# Deep Frying

## Chemistry, Nutrition, and

## Practical Applications

### Second Edition

# Deep Frying

## Chemistry, Nutrition, and

## Practical Applications

### Second Edition

**Editor**
**Michael D. Erickson**

**AOCS**
PRESS

**Urbana, Illinois**

AOCS Press, Urbana, IL 61802

ISBN 978-1-893997-92-9

Library of Congress Cataloging-in-Publication Data

Deep frying : chemistry, nutrition, and practical applications / editor, Michael D. Erickson. -- 2nd ed.
  p. ; cm.
  Includes bibliographical references and index.
  ISBN-13: 978-1-893997-92-9 (alk. paper)
  ISBN-10: 1-893997-92-8 (alk. paper)
  1. Oils and fats, Edible. 2. Deep frying. I. Erickson, Michael D.
  [DNLM: 1. Dietary Fats, Unsaturated. 2. Food Industry--standards. 3. Diet--standards. 4. Fats, Unsaturated--chemistry. 5. Nutritive Value. WA 722 D311 2006]

  TP670.D44 2006
  664'.36--dc22

                                                                                          2006026196
Printed in the United States of America.
11 10 09 08 07 5 4 3 2 1

# Dedication

This book is dedicated to the memory of my friend, Professor Edward Perkins. Both your presence in and contributions to the science and technology of edible fats and oils are greatly missed.

*Michael D. Erickson, Editor*

# Contents

# Foreword

# Tribute to Edward G. Perkins

Edward G. Perkins, Professor Emeritus of Food Chemistry at the University of Illinois, died on Friday, May 19, 2000. He was an internationally recognized authority on the chemistry of fats, oils, and frying. He obtained his B.S. degree in chemistry in 1956, his M.S. in 1957, and his Ph.D. degree in food chemistry in 1958, all from the University of Illinois. After a short time in the U.S. Army and a slightly longer stay at Armour and Co., he returned to the Department of Food Science at the University of Illinois in 1960 as an assistant professor and was quickly promoted to associate professor, and then professor.

His focus was research, teaching, and outreach (which included consulting and giving advice to the food oil processing industry) in fats and oils processing and chemistry, analytical chemistry, and nutritional biochemistry. One special interest was the study of the chemical transformations that occur during heating, the compounds formed and their isolation and characterization, which often included the synthesis of these compounds. He conducted research using animal studies to examine the metabolism and distribution of these same compounds. He worked on the development of gas chromatography-mass spectrometry-computer analysis methods for the detection of biologically important compounds. He also studied phospholipids to understand their function in the degumming process and in food emulsions.

Dr. Perkins was considered one of the leading experts in the application of chromatography and spectrometry, particularly mass spectrometry, to fats and oils. He was the director of the College of Agriculture Mass Spectrometry Laboratory at the University of Illinois. He was the author of more than 140 scientific journal articles, numerous book chapters, and the editor of 10 books. He received 8 major awards, including the A.E. Bailey award and the A.R. Baldwin Distinguished Service award from the American Oil Chemists' Society, the Funk award for excellence in research from the University of Illinois, and the first Steven S. Chang award from IFT. He served as Vice-President and President of AOCS and he was elected a charter Fellow of the Society in 1998. He supervised 22 M.S. theses and 25 Ph.D. dissertations. Dr. Perkins retired from the Department in August of 1999, but he continued to maintain an active research program.

Professor Edward G. Perkins was a good friend and trusted colleague of mine at the University of Illinois for nearly two decades. He was a charismatic and memorable individual. He demanded much of himself and he tended to have high expectations for those around him — students, staff, and faculty (even restaurant chefs). One indication of the true quality of a person's character is how they are perceived by their subordinates. Almost all of the students in his classes, particularly the graduate students under his direction and the staff members working in his office and lab, thought highly of him and were not hesitant to say so. At first he could seem rather gruff and unapproachable to some, but once he learned that one's interest in science and research was sincere, Professor Perkins was extremely collegial, generous with his time, and intent on being helpful. He was a very pro-student professor, much more than average. On numerous occasions I witnessed him go "above and beyond" to assist students with difficult analytical food chemistry problems. He thoroughly enjoyed his role as a mentor. Because of his extensive knowledge, he was consulted often by industry scientists and personnel, and by faculty at universities from all over the world, becoming good friends with many academic and industry leaders. His students, friends, and colleagues miss him.

*William Artz, Ph.D.*
*Professor, Food Sciences and Human Nutrition*
*University of Illinois, Urbana, Illinois*

# Introduction

During the 11 years since the first edition of *Deep Frying* was published, there have been a number of advances in frying technology and improvements in equipment. Batch fryers with automatic timing cycles and basket lifts are now used extensively in fast-food restaurants to consistently produce high quality products. In the snack food industry, fryers with production capacities exceeding 4,000 lbs./hr for tortilla chips and 6,000 lbs./hr for potato chips, supporting improved oil quality through rapid turnover, are now widely used.

The new frying equipment recently introduced includes computerized systems for controlling fryers that support product quality and economy of operation, dry-steam oil-stripping systems to produce reduced-oil snacks, various options for full-flow oil filtration to better maintain oil quality, and new fryers utilizing falling curtains of oil that support the production of new products with unique texture.

These advances in equipment and technology are having a significant impact on the industry. However, perhaps an even greater impact on the industry may be the result of recent changes to FDA Dietary Guidelines and nutritional labeling laws.

The new nutritional labeling law requiring a line item entry for *trans* fat has resulted in a flurry of activity among edible oil refiners to develop low- and no-*trans* oils and shortenings for frying and other applications. Methods for reducing and/or eliminating *trans* fat include reduced *trans* hydrogenation, full hydrogenation and blending, other blending options, traditional (catalyzed) interesterification, enzymatic interesterification, and switching to oils with intrinsic high stability. New, higher stability oils are also being developed through seed breeding technology.

The inclusion of *trans* fat in the Nutritional Facts panel provides additional information for consumers to use in making food choices. While making such choices, consumers need to take into account the fact that *trans* fat constitute 2.6% of calories in the American diet, as compared to some 15% of calories from saturated fat, and guard against increasing saturates to avoid *trans*.

In the Dietary Guidelines for Americans 2005, the FDA addressed the consumption of *trans* and saturates in stating, "Population based studies of American diets show that intake of saturated fat is more excessive than intake of *trans* fats and cholesterol. Therefore, it is most important for Americans to decrease their intake of saturated fat."

The Dietary Guidelines also address the importance of fat in the diet, recommending moderate consumption (between 20 and 35% of calories for adults) and noting that fats are a major source of energy and aid in the absorption of vitamins A, D, E, and K, and cartoenoids. The guidelines also note that, "As a food ingredient, fat provides taste, consistency, and stability and helps us feel full."

The importance of eating a balanced diet and balancing caloric intake with energy expenditure are often stressed to support good health and weight management. In making food choices, consumers regularly demonstrate preferences based on taste, consistency, and freshness and many food ingredients can provide these attributes. But few if any ingredients other than fats and oils help us feel full and provide satiation. With proper planning, fried foods can be included in diets that are nutritious, uniquely palatable, and satisfying to eat.

*Don Banks*
*Edible Oil Technology*
*8155 San Leandro*
*Dallas, Texas 75218*

# About the Editor

Michael D. Erickson is currently a Technical Service Manager for Cargill Dressing, Sauces and Oils in Fullerton, California. He started his career in 1981 with Interstate Foods Corp., Chicago, IL, which, at the time, was the sole supplier of frying fat to the McDonald's system. He worked closely with the largest international quick-service restaurant chains and food processors as primary technical lead for matters related to frying including product development, equipment evaluation and education.

In 1990 he started work for Humko Oils division of Kraft Food Ingredients in Memphis, TN, where, in addition to continued work in the frying industry, he led various projects including alternatives to hydrogenation and identification and development of biotechnology oil through the partnership with Pioneer Hi-Bred International, Inc. He also worked extensively in edible oil processing on projects involving new refining technologies and identification of productivity opportunities in each unit process (refining, bleaching, hydrogenation, deodorization, winterization).

His current role at Cargill includes continued active participation in the frying and frying related industry in addition to pioneering product and service support for unique customer solutions involving any food ingredient. He is a member of the American Oil Chemists' Society having served as the technical chairperson for the 1993 Annual Meeting (Anaheim, CA), and member of the AOCS Governing Board. He developed and chaired two AOCS Laboratory Short Courses on Frying Fats and co-chair for frying fat sessions with Dr. Perkins.

# PART 1

# Overview

# 1

# Production and Composition of Frying Fats

**David R. Erickson**

*DJ Consultants, 507 Woodlake Drive, Santa Rosa, CA 95405-9203*

Production of suitable frying fats and oils starts with crude fats or oils extracted from oil seeds, tree fruits and nuts, and animal or marine fatty tissues. In this chapter the successive steps of extraction, refining and formulation of fats and oils into frying media will be briefly discussed.

## Composition of Crude Fats and Oils

Fats and oils are considered less complex than proteins and carbohydrates because they consist largely of triacylglycerols (triglycerides), which are relatively simple compounds. Nevertheless, there are some common conventional terms used in the edible oils industry that may be somewhat confusing.

The first of these are the actual definitions of fats and oils. The classic definition for fats is that they are solid at room temperature (20°C), while oils are liquid. Table 1.1 shows, however, that classic definitions do not necessarily agree with common usage. More confusion is introduced when fats and oils such as cottonseed, palm, lard, and tallow, are fractionated to give lower-melting, liquid, or soft fractions (olein), and higher-melting, more solid fractions (stearine). Such fractionations and their products are shown in Table 1.2. Further uncertainty is introduced depending on the degree and efficiency of fractionation.

Another variable is that the commonly used fats and oils may vary in their compositions, as shown in Table 1.3. Even this table is now subject to change due to recent and relatively rapid advances in biotechnology that may cause changes in the fatty acid composition of fats and oils from plant sources. These changes will come largely due to economic considerations.

In cattle and sheep (ruminants) and swine, the fat surrounding their internal organs is harder, getting progressively softer toward the skin. This is why the killing fat (internal) is harder than the cutting fat (external) to a discernible degree. Ruminants generally do not change their fat composition to match what is found in their diets, but monogastric animals do. A good example is the extreme softness of lard from peanut-fed swine, which matches the softness and liquidity of peanut oil.

**TABLE 1.1**
**Fats and Oils by Classic Definitions**
**Compared to Common Usage**

| Oil (liquid at 20°C) | Fat (solid at 20°C)[a] |
|---|---|
| Corn | Coconut |
| Cottonseed | Palm |
| Fish | Palm kernel |
| Olive | Butter (butter oil) |
| Peanut | Lard (lard oil) |
| Rapeseed/canola | Tallow (oleo oil) |
| Safflower | |
| Soybean | |
| Sunflower | |

[a]All are solid or partially solid at room temperature.

**TABLE 1.2**
**Fractionated Fats/Oils and Resulting Products**

| Product | Fraction |
|---|---|
| Cottonseed | Winterized cottonseed oil |
| | Cottonseed stearine |
| Palm | Palm olein |
| | Palm stearine |
| Lard | Lard oil |
| | Lard stearine |
| Tallow | Oleo oil |
| | Oleo stearine |

Fish oils, such as those from herring or menhaden, reflect an environmental effect in that they are softer (higher iodine value) in cold water than those from warmer water. Sunflower oil also has a higher degree of polyunsaturation in cooler climates than it does in warm. For instance, the same variety of sunflower grown in North Dakota may be as much as 40% higher in linoleic acid than those grown in Oklahoma.

Rapeseed oil exists as both high erucic acid rapeseed (HEAR) and low erucic acid rapeseed varieties (LEAR, or canola), with large differences in both chemical and physical properties.

Although there are many possible sources of edible fats and oils, those for use as commercial frying fats are relatively few. These are the animal fats, (i.e., lard and

**TABLE 1.3**
**Comparison of Fats and Oils of "Constant" and Variable Compositions**

| Fats and oils with "constant" compositions | Fats and oils of varying compositions |
|---|---|
| Butter | Fish oil (cold and warm water) |
| Coconut | Lard (interior, exterior, and feed effects) |
| Corn | Tallow (interior and exterior) |
| Cottonseed | Rapeseed (HEAR[a] and LEAR[b]) |
| Olive | Safflower (old and high oleic) |
| Palm | Sunflower (temperature effects and high oleic) |
| Peanut | |
| Soybean | |

[a]HEAR = High-erucic acid rapeseed.
[b]LEAR = Low-erucic acid rapeseed.

tallow), marine oil, and vegetable oil. The availability of vegetable fats and oils in descending order are: soybean, palm, sunflower, cottonseed, coconut, peanut, olive, palm kernel, and corn.

Nontriglyceride components of major crude fats and oils are shown in Table 1.4 (1), and their fatty acid compositions are shown in Table 1.5 (2). As a fat or oil is refined, nontriglyceride components are reduced, as is illustrated for soybean oil in Table 1.6 (3). Included in the unsaponifiable matter are pigments and waxes. Pigments are reduced in refining and are not ordinarily of concern in frying fat performance, but they may affect the color of fried products. Sterols may become oxidized and have some physiological effects. Waxes are of little consequence in frying, but are a matter of cosmetics for liquid oil use because they cause cloudiness at lower temperatures. Waxes are usually removed by cooling, crystallization, and filtration after bleaching.

# Processing Fats and Oils for Frying

In discussing the processing of fats and oils, it is helpful to view the overall process from extraction to final product as a system comprised of a series of unit operations carried out consecutively as follows: extraction or rendering, degumming (optional), chemical refining, bleaching and adsorptive treatment, physical refining, hydrogenation and formulation, deodorization, and crystallization. The goal in processing is to produce a salable product processed in the most efficient and economic manner consistent with marketplace quality. Reaching this goal involves balancing maximum quality and yield against the cost of processing. Optimal processing consists of performing each unit process well enough to avoid relying on the next step to overcome any deficiencies. Done correctly, this usually results in the production of the best quality in the most economic manner.

**TABLE 1.4**
**Nontriglyceride Components of Common Crude Fats/Oils**

| Fat/oil | % Unsaponifiables | % Sterols | % Phosphatides | Waxes[a] |
|---------|-------------------|-----------|----------------|----------|
| Canola | 0.8–1.0 | 0.5–0.9 | 1.8–3.5 | trace |
| Corn | 0.8–2.9 | 0.9–1.1 | 0.5–2.0 | present |
| Cottonseed | 0.4–0.6 | 0.2–0.4 | 0.7–0.9 | absent |
| Fish oil | 2.0–5.0 | 0.6–0.8 | 0.8–1.1 | absent |
| Lard | 1.0–1.2 | 0.1–0.2 | 0.04–0.06 | absent |
| Palm | 0.2–0.5 | 0.04–0.06 | 0.05–0.1 | absent |
| Peanut | 0.2–1.0 | 0.2–0.3 | 0.3–0.4 | absent |
| Soybean | 1.4–1.8 | 0.2–0.4 | 1.5–3.0 | absent |
| Sunflower | 1.0–1.5 | 0.5–0.6 | 0.5–1.0 | present |
| Tallow | 1.0–1.5 | 0.08–0.14 | 0.06–0.8 | absent |

[a]Source: Sonntag, N.O.V. (1)

## Extraction

To obtain a crude fat or oil for further refining, it is first necessary to separate or extract the fat or oil from its natural state in either animal or vegetable tissue.

### Animal Fats

Extraction of animal fats is a relatively simple process called rendering. Dry rendering is heating the fat-containing tissue, which solidifies the proteinaceous material and releases the fat, which then pools together in a melted form. Frying bacon is an everyday example of dry rendering. Fat from dry rendering is usually moisture free, may be dark colored, and may have a cooked flavor. For frying purposes, dry-rendered fats are normally further refined, although in some cultures such fats are acceptable for frying without further refining. Wet rendering is using steam under pressure (autoclaves) to cook the fat-containing tissue. It can be done batch wise or in a continuous manner using lower temperatures and centrifuges. The latter system produces an edible protein as well as edible fat. Fats from wet rendering are usually light colored and light flavored and can be made relatively moisture free. Marine oils are wet rendered in a similar manner. The key to the quality of animal and marine fats and oils is directly related to the proper handling of the fatty tissues in a sanitary manner. Such handling prevents microbial growth or enzymatic action. This is readily reflected by the free fatty acid (FFA) content of the fat. A high FFA indicates either mishandling of the raw materials or errors in processing. The odor of crude animal or marine fats and oils is often a reliable indicator of quality.

Animals, both land and water, tend to deposit fat-soluble pesticides in their fatty

**TABLE 1.5**
**Fatty Acid Composition of Common Fats and Oils**

| | Canola | Coco-nut | Corn | Cot-ton-seed | Fish[a] | Lard | Palm | Palm Kernel | Peanut | Soy-bean | Sun-flower | Tallow |
|---|---|---|---|---|---|---|---|---|---|---|---|---|
| | | | | (g/100 g Fat or Oil) | | | | | | | | |
| $C_{6:0}$ | — | 0.6 | — | — | — | — | — | 0.2 | — | — | — | — |
| $C_{8:0}$ | — | 7.5 | — | — | — | — | — | 3.3 | — | — | — | — |
| $C_{10:0}$ | — | 6.0 | — | — | — | 0.1 | — | 3.7 | — | — | — | — |
| $C_{12:0}$ | — | 44.6 | — | — | — | 0.2 | 0.1 | 47.0 | — | — | — | 0.9 |
| $C_{14:0}$ | — | 16.8 | — | 0.8 | 9.3 | 1.3 | 1.0 | 16.4 | 0.1 | 0.1 | — | 3.7 |
| $C_{16:0}$ | 4.8 | 8.2 | 10.9 | 22.7 | 17.1 | 23.8 | 43.5 | 8.1 | 9.5 | 10.3 | 5.9 | 24.9 |
| $C_{18:0}$ | 1.6 | 2.8 | 1.8 | 2.3 | 2.8 | 13.5 | 4.3 | 2.8 | 2.2 | 3.8 | 4.5 | 18.9 |
| $C_{16:1}$ | — | — | — | 0.8 | 12.5 | 2.7 | 0.3 | — | 0.1 | 0.4 | — | 4.2 |
| $C_{18:1}$ | 53.8 | 5.8 | 24.2 | 17.0 | 11.4 | 13.5 | 36.6 | 11.4 | 44.8 | 22.8 | 19.5 | 36.0 |
| $C_{18:2}$ | 22.1 | 1.8 | 58.0 | 51.5 | 1.5 | 41.2 | 9.1 | 1.6 | 32.0 | 51.0 | 65.7 | 3.1 |
| $C_{18:3}$ | 11.1 | 1.8 | 0.7 | 0.2 | 1.6 | 10.2 | 0.2 | — | — | 6.8 | — | 0.6 |
| $C_{20:1}$ | — | — | — | — | 1.6 | — | — | — | 1.3 | 0.2 | — | 0.3 |
| $C_{20:4}$ | — | — | — | — | 2.0 | — | — | — | — | — | — | — |
| $C_{20:5}$ | — | — | — | — | 15.5 | — | — | — | — | — | — | — |
| $C_{22:5}$ | — | — | — | — | 2.4 | — | — | — | — | — | — | — |
| $C_{22:6}$ | — | — | — | — | 9.1 | — | — | — | — | — | — | — |

[a]Source: Agricultural Handbook 8-4 (2).

tissues, and thus these may be present in the separated fat. This may be completely overcome by deodorization; otherwise, such fats should be periodically tested to ensure acceptable levels.

*Palm and Olive Oils*
Palm oil is separated from the flesh of palm fruits by a type of wet rendering. Palm fruits contain very active lipases that require deactivation before tissue disruption (mashing) to avoid high FFA; this is done by autoclaving the fruit bunches as the first step. The autoclaved bunches are then run through equipment to remove the stems and mashed to rupture fat cells and separate out the palm kernels. The pulpy mash is then centrifuged to remove the oil.

Olive oil is obtained by first crushing olives without cooking and removing the oil by pressing and decantation. More modern facilities utilize centrifuges to separate the oil. Olive oil is prized for its natural flavor and is often used without further refining. For frying purposes, olive oil may be susceptible to smoking and hence is

**TABLE 1.6**
**Typical Base Stock Program for Soybean Oil**[a]

| | Shortening Bases | | Margarine Bases | | Shortening or Margarine Base | |
|---|---|---|---|---|---|---|
| Base Stock Number[b] | 1 | 2 | 3 | 4 | 5 | 6 |
| Hydrogenation conditions | | | | | | |
| Initial temperature (°C) | 150 | 150 | 150 | 150–163 | 150–163 | 140 |
| Hydro temperature (°C) | 165 | 165 | 165 | 218 | 218 | 140 |
| Pressure (atm) | 1.0 | 1.0 | 1.0 | 0.3 | 0.6 | 2.7 |
| Catalyst concentration | | | | | | |
| (% Nickel)[c] | 0.02 | 0.02 | 0.02 | 0.05[c] | 0.02[d] | 0.02 |
| Final iodine value[e] | 83–86 | 80–82 | 70–72 | 64–68 | 73–76 | 104–106 |
| Final congeal point (°C) | — | — | 25.5–26 | 33–33.5 | 24–24.5 | — |
| SFI[f] | | | | | | |
| 10.0°C | 16–18 | 19–21 | 40–43 | 58–61 | 26–38 | 4 max |
| 21.1°C | 7–9 | 11–13 | 27–29 | 42–46 | 19–21 | 2 max |
| 33.3°C | — | — | 9–11 | 21 max | 2.0 max | — |

[a]Source: Latondress, E.G., (4).
[b]Properly refined and bleached to remove soap, phosphatides, and peroxides.
[c]Based on weight of oil.
[d]Very selective catalyst.
[e]Final iodine value approximate.
[f]SFI = Solid Fat Index.

**TABLE 1.7**
**Average Compositions for Crude and Refined Soybean Oil**[a]

| Average composition (%) | Crude oil | Refined oil |
|---|---|---|
| Triglycerides | 95–97 | 99+ |
| Phosphatides | 1.5–2.5 | 0.0–0.003 |
| Unsaponifiable matter | 1.6 | 0.3 |
| Plant sterols | 0.33 | 0.13 |
| Tocopherols | 0.15–0.21 | 0.11–0.18 |
| Hydrocarbons (squalene) | 0.014 | 0.01 |
| Free fatty acids | 0.3–0.7 | 0.02–0.03 |
| Trace metals (ppm) | | |
| Iron | 1–3 | 0.1–0.3 |
| Copper | 0.03–0.05 | 0.02–0.06 |

[a]Source: Erickson, D.R. and L.H. Wiedermann, (3).

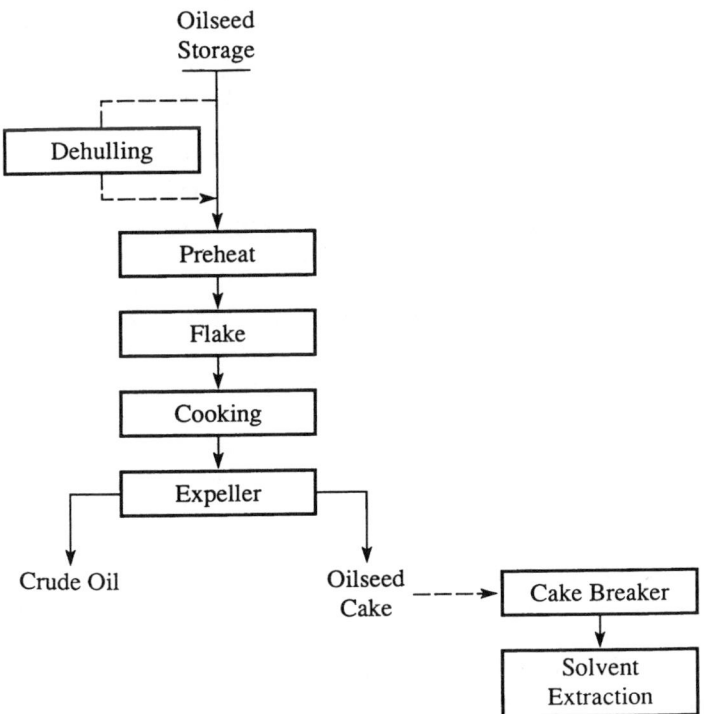

**Fig. 1.1.** Expeller process.

sometimes further refined and/or blended to reduce this tendency for smoking at frying fat temperatures.

### Oil Seeds
Mechanical pressing, solvent extraction, or a combination of the two processes obtains oil from oil seeds.

### Mechanical or Pressure Extraction of Oilseeds
The use of pressure is an old practice. As early as 295 B.C., lever presses in Egypt recovered sesame, linseed, and castor oils. By 184 B.C., screw and wedge presses and filters were in use in the Roman Empire. The first hydraulic press was constructed in England in 1795, followed by perfection of the continuous screw press (Expeller) by Anderson in 1903. The basic principle of the Anderson Expeller remains the same to the present day. (Note: The term "Expeller" is a trademark of Anderson International for their continuous screw press but has become almost a generic name for this type of equipment.)

A general flow diagram for mechanical extraction for oil seeds is shown in Fig. 1.1. Preparation of seeds or oil-bearing materials for expelling usually requires decortication, size reduction by grinding or flaking, and heating. Such preparation is critical because moisture must be reduced to less than 5% and the material must be cooked enough to allow ready release of the oil. Decortication increases protein content of the residual cake and removes the hulls that have little or no oil content and are abrasive.

The drawbacks to mechanical extraction are as follows: low capacity (15–30 tons/day), high residual oil in the press cake (4–7%), high power requirements, and high maintenance and operator skill required. Despite these drawbacks, the use of mechanical extraction is still practiced for some niche markets for specialty oil seeds and in less-developed countries. An example of the first is so-called "cold-pressed" oils, which are specialty products. "Cold pressing" is a loose term that is not defined but appeals to the natural-foods consumer. The most common use of mechanical extraction now is as a preparation step (prepressing) on the higher oil-containing seeds, such as cottonseed, rapeseed (canola), copra, palm kernel, sunflower, and sometimes corn germ, prior to solvent extraction. Soybeans, with lower oil content (18–20%), are solvent-extracted directly without prepressing.

*Solvent Extraction*
Solvent extraction is simple in concept; a suitable solvent is used to dissolve oil out of oil-bearing tissue. The first patent for the process was issued in 1856, and the process was first commercialized as a batch process in 1870. While the principle remains the same, modern continuous processes are complex and represent large capital investments. Plants with capacities of 3,000–4,000 tons/day are now in operation.

Preparation of oilseeds for extraction is a critical step. Preparation of soybeans is typical and is shown in Fig. 1.2. A generic oil seed extraction diagram is shown in Fig. 1.3.

## Refining Fats and Oils

Refining is an overall process for the removal of unwanted constituents from crude fats and oils. The desired quality of the finished product dictates the degree of refining. The second factor in determining processing alternatives is the quality of the crude fats or oils. Fig. 1.4 shows a generic-process flow diagram for refining.

*Crude Fat or Oil Quality*
The quality of a fat or oil depends on the quality of the raw material and the methods used for extraction. Within reasonable limits, modern refiners learned how to assess quality and to adjust their refining practices in relation to both initial crude quality and final product quality targets.

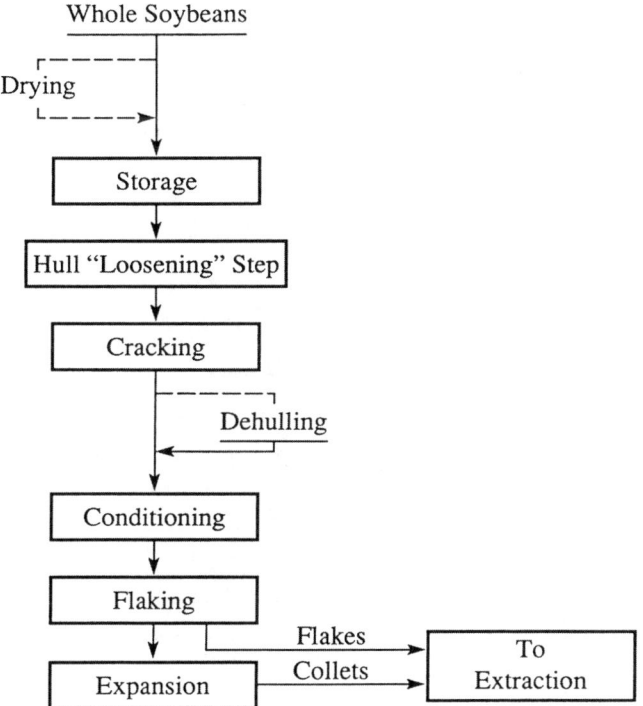

**Fig. 1.2.** Preparation of soybeans for solvent extraction.

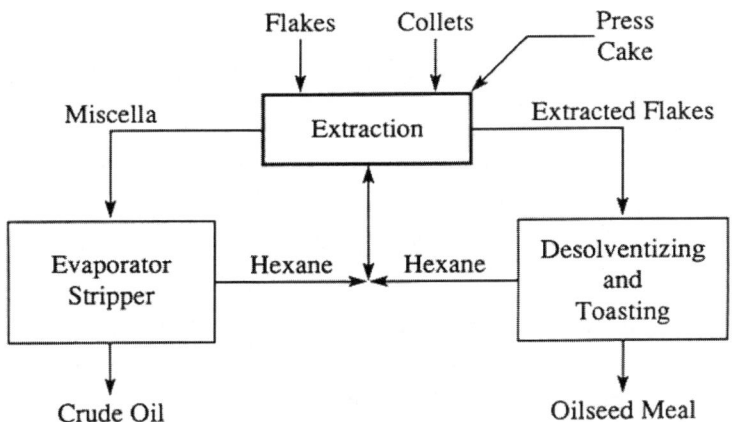

**Fig. 1.3.** Solvent extraction process.

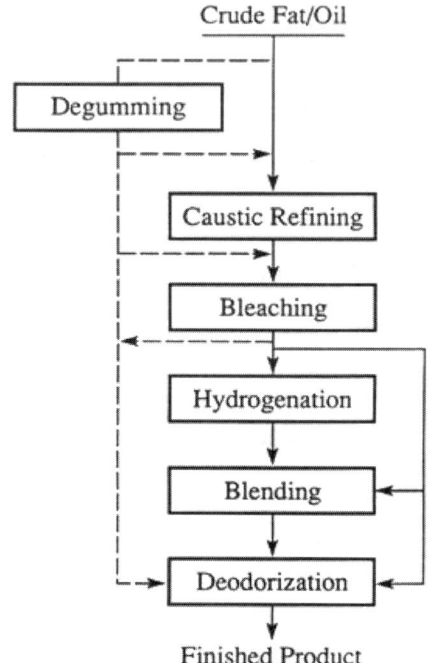

**Fig. 1.4.** Generic fat/oil refining flow diagram.

## *Degumming*

Degumming is the process of hydrating phosphatides present in an oil by adding water followed by centrifugation. There are only three reasons to degum oil: to produce lecithin (phosphatides), to provide degummed oil for long-term storage or transport, and to prepare for physical refining. Degumming is optional when an oil will be caustic-refined since gums are removed in that step. Phosphatides are present in either a hydratable or nonhydratable form (NHP). Nonhydratable phosphatides are particularly troublesome and require either pretreatment with phosphoric acid or use of a long-mix caustic refining system. Phosphatides in soybean oil are of particular importance since they can be present in high quantities (2–4%) in crude oil. Other vegetable oils are lower in quantity (see Table 1.7).

## *Caustic Refining (Neutralization)*

Caustic refining of vegetable oils consists of adding a predetermined amount of a caustic solution (usually sodium hydroxide), mixing thoroughly, and providing contact or retention time, followed by heating and centrifugation. The two basic processes differ in application of heat, caustic strength, and retention time. These are the short- and long-mix systems, and they are compared in Fig. 1.5.

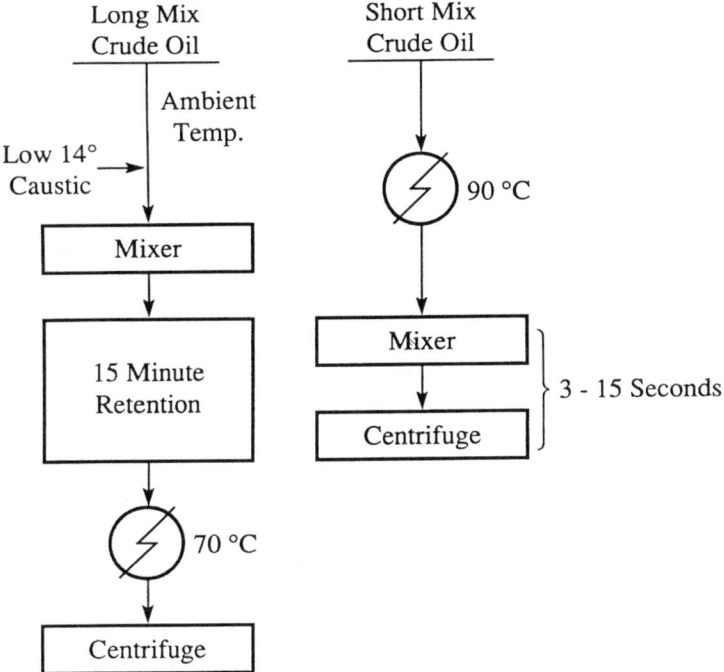

**Fig. 1.5.** Comparison of long- and short-mix caustic refining system.

In the United States, the long-mix system is preferred for soybean and some other oils, while in other parts of the world the short-mix system is used. Since exhaustive removal of phosphatides is required for production of the best-quality soybean oil, U.S. refiners use the long-mix system to ensure adequate time for hydration and removal of both types of phosphatides (3,5).

Following the first centrifugation, the oil is water-washed once or twice by mixing with soft hot water and centrifuging again. A typical degumming and caustic refining schematic is presented in Fig. 1.6, showing the water-washing steps. Typically, fats and oils with low FFA contents and high phosphatides are candidates for caustic refining, and those with the opposite characteristics are best for physical refining.

## Bleaching and Adsorption Treatment

While the term "bleaching" implies color reduction, this may not be the most important function of the process. Depending on the fat or oil, color reduction may be only incidental. If color itself, such as chlorophyll in canola or soybean oil, is a problem, then its removal is paramount. Otherwise, bleaching treatment is guided by the amount needed to improve oxidative flavor stability in the finished products (3,6,7). The functions of a bleaching earth on a normal soybean oil are

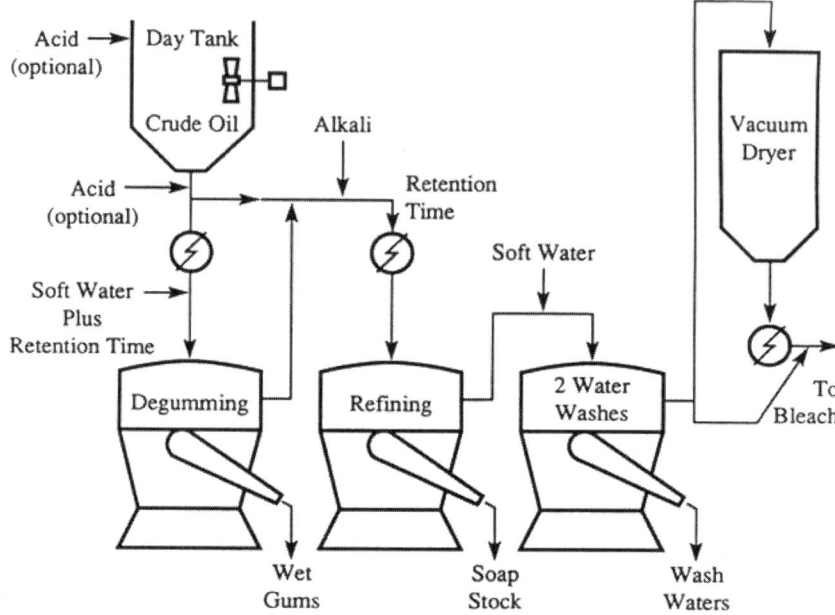

**Fig. 1.6.** Recommended degumming and alkali refining process for soybean oil.

listed in descending order of importance: to decompose peroxides, to remove/change oxidation products, to remove trace gums/soaps, to remove trace metals, and, finally, to decolorize. Properly bleached oils provide maximum flavor stability in finished products and ensure minimum problems in the succeeding steps of hydrogenation and deodorization.

In bleaching, a predetermined amount of bleaching earth is added to the oil, which is then heated and mixed, followed by filtration to remove the earth. Early practice was done at atmospheric pressure in open tanks, but the modern practice is to operate under vacuum in a continuous system. A diagram of the latter system is shown in Fig. 1.7.

## Physical Refining

Physical or steam refining is a deacidifying step similar to deodorization. It has long been used on fats and oils with high FFA contents and/or low phosphatides. High FFA content would lead to unacceptably high refining losses if caustic refining was used, and high phosphatides interfere with the exhaustive degumming necessary for physical refining.

Tallow, coconut, palm kernel, and palm oils are good candidates for physical refining and have been so treated for many years. Physical refining of the normal

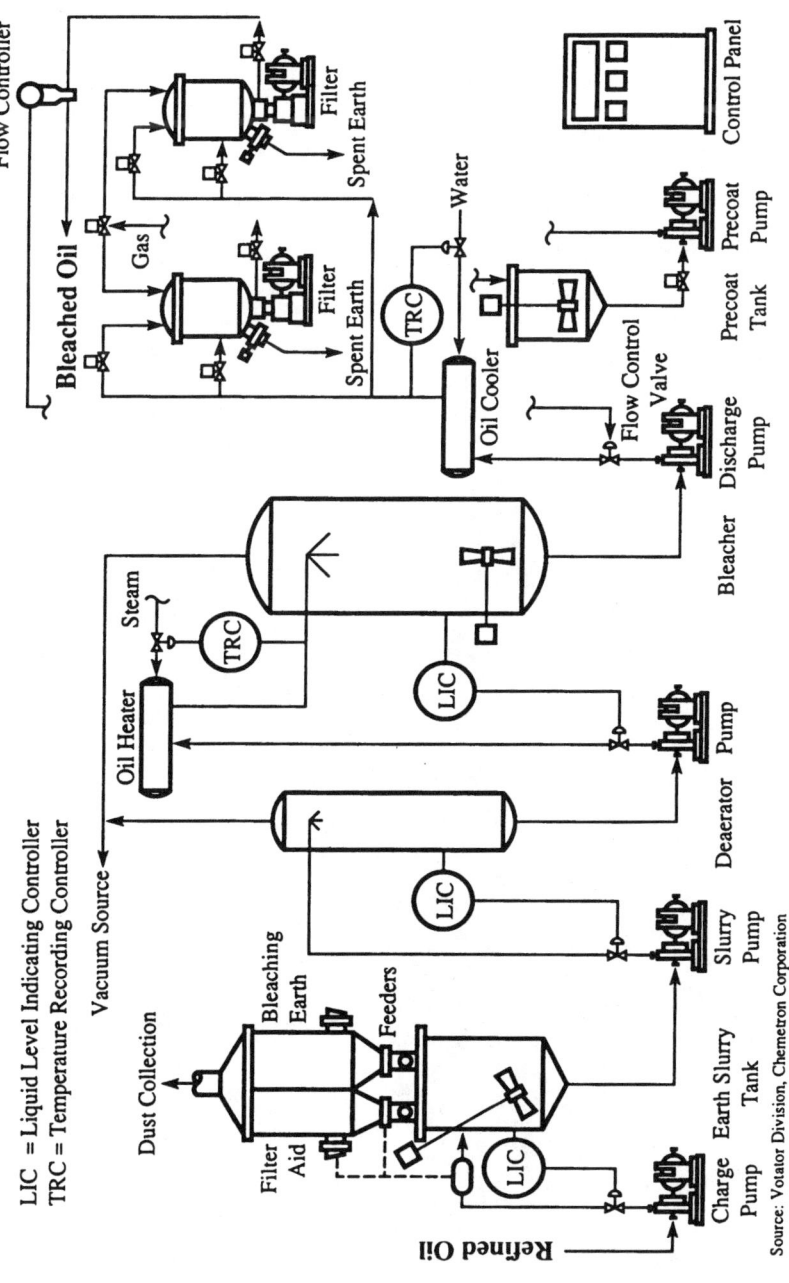

LIC = Liquid Level Indicating Controller
TRC = Temperature Recording Controller

Source: Votator Division, Chemetron Corporation

**Fig. 1.7.** Flowsheet for Votator continuous vacuum bleaching system.

soft vegetable oils is not as satisfactory because of the difficulty in performing the exhaustive degumming necessary to assure maximum quality in the U.S. market. Other markets accept physically refined oils despite their lower quality.

The processing sequence for physical refining is degumming, then bleaching to scavenge remaining phosphatides, followed by steam distillation at a high temperature under vacuum. This is similar to deodorization, with some changes in equipment to accommodate the higher FFA content.

## Hydrogenation and Formulation

Hydrogenation of fats and oils is defined as the addition of hydrogen to the double bonds of unsaturated fatty acids contained in the triglycerides. The actual reaction is not as straightforward, however, because isomerization of the double bonds occurs simultaneously with the hydrogen addition. Although the reaction is complex, experimental results and experience usually allow a refiner to make a wide variety of functional products by blending various hydrogenated products. The purpose of

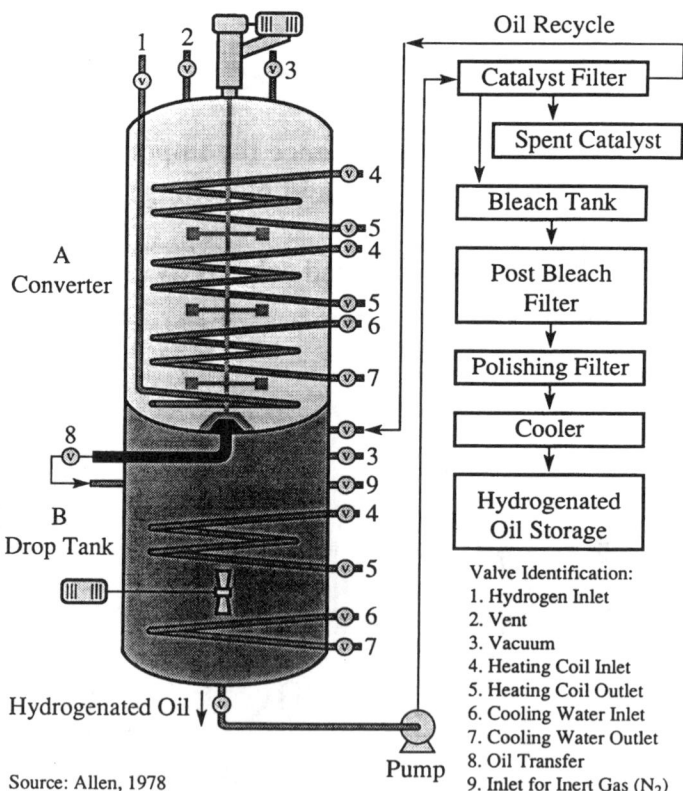

**Fig. 1.8.** Batch hydrogenation apparatus.

**TABLE 1.8**
**Formulation for Margarines from Base Stocks[a,b]**

| Margarine | Required SFI for blend[c] | | | | Formulation | |
|---|---|---|---|---|---|---|
| | 10.0°C | 21.1°C | 33.3°C | 40°C | Base stock | |
| Soft stick | 20–26 | 13–17 | 1.4-3.0 | 0 | #4 | 50 ± 5 |
| | | | | | Unhydrogenated soybean oil | 50 ± 5 |
| Regular stick | 30.0 max | 17.5 min | 3.5 min | 0 | #4 | 38 ± 5 |
| | | | | | #5 | 20 ± 5 |
| | | | | | #6 | 42 ± 5 |
| Tub | 12–13 | 7–8 | 3 max | 0 | #3 | 20–30 |
| | | | | | Unhydrogenated soybean oil | 70–80 |

[a]Source: Latondress, E.G. (4).
[b]These formulations are all soybean oil, suitable for refrigerated products. For maximum resistance to thermal shock, use 5–10% (β′ fats with SFI profile similar to base stock #4. Abbreviation as in Table 1.6).
[c]SFI = Solid Fat Index.

hydrogenating a fat or oil is to increase either its oxidative stability or its solids content. For frying purposes, the increase in oxidative stability is the most important.

Operationally, hydrogenation consists of bringing the oil, catalyst (nickel), and hydrogen gas together in a suitable vessel. Initially, heat is applied until the reaction starts, and from that point the reaction becomes exothermic. The controlling parameters for hydrogenation are type and amount of catalyst, hydrogen gas pressure, and temperature. Agitation is also a variable but most often is fixed. A typical hydrogenation vessel is shown in Fig. 1.8.

Once the reaction is complete, the catalyst is filtered out of the oil. While a refiner may formulate products using a specific hydrogenated oil for each individual formulated product, this is operationally difficult. Such individual formulations rarely use exact multiples of a hydrogenation batch size. This means a greater number of storage tanks or manipulation of batch sizes. The more modern and efficient practice involves the development and use of a base stock program. Such a program results in an unlimited number of hydrogenated products, which allows the following: full use of hydrogenation capacity, a limited number of storage tanks, and accommodation for normal variations found in day-to-day hydrogenation operations. This greatly simplifies scheduling of plant operations because it involves only keeping the base stock tanks full rather than scheduling on the basis of individual orders.

Base stock programs are usually a proprietary matter with an individual refiner; however, the following example for a soybean oil-based base stock program will

**TABLE 1.9**
Formation for Shortenings from Base Stocks[a]

| Shortening type | Required SFI for blend[e] | | | | Formulation | |
|---|---|---|---|---|---|---|
| | 10.0°C | 21.1°C | 33.3°C | 40°C | Base stock | % |
| All purpose[b] | 22–24 | 18–20 | 13–15 | 10–12 | #1 | 88–99 |
| | | | | | Hardfat | 11–12 |
| All purpose[c] | 24–27 | 18–20 | 12–14 | 6–8 | #2 | 92–93 |
| | | | | | Hardfat[c] | 7–8 |
| Frying (heavy duty) | 41–44 | 29–30 | 12–14 | 2–5 | #2 | 97 |
| | | | | | Hardfat[c] | 3 |
| Frying (fluid) | 5–6 | 3–4 | 2–3 | 1–2 | #6 | 98 |
| | | | | | Hardfat[c,d] | 2 |
| Specialty (nondairy) | 43–47 | 27–30 | 6–9 | 1–5 | #4 | 30 ± 5 |
| | | | | | #5 | 70 ± 5 |

[a]Abbreviation and source (4)
[b]Emulsified shortenings made by adding proper type and amount of emulsifier.
[c]Hardfat β' (1–8 iodine value).
[d]Hardfat β' (1–8 iodine value).
[e]SFI = Solid Fat Index

illustrate the principle (4). For other oils, similar results can be achieved based on an identical solids profile (NMR/SFI). It is possible to make a wide variety of products using only six hydrogenated soybean oil base stocks. Suggested beginning parameters of hydrogenation with analytical results are shown in Table 1.6. Note the allowable range in SFI, which is acceptable if the slopes remain the same. Using the base stocks shown in Table 1.6, formulation of shortenings and margarines can be done typically as shown in Tables 1.8 and 1.9.

## Deodorization

Deodorization is a steam distillation process carried out under vacuum (1–5 mm) and at temperatures ranging from 210–270°C. The steam acts as a carrier that removes odoriferous compounds and results in a bland product.

Typical deodorization conditions are shown in Table 1.10. Often there are questions asked as to the so-called "right" temperature, time, pressure, and so on, for deodorization. A simple answer cannot be given because of differences in deodorizing equipment, conditions used in them, and type of product being deodorized. Efficiency

**TABLE 1.10**
Commercial Deodorization Conditions[a]

| | |
|---|---|
| Absolute pressure: | 1–6 mm Hg |
| Deodorization temperature: | 210–274°C |
| | |
| Holding time at elevated temperature | |
| Batch-type: | 3–8 h |
| Continuous and semi-continuous types: | 15–120 min |
| Stripping steam: | wt% of oil |
| Batch-type: | 5–15% |
| Continuous and semi-continuous types: | 1–5% |
| | |
| Product free fatty acid (FFA) | |
| Feed, including steam refining: | 0.05–6% |
| Deodorized oil: | 0.02–0.05% |

[a]Source: Zehnder, C.T. (7)

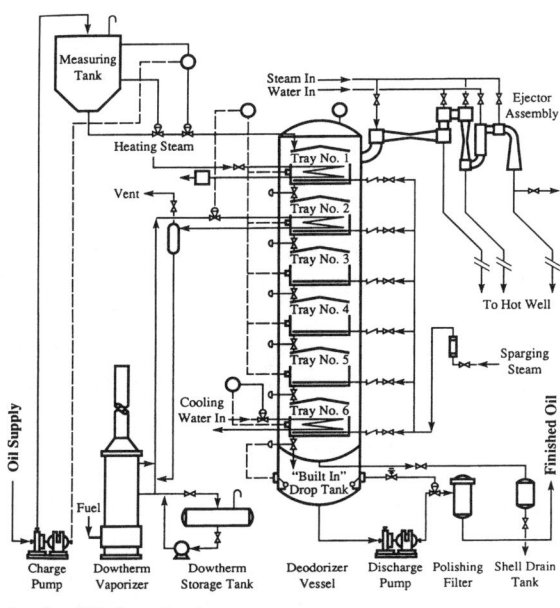

Source: Votator Division, Chemetron Corporation

**Fig. 1.9.** Six-tray semicontinuous deodorizer flow diagram.

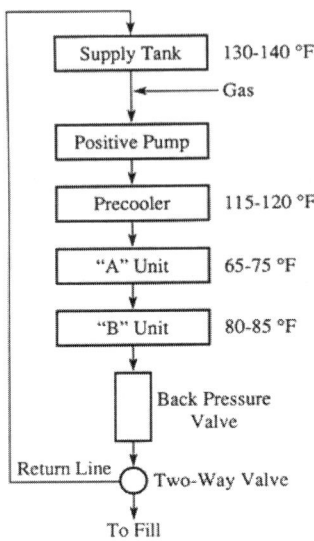

**Fig. 1.10.** Typical shortening crystallization flow diagram.

of deodorization is a function of absolute pressure, top temperature and time, throughput rate, and stripping steam rate. However these functions are varied, the objective is a flavorless product with an FFA content of less than 0.05 for frying fats and preferably 0.02–0.03 for softer products. A typical semi-continuous deodorizer is shown in Fig. 1.9. As a usual practice, citric acid is added to all deodorized fats and oils in the cooling section of the deodorizer.

## Finished Product Handling

For frying fats it is common to add silicone antifoams and antioxidants after deodorization. This is usually done by utilizing a premix that can be metered into the outflow of the deodorizer or added batch-wise to the finished product tanks or transport vessel.

If the frying fat will be shipped out in bulk, this ends the process. The usual precautions are bottom-filling practices or otherwise minimizing contact between the product and air. The product is simply pumped into a tank vehicle and shipped out. Fluidity may be maintained by insulation, or the fat may be allowed to solidify and be remelted for off-loading at its destination.

When frying fats will be handled as solid cubes, or need cooling/chilling to maintain a stable fluid suspension, a scraped surface heat exchanger (SSHE) may be utilized. For solid shortenings that are also used in baking, crystallization and tempering conditions are critical for good baking performance. For frying fats, the only requirement is a reasonably good appearance plus enough softness to allow

**TABLE 1.11**
**Relative Reaction Rates of Unsaturated Fatty Acids with Oxygen**

| Unsaturated fatty acid | Approximate relative reaction rate with oxygen |
|---|---|
| Oleic ($C_{18:1}$, $\Delta 9$) | 1 |
| Linoleic ($C_{18:2}$, $\Delta 9,12$) | 10 |
| Linolenic ($C_{18:3}$, $\Delta 9,12,15$ | 25 |

*Source*: List, G.A., and D.R. Erickson (8).

**TABLE 1.12**
**Sample Calculation of Inherent Stability**[a]

Sample calculation (soybean oil)

| | % Decimal fraction | Approximate relative reaction rate with oxygen | |
|---|---|---|---|
| Oleic ($C_{18:1}$, $\Delta 9$) | 22.0 | $0.228 \times 1$ | $= 0.228$ |
| Linoleic ($C_{18:2}$, $\Delta 9,12$) | 51.0 | $0.510 \times 10$ | $= 5.100$ |
| Linolenic ($C_{18:3}$, $\Delta 9,12,15$) | 6.8 | $0.068 \times 25$ | $= 1.700$ |
| | | Sum | $= 7.028$ |
| | | Inherent stability | $\sim 7.0$ |

[a]*Source*: (8)

digging into, or cutting, of the cube.

Control of SSHE operations is somewhat based on experience. Simultaneous removal of sensible heat and heat of crystallization creates a dynamic situation wherein outlet temperatures alone can be deceiving but when combined with a knowledge of the throughput rate and refrigerant back pressure, control can be reasonably precise.

A typical SSHE system including a working B unit is shown in Fig. 1.10. Injection of gas/air is to enhance the appearance of shortening and to soften it somewhat.

# Composition of Frying Fats

The products used for frying range from unhydrogenated fully refined fats and oils to specially hydrogenated products specifically designed for frying. A useful way to put the different fats and oils in the context of suitability for frying use is to consider their basic, or inherent, stability to oxidation. Inherent stability relates to extent and type of unsaturation of the fatty acids in the fat or oil and relative reaction rates of unsaturated fatty acids with oxygen (9). An example of a sample calculation for soybean oil is shown in Tables 1.11 and 1.12; these tables also show why cottonseed oil represents the first vegetable oil considered for deep frying and why beef tallow was chosen for many years. Table 1.13 shows the results of calculations for other common

**TABLE 1.13**
**Inherent Stability and Calculated Iodine Value (IV) of Common Fats and Oils[a]**

|  | Inherent stability[b] | Calculated IV[c] |
|---|---|---|
| Safflower | 7.6 | 149 |
| Soybean | 7.0 | 132 |
| Sunflower | 6.8 | 136 |
| Corn | 6.2 | 128 |
| Rapeseed (LEAR) | 5.5 | 120 |
| Cottonseed | 5.4 | 110 |
| Rapeseed (HEAR) | 4.1 | 99 |
| Peanut | 3.7 | 100 |
| Lard | 1.7 | 62 |
| Olive | 1.5 | 82 |
| Palm | 1.3 | 50 |
| Tallow | 0.86 | 44 |
| Palm kernel | 0.27 | 13 |
| Coconut | 0.24 | 8 |

[a]Source: (8)
[b]Decimal fraction of fatty acids multiplied by relative rate of reaction with oxygen of each fatty acid (see Table 1.12).
[c]Calculated IV for $C_{18:1}$, $C_{18:2}$, and $C_{18:3}$ using USDA Handbook 8-4 quantities for each.

fats and oils.

Inherent stability as calculated assumes that all oils would be refined, bleached, and deodorized from reasonably good-quality crude oils. Olive oil has good inherent oxidative stability based on its fatty acid composition, but unrefined olive oil may have a smoking problem due to its nontriglyceride constituents. The foregoing comment about the importance of good refining practices applies to all fats and oils destined for use as frying fats.

## Liquid or Pourable Frying Fats

Pourable frying fats have become popular mainly due to their convenience in handling. They range from clear to opaque fluids at room temperature (20°C). An example of an opaque fluid product from soybean oil is shown in Table 1.9. Similar products based on rapeseed (canola) oil are also available in the United States. In other parts of the world, palm olein serves as a pourable frying fat quite adequately.

Clear liquid frying fats can be made from refined, bleached, and deodorized (RBD) oils, such as soybean, corn, cottonseed, and canola, but they are not as stable to oxidation as are hydrogenated versions. Clear liquid frying fats are also made by fractionation of hydrogenated fats.

**TABLE 1.14**
**Properties of Frying Fat/Oils for Foodservice**

| | RBD[a] Soil Oil | Hydroge- nated and Winterized Soil Oil | All Purpose Shortening | Opaque Shortening | Heavy Duty Shortening |
|---|---|---|---|---|---|
| Appearance | Clear Liquid | Clear Liquid | Plastic Solid | Opaque Pourable | Plastic Solid |
| m.p. (°C)[b] | liquid | liquid | 43–48 | 33–37 | 40–43 |
| AOM[c] | 10–25 | 16–30 | 40+ | 35+ | 200+ |
| Polyunsat. % | 34–61 | 54–60 | 15–20 | 35–40 | 4–8 |
| Saturate % | 13–27 | 14–21 | 20–30 | 15–20 | 22–25 |
| Iodine value | 130–135 | 108–112 | 80–85 | 102–108 | 65–75 |

[a]RBD = Refined, bleached, and deodorized.
[b]MP = Melting point.
[c]AOM = Active Oxygen Method (hrs).

## Plastic or Solid Frying Fats

There are a wide variety of plastic or solid frying fats ranging from RBD palm, palm stearine, coconut, and palm kernel oils, and the animal fats lard and tallow. In some cases, these are blended with RBD oils, hydrogenated fats and oils, or with each other, with one important exception. Lauric acid oils (e.g., palm kernel, coconut, and so forth) are special cases because they can only be blended with each other and not with fats and oils containing longer chain fatty acids which lead to premature foaming. Table 1.9 shows two all-purpose shortenings suitable for deep fat frying: a pourable type and a heavy-duty type formulated from hydrogenated soybean oil.

Similar products can be formulated from other hydrogenated oils with similar solids profiles. Such considerations show that a wide variety of products is available for various uses, as shown in Table 1.14, along with their usual specifications.

# Conclusions

A brief overview of the processing of fats and oils was given, followed by formulation of various products used for frying. There are many steps from raw material to finished product, and each must be done correctly to ensure maximum quality in a frying fat. The final user is often far removed from both the processing and the processor. Because of this, reliance must be placed on either their own purchasing specifications or manufacturers' specifications. Unfortunately, adherence to agreed-on specifications does not always mean that the product was processed in an optimal manner. The assurance of expected quality most often is a result of experience with suppliers, working closely with suppliers, an actual performance test, or a combination of the three.

## References

1. Sonntag, N.O.V. In *Bailey's Industrial Oil and Fat Products*, 4th ed.; D. Swern, Ed.; John Wiley & Sons: New York, 1979; Vol. 1, pp. 45–98.
2. U.S. Department of Agriculture (USDA). *Agricultural Handbook 8-4*; USDA, Science and Education Administration: Washington, DC, 1979.
3. Erickson, D.R.; and L.H. Wiedermann. *inform* **1991**, 2, 200.
4. Latondress, E.G. *J. Am. Oil Chem. Soc.* **1981**, 58, 186.
5. Wiedermann, L.H. *J. Am. Oil Chem. Soc.* **1981**, 58, 159.
6. Norris, F.A. In *Bailey's Industrial Oil and Fat Products*, 4th ed.; D. Swern, Ed.; John Wiley & Sons: New York, 1982; Vol. 2, pp. 292–314.
7. Zehnder, C.T. *J. Am. Oil Chem. Soc.* **1976**, 53, 384.
8. List, G.A.; and D.R. Erickson. In *Bailey's Industrial Oil and Fat Products*, 4th ed.; T.H. Applewhite, Ed.; John Wiley & Sons: New York, 1985; Vol. 3, pp. 275–277.
9. Mag, T.K. In *Proceedings of the World Conference on Edible Fats and Oils Processing*; D.R. Erickson, Ed.; AOCS Press: Champaign, IL, 1990; pp. 107–116.

# 2

# Storage and Handling of Finished Frying Oils

**David R. Erickson**
*DJ Consultants, 507 Woodlake Drive, Santa Rosa, CA 95405-9203*

The users of frying fats can best ensure the initial quality of a finished product by specifying the quality parameters needed for their use (see Chapter 1). This may be done simply by using, and relying on, the provider's specifications or by setting a mutually agreed set of specifications with the supplier. Such specifications need to include only those that are critical to the end user in either case.

To better understand what is expected, it is necessary to start by considering what happens during frying: hydrolysis and oxidation. The tendency for either reaction depends first on the composition of the frying fat, and then by the product(s) being fried, the equipment being used, and the actual frying conditions.

Once a frying fat/oil is fully processed and ready for transport, it is at its peak quality the moment it leaves the final step of deodorization. The overall goal of subsequent storage, transport, and handling is to maintain and protect that quality. Further reference to frying fats/oil will be designated FFO for simplicity.

## Deterioration of Frying Fats

Practically, two main reactions lead to degradation of FFO. These are reaction with moisture leading to hydrolysis and reaction with oxygen (air) resulting in oxidation.

### Reaction of FFO with Moisture (Water)

The most usual reaction of FFO with water is hydrolysis. It occurs in the triglyceride fraction resulting in formation of free fatty acids:

$$\text{Triglyceride} + \text{water} = \text{Diglyceride} + \text{free fatty acids}$$
$$\text{Diglyceride} + \text{water} = \text{Monoglyceride} + \text{free fatty acids}$$
$$\text{Etc.}$$

For users of FFO, the moisture content of purchased products is essentially nil, barring any introduction of water during storage or transport after deodorization. A typical maximum moisture specification for fully processed fats and oils is < 0.10%. The solubility of water in oil is about 0.1% at 20°C (68°F) that will increase with temperature and decrease as the average chain length of the component fatty acids increase. For FFO with fatty acids having chain lengths of $C_{16}$, or greater, flavor deterioration due to hydrolysis is of little concern since these fatty acids are essentially flavorless. Fats and oils with component fatty acids < $C_{16}$ will develop off flavors often described as "hydrolytic rancidity." For very short chains, such as those found in butter, the resulting off flavors are quite pronounced. Hydrolysis of the longer chains such as the $C_{12}$ and $C_{14}$ found in coconut and palm kernel oils lead to "soapy" flavors.

In addition to the potential for soapy flavors due to fatty acids < $C_{16}$ there is another important factor to consider in the use of FFO. "Lauric" acid oils such as coconut and palm kernel oils, for reasons not well understood, but nevertheless practically evident, will foam excessively during frying when these oils are mixed with oils of longer chain length.

Contact of FFO with water is, of course, unavoidable when frying moisture-containing products. This may contribute to FFO breakdown due to some hydrolysis at the high temperatures of frying and is often said to occur because of the inevitable increase in FFA over time in a fryer. It should be noted that the determination of FFA is a simple titration to an endpoint of pH 8.3 with sodium hydroxide and the results expressed as %FFA (AOCS Method Ca5a-40).

As purchased, moisture in FFO is usually of little concern. Subsequent handling practices should be monitored to insure there are no opportunities to introduce water into the FFO. Most often this happens during cleaning operations and sometimes occurs by reflux, or "drip back" from steam exhausting systems, with the latter being especially detrimental to fry life from both hydrolysis and oxidation. This will be discussed later in the chapter.

Introduction of excess or free water to hot FFO decreases fry life. It is also extremely dangerous to personnel and equipment due to explosive spattering. It also requires unnecessary and costly expenditure of heat energy to evaporate and dissipate the excess moisture. It is recommended that every precaution be taken to remove free moisture from products before being placed into hot FFO. This applies to both small retail frying operations as well as in larger commercial operations. Excess moisture may also contribute unnecessarily to the reflux mentioned above.

As shown in Fig. 2.1, oxidation also leads to the formation of acidic products. As a result, what is expressed as FFA may include both actual FFA and acidic products produced by oxidation.

This is simply a point of clarification and does not lessen the usefulness of using FFA as a measure of FFO deterioration in a frying vessel. In fact, FFA is a simple and practical method of following and/or estimating such deterioration.

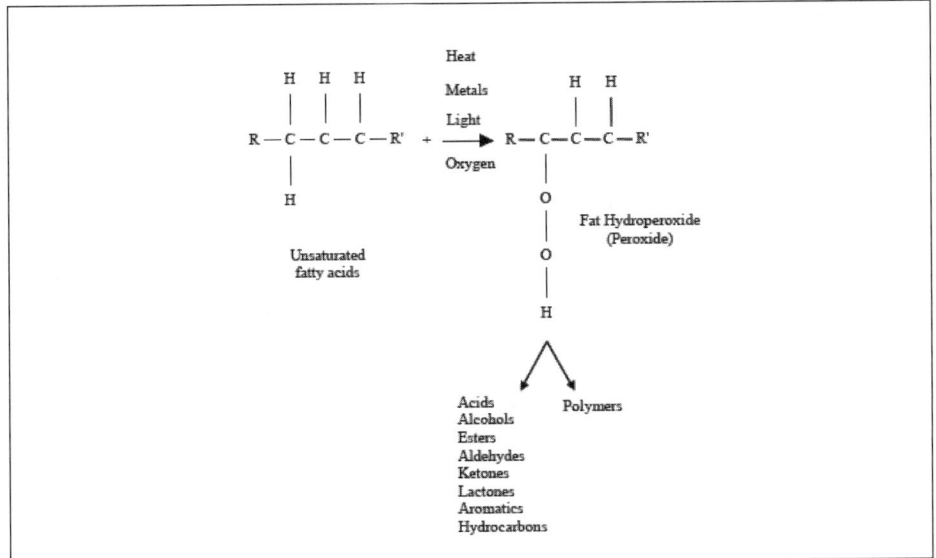

**Fig. 2.1.** *Source*: Bailey's *Industrial Oil and Fat Products*, Vol. 3, 4th edn, p. 275.

## Reaction of Frying Fats with Oxygen

Detailed mechanisms will be addressed in subsequent chapters. For purposes of this chapter the simplified diagram in Fig. 2.1 illustrates the oxidative reactions of FFO.

Practically speaking, the propensity for FFO to oxidize is related to their fatty acid composition (FAC), and more specifically to the amount and type of unsaturation. There may be some effect due to position of the fatty acid on the glycerol, but this is probably of minor consequence.

Relative rates of reaction of the three most common unsaturated fatty acids with oxygen are approximated and shown in Table 1.11. The relative rates of reaction for the type and amount of individual fatty acids can be useful for assessing the suitability of common fats and oils when formulating FFO. For any fat or oil, this is done by multiplying the decimal fraction of each unsaturated present by its relative rate of reaction and then summing to arrive at a relative measure of Inherent Stability (IS). The higher the IS, the more susceptible the fat or oil is to oxidation. IS is a convenient tool to calculate and assess the basic oxidative stability of a FFO when the fatty acid composition (FAC) is known. Using the FAC listed in *USDA Handbook 8-4* (1) the inherent stability of common fats and oils is shown in Table 1.13.

Inherent Stability also supports the need for better handling of those fats and oils with higher numbers. Use of unhydrogenated refined, bleached, and deodorized (RBD) cottonseed and peanut oils for frying is possible because of their lower IS number. Greater use of the more unsaturated oils such as soybean oil with higher IS required the industry to develop better handling practices. This included use of stainless steel equipment, reconsideration of "standard" refining practices, and improved finished product handling. It also invited the use of hydrogenation to improve the oxidative stability of finished products, especially for use as FFO. The inherent instability of any FFO can be easily and simply calculated from the fatty acid composition and provides a basis for comparing FFO in relation to their expected susceptibility to oxidation.

Catalytic factors, as shown in Fig. 2.1, also affect oxidative stability. Avoidance of copper or iron in finished product storage and in frying equipment is mandatory. Exposure to light is usually not of concern during the frying process but may be in packaged products such as fried snacks where large surface areas are exposed. Both opaque packaging and inert gas packing are particularly effective in preventing both light and oxidative induced off flavor development.

## Finished Product Handling

Finished FFO can be of two basic types that are either solid or fluid at normal room temperatures, i.e., 20–25°C.

### Solid FFO

Solid FFO are received by smaller fryers as cubes in boxes with distinctively colored plastic liners, which are usually blue. This is to assure that such plastic liners are easily distinguished from the normally white shortening, thereby reducing the chance of the plastic being introduced into the frying equipment.

Handling of such cubed product is straightforward as to maintaining quality. Storage should be at or near normal room temperature (20°C). Depending on the product, it may be necessary to warm the cube to a higher temperature to facilitate easier scooping or cutting portions for addition to frying equipment. This should be done for each cube as needed rather than the whole stock on hand, however. Use of cubes should be on a first-in, first-used basis and product storage on site should be kept at a minimum.

Larger commercial fryers may use solid shortenings that are shipped in tank cars or trucks and are shipped as melted product in insulated vessels or as solid product requiring re-melting at the plant site for off-loading. Product specifications for such bulk shipments may include inert gas (nitrogen) purging and blanketing during loading. Gas (nitrogen) blanketing should be considered at the off-loading site depending on the turnover rate at the frying site. Normally, the stability of solid FFO and the turnover rate is rapid enough to preclude the need for such inert gas use by the user.

It is the supplier's responsibility to assure use of properly cleaned tank cars and/or tank trucks and prevent commingling of finished FFO with other oils or chemicals. Dedication of tank cars or trucks to finished products is one method used by suppliers to avoid problems and could be a point of negotiation with potential suppliers.

## *Liquid and/or Pourable FFO*

Liquid and/or pourable FFO are used both in small frying operations and in larger commercial operations. Their use has increased due to their ease of handling, allowing simple pouring or pumping into frying vessels with no requirement for re-melting or introducing solid product into frying vessels.

Some fluid FFO are produced by fractionation of oils containing triglycerides that are solid at temperatures higher than normal room temperature. Such oils are usually more expensive due to the cost of fractionation but have been used when convenient outlets are available for use of the solid residuals. Such shortenings may be clear at room temperature as received but temperature fluctuations below room temperature can cause some settling out of the harder fractions, thereby requiring remixing to assure uniformity of product going to the frying vessels.

The obvious advantage of handling fluid and/or pourable FFO led to the development of opaque, or cloudy, pourable FFO and other shortenings that depend on the ability of producers to add harder fractions to hydrogenated products. Fluidity is then controlled by crystallization conditions. Such products will remain uniform and fluid at or near room temperature. Over time, some settling out of the solid fraction may occur but simple remixing can usually restore uniformity of the product. This will be reviewed again when considering bulk handling of pourable FFO.

These pourable FFO can be received and handled as fluids by the user, thus eliminating the cubes and packaging disposal related to solid products and the labor and danger of adding solid product to frying vessels. Instead of handling 50-lb cubes, the operator may use plastic containers, usually #35 "jug-in-a-box," and simply pour out the product. The product should be remixed as frequently as necessary to assure uniformity of product going into the frying vessel.

## Bulk Handling of Pourable FFO for Industrial Frying

Even for smaller frying operations there may be merit in considering bulk handling of pourable FFO thereby eliminating all packaging on-site. This is possible, and is being done, but care needs to be taken in designing a suitable system as outlined below:

1.  When receiving product, all lines, pumps, valves, tanks, etc. should ideally be constructed of stainless steel, with no opportunity for incorporation of air through pump suction, avoiding cascading by bottom filling. Plastic or epoxy coated metal equipment may be considered but may not be acceptable due to sanitary reasons and local requirements.
2.  Agitation in the holding vessel(s) is important and should not be overlooked due to the tendency for the more solid fractions to settle out over time. Uniformity of product can be assured by programming periodic gentle agitation or determining when agitation is needed by other means (see 3 below).
3.  Provision for sampling product in the holding vessel to determine whether

fractionation and/or degradation is occurring is critical.

4. Provision of two independent receiving vessels to allow complete emptying, subsequent inspection, and cleaning, if and when necessary, without interrupting operations.

5. Consideration should be given to both purging and blanketing incoming products to maintain maximum quality of purchased product. If turnover and/or oxidative stability of the FFO are high enough such use of inert gas may be unnecessary.

This has been a brief discussion of finished product handling related to finished FFO. A more extensive treatment of the overall subject of storage, handling, and stabilization of edible oils has been published by List and Erickson (2).

## Bulk Handling of Pourable FFO for Foodservice Frying

A relatively new service in the foodservice frying industry is bulk delivery of pourable frying oils to restaurants. Though the concept is not new, only recently has the economic justification been favorable because of the current industry trend of switching to oils with healthier perceptions that, at the same time, tend to be less stable and more expensive. Whereas frying fat was once considered an inexpensive part of doing business, its total management now commands the need to revisit all opportunities to stretch food/oil ratios without compromising quality. Moreover, the cost of energy, freight/transportation, and packaging all contribute significantly to the cost of traditional packaged frying fats and oils. To that end, bulk deliveries to restaurants continue to attract the attention of not only the major frying fats and

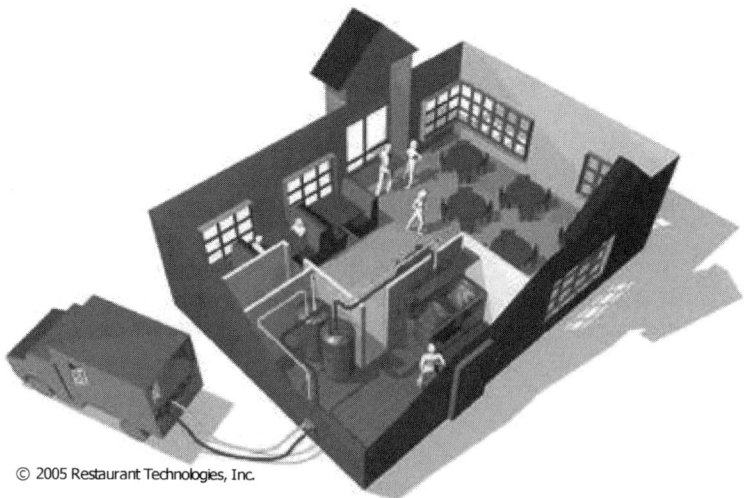

© 2005 Restaurant Technologies, Inc.

**Fig. 2.2.** Bulk oil concept (used by permission by Restaurant Technologies, Inc., Minneapolis, MN).

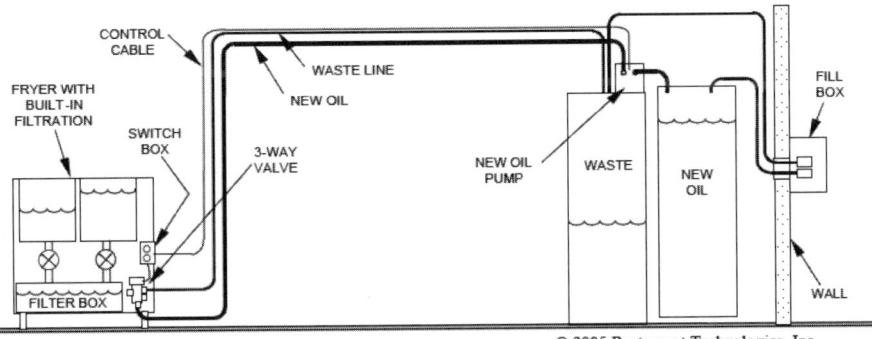

**Fig 2.3.** Bulk oil schematic overview (used by permission – Restaurant Technologies, Inc., Minneapolis, MN).

oils users, such as quick service restaurants, but even the more upscale "table cloth" restaurants. Figures 2.2 and 2.3 show a typical bulk delivery concept and system currently available by the pioneers and leaders in this area, Restaurant Technologies, Inc. (RTI).

## References

1. U.S. Department of Agriculture (USDA). *Agricultural Handbook 8-4*; USDA, Science and Education Administration: Washington DC, 1979.
2. List, G.L.; and D.R. Erickson. In *Bailey's Industrial Oil and Fat Products*, 4th ed.; T.H. Applewhite, Ed.; John Wiley & Sons: New York, 1985; Vol. 3, pp. 273–309.

# ·3·

# Initial Quality of Frying Oil

**Frank T. Orthoefer**[a] **and Gary R. List**[b]

[a]Germantown, TN, [b]USDA ARS NCAUR, Peoria, IL

Fried food has grown in popularity despite the low-fat/no-fat health trend. For example, between 1979 and 1988, the snack food industry in the United States increased by about 88% (1). The deep-frying process is commonly used by the snack food industry. The consumer, obviously, prefers the flavor and texture of fried food, especially when eating out (2). With the growth of fried food, there also has been continual improvement in the quality of food prepared by frying. Higher quality food ingredients, better frying oil, and improved frying equipment and frying practices have contributed to the improvement of fried food.

Fat or oil used for frying often determines the acceptability of food prepared with them. Although frying oil serves primarily as a heat exchange medium, oil often makes up a significant portion of the final food product, as much as 45% of the total product (3). Both physical and chemical changes occur in oil as a result of frying. These are due to partial oxidation as well as interaction between oil, water, and food components (4). Oil varies widely in eating quality, functionality, and rate of deterioration depending on source, processing, or formulation. An ingredient specification combined with a total quality management perspective for the oil component ensures production of high-quality frying oil and, subsequently, fried food.

## Definition of Quality

Quality is defined as the essential characteristic, distinguishing property, or degree or grade of excellence. The quality of a food may be defined as the composite of those characteristics that differentiate individual units of a product having significance in determining the degree of that product's acceptability (5,6).

The overall quality of a product or an ingredient is analyzed for its component attributes. For a fat or oil, quality comprises those attributes that affect the acceptability of food prepared with it, such as flavor, color, and texture. Commercially, quality is based on the average of those characteristics acceptable in the marketplace, and not necessarily on the highest quality available. Uniformity of oil is determined by whether it falls within the upper and lower control limits set for each characteristic. Thus, maintenance of oil quality acceptable to the buyer is the function of quality control at the food oil refinery.

## Quality in the Edible Oil Refinery

Traditionally, two terms have been used to describe how quality is monitored in processing facilities: quality control and quality assurance. Quality control is a process function with evaluations at each step of the process (7) with specifications set at each step. Interpretation of results followed by process adjustments ensure that the final product meets product specifications.

The function of quality assurance encompasses all aspects of oil processing including specifications, packaging, shipping, receiving, purchasing, sales, service, storage, and sanitation (8). In today's environment, there is a "partnering" between oil processor (supplier) and user (customer). Partnering implies that both are in business together and have a stake in the outcome. Since the early 1980s, quality assurance in processing industries has been referred to as a "total quality movement." No longer is quality concerned only with controlling the process to meet minimum product specifications. Quality is now assured through all aspects of production from management to shipping to product characteristics. Terms frequently used to describe today's quality programs are select supplier programs, ISO 9000, statistical process control (SPC), good manufacturing practices (GMP), and total quality management (TQM).

Within the edible oil refinery, some tools of TQM are flowcharts, Pareto diagrams, histograms, run charts, control charts, and process capability charts. All are part of SPC and enhance problem-solving capabilities. Each addresses preventing, detecting, and correcting problems during production. Some quality programs prescribe continuous improvement of products and processing.

Several factors contribute to producing poor quality oil products, thus increasing cost. Some causes of inadequate quality include management focused on quantity rather than quality of oil produced and personnel inadequately trained to perform required tasks and adjustments. In some instances, the cost of poor quality is hidden by reblending out-of-spec products, reprocessing "mistakes," blending low-quality raw materials, or ingredient substitutions. If the user does not understand the quality parameters for frying oil or insists on the lowest price without regard for quality, the real costs may be magnified through decreased consumer acceptance of fried food and loss of sales by the user. It is critical to have a reputable and credible oil supplier.

## Frying Oil Specifications

Specifications are prepared for oil to ensure that quality is acceptable. Specifications may originate from the producer or be part of buyers' specifications. Specifications are guidelines for the oil processor to purchase raw materials, process intermediates, and formulate products (9). A typical production specification at the oil refinery is shown in Table 3.1. The specification includes the source or composition of the oil as well as minor ingredients that may be added. Both subjective and objective characteristics and their numerical values are identified.

**TABLE 3.1**
**Typical Specifications for Winterized Soybean Oil[a]**

| Specifications | | Specifications | |
|---|---|---|---|
| Color (5 1/4" Lovibond) | 1.0 max | Iodine value (Wijs) | 108–112 |
| Free fatty acid (% as oleic) | 0.05 max | Drop point (°F) | |
| Peroxide value (meg/kg) | 0.5 max | SFI[b] 50°F | |
| Flavor/odor | 7 min | 70°F | |
| Filter test | 6 | 80°F | |
| Moisture (%) | 0.05 max | 92°F | |
| Appearance | clear | 104°F | |
| Chlorophyll (ppb) | 75 max | AOM[c] | |
| Cold test (h) | 10 min | Linolenic acid (%) | 3.0 max |

[a]Composition: ingredients, winterized SB–110W 100%; additives, none.
[b]SFI = Solid Fat Index.
[c]AOM = Active Oxygen Method.

Typical characteristics for an oil are shown in Table 3.2. Determinations of chemical and physical characteristics of an oil are made during oil refining, in storage after processing, and prior to using the finished oil.

# Quality Analysis

Numerous analytical determinations are performed at the oil refinery laboratory (10). Technicians are trained in the use of special methods and equipment to perform these determinations. Generally, the methods used to assess quality have been developed by professional organizations, particularly the American Oil Chemists' Society (AOCS), Association of Official Analytical Chemists (AOAC), and American Society for Testing Materials (ASTM). Industries, universities, and private laboratories also collaborated with these organizations in developing relevant and accurate methods.

# Refinery Quality

With oil, there is a natural fluctuation in its composition and characteristics. In the oil refinery, variations in crude oil often lead to process upsets and product quality deviations. Grading systems have been developed for many crude oils (11). These systems specify grades with premiums or discounts based on refining tests. Through timely assessment and continual monitoring of the oil at each processing step, the process may be adjusted to produce an oil that meets particular grades and product specifications. The objective of oil processing is to achieve an end product within the chemical and physical characteristics defined by product specifications.

**TABLE 3.2**
**Quality Characteristics Determined During Processing, After Storage and Use**[a]

| Characteristic | Oil refining | Storage | Use |
|---|:---:|:---:|:---:|
| | | Tests performed during: | |
| Color | ✓ | | |
| Free fatty acid | ✓ | ✓ | ✓ |
| Peroxide value | ✓ | ✓ | ✓ |
| Filter test | ✓ | | |
| Moisture | ✓ | ✓ | |
| Chlorophyll | ✓ | | |
| Cold test | ✓ | | |
| Iodine | ✓ | | |
| Drop point | ✓ | | |
| AOM | ✓ | | |
| SFI @10°C | ✓ | | |
| 21.1°C | | | |
| 26°C | | | |
| 33.3°C | | | |
| 40°C | | | |
| Flavor/odor | ✓ | ✓ | ✓ |
| Appearance | ✓ | ✓ | ✓ |

[a]Abbreviations are as in Table 3.1.

# Processing of Crude Oil

Edible fats and oils are derived from several plants (soybean, corn, sunflower, rapeseed, or canola) and animal sources (tallow, lard, fish, and butterfat) (12). They are differentiated by source, fatty acid chain length, and physical form. Each has a unique fatty acid composition that influences its suitability for various applications (Table 3.3). In addition to the identity of oil, geographical origin or the method of recovery may be identified.

Crude oil refers to any oil in its native state. Components that adversely impact flavor, odor, appearance of finished product, or performance of an oil during use are removed during processing. The composition of crude soybean oil is shown in Table 3.4. In addition to the undesirable flavor, odor, and color, the presence of free fatty acids (FFA), partial glycerides, and phospholipids make the crude oil unsuitable for deep-frying use.

The common steps of oil processing include (Table 3.5): i) filtration and degumming, ii) alkaline or physical refining, iii) bleaching, iv) hydrogenation, v) winterization or fractionation, vi) deodorization, and vii) packaging. The objectives of oil refining are to isolate, in a commercially practical sense, pure triglycerides from crude oil. The triglycerides may then be modified or combined with additives to obtain desirable performance and packaged in a usable form.

## Objectives of Processing Steps

### Filtration and Degumming

This step removes small amounts of nontriglyceride components, such as residual meal, metal fragments, dirt and gums (hydrated phospholipids), and other insolubles from crude oil (13). A filter aid, such as diatomaceous earth (DE), is slurried with the crude oil prior to filtration to assist the flow of oil through the filter.

After filtration, the oil is mixed with small quantities of softened water. Hydratable components are allowed to hydrate; then the oil–water mixture is centrifuged to remove the higher density wet gums from the less dense oil. Properly filtered and degummed oil is clear and brilliant with less than 20 ppm phosphorus. The adequacy of filtration and degumming are determined by clarity or filter tests and residual phosphorus analysis.

### Alkali or Physical Refining

This step is utilized to remove FFA from oil (14). Alkali refining consists of the reaction of dilute sodium or potassium hydroxide with the FFA to form water-soluble soaps (15). The amount of alkali or caustic to be added is optimized to remove only FFA and prevent further hydrolysis of triglycerides (16). The reaction of the alkali with FFA forms a thick stable emulsion, called soapstock, consisting of soaps, triglycerides, and water. The emulsion is passed through a refining centrifuge to separate the high-density soapstock and water mixture from oil. Final soap contents are normally less than 50 ppm, and FFA is near zero.

Physical refining does not involve caustic addition (17). Oil is steam sparged under a 4- to 6-mm Hg vacuum at more than 200°C. All volatile components, including FFA, are stripped from oil. Free fatty acids in soybean oil, for example, are reduced from about 1.5% in crude oil to less than 0.05% in finished oil (18).

### Bleaching

Bleaching removes color bodies, residual soaps, and phosphatides from refined oil (19). Bleaching is an adsorption process. The adsorbents, referred to as bleaching clays, consist of calcium montmorillonite, natural hydrated alumina silicate, silicon dioxide, or activated carbon (20). Bleaching consists of adding adsorbents to oil followed by heating under a vacuum. The adsorbents are then filtered from the oil. Filter tests ensure complete removal of clay adsorbents.

### Hydrogenation

This is the process of chemically reacting an oil with hydrogen gas in the presence of

**TABLE 3.3**
**Typical Composition and Chemical Constants of Common Edible Fats and Oils**

| Oil/Fat | Iodine Value Range | Saponification Value Range | Butyric 4:0 | Caproic 6:0 | Caprylic 8:0 | Capric 10:0 | Undecanoic 11:0 | Lauric 12:0 | Tridecanoic 13:0 | Myristic 14:0 | Myristoleic 14:1 | Pentadecanoic 15:0 | Pentadecenoic 15:1 |
|---|---|---|---|---|---|---|---|---|---|---|---|---|---|
| Babassu | 13–18 | 247–254 | 0.4 | 5.3 | 5.9 | | | 44.2 | | 15.8 | | | |
| Butterat | 25–42 | 210–240 | 3.8 | 2.3 | 1.1 | 2.0 | 0.1 | 3.1 | 0.1 | 11.7 | 0.8 | 1.6 | |
| Chicken Fat | 76–80 | 194–204 | | | | | | 0.2 | | 1.3 | 0.2 | | |
| Citrus Seed Oil | 99–106 | 192–197 | | | | | | 0.1 | | 0.5 | | | |
| Cocoa Butter | 32–40 | 190–200 | | | | | | | | 0.1 | | | |
| Coconut Oil | 7–13 | 248–264 | 0.5 | 8.0 | 6.4 | | | 48.5 | | 17.6 | | | |
| Cohune Oil | 8–14 | 250–260 | 0.3 | 8.7 | 7.2 | 0.1 | | 47.3 | | 16.2 | | | |
| Corn Oil | 110–128 | 186–196 | | | | | | | | | | | |
| Cottonseed Oil | 99–121 | 189–199 | | | | | | | | 0.9 | | | |
| Lard | 53–68 | 192–203 | | | | 0.1 | | 0.1 | | 1.5 | | 0.2 | |
| Murumura Tallow | 8–13 | 237–247 | 0.1 | 1.3 | 1.5 | | | 46.2 | | 32.4 | | | |
| Oat Oil | 105–110 | 180–198 | | | | | | | | 0.2 | | | |
| Olive Oil | 76–90 | 188–196 | | | | | | | | | | | |
| Palm Oil | 45–56 | 195–205 | | | | | | 0.3 | | 1.1 | | | |
| Palm Kernel Oil | 14–24 | 243–255 | 0.3 | 3.9 | 4.0 | | | 49.6 | | 16.0 | | | |
| Peanut Oil | 84–102 | 188–196 | | | | | | | | 0.1 | | | |
| Rapeseed Oil | 97–110 | 168–183 | | | | | | | | 0.1 | | | |
| Canola Oil | 110–115 | – | | | | | | | | | | | |
| Rice Bran Oil | 92–109 | 181–195 | | | 0.1 | 0.1 | | 0.4 | | 0.5 | | | |
| Safflower Oil | 138–151 | 186–198 | | | | | | | | 0.1 | | | |
| Safflower Oil (High Oleic) | 85–93 | 185–195 | | | | | | | | 0.1 | | | |
| Sesame Seed Oil | 104–118 | 187–196 | | | | | | | | | | | |
| Soybean Oil | 125–138 | 188–195 | | | | | | | | 0.1 | | | |
| Sunflower Oil | 122–139 | 186–196 | | | | | | 0.5 | | 0.2 | | | |
| Tallow (Beef) | 33–50 | 190–202 | | | | 0.1 | | 0.1 | | 3.3 | 0.2 | 1.3 | 0.2 |
| Tallow (Mutton) | 35–46 | 192–198 | | | | 0.2 | | 0.3 | | 5.2 | 0.3 | 0.8 | 0.3 |
| Tucum Oil | 10–14 | 240–250 | 0.2 | 2.9 | 2.3 | | | 51.8 | | 22.0 | | | |
| Uchuhuba Tallow | 6–17 | 215–232 | | | 0.3 | 0.8 | | 16.3 | 0.3 | 70.8 | 1.3 | | |

a catalyst (21). The catalyst is generally reduced metallic nickel. The hydrogen reacts at the unsaturated double bonds of fatty acids as these bonds are the most reactive groups in the triglyceride. Hydrogenation reduces the degree of unsaturation. This dramatically changes the chemical and physical attributes of fat or oil. Both oxidative and thermal stability of oil are increased with increased hydrogenation. Oil may be physically converted from a liquid oil to semisolid or solid fat during hydrogenation.

When the desired degree of hydrogenation is achieved, as represented by iodine value (IV), the oil is slurried with DE and passed through a filter to remove the catalyst. Citric acid may be added to chelate and remove any residual nickel catalyst or acid-activated clay.

The extent of hydrogenation is readily determined by IV, which is defined as the

**TABLE 3.3, CONT**
**Typical Composition and Chemical Constants of Common Edible Fats and Oils**

| Palmitic | Palmitoleic | Margaric | Margaroleic | Stearic | Oleic | Linoleic | Linolenic | Nonadecanoic | Arachidic | Gadoleic | Eicosadienoic | Arachidonic | Behenic | Erucic | Docosadienoic | Lignoceric |
|---|---|---|---|---|---|---|---|---|---|---|---|---|---|---|---|---|
| 16:0 | 16:1 | 17:0 | 17:1 | 18:0 | 18:1 | 18:2 | 18:3 | 19:0 | 20:0 | 20:1 | 20:2 | 20:4 | 22:0 | 22:1 | 22:2 | 24:0 |
| 8.6 | | | | 2.9 | 15.1 | 1.7 | | | 0.1 | | | | | | | |
| 26.2 | 1.9 | 0.7 | 0.2 | 12.5 | 28.2 | 2.9 | 0.5 | | | 0.2 | | 0.1 | | | | |
| 23.2 | 6.5 | 0.3 | 0.1 | 6.4 | 41.6 | 18.9 | 1.3 | | | | | | | | | |
| 28.4 | 0.2 | | | 3.5 | 23.0 | 37.8 | 5.7 | | 0.8 | | | | | | | |
| 25.8 | 0.3 | | | 34.5 | 35.3 | 2.9 | | | 1.1 | | | | | | | |
| 8.4 | | | | 2.5 | 6.5 | 1.5 | | | 0.1 | | | | | | | |
| 7.7 | | | | 3.2 | 8.3 | 1.0 | | | | | | | | | | |
| 12.2 | 0.1 | | | 2.2 | 27.5 | 57.0 | 0.9 | | 0.1 | | | | | | | |
| 24.7 | 0.7 | 0.1 | | 2.3 | 17.6 | 53.3 | 0.3 | | 0.1 | | | | | | | |
| 24.8 | 3.1 | 0.5 | 0.3 | 12.3 | 45.1 | 9.9 | 0.1 | | 0.2 | 1.3 | 0.1 | 0.4 | | | | |
| 5.6 | 0.1 | | | 2.2 | 8.9 | 1.5 | | | 0.2 | | | | | | | |
| 17.1 | 0.5 | | | 1.4 | 33.4 | 44.8 | | | 0.2 | 2.4 | | | | | | |
| 13.7 | 1.2 | | | 2.5 | 71.1 | 10.0 | 0.6 | | 0.9 | | | | | | | |
| 45.1 | 0.1 | | | 4.7 | 38.8 | 9.4 | 0.3 | | 0.2 | | | | | | | |
| 8.0 | | | | 2.4 | 13.7 | 2.0 | | | 0.1 | | | | | | | |
| 11.6 | 0.2 | 0.1 | | 3.1 | 46.5 | 31.4 | | | 1.5 | 1.4 | 0.1 | | 3.0 | | | 1.0 |
| 2.8 | 0.2 | | | 1.3 | 23.8 | 14.6 | 7.3 | | 0.7 | 12.1 | 0.6 | | 0.4 | 34.8 | 0.3 | 1.0 |
| 3.9 | 0.2 | | | 1.9 | 64.1 | 18.7 | 9.2 | | 0.6 | 1.0 | · | | 0.2 | | | 0.2 |
| 16.4 | 0.3 | | | 2.1 | 43.8 | 34.0 | 1.1 | | 0.5 | 0.4 | | | 0.2 | | | 0.1 |
| 6.5 | | | | 2.4 | 13.1 | 77.7 | | | 0.2 | | | | | | | |
| 5.5 | 0.1 | | | 2.2 | 79.7 | 12.0 | 0.2 | | 0.2 | | | | | | | |
| 9.9 | 0.3 | | | 5.2 | 41.2 | | | | | | | | | | | |
| | | | | | | 43.2 | 0.2 | | | | | | | | | |
| 11.0 | 0.1 | | | 4.0 | 23.4 | 53.7 | 7.8 | | 0.3 | | | | 0.1 | | | |
| 6.8 | 0.1 | | | 4.7 | 18.6 | 68.2 | 0.5 | | 0.4 | | | | | | | |
| 25.5 | 3.4 | 1.5 | 0.7 | 21.6 | 38.7 | 2.2 | 0.6 | 0.1 | 0.1 | | | 0.4 | | | | |
| 23.6 | 2.5 | 2.0 | 0.5 | 24.5 | 33.3 | 4.0 | 1.3 | 0.8 | | | | 0.4 | | | | |
| 6.8 | | | | 2.3 | 9.3 | 2.4 | | | | | | | | | | |
| 4.3 | 0.5 | | | 0.7 | 9.3 | 0.6 | 0.3 | | 0.1 | | | | | | | |

grams of iodine that combine with 100 g of oil. A lightly hydrogenated soybean oil has an IV of 110 and is a cloudy fluid at room temperature. Continued hydrogenation of oil to less than 90 IV produces a semisolid fat. Oil with less than 70 IV has a hard-to-brittle consistency.

Hydrogenation may result in the formation of chemical isomers, particularly *trans* fatty acid isomers (22). Native vegetable oil contains only *cis* isomers. *Trans* fatty acid isomers are present in low quantities in animal fat. The amount of *trans* isomers formed during hydrogenation depends on hydrogenation conditions of temperature, time, hydrogen gas pressure, and type of catalyst.

**TABLE 3.4**
**Average Compositions for Crude and Refined Soybean Oil**

|  | Crude Oil | Refined Oil |
|---|---|---|
| Triglycerides % | 95–97 | >99 |
| Phosphatides % | 1.5–2.5 | 0.00–0.045 |
| Unsaponifiable matter % | 1.6 | 0.3 |
| Plant sterols % | 0.3 | 0.13 |
| Tocopherols % | 0.15–0.21 | 0.11–0.18 |
| Hydrocarbons (squalene) % | 0.014 | 0.01 |
| Free fatty acids % | 0.3–0.7 | <0.05 |
| Trace metals |  |  |
| Iron ppm | 1–3 | 0.1–0.3 |
| Copper ppm | 0.03–0.05 | 0.02–0.06 |

**TABLE 3.5**
**Processing Steps in Production of Finished Oils**

| Crude Oil | | |
|---|---|---|
| Filtration | | |
| Water degumming | Lecithin (wetgums) | |
| Alkali refining | | Soapstock |
| Water washing | | |
| Bleaching | | |
| Hydrogenation | | |
| Winterization | | |
| Oil blending | | |
| Deoderization | Deodorizer distillate | |
| Finished oil | | |

*Winterization or Fractionation*

This step is the removal of high-melting triglycerides from lower-melting components. For example, a lightly hydrogenated 110 IV soybean oil is cloudy at room temperature because of the presence of high-melting triglycerides. In winterization, oil is slowly cooled to force crystallization of higher-melting glycerides (23). The crystallized components are then removed by filtration, producing a clear fluid oil at room temperature. Soybean oil hydrogenated to less than 95 IV, then winterized, produces a high-stability frying shortening that is fluid at room temperature. Palm oil is generally fractionated into low-melting olein and high-melting stearin components. Winterized oil is tested for resistance to crystallization at ice water temperatures. A minimum number of hours, such as 10, is commonly used for hydrogenated, winterized soybean oil.

*Deodorization*

Deodorizing oil removes the final traces of volatile components, primarily those that contribute to flavor and odor (24,25). The goal of deodorization is to produce a nearly flavorless, odorless oil with a light color. Deodorization consists of steam-sparging the oil under high vacuum (< 10 mm Hg) at high temperatures (> 200°C). After deodorization and cooling of the oil, a chelating agent, such as citric acid, may be added to deactivate trace metals. Antioxidants may also be added to enhance stability. Commonly used antioxidants are TBHQ (tertiary butyl hydroquinone), BHA (butylated hydroxy aniline), BHT (butylated hydroxy toluene), ascorbyl palmitate, natural antioxidants, and tocopherols.

*Packaging*

Packaging oil may include additional processing steps. For bulk storage of frying oil, vertical tanks are preferred. Agitation or circulation and nitrogen blanketing prevents stratification and contact of oil with air during storage. Maintaining oil temperatures above the melting point is necessary to prevent solidification. Stainless steel storage tanks are preferred to prevent contact of oil with iron or other proxidant metals.

Oil for packaging is nitrogen flushed to prevent air contact. Saturating the oil with nitrogen also prevents partial collapse of plastic containers upon cooling of the oil in the container. Specially formulated frying oils, such as liquid frying shortening, require a positive nitrogen headspace. Votated plastic shortening in cartons contains 10–15% nitrogen gas to improve handling and appearance.

# Finished Product Quality

Routine tests performed at the edible oil refinery are shown in Table 3.2 (26,27). Finished product specifications provide information on composition of the oil, chemical characteristics of the processed product, and relative product performance.

## Oil Composition

Frying oil generally consists of a single source. Common additives to frying oil are antioxidants and antifoams. In some instances, blends of source oils are utilized to impart a desired flavor (e.g., peanut oil with soybean oil), improve stability (e.g., cottonseed oil with soybean oil), or modify texture and appearance (e.g., cottonseed oil and tallow). Chemically combined oils produced by interesterification are also available (e.g., coconut oil with soybean oil). Oil may also be partially hydrogenated. The composition of the blend is determined by fatty acid composition studies using gas–liquid chromatography (GLC). Other indicators of composition are melting point and IV.

The presence of antioxidants in an oil requires analysis for the individual types present. TBHQ, BHA, and BHT are the most commonly used antioxidants. The maximum permitted level is limited to 200 ppm for any single antioxidant or 200 ppm for combinations of antioxidants (28).

The antifoam used for frying oil is dimethylpolysiloxane, often referred to as DMS or silicone. DMS is limited to 10 ppm (28). Analysis for DMS requires analysis for silicon in oil and is often performed by atomic absorption spectrometry.

## Chemical Characteristics of the Processed Product

Quality characteristics that relate to the efficiency of the refining process are color, FFA, peroxide value (PV), flavor and odor, filter, and moisture (Table 3.6).

### Color

The color of an oil depends on its source and degree of processing. Carotenoid and chlorophyll pigments are responsible for color. Chlorophylls may also be a proxidant that limits the useful life of an oil. Color is removed by bleaching. Final oil color is determined by a color comparison technique, such as the Lovibond system. In this system, color is expressed as red and yellow components. For example, the color of fully refined soybean oil may be 0.8 R, 8.0 Y (0.8 red, 8.0 yellow). Chlorophyll may be determined separately by spectrophotometry.

### FFA

Free fatty acid analysis indicates efficiency of alkali refining and deodorization. Less than 0.05% FFA is the industry standard for finished oil. Free fatty acid may be easily determined by titration with an acid or base indicator. Newly developed indicator test strips are also used for FFA analysis.

### Peroxide Value

Peroxide value (PV) is primarily an indication of oil oxidation (hydroperoxide formation). Fully refined oil, after deodorization, has a PV of zero. Peroxide value steadily increases post deodorization. Peroxide value is determined by reacting the oil with iodine followed by titration with sodium thiosulfate. Oil from a refinery should

**TABLE 3.6**
**Routine Quality Control Tests for Soybean Oil**

| Determination | AOCS[a] Method |
|---|---|
| Lovibond color | Cc13e-92 |
| Karl Fischer moisture | Ca 2e-84 |
| Oven moisture (moisture and volatile matter-air oven method) | Ca 2c-25 |
| Free fatty acids | Ca 5a-40 |
| Soap in oil (titrimetic method) | Cc 17-79 |
| Peroxide value (acetic acid-chloroform method) | Cd 8-53 |
| Solids (sediment; insoluble impurities) | Ca 3a-46 |
| Phosphorus content in oil | Ca 12-55 |
| Phospholipids in oil (nephelos) | Ca 19-86 |
| Iodine value | Cd 1-25 |
| Refractive index | Cc 7-25 |
| Capillary melting point | Cc 1-25 |
| Dropping point | Cc 18-80 |
| Solid fat index | Cd 10-57 |
| Solid fat content (by NMR[b]) | Cd 16b-93 |
| Active oxygen method | Cd 12-57 |
| Oxidative stability (by OSI[c]) | Cd 12b-92 |
| Fatty acid composition (by GLC[d]) | Ce 1-62 |
| Metals by atomic absorption | Ca 15-75 |
| | Ca 18-79 |
| Flavor/odor | Cg 1-83 |
| Metals (by ICP)[e] | AOAC 985.01[f] |

[a]AOCS = American Oil Chemists' Society.
[b]NMR = Nuclear magnetic resonance.
[c]OSI = Oil stability instrument.
[d]GLC = Gas-liquid chromatography.
[e]ICP = Inductively couples plasma.
[f]Association of Official Analytical Chemists.

be less than 1.0 PV.

## Oil Flavor and Odor

These factors are influenced by partial oxidation. Oxidation can produce off flavor and odor in an oil. Flavor analysis in the refinery is commonly performed by a laboratory analyst on a pass/fail basis. Formal taste analyses may be conducted by trained panelists (Fig. 3.1) (29). Flavor is graded against descriptors of known standards. A numerical grade between 1 (lowest) and 10 (highest) is assigned. A minimum score of 7.0 is often required before shipping oil from refineries. Descriptors of a satisfactory oil flavor are "nutty," "bland," or "buttery." Unsatisfactory descriptors of oil flavor are "rancid," "burnt," "painty," or "fishy."

## Filter Test

A filter test is performed to ensure clarity and absence of particulates in oil. Particulates may consist of bleaching earth, filter aid, or insoluble material from storage tanks suspended in oil. Particulates are determined by filtering a defined quantity of oil through a filter and comparing the filter disk to acceptable standards.

## Moisture Analysis

Moisture analysis is an indicator of refining upsets or equipment failure. The industry standard for moisture content in a finished oil is a maximum of 0.05% water. Analysis

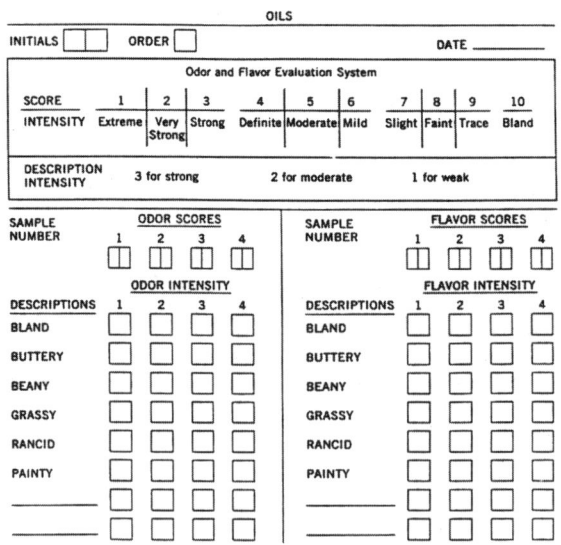

**Fig. 3.1.** Flavor Evaluation Score Sheet

is performed by the Karl Fischer potentiometric technique. High moisture levels are visible, however, and are unacceptable.

## Relative Product Performance

Quality characteristics that relate to product performance and fried product acceptability are solid fat index (SFI), active oxygen method (AOM), melting point (drop point), and iodine value (IV).

### Solid Fat Index (SFI)

Solid fat index, or solid fat content (SFC), generally characterizes "plastic" fat such as in margarine or shortening. Solid fat index measures solid fat present over a defined temperature range. Solid fat index is determined by dilatometry, and SFC by nuclear magnetic resonance (NMR) (30). The water bath temperatures used are 10, 21.1, 26.7, 33.3, and 40°C. With frying oil, plasticity is important, particularly in the frying of doughnuts as well as in other bakery products. Solid content also affects appearance of the fried product. Products fried in oil with higher solids have a drier, less greasy appearance compared to products prepared with liquid oil.

### AOM

The active oxygen method (AOM) measures resistance of oil to oxidation under defined conditions. An instrumental version called an oil stability instrument (OSI) is also available (31,32). The method consists of subjecting oil to high temperatures and air exposure, similar to the conditions of a deep fryer. The volatiles generated over time are monitored. The time to reach a specific end point is a reproducible characteristic of oil. An oil stability instrument provides a relative correlation to the fry life of oil.

### Melting Point

There are various methods for melting point determination. Most have been replaced by the automated dropping point method. This method measures the minimum temperatures for oil to flow through a defined orifice. Melting point is an important characteristic to monitor for maintenance of product consistency.

### Iodine Value (IV)

Iodine value is a measure of the average degree of unsaturation of oil determined by titration with active iodine. Refractive index is a rapid technique used to correlate to IV. The greater the unsaturation (or high IV), the more rapid the oil tends to oxidize, particularly in deep frying. Oil product consistency in terms of IV relates to susceptibility to oxidative deterioration and potential frequency of oil discard.

## Determination of Purchased Oil Quality

The quality of frying oil depends on the purchase specification for oil and the reliability and credibility of the supplier or processor to produce oil that meets this specification. It is important for the purchaser to incorporate only essential requirements into a specification. Analytical variance must be taken into account when writing a specification.

The evaluations most frequently determined by the purchaser include FFA content, PV, flavor, color, IV, and fatty acid profile. Less often determined quality characteristics are AOM (or OSI), IV, and smoke point. The fry life of oil, perhaps most important to the operator of a deep frying operation, is seldom determined because actual frying is the only way to measure frying performance (33).

Specialized techniques are used to evaluate modified frying oil, such as fluid or plasticized shortenings. Fluid shortening must remain pourable, and the solids must remain suspended. Measuring viscosity using a rotational viscometer and noting product appearance are generally sufficient for evaluating an oil. A partial separation in the container of solid fat, appearing as clumps, indicates improper formulation or inadequate tempering of fluid shortening.

Plastic shortening consistency is dependent on formulation, votation conditions, and tempering prior to shipment. Consistency is generally determined by a penetrometer.

## Summary

The initial quality of a frying oil may have a significant impact on the quality of fried food prepared with it. The initial quality of an oil, as well as its durability during frying, is influenced by all steps involved in processing the oil.

The quality of an oil is assessed against standards or specifications. Various methods are available to evaluate the initial oil quality and durability during use. Specifications, however, must be meaningful and realistic to the oil processor and purchaser.

## Suggested Reading

Since publication of the first edition of this book in 1996, a number of excellent reviews have been published. Gupta et al. (34) recently published a book entitled *Frying Technology and Practices* that presents an overview of the frying industry, selection of frying fats, oil handling, frying operations, design variables in industrial fryer construction, coatings used in fried food products, packaging of fried foods, toxicology of frying fats, and regulatory requirements for the frying industry. Mounts (35) presents an overview of frying oils including deterioration, the use of metal chelating agents, antioxidants, and nitrogen-blanketing to protect and extend the life of frying fats. Methods to establish the quality of used frying fats are also reviewed. Another excellent review of all aspects of frying oils has been published in *Bailey's Industrial Oil and Fat Products*, 6th edition, and should be consulted for further information (36).

## References

1. Tettweiler, P. *Food Technol.* **1991,** *45*, 58–62.
2. Singh, P. *Food Technol.* **1995,** *49*, 134–137.
3. Saguy, I.S.; and E.J. Pinthus. *Food Technol.* **1995,** *49*, 142–145.
4. Fritsch, W.W. *J. Am. Oil Chem. Soc.* **1981,** *58*, 272.
5. Kramer, A.; and B.A. Twigg. *Quality Control for the Food Industry*, 3rd ed.; AVI Publishing: Westport, CT, 1970.
6. Lawson, H.W. In *Standards for Fats & Oils*; AVI Publishing: Westport, CT, 1985; Vol. 5.
7. Croy, C. *inform* **1993,** *4*, 1034–1040.
8. Croy, C. *inform* **1993,** *4*, 884–895.
9. Brekke, O.L. In *Handbook of Soy Oil Processing and Utilization*; D.R. Erickson, E.H. Pryde, O.L. Brekke, T.L. Mounts, and R.A. Falb, Eds.; American Oil Chemists' Society: Champaign, IL, 1980; pp. 377–381.
10. Podmore, J. In *Edible Fats and Oils Processing: Basic Principles and Modern Practices*; D.R. Erickson, Ed.; American Oil Chemists' Society: Champaign, IL, 1990; pp. 374–389.
11. National Oilseed Processors' Association. *Yearbook and Trading Rules: 1995–1996*; National Oilseed Processor's Association: Washington, DC, 1995.
12. Weiss, T.J. In *Food Oils and Their Uses*; AVI Publishing: Westport, CT, 1983; pp. 65–101.
13. Hunter, J.E. *J. Am. Oil Chem. Soc.* **1981,** *58*, 283–287.
14. Young, F.V.K. In *Proceedings of the American Soybean Association Symposium on Soybean Processing*; R. Leysen, Ed.; American Soybean Association: St. Louis, MO, 1981; Paper No. 4.
15. Tatum, V.; and C.K. Chow. *J. Food Sci. Technol.* **1992,** *53*, 337–351.
16. Mounts, T.L.; and R.A. Anderson. In *Lipids in Cereal Technology*; P.J. Barnes, Ed.; Academic Press: New York, 1983; pp. 373–387.
17. Thiagarajan, T. In *Edible Fats and Oils Processing: Basic Principles and Modern Practices*; D.R. Erickson, Ed.; American Oil Chemists' Society: Champaign, IL, 1990; pp. 362–365.
18. Mounts, T.L.; and F.P. Khym. In *Handbook of Soy Oil Processing and Utilization*; D.R. Erickson, E.H. Pryde, O.L. Brekke, T.L. Mounts, and R.A. Falb, Eds.; American Oil Chemists' Society: Champaign, IL, 1980; pp. 89–105.
19. Mag, T. In *Edible Fats and Oils Processing: Basic Principles and Modern Practices*; D.R. Erickson, Ed.; American Oil Chemists' Society: Champaign, IL, 1990; pp. 107–116.
20. Hastert, R.C. *Introduction to Fats and Oils Technology*; American Oil Chemists' Society: Champaign, IL, 1991; pp. 95–104.
21. Hastert, R.C. *Introduction to Fats and Oils Technology*; American Oil Chemists' Society: Champaign, IL, 1991; pp. 114–136.
22. Mounts, T.L., In *Handbook of Soy Oil Processing and Utilization*; D.R. Erickson, E.H. Pryde, O.L. Brekke, T.L. Mounts, and R.A. Falb, Eds.; American Oil Chemists' Society: Champaign, IL, 1980; pp. 131–144.
23. Latondress, E.G. *J. Am. Oil Chem. Soc.* **1983,** *60*, 257–261.
24. Mounts, T.L. *J. Am. Oil Chem. Soc.* **1981,** *58*, 51A–54A.
25. Brekke, O.L. In *Handbook of Soy Oil Processing and Utilization*; D.R. Erickson, E.H.

Pryde, O.L. Brekke, T.L. Mounts, and R.A. Falb, Eds.; American Oil Chemists' Society: Champaign, IL, 1980; pp. 105–130.

26. Roden, A.; and G. Ullyot. *J. Amer. Oil Chem. Soc.* **1984,** *61*, 1109–1111.

27. Thomas, A. In *Proceedings of the American Soybean Association Symposium on Soybean Processing*; R. Leysen, Ed.; American Soybean Association: St. Louis, MO, 1981; Paper No. 4.

28. *Code of Federal Regulations*, 173.34.

29. Brekke, O.L. In *Handbook of Soy Oil Processing and Utilization*, D.R. Erickson, E.H. Pryde, O.L. Brekke, T.L. Mounts, and R.A. Falb, Eds.; American Oil Chemists' Society: Champaign, IL, 1980; pp. 67–70.

30. Sleeter, R. *J. Am. Oil Chem. Soc.* **1983,** *60*, 343–349.

31. Matlock, M. *J. Am. Oil Chem. Soc.* **1988,** *65*, 530; Abstract KK10.

32. Matlock, M. In *Edible Fats and Oils Processing: Basic Principles and Modern Practices*; D.R. Erickson, Ed.; American Oil Chemists' Society: Champaign, IL, 1990; pp. 385–389.

33. Lawson, H.W. *Standards for Fats & Oils*; AVI Publishing: Westport, CT, 1985; Vol. 5, pp. 44–48.

34. Gupta, M.K.; K. Warner; and P.J. White, Eds. *Frying Technology and Practices*; AOCS Press: Champaign, IL, 2004.

35. Mounts, T.L. In *Lipid Technologies and Applications*; F.D. Gunstone and F.B. Padley, Eds.; Marcel Dekker: New York, 1997; pp. 433–451.

36. Gupta, M.K. In *Bailey's Industrial Oil and Fat Products*, 6th ed.; F. Shahidi, Ed.; Wiley Interscience: New York, 2005; Vol. 4, pp. 1–31.

# PART 2

# Physical Characteristics

# 4

# Volatile Odor and Flavor Components Formed in Deep Frying

**Edward G. Perkins**

*Department of Food Science, Burnsides Research Laboratory, University of Illinois, Urbana, Illinois 61801*

The presence of flavor and odor components in frying oil and fried-products oils are important since they add to both the desirable and undesirable flavor aspects of the fried food. Flavor and odor components in frying fats and fried products arise through autoxidation and decomposition at frying temperatures as well as by hydrolysis of fat by steam generated by water present in the food product.

## Fat Hydrolysis

Hydrolysis of frying fat by moisture produces diacyl glycerides and free fatty acids. The diacyl glycerides formed can further be hydrolyzed into monoglycerides and ultimately glycerol and fatty acid. Such fatty acids are themselves very reactive and exhibit flavor characteristics. A majority of the fatty acids thus formed, however, are volatilized and removed from the oil by steam generated by deep frying (Fig. 4.1).

## Autoxidation

Autoxidation takes place at deep-frying temperatures (160°F), but hydroperoxides formed immediately decompose into primary and secondary decomposition products, which themselves are responsible for the flavor of oils used for deep frying and the resulting fried products.

Numerous studies have described this process as a combination of initiation, propagation, and termination processes (7–9,22):

$$\text{Initiation } RH \rightarrow R^*$$
$$\text{Propagation } R^* + O_2 \rightarrow ROO^*$$
$$ROO^* + RH \rightarrow ROOH + R^*$$
$$\text{Termination } R^* + R^* \rightarrow \text{Nonradical Products}$$

Lipid oxidation has been repeatedly reviewed (3,9). Autoxidation occurs when a lipid containing fatty acids, especially linoleic and linolenic acids as well as more

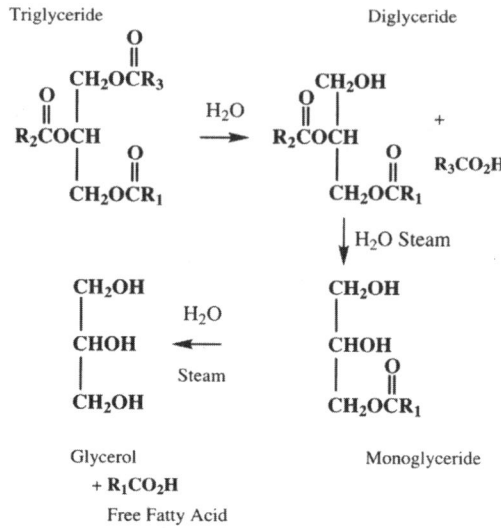

**Fig. 4.1.** Hydrolysis of triglycerides by steam.

polyunsaturated fatty acids is exposed to oxygen.

As an example, the free radical previously described in the case of linoleate is formed at the 11 position, and then resonance stabilization occurs to form conjugated free radicals. These react with oxygen and a hydrogen source to form the corresponding hydroperoxide, which then decomposes with heat into volatile products (Fig. 4.2).

The presence of heavy metals accelerates oxidation. Such metals are (in decreasing order of activity) Fe > Mn > Cu > Co > Pb > Ni > Zn. In general, the presence of brass, as in thermo switches, and any of the other metals as components of a fryer increase the rate of fat deterioration.

Although stable lipid hydroperoxides form at lower temperatures, they immediately decompose if formed at frying temperatures. A variety of compounds consisting of a homologous series of alkanes, alkenes, alkanals, mono and diunsaturated alkenals, ketones, and methyl ketones forms (Table 4.1). Several lactones and pentyl furans form. In addition, when potatoes are fried in such oils, additional compounds such as pyrazines form (4). The volatile components formed in frying oils were previously reported (4,16). These studies were confirmed by a more recent study in which dynamic headspace analysis was coupled with gas chromatography/mass spectrometry (GC/MS) to directly separate and identify components present in both french fries and the corresponding frying oil. A list of these components is in Table 4.1. Amounts of these materials vary, depending on the length of time the oil was heated and the fries prepared (20). Ill-cared for oil may build up peroxides, which decompose when heated. The volatile decomposition products formed from peroxidized oils, however, have approximately the same composition as those formed during deep frying, except

**TABLE 4.1**
**Volatile Compounds Found in French Fries and Frying Oils**[a]

| Component | |
|---|---|
| Frying Oil | French Fries |
| Butanal | Hexanal |
| Hexane | Methyl pyrazine |
| 1-Butanol | 2-Hexenal |
| Pentanol | 2-Heptanone |
| Heptane | Nonane |
| Hexanal | Heptanal |
| 2-Hexenal | 2,5-Dimethylpyrazine |
| 2-Heptanone | 2-Heptenal |
| Nonane | 2-Pentylfuran |
| Heptanal | Octanal |
| 2-Heptenal | t2,t4-Heptadienal |
| 2-Pentylfuran | 2-Octenal |
| Octanal | Nonanal |
|  | 2-Nonenal |
| t2 t4-Heptadienal |  |
| 2t 4-Heptadienal | Decanal |
| 2-Octenal | 2-Decenal |
| Nonanal | 2,c 4-Decadienal |
| 2-Nonenal | Undecanal |
| Decanal | 2,t 4-Decadienal |
| 1-Decene |  |
| 3-Octanone |  |
| 2-Decenal |  |
| 2,c 4-decadienal |  |
| Undecanal |  |
| 2,t4-Decadienal |  |
| 2-Octen-1-ol |  |
| 2-Undecenal |  |
| Dodecanal |  |

[a]In order of elution on a nonpolar capillary column.

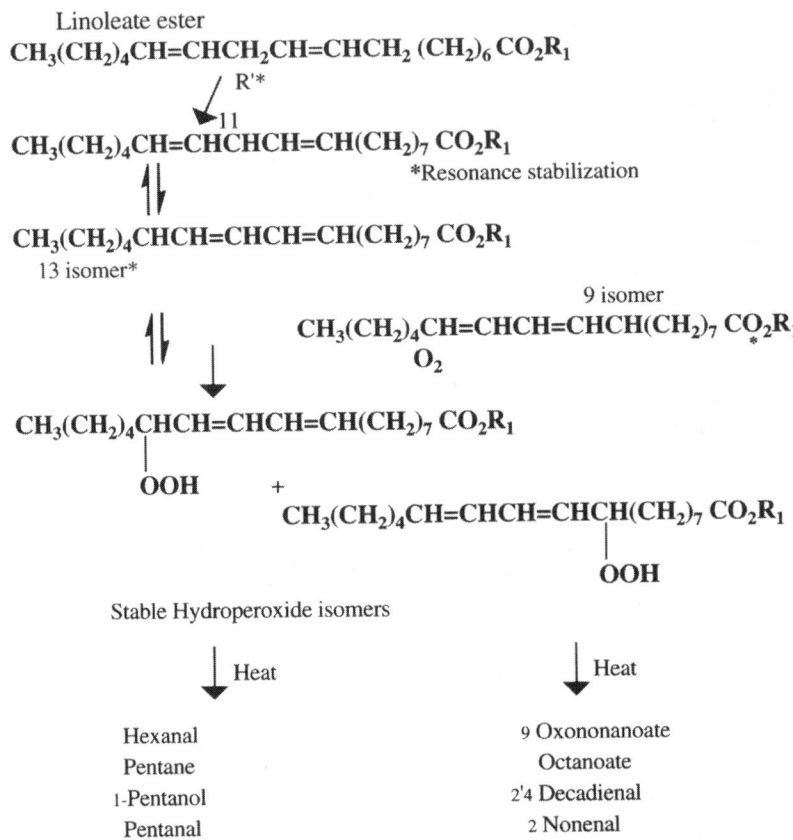

**Fig. 4.2.** Formation and decompostion of linoleate hydroperoxides.

that there are no pyrazines.

In addition to primary oxidation products, such as hexanal, formed directly by hydroperoxide decomposition, secondary oxidation products form (19).

## Isolation and Separation of Volatiles

Complex mixtures of differing structures and polarity are difficult to separate and identify. The classical work of Chang employs a distillation fractionation technique (4). Using this method, most of the compounds listed in Table 4.1 were determined.

With the advent of high-resolution GC and MS, however, it became much less labor intensive and relatively simple to analyze such complex mixtures.

Currently, the three principal methods primarily used are static headspace analysis, dynamic headspace analysis, and direct headspace analysis by direct injection into the gas-liquid chromatograph.

Static headspace depends on equilibrium formed between volatile compounds and oil at any given overall temperature and concentration. Static headspace analysis involves removing a sample of the headspace gas with a syringe, either manually or automatically, and injecting the contents into the GC column. Sample size is somewhat restricted, but it is common to use either 200 or 500 µL sample loops (11).

Dynamic headspace analysis involves volatile collection by "sparging" the sample with a purified gas. Passing the gas through the sample gives good yields of volatile components without problems involving sample equilibration. The volatile compounds collected are transported to a trap or directly to the GC (21).

Direct headspace analysis may be carried out by adding material to be sampled to the head of a purge trap, followed by desorption into the GC. It is also accomplished by placement of a sample directly into a column injection liner or modified injector port (6,13). Purge traps are tubes about ¼″ in diameter and 4″ long packed with either Tenax GC or Carbopac B (10,17). These materials are advantageous because they do not adsorb water, only organic material. They can form artifacts in the gas chromatogram, however, as a result of thermal decomposition, especially if oxygen contaminated the adsorbed compound and adsorbent (14,15). Furthermore, regardless of how the sample is collected, better separations and quantitation are obtained if the volatile fraction is concentrated on the head of the column. This may be accomplished by cooling the GC column head or instrument oven to low temperatures with liquid nitrogen. This then also allows cryogenic or subambient temperature programming for optimal separation (5,12,18).

## Odor and Flavor Threshold Levels

Berger showed that four major flavor components of soybean oil have threshold limits ranging from 2 – 0.0001 (2). An extensive compilation of the threshold levels of a large number of compounds generally appearing in autoxidative mixtures was published by the American Society for Testing Materials (1). Much effort went into the development of accurate descriptors to characterize lipid flavor. Some of these are: acrid (acrolein), beany (cooked soybean odor), fishy, fruity (olive oil), grassy (hexanal), and green (raw soybeans) (24). The quality of an oil is related to the appearance of materials with these descriptors (24).

## Summary

Both the fried product and frying oil absorb odor and flavor compounds formed during heating of the initial frying oil. Such compounds are responsible for many aspects of desirable fried food flavor, although their presence may also cause quality deterioration of the product.

## References

1. American Society for Testing Materials (ASTM). ASTM DS48. In *Compilation of Odor and Taste Threshold Values Data*; W.H. Stahl, Ed.; ASTM: Baltimore, MD, 1973.
2. Berger, K.G. *Chem. Ind.* **1975,** *3*, 194–199.
3. Chan, H.W.S., Ed. *Autoxidation of Unsaturated Lipids*; Academic Press: London, 1987.
4. Chang, S.S.; R.J. Peterson; and C.-T. Ho. In *Chemistry of Deep Fried Flavor in Lipids as a Source of Flavor*; M.K. Supran, Ed.; American Chemical Society: Washington, DC, 1978; pp. 18–42.
5. Charalambous, G, Ed. *Analysis of Foods and Beverages: Headspace Techniques*; Academic Press: New York, 1978.
6. Dupuy, H.; E.T. Rayner; J.I. Wadsworth; and M.C. Legendre. *J. Am. Oil Chem. Soc.* **1977,** *54*, 445.
7. Eriksson, P.C.E. In *Autoxidation of Unsaturated Lipids*; H.W.S. Chan, Ed.; Academic Press: London, 1987; p. 207.
8. Forrs, D.A. *Prog. Chem. Fats Other Lipids* **1972,** *13*, 181.
9. Frankel, E.N. *Prog. Lipid Res.* **1980,** *19*, 1.
10. Hinshaw, J.V. *LC–GC* **1990,** *6*, 873.
11. Ioffe, B.V.; and A.G. Vitenberg. *Head-Space Analysis and Related Methods in Gas Chromatography*; John Wiley & Sons: New York, 1983; p. 9.
12. Jennings, W. *Analytical Gas Chromatography*; Academic Press: New York, 1987.
13. Legendre, M.C.; H.P. Dupuy; E.T. Rayner; and W.H. Schuller. *J. Am. Oil Chem. Soc.* **1980,** *62*, 1657.
14. MacLeod; and J.M. Ames. *J. Chromatogr.* 1986 *355*, 393.
15. Midleditch, B.S.; and A. Zlatkis. *J. Chromatogr. Sci.* **1987,** *25*, 547.
16. Nawar, W.W.; S.J. Bradley; and S.S.L. Manno. In *Lipids as a Source of Flavor*; M.K. Supran, Ed.; American Chemical Society: Washington, DC, 1978; pp. 42–55.
17. Nunez, A.J.; L.F. Gonzales; and J. Janak. *J. Chromatogr.* **1984,** *300*:127.
18. Perkins, E.G., In *Flavor Chemistry of Lipid Foods*; D.B. Min and T.H. Smouse, Eds.; American Oil Chemists' Society: Champaign, IL, 1989; pp. 43–45.
19. Pokorny, J. In *Flavor Chemistry of Lipid Foods*; D.B. Min and T.H. Smouse, Eds.; American Oil Chemists' Society: Champaign, IL, 1989.
20. Qian, C. Characterization of Deep Fried Flavor: Preliminary Study Using French Fries; M.S. Thesis, University of Illinois, Urbana, 1990.
21. Raghavan, S.K.; S.K. Reeder; and A. Khayat. *J. Am. Oil Chem. Soc.* **1989,** *66*, 942.
22. Schieberle, P.; and W. Grosch. *J. Am. Oil Chem. Soc.* **1981,** *58*, 602.
23. Warner, K. In *Flavor Chemistry of Fats and Oils*; D.B. Min and T.H. Smouse, Eds.; American Oil Chemists' Society: Champaign, IL, 1985.
24. Warner, K.; and N.A.M. Eskin. *Methods to Assess Quality and Stability of Oils and Fat-Containing Foods*; American Oil Chemists' Society: Champaign, IL, 1995; Appendix b.

# Isomeric and Cyclic Fatty Acids as a Result of Frying

**Jean-Louis Sébédio[a] and Pierre Juaneda[b]**
[a]Human Nutrition Unit, Mass spectrometry platform UMR INRA- Université D'auvegrne, Centre de Clermont-Theix, 63122 Saint Genes Champanelle, France and [b]INRA UMR Flavic, 17 rue Sully, 21069 Dijon Cedex, France

## Introduction

Deep fat frying is one of the most common procedures for the preparation of food. In this process, fat is exposed to high temperatures in the presence of air and food containing water. Consequently, many chemical reactions take place during this process (Fig. 5.1). Hydrolysis results in the formation of fatty acids, monoglycerides, and diglycerides while the presence of air and high temperatures gives rise to the formation of thermal and oxidation alteration products. Oxidation products may include oxidized monomeric, dimeric, and oligomeric triglycerides, as well as volatiles components such as aldehydes, ketones, hydrocarbons, etc. Thermal alteration products include cyclic fatty acid monomers (CFAM), fatty acid geometrical isomers, nonpolar dimeric, and oligomeric triglycerides.

This chapter will review the structural analysis and identifications of geometrical fatty acids isomers including those having a conjugated system and the cyclic fatty acid monomers formed from both essential fatty acids linoleic and linolenic acids. Methods of analysis, quantification, and levels in food products will be described when available. Emphasis will be given to the CFAM considering the great number of studies carried out on these molecules.

## Cyclic Fatty Acid Monomers

### Structural Analysis and Identification

Evidence for the presence of cyclic monomers in heated fat was found as early as 1953 (1,2). It was a few more years, however, before work began on structural analyses using urea adduction that permitted isolation of fractions enriched in cyclic components. The first experiments on heated fats were carried out with linseed oil containing a significant quantity of 18:3(n-3). The first gas–liquid chromatographic fractionations

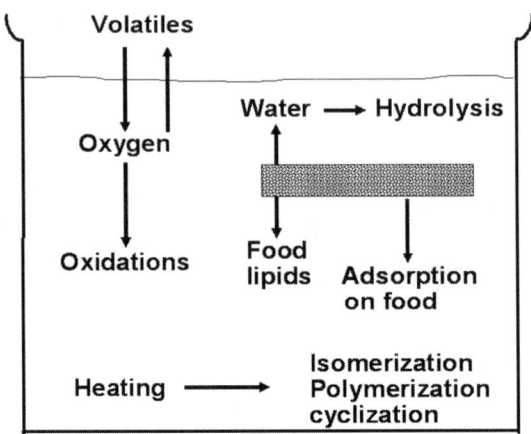

**Fig. 5.1.** Major reactions which are taking place during deep frying operations. Adapted from Fritsch, C.W., *JAOCS 58*,1981.

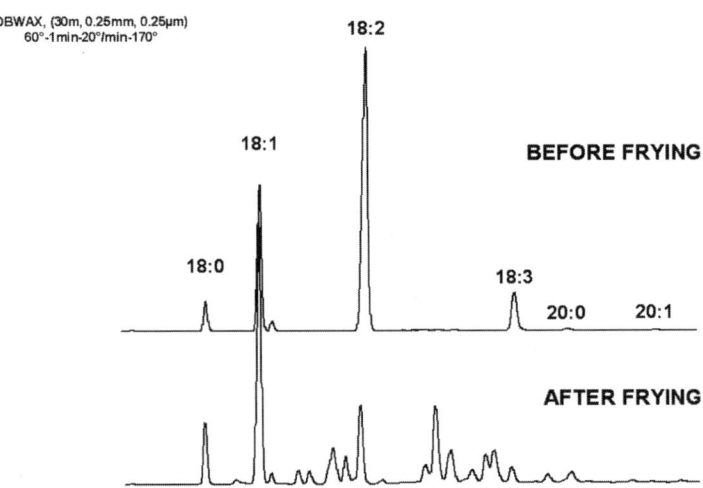

**Fig. 5.2.** Part of GC analyses of total fatty methyl esters of rapeseed oil (top) and methyl esters of heated rapeseed oil (275°C, 12 h, under N₂); column: Carbowax 20 M, 25 m 0.3 mm i.d., isothermal 170°C.

permitted McInnes et al. (3) in 1961 to propose that cyclic fatty acids formed after heat treatment were dienoic components with 18 carbon atoms and a cyclohexenic substructure.

These components were supposed to be disubstituted cyclic fatty acids (one substituent being a propyl group) having one ethylenic bond in the 6-carbon-membered ring and the other on the substituent having the acid function. Later, a study by Saito and Kaneda (4) confirmed the existence of the same type of components, the major difference being location of the double bonds. Hutchinson and Alexander (5), however, showed that the number of carbons of both substituents could also be different. One component was isolated and identified as being ethyl 11-(2-methylcyclohex-2-en-1-yl)undec-*trans*-9-enoate. It was suggested by Gast et al. (6) that some of the cyclic fatty acids formed from heated linseed oil could be C18 disubstituted 5-carbon-membered ring acids. This was confirmed later by Potteau et al. (7) on heated linseed oil. In that study, a fraction enriched in cyclic fatty acids was obtained using urea adduct fractionation. The fraction was further hydrogenated and analyzed by gas chromatography/mass spectrometry (GC/MS). Following studies of Michael (8,9) and Zeman and Scharmann (10) on MS of cyclic fatty acids, major peaks were identified as *cis* and *trans* isomers of methyl 9-(2′-n-propyl-cyclohexyl)-nonanoate. Minor peaks were tentatively identified as methyl 10-(2′-n-propylcyclopentyl)-decanoate, methyl 8-(2′-n-butyl cyclohexyl)-octanoate, methyl 9-(2′-n-butyl cyclopentyl)-nonanoate, and methyl 8-(2′-n-butyl cyclohexyl)-octanoate.

Some fatty acids with a 6-carbon-membered ring were synthesized (11,12). It was possible to compare the spectra of molecules isolated to those of authentic standards. This was not possible, however, for cyclopentyl derivatives, as monosubstituted fatty acids having a 6-carbon-membered ring were the only ones available (13).

Vegetable oils used for frying are composed of five major fatty acids: 16:0, 18:0, 18:1, 18:2(n-6), and 18:3(n-3) (14). It is possible to divide these oils into two families. The first family is composed of oil containing 18:2(n-6) as the major polyunsaturated fatty acid and only traces of 18:3(n-3). These are peanut, sunflower, olive, and corn oils. The second family not only contains linoleic acid but also appreciable amounts of the more unsaturated fatty acid, linolenic acid. These are mainly rapeseed (canola), and soybean oils.

However, some countries, such as France, stipulate that an oil is suitable for frying only if it contains less than 2% linolenic acid. It is therefore necessary to distinguish those cyclic fatty acids formed from linolenic acid. For that purpose, extensive studies were carried out in order to produce, isolate, and characterize CFAM fractions that were representative of linoleic and of linolenic acids. Sunflower and linseed oils were chosen as model oils. Sunflower oil contains only trace amounts of linolenic acid but a high amount of linoleic acid, around 65%, while the major fatty acid of linseed oil is α-linolenic acid representing more than 55% of the total fatty acids, approximately. These were heated at different temperatures and under different experimental conditions in order to isolate different CFAM fractions.

As an example, gas chromatography analyses (Fig. 5.2) of both original and heated canola oil showed the extent of transformations during heat treatment. The resulting esters are a complex mixture of "natural fatty acids" present in the original oil, some 18:2 and 18:3 geometrical isomers (15), as well as CFAM with retention times ranging from that of 18:1 to that of 18:3 geometrical fatty acid isomers. Even considering the complexity of the samples, it is rather simple to isolate a pure fraction of CFAM. A combination of sound techniques, such as column chromatography on silicic acid, and urea fractionation permitted isolation of relatively pure CFAM fraction from heated linseed oil, for example (16). Isolating a CFAM fraction from a heated sunflower oil was a difficult task as it was necessary to add one step in the fractionation process (16), since the second urea fractionation, the nonadduct fraction, still contained some 18:2n-6.

Preparative high-performance liquid chromatography (HPLC) on a $C_{18}$ reversed-phase column permitted isolation of a relatively pure CFAM fraction that still contained about 4% 18:2n-6. CFAM fractions are usually complex mixtures and, to elucidate the structures of the newly formed molecules, one must determine the skeleton of the molecule, the degree of unsaturation, and positions and geometry (Z or E) of double bond(s), as well as geometry of the ring substitution (cis or trans). For GC analyses, polar columns, such as 100% cyanopropyl polysiloxane, give very good separation, as illustrated in Fig. 5.3. The top of the figure represents a mixture of CFAM isolated from linseed oil heated at 275°C for 12 h under nitrogen. On this column, at 160°C, ECL values of major CFAM ranged from 18.94 to 20.33 under the experimental conditions described in the figure.

The easiest and most commonly used method to determine the skeleton of the molecule is total catalytic hydrogenation. The result of hydrogenation, which eliminates positional and geometrical isomers, is to simplify complex mixtures of unsaturated compounds into a simpler mixture of saturated ones (see Fig. 5.3B as an example for linseed oil). It is then possible to study the structure of these hydrogenated components by GC coupled with MS. For example, Fig. 5.4 (top) represents (after hydrogenation) total ion current of the CFAM mixture (as methyl esters) isolated from sunflower oil heated at 240°C for 10 h under nitrogen. On the bottom is CFAM (as methyl ester after hydrogenation) isolated from linseed oil heated at 240°C for 10 h under nitrogen.

Gas chromatography/mass spectrometry is, so far, the method of choice for identifying hydrogenated CFAM in heated fats and oils. Interpretation of mass spectra when using methyl ester derivatives as in early studies (7), however, is often very difficult, especially when cyclopentyl isomers are involved. It is generally accepted that $C_{18}$ cyclic fatty acid methyl esters give four characteristic fragments A, B, C, and D with electron impact MS (7), corresponding to the fragmentation in the α positions of the ring. Additional ion fragments are D-32, which results from loss of methanol, and D-32-18, from further loss of water. A fragment of B+1 is present

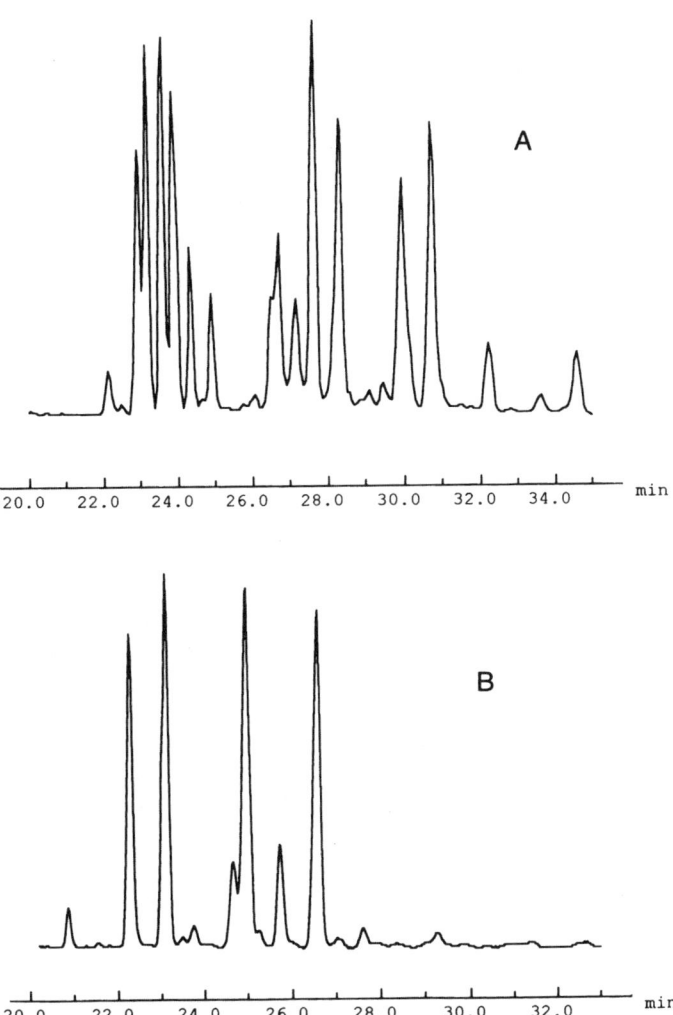

**Fig. 5.3.** GC analyses on a polar fused silica column (BPX 70), 50 m 0.32 mm i.d., isothermal conditions 160°C, of CFAM isolated from a) linseed oil heated at 275°C for 12 h under $N_2$, and b) the corresponding hydrogenated fraction (J.-L. Sébédio and P. Juaneda, unpublished data).

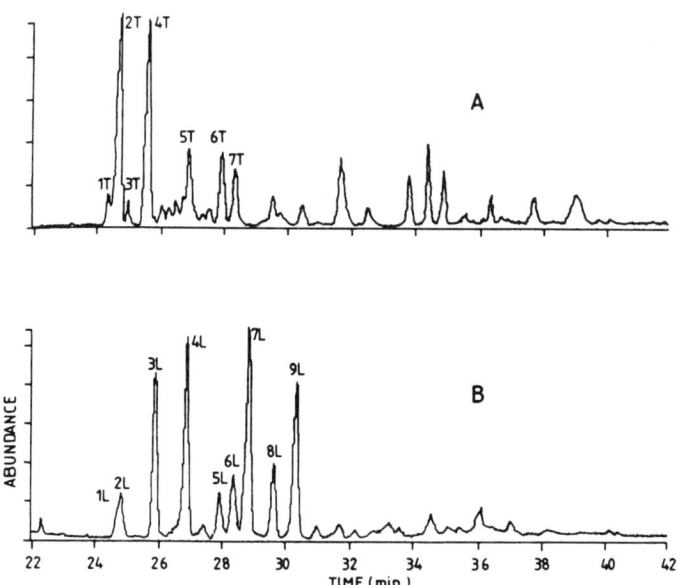

**Fig. 5.4.** Total ion currents of a) CFAM isolated from heated sunflower oil after total hydrogenation and b) CFAM isolated from heated linseed oil. See Sébédio et al. (18).

in most spectra due to protonation of fragment B. This type of fragmentation is usually observed for cyclohexyl and cyclohexenyl isomers (17,18). Cyclohexenyl isomers are also characterized by an important fragment resulting from a retro Diels-Alder reaction, from which there is a further loss of methanol. Also, a McLafferty rearrangement gives rise to other characteristic fragments (17).

The fragmentation pattern, however, is much more complex for cyclopentyl and the cyclopentenyl isomers. In this case, $\alpha$- and $\beta$-fragmentations take place. Furthermore, the fragmentation pattern seems to get increasingly more complex with increasing numbers of carbon atoms from the alkyl moiety. The existence of a $\beta$-cleavage for monosubstituted cyclopentenyl esters was originally proposed by Christie et al. (13), and this trend was later verified by synthesis of cyclopentyl and cyclopentenyl isomers (19,20).

Later, the utilization of other derivatives such as dimethyloxazoline (DMOX) or picolinyl esters as described by Christie et al. (21), permitted a much faster and easier interpretation of the fragmentation pattern. For example, we have reported in Fig. 5.5 spectra of two CFAM isolated from heated linseed oil and the total chromatogram. Separation of DMOX derivatives is as good as the corresponding methyl esters. Furthermore, interpretation of the fragmentation pattern of $C_5$-membered ring is

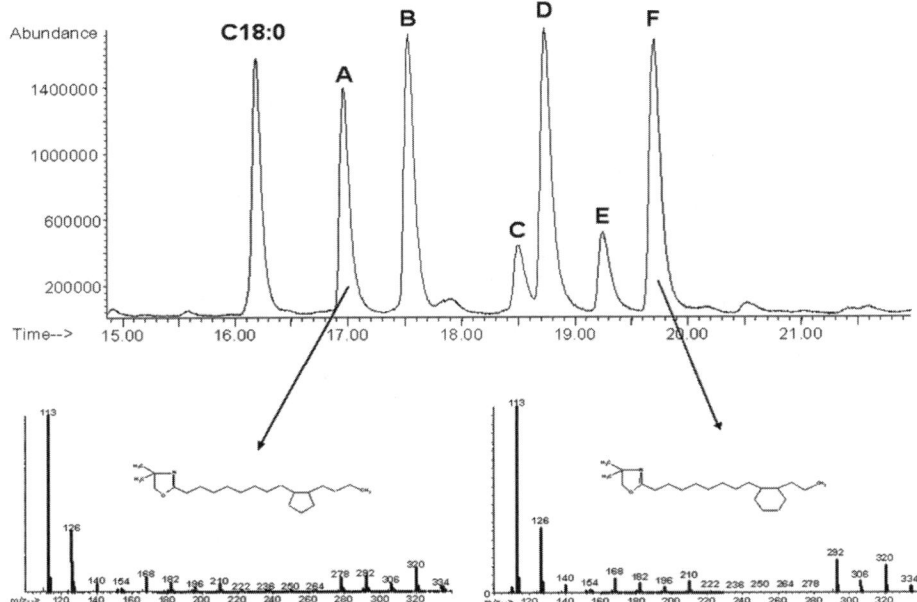

**Fig. 5.5.** Total ion current of dimethyloxazoline derivatives of hydrogenated CFAM isolated from heated linseed oil and mass spectra of components A and F (BPX-70 column, 50 m × 0.32 mm i.d., 0.25 μm film).

as easy as that of a $C_6$-membered ring isomer. In the mass spectra of peak A, there were regular gaps of 14 amu from the M-15 ion ($m/z$=320) down to 278, indicating a saturated straight chain between $C_{18}$ and $C_{14}$. A gap of 68 amu between $m/z$ 278 and $m/z$ 210 was due to a $C_5$ ring between $C_{14}$ and $C_{10}$, and the gaps of 14 amu from 210 to 126 indicated the saturated straight chain. Similarly, a close examination of the fragmentation pattern of peak F permitted localization of a $C_6$-membered ring (fragments 292 and 210) between $C_{15}$ and $C_{10}$.

Extensive work established structures of CFAM. The intact unsaturated species isolated from heated linseed and sunflower oils were characterized by GC/MS and Fourier transform infrared (FT-IR) spectroscopy (17). In this study, the degree of unsaturation of CFAM was directly determined by GC/MS. CFAM isolated from heated linseed oil were mostly dienoic isomers, while CFAM isolated from heated sunflower oil were mostly monoenoic fatty acids (17). The same work also reported the geometry of double bonds studied with GC coupled with FT-IR spectroscopy. Most of CFAM isolated from linseed oil were Z,E isomers, while major CFAM formed from $C_{18:2}$ (sunflower oil) were Z isomers (17). A close examination of mass and infrared spectra showed that the four CFAM isolated from heated linseed oil (50% of total CFAM) were diunsaturated cyclic ester isomers with a cyclohexenyl ring (shift of about 50 cm$^{-1}$ in the out-of-plane CH deformation IR absorption from

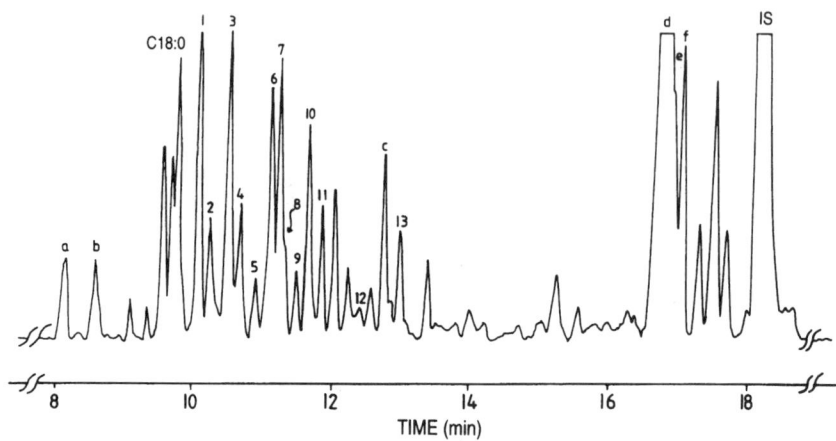

**Fig. 5.6.** Partial chromatogram of a fraction of heated partially hydrogenated soybean oil; internal standard (IS) was phenanthrene. Reproduced by permission from Rojo and Perkins (23).

**TABLE 5.1**

**Hydrogenated CFAM[a] Identified by Gas Chromatography/Mass Spectrometry in Heated Sunflower and Linseed Oils (240°C, 10 h, N$_2$)**

| Peak code | Configuration | | ECL (Fig. 5.4) |
|---|---|---|---|
| *Heated linseed oil* | | | |
| 3L | methyl 9-(2´-butylcyclopentyl)nonanoate | *trans* | 18.12 |
| 4L | methyl 10-(2´-propylcyclopentyl)decanoate | *trans* | 18.28 |
| 6L | methyl 9-(2´-butylcyclopentyl)nonanoate | *cis* | 18.51 |
| 7L | methyl 9-(2´-propylcyclohexyl)nonanoate | *trans* | 18.54 |
| 8L | methyl 10-(2´-propylcyclopentyl)decanoate | *cis* | 18.70 |
| 9L | methyl 9-(2´-propylcyclohexyl)nonanoate | *cis* | 18.76 |
| *Heated sunflower oil* | | | |
| 2T | methyl 7-(2´-hexylcyclopentyl)heptanoate | *trans* | 17.96 |
| 4T | methyl 9-(2´-butylcyclopentyl)nonanoate | *trans* | 18.12 |
| 5T | methyl 7-(2´-hexylcyclopentyl)heptanoate | *cis* | 18.34 |
| 6T | methyl 9-(2´-butylcyclopentyl)nonanoate | *cis* | 18.51 |
| 7T | methyl 9-(2´-propylcyclohexyl)nonanoate | *trans* | 18.54 |

[a]CFAM = Cyclic Fatty Acid Monomers.
[b]*Source:* Sébédio, J.-L., et al., *J. Am. Oil Chem. Soc.* 66:704–709 (1989).

**TABLE 5.2**
Some C$_{18}$ Cyclic Monomers Identified by Gas Chromatography/Mass Spectrometry in Heated, Partially Hydrogenated Soybean Oil

| Peak (Fig. 5.6) | Main compound | Configuration |
|---|---|---|
| 1 | methyl 9-(2′-*n*-butylcyclopentyl)nonanoate | *trans* |
| 2 | methyl 7-(2′-*n*-pentylcyclohexyl)heptanoate | *trans* |
| 3 | methyl 10-(2′-*n*-propylcyclopentyl)decanoate | *trans* |
| 4 | methyl 8-(2′-*n*-butylcyclohexyl)octanoate | *trans* |
| 5 | methyl 7-(2′-*n*-pentylcyclohexyl)heptanoate | *cis* |
| 6 | methyl 9-(2′-*n*-butylcyclopentyl)nonanoate | *cis* |
| 7,8 | methyl 9-(2′-*n*-propylcyclohexyl)nonanoate | *trans* |
| 9 | methyl 8-(2′-*n*-butylcyclohexyl)octanoate | *cis* |
| 10 | methyl 10-(2′-*n*-propylcyclopentyl)decanoate | *cis* |
| 11 | methyl 9-(2′-*n*-propylcyclohexyl)nonanoate | *cis* |
| 12 | methyl 10-(2′-*n*-ethylcyclohexyl)decanoate | *trans* |
| 13 | methyl 10-(2′-*n*-ethylcyclohexyl)decanoate | *cis* |

*Source:* Rojo, J.A., and E.G. Perkins, *J. Am. Oil Chem. Soc.* 64:414–421 (1987).

710 to 660 cm$^{-1}$ and a retro Diels-Alder fragment at *m/z* 238 in the mass spectra). Two of these CFAM were tentatively identified as *cis/trans* isomers (ring substitution) of 9-(2′-propyl-4′-cyclohexenyl)-8-nonenoate by comparison of their mass spectra with mass spectra published by Awl and Frankel (12) for the synthetic compound. The other two were tentatively identified as positional isomers for the E-ethylenic bond on the carbon chain (17). Gas chromatography/mass spectrometry of CFAM isolated from heated sunflower oil suggested that the major acids were disubstituted cyclopentenic isomers (17).

Structures of CFAM depend very little on the temperatures used in the range of 200–275°C (17). Under low or high temperatures, the same major CFAM are formed. Differences were observed only in their relative proportions and amounts in the oil. The determining factor, however, is the nature of the polyunsaturated fatty acid present in the original oil. An oil containing linoleic acid (sunflower) and only traces of 18:3(n-3) gave a mixture of CFAM having a 5-carbon-membered ring, while only minor amounts of CFAM having a 6-carbon-membered ring were observed. In contrast, an oil rich in α-linolenic acid gave a mixture with approximately the same amount of cyclopentyl and cyclohexyl isomers. The complete structure of the isomers could not be elucidated at the time the study was carried out due to the complexity of CFAM mixtures and the individual structures under investigation.

As previously mentioned, a GC/MS study of CFAM isolated from heated linseed and sunflower oils was undertaken after complete hydrogenation of isolated CFAM

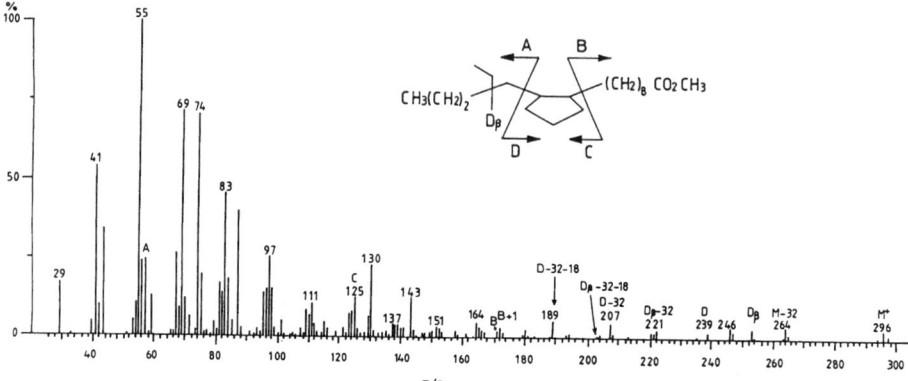

**Fig. 5.7.** Mass spectrum of peak 3L (Fig. 5.4) identified as *trans*-methyl 9(2′-butylcyclopentyl)-nonanoate. Sébédio et al. (18).

mixtures (18). This study confirmed previous ones, while extending the number of identified species. For linseed oil, the main peaks in Fig. 5.4b (3L, 4L, 6L, 7L, 8L, and 9L) were unambiguously identified (Table 5.1). Peak 5L was identified as methyl 8-(2′-butylcyclohexyl) octanoate (*trans*). For sunflower oil, the main peaks were also unambiguously identified (Table 5.1), confirming that the major parts of the compounds are disubstituted cyclopentyl isomers in this case, some with rather long hydrocarbon chains. Thus, peak 1T (Fig. 5.4A) was tentatively identified as *trans*-methyl 4-(2′-nonylcyclopentyl)-butanoate (18).

The *cis* and *trans* configurations of these isomers were assigned according to what was already published on cyclohexyl isomers and also after syntheses of the cyclopentyl ones (19,20). With total catalytic hydrogenation, positional and geometrical isomers give the same saturated product. Therefore, for these isomers, only one peak is detected in a GC after hydrogenation. To establish clear correlations between unsaturated CFAM isomers and their hydrogenated analogs, Le Quéré et al. (22) developed an on-line hydrogenation method. In this system, the sample was injected in GC using hydrogen as the carrier gas. The analytical column was connected to a hydrogenation capillary reactor. The capillary reactor consisted of a deactivated fused silica capillary column (60 cm × 0.32 mm i.d.) coated with palladium acetylacetonate. This reactor was connected through 60 cm of deactivated fused silica capillary tubing to the ion source of a mass spectrometer (22). For heated linseed oil, total ion currents obtained with and without the Pd reactor could be superimposed. Under the same chromatographic conditions, retention times were not altered by inserting the capillary reactor into the system.

Similar structures were proposed by Rojo and Perkins (Table 5.2, Fig. 5.6) in a

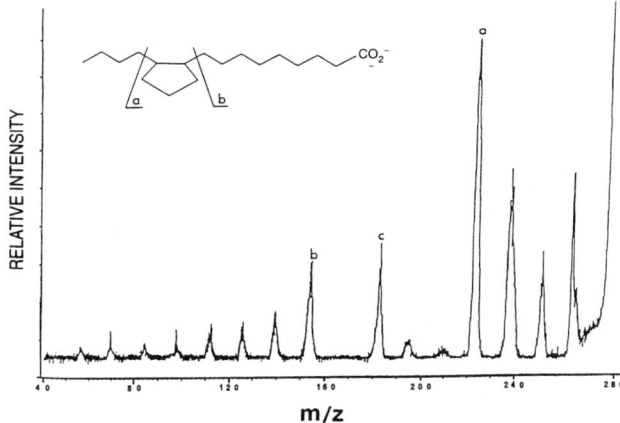

**Fig. 5.8.** CAD-MIKE spectrum of the carboxylate anion (*m/z* 281) of synthetic 9-(2′-butylcyclopentyl)-nonanoic acid, measured from negative FAB-generated parent ion. Le Quéré et al. (27).

study on partially hydrogenated soybean oil (23) heated under different conditions. The fresh oil which contained about 2.4% 18:3n-3 was heated at 195°C for 80 h under simulated deep-fat frying using cotton balls containing 75% water by weight. In that study, a large variety of cyclohexyl isomers were found. Methyl 9-(2′-n-propylcyclohexyl)-nonanoate (*cis* and *trans*) were the major isomers. Ethyl, butyl, and pentyl cyclohexyl isomers (peaks 2, 4, 5, 12, and 13; Fig. 5.6) were also reported. Butyl- and pentyl-cyclohexyl isomers were also found later in heated linseed oil (18).

At this point it is important to stress the utilization of DMOX derivatives for elucidation of structural properties of saturated cyclic fatty acids. A close examination of Figs. 5.5 and 5.7 indicates the potentiality of these derivatives that give simpler spectra than methyl esters that enable the determination of the size of the ring and the substituants without ambiguity.

An interesting approach to structural elucidation of fatty acids was later introduced, which used tandem mass spectrometry (MS/MS), and especially fast atom bombardment MS/MS (FAB-MS/MS) (24,25). Briefly, mass-analyzed ion kinetic energy (MIKE) spectra, or the B/E-linked scan spectra, obtained after collisional activation of the carboxylate anion of mono- or polyunsaturated fatty acids desorbed by FAB display characteristic features from which the position of the carbon–carbon double bond is determined unambiguously.

The fragmentation pattern, known as charge-remote fragmentation (24), results from characteristic allylic cleavages. This charge-remote fragmentation process was also successful in locating the three-membered rings in cyclopropane and cyclopropene fatty acids (26).

Using the same fragmentation process, Le Quéré et al. (27) investigated the behavior of 1,2-disubstituted cyclopentane rings on model compounds of hydrogenated CFAM (27). Collisionally activated dissociation (CAD)-MIKE

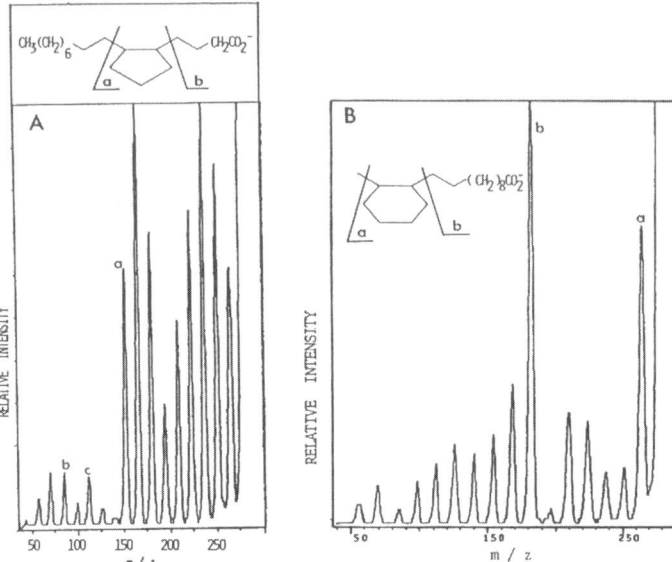

**Fig. 5.9.** CAD-MIKE spectra of the carboxylate anions (*m/z* 281) of hydrogenated 1,2-disubstituted cyclopentyl and cyclohexyl acids isolated from heated sunflower oil, obtained in a GC-MS/MS analysis of their pentafluorobenzyl esters. Le Quéré et al. (27).

**TABLE 5.3**
**Main Hydrogenated Cyclic Fatty Acid Monomers (CFAM) Identified in Heated Linseed and Sunflower Oils. Identifications Made from CAD-MIKE Spectra of Carboxylate Anions Obtained by GC-MS/MS of Pentafluorobenzyl Esters**

| Main component | Referenced figure |
|---|---|
| *Linseed oil* | Fig. 5.8 |
| 9-(2´-butylcyclopentyl)nonanoic acid | |
| 10-(2´-propylcyclopentyl)decanoic acid | |
| 9-(2´-propylcyclohexyl)nonanoic acid | |
| *Sunflower oil* | |
| 4-(2´-nonylcyclopentyl)butanoic acid | Fig. 5.9 |
| 7-(2´-hexylcyclopentyl)heptanoic acid | |
| 9-(2´-butylcyclopentyl)nonanoic acid | Fig. 5.8 |
| 12-(2´-methylcyclopentyl)dodecanoic acid | |
| 9-(2´-propylcyclohexyl)nonanoic acid | |
| 10-(2´-ethylcyclohexyl)decanoic acid | |
| 11-(2´-methylcyclohexyl)undecanoic acid | Fig. 5.9 |

*Source:* Le Quéré, J.-L. et al., *J. Chromatogr. 562*:659–672 (1991).

spectrum of the carboxylate anion, generated by FAB from 9-(2′-butylcyclopentyl) nonanoic acid, displayed charge-remote fragmentation, with easier cleavage at points a and b (Fig. 5.8), corresponding to α-fragmentation occurring to the cyclopentane ring. Another fragment (labeled c in Fig. 5.8) of enhanced intensity was found in the spectra between the characteristic fragment ions a and b. Its origin may involve cleavage unique to the cyclopentane ring. It is noteworthy that no signal appears between fragment ions b and c. This pattern of three enhanced signals a, b, and c, with two minor peaks between a and c, and no fragment between c and b, appears characteristic of cyclopentane-disubstituted acids. The main fragmentations of the ring correspond to the known weak characteristic fragmentation of the disubstituted cyclopentyl acids in electron ionization MS. The characteristic peaks obtained with charge-remote fragmentation, however, are of considerably enhanced intensity (27). This fragmentation appears different from that of cyclopropane acids, where cleavage β to the ring occurs (26).

When complex mixtures of isomeric forms of fatty acids are analyzed, chromatographic separation is necessary. Promé and co-workers (28) developed an elegant GC–MS/MS method to generate high yields of gas-phase carboxylate anions from electron capture ionization of pentafluorobenzyl fatty acid esters. Applied to characterize hydrogenated CFAM isolated from heated linseed and sunflower oils, this method showed some success (27). Typical spectra with important characteristic fragments defining structural features are shown in Fig. 5.9 for cyclopentyl and cyclohexyl acids. The main hydrogenated CFAM identified by this technique in heated linseed and sunflower oils are indicated in Table 5.3. This study confirmed earlier works (18) conducted with GC/MS of CFAM methyl esters and extended the number of identified compounds. In particular, CFAM with very short alkyl moiety (methyl, for example) were more easily characterized (27).

At that point, the carbon skeleton of CFAM could be considered established, but considerable work was still needed to locate the double bond(s) in these CFAM isolated from oil containing polyunsaturated fatty acids. Direct localization of unsaturation sites of methyl esters via MS is not possible due to facile migration of the double bonds on electronic impact.

Among available methods, chemical degradation by means such as oxidative ozonolysis in BF3-MeOH, developed by Ackman et al. (29) for straight-chain fatty acids, was applied to diunsaturated CFAM isolated from heated linseed oil (27). Oxidative ozonolysis of a mixture of cyclic monomer methyl esters isolated from heated linseed oil gave essentially seven products, easily separated by GC, among which octanedioate and nonanedioate dimethyl esters were the major products (27). However, other di- or trimethyl esters were suspected in the reaction mixture, owing to their mass spectra. The complex nature of the reaction mixture, some side reaction, such as in situ decarboxylations, and the necessity to confirm all hypotheses concerning the breakdown products by chemical synthesis, precluded a complete assignment of the structures of unsaturated CFAM isolated from heated linseed and sunflower oils

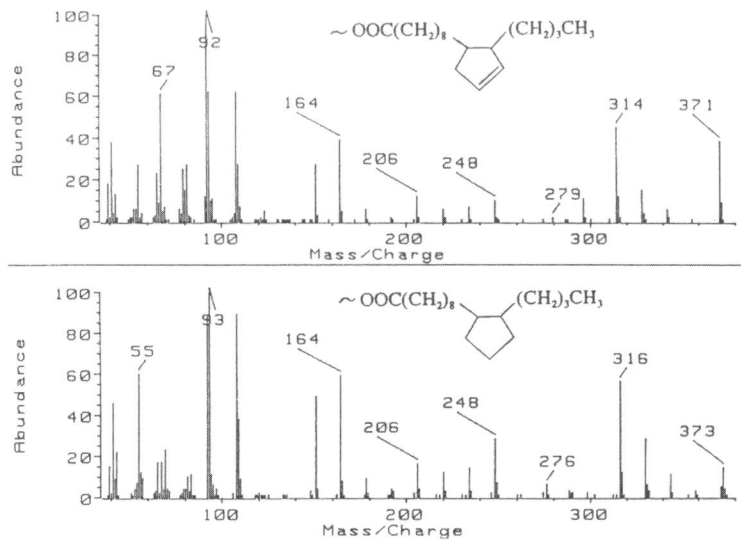

**Fig. 5.10.** Mass spectra of picolinyl ester derivatives of the cyclic monoenoic acid isolated from heated sunflower oil as a pure silver-ion high-performance liquid chromatography (HPLC) fraction, before (top) and after (bottom) hydrogenation. Reproduced with permission from Christie et al. (21).

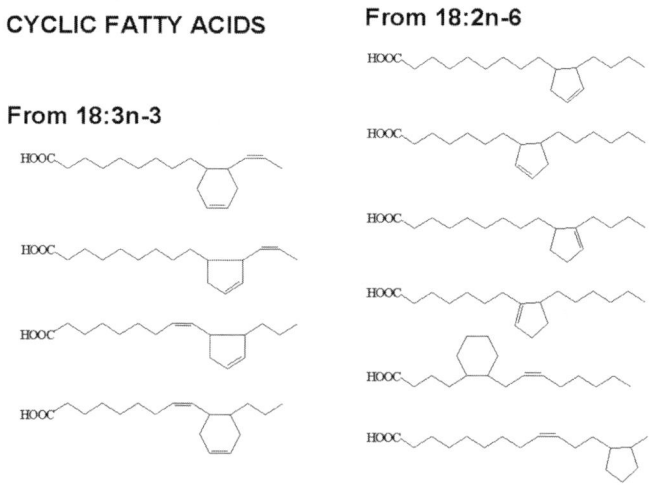

**Fig. 5.11.** Main structures of the cyclic fatty acids formed from linoleic and linolenic acids in heated sunflower and linseed oils, respectively. From (21) and (37).

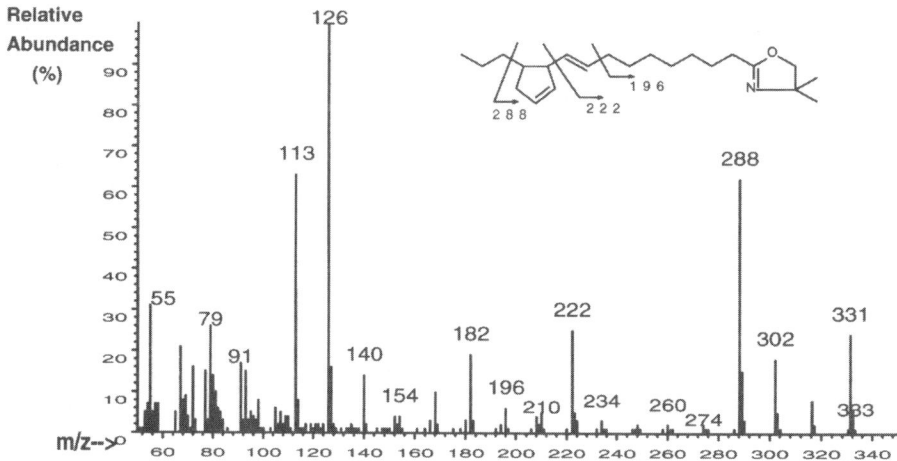

**Fig. 5.12.** Mass spectrum of the dimethyloxazoline derivative of (2′-propyl-4′-cyclopentenyl)9-decenoic acid found in the CFAM mixure isolated from heated linseed oil. Le Quéré unpublished data.

(27) so that this method cannot be used routinely for structural identification.

Fast atom bombardment-desorbed CAD-MIKE spectra of synthetic unsaturated cyclopentyl and cyclopentenyl acids were recorded to investigate the charge-remote fragmentation process for locating rings and double bonds in a single run. All acquired spectra displayed the same features. The main characteristic fragment ion corresponds to cleavage of the alkyl substituent α to the cyclopentane or cyclopentene moiety. The second important fragment is formed by an allylic cleavage on the carboxylate side, and this behavior corresponds to the allylic cleavage observed for monounsaturated fatty acids (28,30). Fragmentation of the carboxylate moiety in the α-position of the ring occurs with low intensity, probably because the corresponding bond is adjacent to the double bond. Thus charge-remote fragmentation was observed for cyclopentyl- and cyclohexyl-disubstituted acids after collisional activation of their carboxylate anions. Generation of carboxylates anions by electron-capture ionization of pentafluorobenzyl esters allowed capillary GC to be performed prior to MS/MS analyses. Results for synthetic unsaturated models make this an interesting method for elucidation of CFAM formed in heated fat, but simpler methods for CFAM structural elucidation present in complex mixtures were further developed.

Pyridine-containing derivatives, such as picolinyl esters, were shown to be suitable for direct MS structural analysis of acids containing straight, branched, unsaturated, cyclic, or oxygenated chains (31–33). In electron impact conditions, these fatty acid derivatives stabilize the charge on the nitrogen atom far from the site of structural interest during ionization, and radical-induced cleavage of the hydrocarbon chain

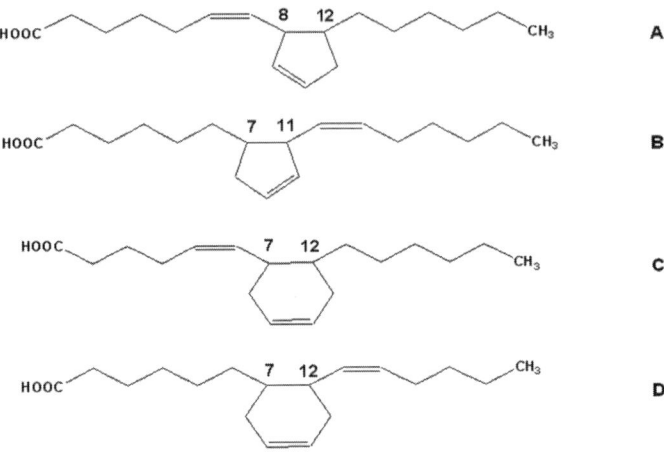

**Fig. 5.13.** Cyclic fatty acids formed from γ-linolenic acid in evening primrose oil. Adapted from Dobson and Sébédio (42).

predominates. Distinctive fragmentations, according to the position of structural features, produce fragment ions of diagnostic value. Particularly significant, a gap of 26 amu represents cleavages on either side of a given double bond.

4,4-Dimethyloxazoline (DMOX) derivatives, first introduced by Zhang et al. (34), also lead to fragmentation patterns suitable for structural investigations in GC/MS. The method was applied successfully to the structural determination of polyunsaturated fatty acids (35) and of cyclopropenoid fatty acids (36). Both methods were used to obtain information on the nature of cyclic monoenes produced from linoleic acid in heated sunflower oil (21) and of cyclic dienes produced from linolenic acid in heated linseed oil (37,38).

For CFAM mixtures isolated from heated oils, however, simplification of mixtures was advisable. CFAM mixtures isolated from heated sunflower and linseed oils were separated, as phenacyl esters by silver-ion HPLC, into several fractions according to degree of unsaturation, size and configuration of the ring, and configuration of double bonds (21). The simplified mixtures were then examined by GC/MS as picolinyl esters and their DMOX derivatives. This approach, combined with GC/FT-IR spectroscopy of the methyl esters of silver-ion HPLC fractions for double bonds configuration assessment, proved highly valuable for complete elucidation of the structures of CFAM isolated from heated oils (21).

As already suggested (17), most CFAM formed from linoleic acid in heated sunflower oil contained a disubstituted cyclopentenyl ring (21). Spectra of a major compound (picolinyl ester), isolated as a single component in a silver-ion HPLC

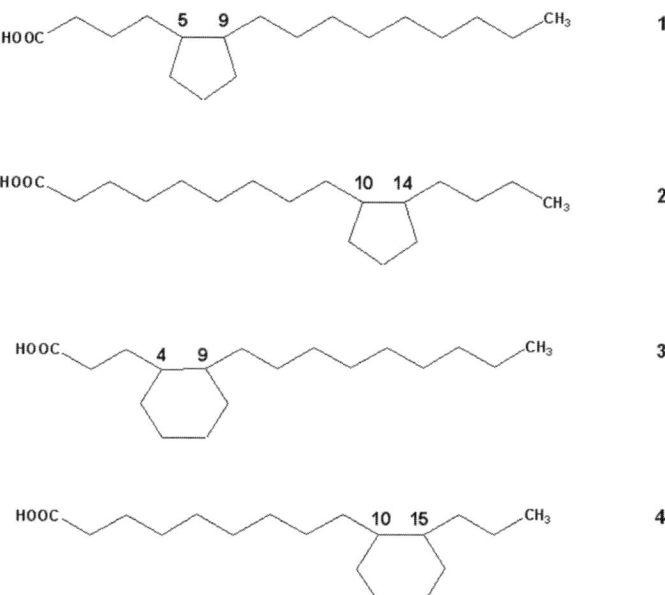

**Fig. 5.14.** Cyclic fatty acids formed from oleic acid. Adapted from Dobson et al. (43).

fraction, are presented in Fig. 5.10. In the spectrum of the unsaturated derivative, a gap of 66 amu between $m/z$ 248 and 314 indicates a cyclopentene ring, incorporating $C_{10}$ to $C_{14}$ of the original $C_{18}$ chain. In the spectrum of the hydrogenated product, the gap for the cyclopentane ring of 68 amu was evident between $m/z$ 248 and 316. In both spectra, regular gaps of 14 amu on either side of the ring confirmed the absence of double bonds in the side chains. The position of the double bond in the ring could not be determined from the spectra, but it presumably remained between $C_{12}$ and $C_{13}$ of the original $C_{18}$ chain. Some cyclohexane and cyclopentane rings were also present in the complex mixture. Examination of the spectra of CFAM picolinyl esters, before and after hydrogenation, allowed the proposal of distinct structural formulae (Fig. 5.11) for CFAM isolated from heated sunflower oil (21).

A study by Mossoba and co-workers (38) published a year later using a similar methodology based on GC-MS of DMOX derivatives and GC-FTIR analyses but not using a prefractionation step demonstrated the presence of $C_5$ and $C_6$ membered ring isomers as the major cyclic fatty acid monomers in heated linseed oils. A year later, Dobson et al. (37) fractionated the CFAM mixture formed from linolenic acid in heated linseed oil, by silver-ion HPLC as phenacyl esters, before conversion to picolinyl esters and DMOX derivatives for examination by GC/MS (37). Hydrogenated compounds and deuterated derivatives, prepared using deuterium on Wilkinson's catalyst, were also examined to locate double bonds. The configuration of double bonds of various isomers was confirmed by GC/FT-IR spectroscopy (37). Identified

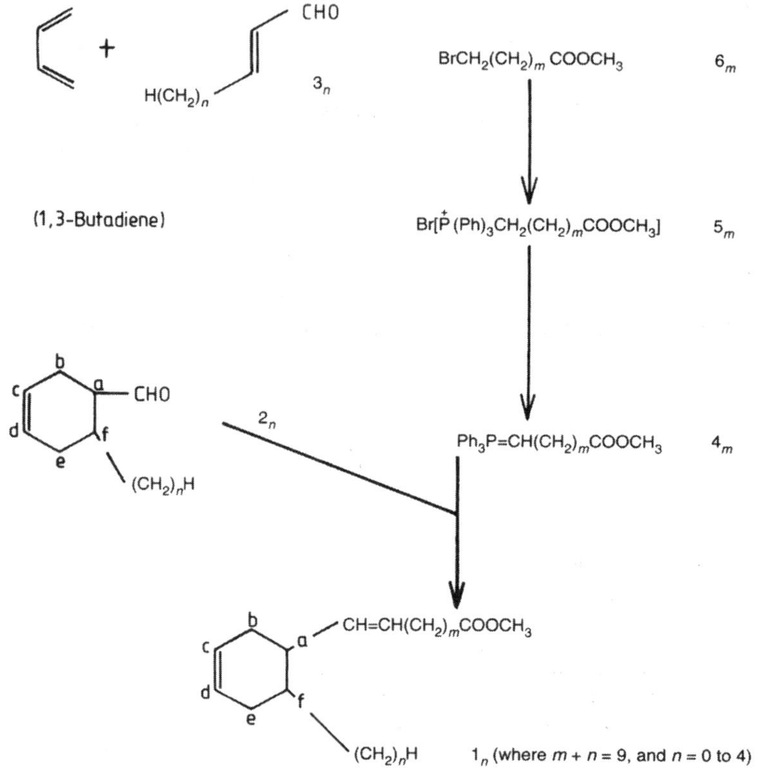

**Scheme 5.1.** Synthesis of cyclohexenyl fatty acids. From Awl and Frankel (12).

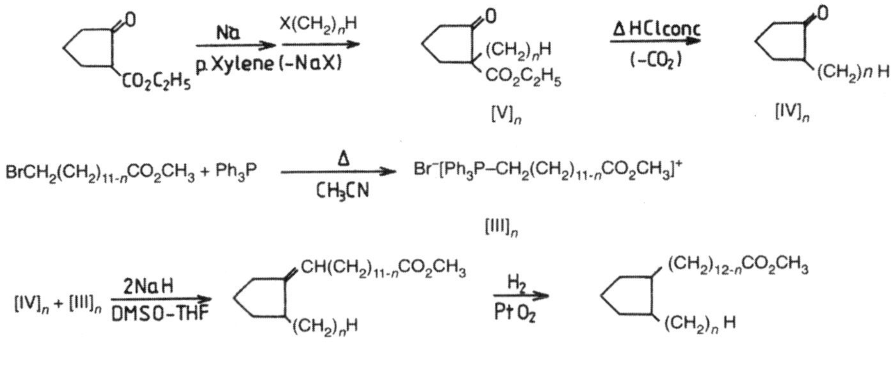

**Scheme 5.2.** Synthesis of $C_5$ CFAM. From Rojo and Perkins (20).

CFAM are shown in Fig. 5.11, and the mass spectrum of the DMOX derivative of 10-(2′-propyl-4′-cyclopentenyl)-9-decenoic acid is presented in Fig. 5.12. A gap of 66 amu between $m/z$ 288 and 222 fixed the presence of a cyclopentenyl ring between $C_{11}$ and $C_{15}$, and the position of the double bond in the carboxylate moiety at $C_9$ was confirmed by a gap of 26 amu between $m/z$ 222 and 196. The position of the double bond in the ring could not be determined from the mass spectra of the picolinyl esters or the DMOX derivatives. As argued for cyclic monoene fatty acids (21), however, double bond migration is unlikely to occur simultaneously with cyclization, and the double bond originally at $C_{12}$ probably remains at that position. This was confirmed by an important retro Diels-Alder fragment on the mass spectra of cyclohexenyl fatty acids [retro Diels-Alder fragment also present in the mass spectra of the methyl esters (17)] and by mass spectra data of picolinyl esters and DMOX derivatives of two model cyclopentenyl dienes. Fourier transform infrared spectroscopic data showed that substantial stereomutation took place in the chain with an overall E/Z ratio of 2.0 (37). Stereomutation of double bonds is known to occur in heated oil (39). Most of the identified molecules corresponded to the structures developed by Mossoba et al. (38,40). However, these authors detected the presence of cyclohexadienyl isomers that were not confirmed in the work of Dobson et al. (37).

Other minor compounds such as bicyclic fatty acids were also detected (7,41) in heated sunflower and linseed oils. In heated sunflower oils, these saturated bicyclic

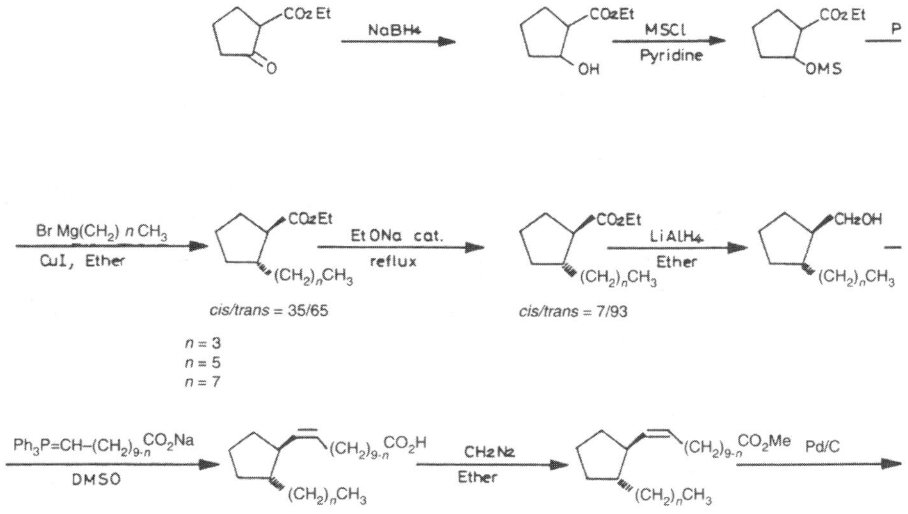

**Scheme 5.3.** Synthesis of $C_5$ CFAM. From Vatèle et al. (19).

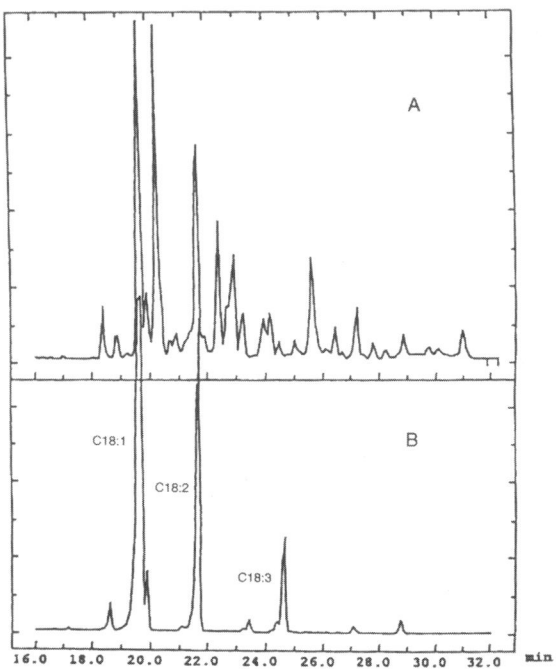

**Fig. 5.15.** Gas chromatography (BPX-70 column) of a) CFAM fraction isolated from heated sunflower oil, and b) fatty acid methyl esters (FAME) of canola oil.

fatty acids were a mixture of a $C_6$-membered ring either from $C_{10}$ to $C_{15}$ or from $C_7$ to $C_{12}$ of the parent fatty acid and of a $C_5$-membered ring with either from $C_{10}$ to $C_{14}$ or from $C_8$ to $C_{12}$. A bond across the ring gave the bicyclic structure but the position of the bond could not be determined.

From the previously mentioned studies, cyclization from linoleic and linolenic acids was rationalized (37). Products from reaction of either the $C_9$ and $C_{12}$ double bonds to give rings between $C_5$ and $C_9$, $C_5$ and $C_{10}$, $C_8$ and $C_{12}$, $C_{12}$ and $C_{17}$, and $C_{13}$ and $C_{17}$ were observed for the cyclic monoenes isolated from heated sunflower oil. For cyclic dienes isolated from heated linseed oil, cyclization was always directed internally toward other double bonds and never involved the double bond at $C_{12}$. Migration of the ring double bond to the $C_{13}$ position in $C_{10}$–$C_{14}$ cyclopentenyl fatty acids was observed for cyclic monoenes (21) but never observed for cyclic dienes (37).

Other structural studies on evening primrose oil (42) and of high oleic sunflower oil (43) also showed that oleic acid and γ-linolenic acid could, upon heat treatment, give cyclic fatty acid monomers. Sixteen cyclic dienoic fatty acids (Fig. 5.13) were identified (42) from γ-linolenic acid. The structures were analogous to those previously reported for α-linolenic acid in the sense that the trienoic unit reacted the same way irrespective of its position on the carbon chain. Oleic acid (43) was found to give a mixture of eight saturated fatty acids comprised of four basic structures of cyclopentyl and cyclohexyl fatty acids (Fig. 5.14). The cyclopentyl fatty acids were twice as abundant as those with a cyclohexane ring.

Recently, by looking at the components formed during the cyclization of oleic, linoleic, and linolenic acids, Destaillats and Angers presented rearrangement mechanisms that may be involved in the formation of CFAM and their further evolution into bicyclic fatty acid monomers (44). A concerted cycloaddition mechanism undergoing [1,6]-or [1,7] prototropic migration was found consistent with the CFAM identified.

## Syntheses of Model CFAM

Earlier work on the synthesis of model CFAM was carried out on disubstituted cyclohexenyl fatty acids. Graille et al. (11) reported synthesis of a monounsaturated disubstituted cyclohexenyl fatty acid. Awl and Frankel (12) reported synthesis of some diunsaturated compounds having one ethylenic bond in the $C_6$ ring and the other one on the substituent having the longer chain length. Scheme 5.1 shows the synthesis of disubstituted cyclohexenyl CFAM (12). According to the synthetic pathway, the major compounds obtained had a double bond on the side chain and ring substituents in a *trans* position. 1H-NMR and 13C-NMR chemical shifts were reported for the major components (12). Mass spectra data given for one cyclic ester 1n [n = 3 and m = 6, i.e., methyl 9-(2′-propyl-4′-cyclohexenyl)-8-nonenoate] were particularly useful for assigning the structure of one cyclohexenyl diene formed from linolenic acid in heated linseed oil (17).

Vatèle et al. (19) and Rojo and Perkins (20) also proposed two different approaches for synthesizing CFAM having a 1,2-disubstituted 5-carbon-membered ring. Three monounsaturated CFAM having the ethylenic bond in the α-position to the ring and four saturated isomers were synthesized.

The synthetic route proposed by Rojo and Perkins (20) (Scheme 5.2) is very similar to that used by Awl and Frankel (12). The synthetic route followed by Vatèle et al. (19) was quite different. The starting product was ethyl 2-oxo-cyclopentanecarboxylate. The first important step of the synthesis (Scheme 5.3) was a Michaël addition of alkylmagnesium bromide to an unsaturated ester (I) having a 5-carbon-membered ring to give a mixture of *trans* and *cis* ethyl-2-alkylcyclopentane carboxylate (II) in 77–83% yield.

Both synthetic routes (19,20) give the possibility of preparing saturated and unsaturated CFAM, the latter ones being intermediates. It is also very easy to obtain different reaction products with different substituents using the same sequence of reactions. Mass spectra data of these saturated models were used to confirm the carbon skeletons of CFAM formed in heated partially hydrogenated soybean oil (21) and in heated sunflower and linseed oils (18).

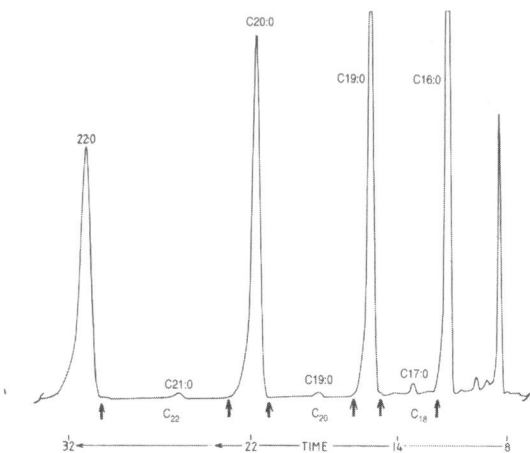

**Fig. 5.16.** HPLC (C$_{18}$ reversed phase; CAN, acetone (90:10), 4 mL min-1) analysis of fully hydrogenated FAME of fish oil capsule.

## Quantitative Analysis

As of now, all methods proposed to determine the amount of cyclic fatty acids (CFAM) in heated fats and oils are based on the analysis of fatty acid methyl esters by gas–liquid chromatography (GC). As shown in Fig. 5.15, cyclic fatty acids formed from either linoleic or linolenic acid have, even on highly polar columns, ECL values close to those of the fatty acids, such as 18:1, 18:2, and 18:3 geometrical isomers, which are present in deodorized and/or vegetable oils subjected to heat (39,45,46). Therefore it is not possible by this direct GC method to determine the quantity of CFAM precisely, due to the large overlap with unsaturated fatty acids as previously mentioned. One solution is to fully hydrogenate the sample and convert all polyenes to 18:0, which is then separated from hydrogenated CFAM, which have longer retention times on polar phases (15,47). This method was developed by Black and Eisenhauer in 1963 (48). Only samples containing high quantities of cyclic fatty acids (> 40%) were analyzed. This method was further improved by Gente and Guillaumin in 1977 (49) using glass capillary columns and an internal standard (IS), which has an ECL value close to C$_{21:0}$. These authors concluded that cyclic components corresponded to peaks having retention times between C$_{18:0}$ and C$_{20:0}$, the detection limit being 0.1%.

One major drawback of the method, however, was that, even on capillary columns, CFAM were present as tails on the major peak of stearic acid, so that one would wonder if some CFAM did not have retention times close to 18:0. This could be partially avoided by using more polar columns such as cyanosilicon phases (15,47). Gas chromatographic analysis on silar-10c of a mixture of 18:0 and CFAM isolated from heated linseed oil after hydrogenation (15) showed four major CFAM having

**TABLE 5.4**
**Quantification of Cyclic Fatty Acids (wt%) in Vegetable Oils**

| Sample no. | After urea enrichment (51) | Direct GLC analysis (49) |
|---|---|---|
| 1 | 0.04 | 0.2 |
| 2 | 0.07 | 0.2 |
| 3 | 0.04 | 0.2 |
| 4 | 0.07 | 0.2 |
| 5 (linseed) | 11.46 | 8.8 |
| 6 | 4.60 | 3.5 |

**TABLE 5.5**
**Quantification of CFAM in Heated Oils**

| Method | Heated sunflower | Heated rapeseed |
|---|---|---|
| A. Gente and Guillaumin (49) | $X = 0.43$ ($n = 9$) | $X = 0.29$ ($n = 9$) |
| B. Gente and Guillaumin (49) with 17:0 added before hydrogenation | 0.55 | 0.28 |
| C. Gente and Guillaumin (49) with urea inclusion + 17:0 before GLC | $X = 0.07$ ($n = 3$) | $X = 0.09$ ($n = 4$) |
| D. Gere et al. (52) + phenanthrene + urea inclusion | $X = 0.11$ ($n = 7$) | $X = 0.14$ ($n = 6$) |

Sources: Gente, M., and R. Guillamin, *Rev. Fr. Corps Gras* 24:211–218 (1977).
Gere, A. et al., *Rev. Fr. Corps Gras* 31:341–346 (1984).

ECL values of 18.43, 18.63, 19.05, and 19.39, and two minor components with ECL values of 19.00 and 19.23. The same analysis on a Carbowax column revealed the presence of only three major CFAM isomers, the fourth having the same ECL value as that of 18:0. Therefore the quantity of CFAM could only be calculated after the elimination of $C_{18:0}$. This may be realized using urea adduction as described by Firestone et al. (50) and Potteau (51). Consequently, considering the overlap between 18:0 and some CFAM, discrepancies between results given by two methods (the first one being direct GC analysis of the totally hydrogenated sample, and the second being GC analysis of a purified fraction after urea adduction to complex $C_{18:0}$ was noted (47). Such discrepancies could be very important (Table 5.4), especially when the first CFAM peak [methyl 9-(2-butylcyclopentyl) nonanoate] was important (samples 5 and 6). This is the case when the original oil contains high quantities of 18:3(n-3).

As outlined by Gere et al. (52), however, utilization of urea for enrichment of CFAM results in a loss of approximately 20% of the cyclic fatty acids during complexation and non-complete elimination of 18:0. To account for this loss of CFAM, utilization of phenanthrene as the IS instead of 17:0 was proposed. Phenanthrene would be added to methyl esters prior to complexation and would behave in the same manner as cyclic monomers. Consequently, higher CFAM values

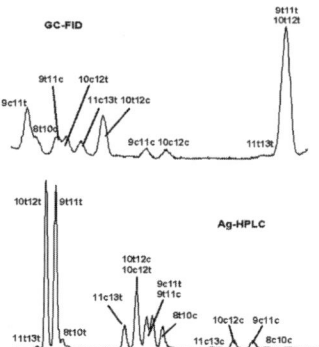

**Fig. 5.17.** Analysis of CLA methyl esters fraction isolated from a used sunflower oil: on top by GLC on CP-sil 88 and on bottom by silver ion HPLC. From Juaneda et al. (70).

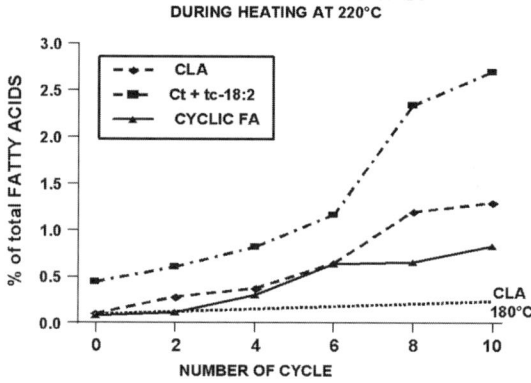

**Fig. 5.18.** Evolution of modified fatty acids during heat treatment of sunflower oil at 220°C. From Juaneda et al. (71).

would be obtained (comparison of methods C and D in Table 5.5). In each case, if we compare the four methods described in Table 5.5, quantities of CFAM are always smaller when using urea adduction (where the peak of 18:0 is smaller) as compared to direct GC analysis.

Urea adduction is not the only technique for concentrating cyclic fatty acid fractions prior to GC analysis. This can also be done by low-temperature crystallization to determine cyclic acids in an oil at levels below 0.5%. This method uses apparatus developed by Friedrich (53) and removes some of the saturated components (77–97%), as reported by Meltzer et al. (54).

However, even if an enrichment step is necessary to concentrate the CFAM fraction prior to GC analyses, urea adduction or low-temperature crystallization does

**TABLE 5.6**
**CFAM Content in Phospholipids (PL) and Nonphosphorus Lipids (NL) of Cardiomyocytes Incubated in a CFAM-Containing Medium**

| Sample | Concentration of CFAM in medium (mg/L) | PL | NL |
|---|---|---|---|
| Heated linseed | 2.5 | $0.5 \pm 0.08^a$ | $0.6 \pm 0.05$ |
| | 5.0 | $0.8 \pm 0.01$ | $1.1 \pm 0.17$ |
| Heated sunflower | 2.5 | $1.0 \pm 0.07$ | $1.5 \pm 0.10$ |
| | 5.0 | $1.5 \pm 0.18$ | $3.2 \pm 0.11$ |

Source: Sébédio, J.-L. et al., J. Chromatogr. 659:101–109 (1994).
[a]Average of 3 separate cultures (in mg esters/mg proteins).

**TABLE 5.7**
**Cyclic Fatty Acids in Commercial Frying Oils**

| | Amounts (%) | | | |
|---|---|---|---|---|
| | Minimum | Maximum | Origin | No. of samples |
| Frankel et al. (59) | 0.02 | 0.50 | USA | 25 |
| | 0.17 | 0.66 | Egypt | 8 |
| | | | Israel | |
| Gere et al. (60) | 0.02 | 0.16 | Hungary | 8 |
| Sébédio et al. (61) | 0.01 | 0.25 | France | 31 |
| Poumeyrol (62) | 0.01 | 0.09 | France | 21 |
| Juaneda et al. (70 ) | 0.15 | 0.17 | France | 4 |

not completely eliminate 18:0. Further, the utilization of urea for the analysis of small sample sizes brings out many impurities that can be detected during GC analyses. A clean-up procedure was developed by Rojo and Perkins (55) using solid-phase extraction. The authors, however, reported inconsistent recoveries of phenanthrene (IS), and utilization of either a naturally occurring or a synthetic cyclic fatty acid for the IS was recommended.

In 1994, Sébédio et al. (56) described a method utilizing HPLC for the enrichment step. Briefly, total fatty acid methyl esters were hydrogenated after addition of ethyl hexadecanoate as the IS. A fraction containing the IS and CFAM was isolated by HPLC on a $C_{18}$ reversed-phase column. The quantity of CFAM was then further calculated after analyses by GC on a polar phase, and CFAM structures were determined by GC/MS. Good reproducibility was obtained not only for oil samples but also in the biological assays (rat tissues and lipids from heart cell cultures), as reported in Table 5.6. Furthermore, no modification of the CFAM profile was obtained, which was not the case when urea adduction was carried out on a small sample size, such as lipids isolated from animal tissue.

Cyclic fatty acid methyl esters are present not only in heated vegetable oil but also in encapsulated fish oils, which may indicate that the oil was submitted at one point to heat treatment (57). As $C_{20:5}$(n-3) (eicosapentaenoic acid, EPA) and $C_{22:6}$(n-3) (docosahexaenoic acid, DHA) are the major fatty acids of fish oil (58), cyclic fatty acids with 20 and 22 carbons were also found. The method developed to quantify cyclic fatty acids also included HPLC as an enrichment step. Briefly, the total lipid sample is converted to methyl esters, which are further hydrogenated on $PtO_2$. The total hydrogenated methyl esters are then fractionated on a $C_{18}$ reversed-phase column. Three fractions are collected. The fraction collected between 16:0 and 18:0 contains the $C_{18}$ CFAM and 17:0 (which is used as the IS), while that collected between 18:0 and 20:0 contains the $C_{20}$ CFAM and 19:0, and that between 20:0 and 22:0 is a mixture of $C_{22}$ CFAM and 21:0. Each fraction is then analyzed by GC on a polar column, as shown in Fig. 5.16, using 17:0, 19:0, and 21:0 as the IS. Gas chromatography/mass spectrometry studies showed that all these fractions contained only CFAM with the fragmentation pattern already observed for CFAM isolated from heated vegetable oil (7). Thirty to forty ppm CFAM was considered the minimum quantity that could be quantified.

### Presence in Food Products

Many studies on cyclic fatty acids, whether structural or concerned with their formation due to temperature, type of oil, or length of heat treatment, were carried out on oil heated in the laboratory or on oil heated under simulated frying operations (16). Unfortunately, few results for oil collected from actual restaurants or industrial operations are available (Table 5.7). In the study by Frankel et al. (59), fat and oil were a mixture of vegetable shortening, animal-vegetable shortening, partially hydrogenated vegetable oil, cottonseed oil, and soybean oil, while Gere et al. (60) described experiments carried out with mixtures of sunflower oil and lard. In the study by Sébédio et al. (61), peanut, sunflower, and soybean oils, as well as some unknown mixtures of oil, were used. Samples studied by Poumeyrol (62) were palm and peanut oils. CFAM levels found in the different studies were in good agreement. They ranged from 0.01–0.66% in the most altered samples.

## Methylene and Conjugated Polyunsaturated Fatty Acid Isomers

*Trans* polyunsaturated fatty acid isomers were first reported by Ackman et al. (45) as a result of heat treatment that was used during the deodorization of oils. Later Grandgirard et al. (39) characterized geometrical isomers of linoleic and linolenic acids in oils heated in the laboratory. Linolenic acid was shown to be more sensitive to isomerization than linoleic acid, and the type of geometrical isomers detected (mono-*trans* vs. di-*trans*) depended on the temperature used for the process. For linolenic acid the major isomers were the 18:3 9c,12c,15t, the 18:3 9t,12c, 15c and the 18:3 9t,12c,15t. Very little isomerization took place in the central position (39). Similarly, linoleic acid gave three isomers upon heat treatment: the 18:2 9c,12t, the 18:2 9t,12c and only minor quantities of the di-*trans* isomer at high temperatures. The position of linolenic acid on the triacylglycerol was also shown to be of importance for geometrical isomerization (63). For example, at high temperatures, 18:3n-3 acylated in the central position of the TG showed the highest sensitivity to geometrical isomerization. Geometrical isomerization was further confirmed in a study where frozen prefried french fries were deep fried in soybean and in peanut oils (64). Thirty frying operations were conducted at 180, 200, and 220°C for 5 min. For both acids, the quantities of *trans* fatty acids increased after 10 frying treatments at 220°C while no differences were found at 180 and 200°C. Furthermore, the di-*trans* isomers of linolenic acid were formed above 200°C. Further studies also confirmed the formation of *trans* fatty acids during frying (65). Interestingly, in a study by Capiono et al. (66), *trans* isomers of linoleic and of linolenic acid were found to increase more after microwave heating than after conventional heating.

Older publications report the presence of *trans* polyunsaturated fatty acids in food products such as deodorized vegetable oils, human milk, and infant formulas, for example (67). However, the data are over 10 years old and considering the emphasis placed on *trans* fatty acids lately, one might think that their quantities in food products should have decreased. Further work is needed in this field in order to complete food composition tables.

While many papers report the presence of conjugated fatty acids in products from ruminants (68), much less data are available on the formation of conjugated fatty acids (CLA) during food processing such as frying. In milk and dairy products the CLA mixture is mainly composed of one major isomer, the 18:2 9c,11t (69). Other isomers such as 7t,9c and 11c,13t have also been observed. The situation of used frying oils is not as simple. The used oils are characterized by a high proportion of the di-*trans* isomers that cannot be separated, even on long polar columns using hydrogen as the carrier gas. Examination of the mixture by the powerful silver nitrate high performance liquid chromatography (Fig. 5.17) revealed that the 10t,12t and the 9t,11t are the major isomers. These are accompanied by a mixture of c,t t,c and c,c 8,10 10,12 and 11,13 isomers. For the five samples collected from market vendors and restaurants (70) the quantity of CLA ranged from 0.3–0.5% while the quantity of polar compounds ranged from 30–53%. No relations between the amounts of polar components and CLA were reported. The evolution of the different isomeric fatty acids described in this chapter (CFAM, 18:2 isomers, CLA) was reported in a study using sunflower oil that was submitted to heating cycles at different temperatures. The evolution of these molecules at 220°C is reported in Fig. 5.18. The quantities of CLA, CFAM, and isomeric fatty acids increased slowly up to cycle 6, where the amount of polar components reached 30%. At this point the amount of 18:2 isomers increased sharply while the other two types of alteration products showed only a moderate increase (71). In any case, the quantity of CLA formed during frying is low and is mainly constituted by two di-*trans* isomers. However, no studies have been carried out on these isomers so far.

## Conclusion

As this chapter has shown, frying is a complex situation where many reactions are taking place. During recent years many studies were carried out to identify the major components due to the heat process, such as cyclic fatty acid monomers and polyunsaturated fatty acid geometrical isomers, including those having conjugated double bonds. The three unsaturated fatty acids with 18 carbons can give rise to CFAM, but linolenic acid is the most sensitive to temperature for isomerization as has been demonstrated by Ackman for the deodorization of oils and for cyclization. Temperature is the most important frying parameter as far as these compounds are concerned; around 200°C seems to be the critical point at which to find these components in appreciable quantities, not only in the frying medium but also in the food matrix.

## References

1. Wells, A.F.; and R.H. Common. *J. Sci. Food Agric.* **1953**, *4*, 233–237.
2. MacDonald, J.A. *J. Am. Oil Chem. Soc.* **1956**, *33*, 394–396.
3. McInnes, A.G.; F.P. Cooper; and J.A. MacDonald. *Can. J. Chem.* **1961**, *39*, 1906–1914.
4. Saito, M.; and T. Kaneda. *Yukagaku* **1976**, *25*, 79–86.
5. Hutchinson, R.B.; and J.C. Alexander. *J. Org. Chem.* **1963**, *28*, 2522–2526.
6. Gast, L.E.; W.J. Schneider; C.A. Forest; and J.C. Cowan. *J. Am. Oil Chem. Soc.* **1963**, *40*, 287–289.
7. Potteau, B.; P. Dubois; and J. Rigaud. *Ann. Technol. Agric.* **1978**, *27*, 655–679.
8. Michael, W.R. *Lipids* **1967**, *1*, 359–364.
9. Michael, W.R. *Lipids* **1967**, *1*, 365–368.
10. Zeman, A.; and H. Scharmann. *Fette Seifen Anstrichm.* **1973**, *75*, 32–44.
11. Graille, J.; A. Bonfand; P. Perfetti; and M. Naudet. *Chem. Phys. Lipids* **1980**, *27*, 23–41.
12. Awl, R.A.; and E.N. Frankel. *Lipids* **1982**, *17*, 414–426.
13. Christie, W.W.; D. Rebello; and R.T. Holman. *Lipids* **1969**, *4*, 229–231.
14. Sébédio, J.-L.; and A. Grandgirard. *Prog. Lipid Res.* **1989**, *28*, 303–336.
15. Sébédio, J.-L. *Fette Seifen Anstrichm.* **1985**, *87*, 267–273.
16. Sébédio, J.-L.; J. Prévost; and A. Grandgirard. *J. Am. Oil Chem. Soc.* **1987**, *64*, 1026–1032.
17. Sébédio, J.-L.; J.-L. Le Quéré; E. Sémon; O. Morin; J. Prévost; and A. Grandgirard. *J. Am. Oil Chem. Soc.* **1987**, *64*, 1324–1333.
18. Sébédio, J.-L.; J.L Le Quéré; O. Morin; J.M. Vatèle; and A. Grandgirard. *J. Am. Oil Chem. Soc.* **1989**, *66*, 704–709.
19. Vatèle, J.M.; J.-L. Sébédio; and J.-L. Le Quéré. *Chem. Phys. Lipids* **1988**, *48*, 119–128.
20. Rojo, J.A.; and E.G. Perkins. *Lipids* **1989**, *24*, 467–476.

21. Christie, W.W.; E.Y. Brechany; J.-L. Sébédio; and J.-L. Le Quéré. *Chem. Phys. Lipids* **1993,** *66,* 143–153.
22. Le Quéré, J.-L.; E. Sémon; B. Lanher; and J.-L. Sébédio. *Lipids* **1989,** *24,* 347–350.
23. Rojo, J.A.; and E.G. Perkins. *J. Am. Oil Chem. Soc.* **1987,** *64,* 414– 421.
24. Gross, M.L. In *Advances in Mass Spectrometry;* P. Longevialle, Ed.; Heyden and Sons: London, 1989; Vol. 11A, pp. 792–811.
25. Gross, M.L. *Int. J. Mass Spectrom. Ion Proc.* **1992,** *118/119,* 137–165.
26. Tomer, K.B.; N.J. Jensen; and M.L. Gross. *Anal. Chem.* **1986,** *58,* 2429–2433.
27. Le Quéré, J.-L.; J.-L. Sébédio; R. Henry; F. Couderc; N. Demont; and J.C. Promé. *J. Chromatogr.* **1991,** *562,* 659–672.
28. Promé, J.-C.; H. Aurelle; F. Couderc; and A. Savagnac. *Rapid Comm. Mass Spectrom.* **1987,** *1,* 50–52.
29. Ackman, R.G.; J.-L. Sébédio; and W.N. Ratnayake. *Methods Enzymol.* **1981,** *72,* 253– 276.
30. Tomer, K.B.; F.W. Crow; and M.L. Gross. *J. Am. Chem. Soc.* **1983,** *105,* 5487–5488.
31. Harvey, D.J. *Spectroscopy (Ottawa)* **1990,** *8,* 211–244.
32. Harvey, D.J. In *Advances in Lipid Methodology—One;* W.W. Christie, Ed.; The Oily Press: Ayr, U.K., 1992; pp. 19–80.
33. Christie, W.W. *inform* **1993,** *4,* 85–91.
34. Zhang, J.Y.; Q.T. Yu; B.N. Liu; and Z.H. Huang. *Biomed. Environ. Mass Spectrom.* **1988,** *15,* 33–44.
35. Luthria, D.L.; and H. Sprecher. *Lipids* **1993,** *28,* 561–564.
36. Spitzer, V. *J. Am. Oil Chem. Soc.* **1991,** *68,* 963–969.
37. Dobson, G.; W.W. Christie; E.Y. Brechany; J.-L. Sébédio; and J.-L. Le Quéré. *Chem. Phys. Lipids* **1995,** *75,* 171–182.
38. Mossoba, M.M.; M.P. Yurawecz; J.A.G. Roach; H.S. Lin; R.E. McDonald; B.D. Flickinger; and E.G. Perkins. *Lipids* **1994,** *29,* 893–896.
39. Grandgirard, A.; J.-L. Sébédio; and J. Fleury. *J. Am. Oil Chem. Soc.* **1984,** *61,* 1563– 1568.
40. Mossoba, M.M.; M.P. Yurawecz; J.A.G. Roach; H.S. Lin; R.E.McDonald; B.D. Flickinger; and E.G. Perkins. *J. Am. Oil Chem. Soc.* **1995,** *72,* 721–727.
41. Dobson, G.; W.W. Christie; and J.L. Sébédio. *Chem. Phys. Lipids* **1997,** *87,* 137–147.
42. Dobson, G.; and J.-L. Sébédio. *Chem. Phys. Lipids* **1999,** *97,* 105–118.
43. Dobson, G.; W.W. Christie; and J.-L. Sébédio. *Chem. Phys. Lipids* **1996,** *82,* 101–110.
44. Destaillats, F.; and P. Angers. *Eur. J. Lipid Sci. Technol.* **2005,** *107,* 767–772.
45. Ackman, R.G.; S.N. Hooper; and D.L. Hooper. *J. Am. Oil Chem. Soc.* **1974,** *51,* 42–49.
46. Wolff, R.L. *J. Am. Oil Chem. Soc.* **1993,** *70,* 219–224.
47. Grandgirard, A.; and F. Julliard. *Rev. Fr. Corps Gras* **1983,** *30,* 123–128.
48. Black, L.T.; and R.A. Eisenhauer. *J. Am. Oil Chem. Soc.* **1963,** *40,* 272–274.
49. Gente, M.; and R. Guillaumin. *Rev. Fr. Corps Gras* **1977,** *24,* 211– 218.
50. Firestone, D.; S. Neisheim; and W. Horwitz. *J. Assoc. Off. Anal. Chem.* **1961,** *44,* 465–474.
51. Potteau, B. *Ann. Nutr. Aliment.* **1976,** *30,* 89–93.
52. Gere, A.; C. Gertz; and O. Morin. *Rev. Fr. Corps Gras* **1984,** *31,* 341–346.
53. Friedrich, J.P. *Anal. Chem.* **1961,** *33,* 974–975.

54. Meltzer, J.B.; E.N. Frankel; T.R. Bessler; and E.G. Perkins. *J. Am. Oil Chem. Soc.* **1981,** *53,* 779–784.
55. Rojo, J.A.; and E.G. Perkins. *J. Am. Oil Chem. Soc.* **1989,** *66,* 1593–1595.
56. Sébédio, J.-L.; J. Prévost; E. Ribot; and A. Grandgirard. *J. Chromatogr.* **1994,** *659,* 101–109.
57. Sébédio, J.-L.; and A. De Rasilly. Analysis of Cyclic Fatty Acids in Fish Oil Concentrates. In *Proceedings of the 17th Nordic Lipid Symposium*; Imatra, Finland, 1993; pp. 212–216.
58. Ackman, R.G. *Food Rev. Int.* **1990,** *6,* 617–646.
59. Frankel, E.N.; L.M. Smith; C.L. Hamblin; R.K. Creveling; and A.J. Clifford. *J. Am. Oil Chem. Soc.* **1984,** *61,* 87–90.
60. Gere, A.; J.-L. Sébédio; and A. Grandgirard. *Fette Seifen Anstrischm.* **1985,** *87,* 359–362.
61. Sébédio, J.-L.; A. Grandgirard; C. Septier; and J. Prévost. *Rev. Fr. Corps Gras* **1987,** *34,* 15–18.
62. Poumeyrol, G. *Rev. Fr. Corps Gras* **1987,** *34,* 543–546.
63. Martin, J.C.; F. Lavillonniere; M. Nour; and J.L. Sebedio. *J. Am. Oil Chem. Soc.* **1998,** *75,* 1691–1697.
64. Sébédio, J.-L.; M. Catte; M.A. Boudier; J. Prevost; and A. Grandgirard. *Food Res. Int.* **1996,** *29,* 109–116.
65. Sébédio, J.-L.; A. Grangirard; and J. Prevost. *J. Am. Oil Chem. Soc.* **1988,** *65,* 362–366.
66. Caponio, F.; A. Pasqualone; and T. Gomes. *Int. J. Food Sci. Technol.* **2003,** *38,* 481–486.
67. Ratnayake, W.M.N.; J.M. Chardigny; R.L. Wolff; C.C. Bayard; J.-L. Sébédio; and L. Martine. *J. Pediatr. Gastroenterol. Nutr.* **1997,** *25,* 400–407.
68. Parodi, P. Conjugated Linoleic Acid in Food. In *Advances in Conjugated Linoleic Acid Research*; J.-L. Sébédio, W.W. Christie, and R. Adlof, Eds.; AOCS Press: Champaign, IL, 2003; Vol. 2, pp. 101–122.
69. Sehat, N.; M.P. Yurawecz; J.A.G. Roach; M.M. Mossoba; J.K.G. Kramer; and Y. Ku. *Lipids* **1998,** *33,* 217–221.
70. Juaneda, P.; O. Cordier; S. Gregoire; and J.-L. Sébédio. *OCL* **2001,** *8,* 94–97.
71. Juaneda, P.; S. Brac de la Perriere; J.-L. Sébédio; and S. Gregoire. *J. Am. Oil Chem. Soc.* **2003,** *80,* 937–940.

# 6

# Formation and Analysis of Oxidized Monomeric, Dimeric, and Higher Oligomeric Triglycerides

## M. Carmen Dobarganes and Gloria Márquez-Ruiz
*Instituto de la Grasa (C.S.I.C.), Avda., Padre García Tejero, 4, 41012 Sevilla, Spain*

## Introduction

During frying, it is well known that a wide variety of chemical reactions results in the formation of new compounds that differ in molecular weight and polarity.

Table 6.1 summarizes the main groups of alteration compounds formed during frying, resulting from the oil or fat being exposed to high temperatures in the presence of air and moisture. Hydrolysis occurs due to the presence of moisture in the food. This involves breaking ester bonds and releasing free fatty acids, monoglycerides, and diglycerides. These compounds have higher polarity and lower molecular weight than the original triglycerides (TG). Additionally, due to the presence of air and exposure to high temperatures, oxidation and thermal alterations take place in the unsaturated fatty acids, leading mainly to modified TG with at least one of the three fatty acyl chains altered.

With the exception of the volatiles resulting from the breakdown of peroxides, oxidative and thermal alteration products have similar or higher molecular weight than that of the original TG. Many of the volatile decomposition products have been identified (1–4), whereas detailed information on the nonvolatile compounds is only partial because of the complexity of this fraction, and the limitations of the analytical methodologies used for the isolation and quantitation of such compounds.

While diglycerides, monoglycerides, and fatty acids likewise originate in the stage before intestinal absorption and hence have no relevance from a nutritional point of view, nonvolatile alteration products do modify the nutritional properties of oils and fats (5). Therefore, the evaluation of the complex mixture of nonvolatile compounds, ingested as a part of fried food, and the understanding of its dependence on the main variables of the frying process are a subject of great interest not only for processors and food technologists, but also for nutritionists and consumers.

This chapter is dedicated to the major nonvolatile alteration compounds in used frying oils and fats, i.e., oxidized monomeric, dimeric, and higher oligomeric TG, also named oxidized monomers, dimers, and polymers throughout this chapter. The contents of this chapter are divided into two major sections. The first part covers their formation, while the second one focuses on the analytical techniques used for their quantitation as well as their occurrence in used frying fats.

# Formation of Oxidized Monomers, Dimers, and Higher Oligomers

The formation of new compounds during frying is clearly associated with the autoxidation process, which proceeds via a free radical mechanism summarized in Fig. 6.1. Here RH represents the triglyceride molecule undergoing oxidation in one of its unsaturated fatty acyl groups. Three triglyceride radicals, -alkyl radicals (R•) formed in the initiation reaction, alkylperoxyl radicals (ROO•) formed by the addition of oxygen, and alkoxyl radicals (RO•) formed during hydroperoxide decomposition, are involved in the formation of hydroperoxides (ROOH) and/or in the set of termination reactions. This leads to a great variety of compounds of different polarity, stability, and molecular weight. Among them, three main groups of compounds are noteworthy:

- Compounds with molecular weights similar to those of the triglycerides (RH) undergoing oxidation in one of their unsaturated fatty acyl groups (6)
- Volatile compounds with molecular weights lower than those of RH. These volatile compounds are produced by alkoxyl radical breakdown (7)
- Polymerization compounds formed through interaction of two triglyceride radicals and thus, with molecular weights higher than those of RH (8)

**TABLE 6.1**
**Main Groups of New Compounds Formed During Frying**

| Alteration | Causative Agent | Resulting Compounds |
|---|---|---|
| Hydrolysis | Moisture | Fatty acids |
| | | Diglycerides |
| | | Monoglycerides |
| Oxidation | Air | Oxidized monomeric triglycerides |
| | | Oxidized dimeric and oligomeric triglycerides |
| | | Volatile compounds (aldehydes, ketones, alcohols, hydrocarbons, etc.) |
| Thermal alteration | Temperature | Cyclic monomeric triglycerides |
| | | Isomeric monomeric triglycerides |
| | | Nonpolar dimeric and oligomeric triglycerides |

At atmospheric pressure, and low or moderate temperatures, solubility of oxygen is high and alkylperoxyl radicals (ROO•) are by far the most common radical species. Once the oil oxidation has been initiated, reaction with oxygen is very rapid and ROOH is the major product originated. However, the chemistry of oxidation at the high temperatures of frying is much more complex since both thermal and oxidative reactions are simultaneously occurring. As temperature increases, the solubility of oxygen decreases drastically while all the oxidation reactions are accelerated. Formation of hydroperoxides is extremely rapid but their decomposition is even faster, such that the amount of hydroperoxides tends to be zero (9). As the oxygen pressure is reduced, the initiation reaction becomes more important, the concentration of alkyl radicals (R•) with respect to alkylperoxyl radicals (ROO•) increases, and polymeric compounds are formed through reactions mainly involving alkyl (R•) and alkoxyl (RO•) radicals (10).

As previously stated, all these reactions take place in the unsaturated fatty acyl groups attached to the glyceridic backbone and, therefore, the stable final products are monomeric, dimeric, and higher oligomeric TG including modified and nonmodified acyl groups.

The following section covers the main mechanisms for the formation of oxidized monomers, dimers, and higher oligomers. In order to decrease the complexity of the

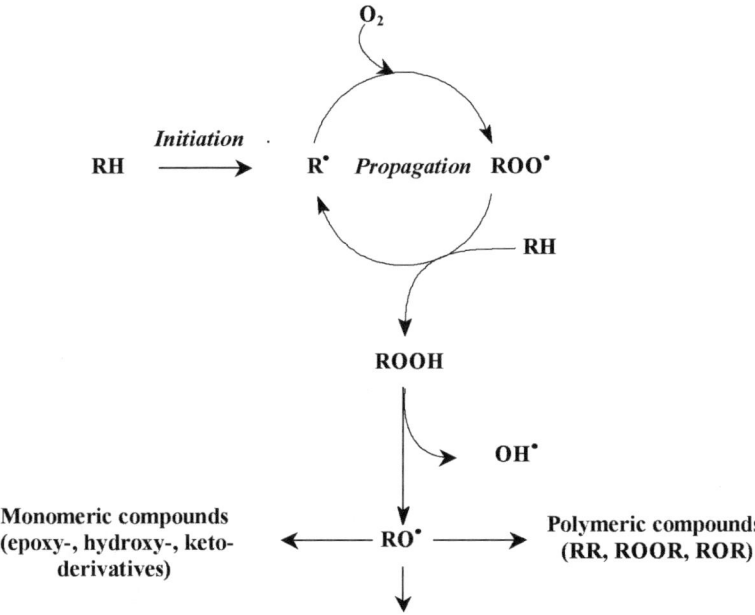

**Fig. 6.1.** Simplified scheme of the autoxidation process.

new compounds formed, most of the studies concerning formation of these groups of compounds at high temperatures started with fatty acid methyl esters (FAME), the simplest TG derivatives.

## Oxidized Monomers

Oxidized monomeric TG, characterized by the presence of extra oxygen in at least one of the fatty acyl groups of the molecule, are final stable products resulting from the breakdown or decomposition of primary oxidation compounds (hydroperoxides). Thus, this group includes TG containing short-chain fatty acyl and short-chain n-oxo fatty acyl groups as the main products (11,12), as well as different oxygenated groups, mainly hydroxy, keto, and epoxy (6). Considering the number of oxygenated forms, which may be present in one or more fatty acyl group of TG, it is easy to imagine the variety of compounds formed and the difficulties encountered during analysis, even if FAME derivatives are used for this purpose (13).

Figure 6.2 summarizes the formation of major short-chain compounds attached to the glyceridic backbone from the 9-hydroperoxide of oleyl, linoleyl, and linolenyl acyl groups. Homolytic β-scission of the alkoxyl radical from allylic hydroperoxide would involve C-C cleavage on either side of the carbon bearing the oxygen. Bound $C_{8:0}$, $C_{9:0}$ aldehyde, and $C_{9:0}$ acid were the major compounds formed in thermoxidized TG (14), besides lower amounts of bound, $C_{7:0}$, $C_{8:0}$ aldehyde, and $C_{8:0}$ acid probably coming from 13-hydroperoxide breakdown of linoleic acid and further oxidative reactions (15) and from 8-hydroperoxide of oleic acid (16).

Concerning oxidized compounds of molecular weight similar to that of the starting TG, the main groups present corresponded to epoxy, keto, and hydroxy fatty acyl groups attached to TG. The route of formation suggested is summarized in Fig. 6.3 for epoxides. Two distinct mechanisms have been proposed for epoxide formation either at the site of the double bond or near the double bond. In the latter case, the original double bond remains (17). However, only the compounds formed when the oxygen added across an existing double bond were detected by a combination of gas chromatography and mass spectrometry at 180°C. Thus, two saturated epoxides, *trans*-9,10- and *cis*-9,10-epoxystearate, were formed in methyl oleate and triolein samples and four monounsaturated epoxides, *trans*-12,13-, *trans*-9,10-, *cis*-12,13-, and *cis*-9,10-epoxyoleate, were formed in methyl linoleate and trilinolein samples (18). The suggested mechanism for the formation of ketones and hydroxides is shown in Fig. 6.4 (6).

Depending on the unsaturation degree of the fatty acyl groups involved, more than one oxygenated function may be present in the same fatty acyl group and more than one oxidized fatty acyl group may be present in one TG molecule.

## Dimers and Higher Oligomers

Two characteristics of the frying process, high temperature and presence of air, favor the development of polymerization reactions accounting for the most complex group

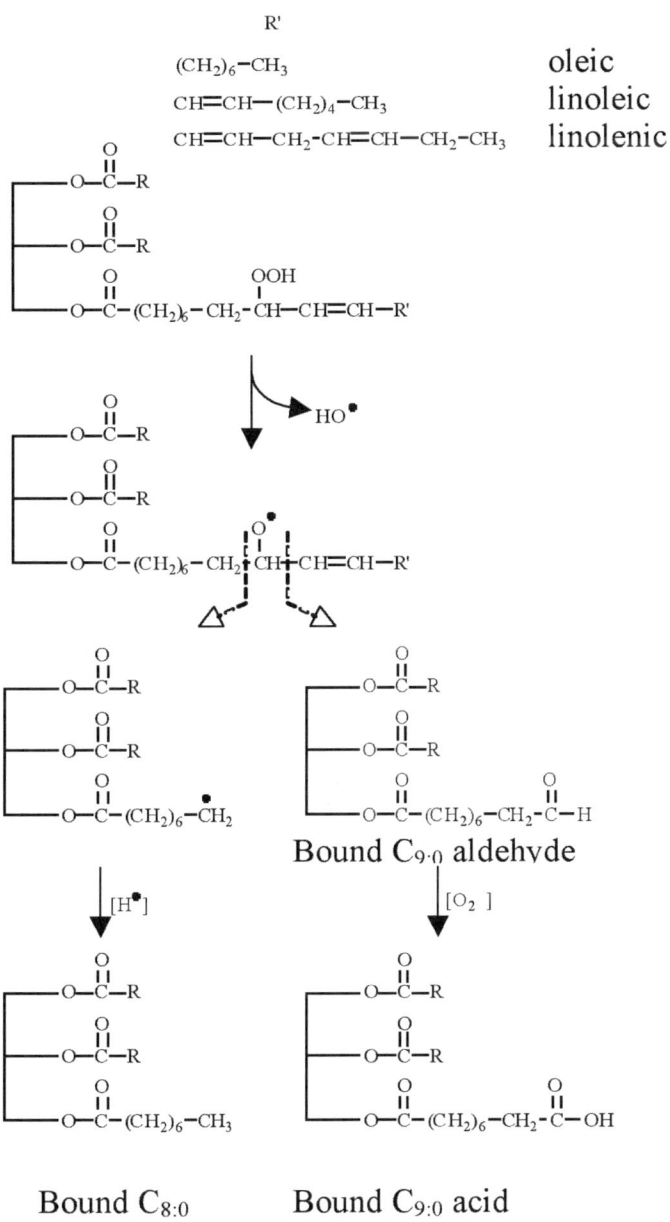

**Fig. 6.2.** Formation of major short-chain oxidation compounds from 9-hydroperoxide of major fatty acyl groups.

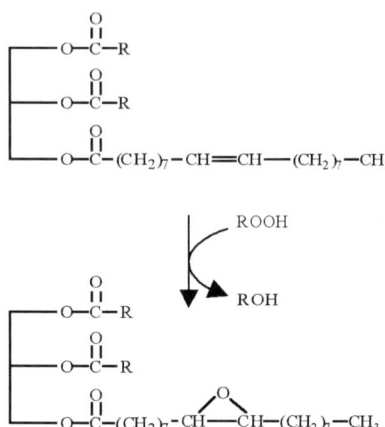

**Fig. 6.3.** Formation of the epoxide ring via external hydroperoxides.

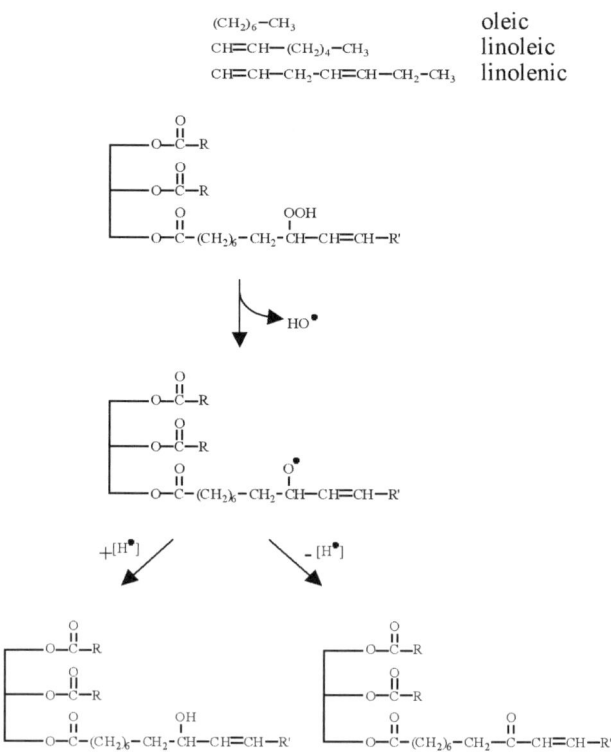

**Fig. 6.4.** Suggested formation of hydroxy and keto functions from 9-hydroperoxide of major fatty acyl groups.

of new compounds among those found in used frying fats and oils.

This complexity results from the different functions available for the oxidation of unsaturated fatty acids, along with the composition of the fat with a high proportion of TG containing more than one unsaturated acyl group per molecule. It also accounts for the lack of studies on the structure and formation of dimeric and higher oligomeric TG.

Significant information on mechanisms of polymerization reactions has been limited to the formation of the dimers obtained in the first step of polymerization. The studies were carried out starting from FAME heated under well-defined conditions, either in the absence or in the presence of air. Detailed results can be found in general reviews (19,20).

*Nonpolar Dimers*

Mechanisms and reactions participating in the formation of nonpolar dimeric FAME, i.e., compounds formed through C-C linkages without any extra oxygen in the molecule, were studied by using FAME subjected to high temperatures, normally between 200 and 300°C, in the absence of air to inhibit oxidative reactions (21,22). Mass spectra of isolated dimers before and after hydrogenation gave clear evidence of the number of double bonds and rings present in the original structures by determining the parent mass peaks. Additional evidence of isomeric forms was obtained from the pattern of fragments observed. From a large series of experiments the following was concluded:

1. The major compounds formed during thermal treatment of FAME are generated through radical reactions from the allyl radicals. Three main reactions have been proposed: a) formation of dehydrodimers by the combination of two allyl radicals, b) formation of noncyclic dimers by intermolecular addition of the allyl radical to a double bond of an unsaturated molecule, and c) formation of cyclic dimers by intramolecular addition of an intermediate dimeric radical to a double bond in the same molecule. Fig. 6.5 shows the three main routes described above.

2. The proportion of the different dimers depends on the conditions used. For example, at 140°C, only dehydrodimers were found in significant amounts. Conversely, above 250°C, mono-, bi-, and tricyclic dimers were mainly generated while dehydrodimers were not found.

3. When conjugated FAME are present, thermal dimers result from Diels Alder reactions, that is, a reaction between two molecules, one of them with a double bond acting as dienophile, added to the conjugated diene of the second molecule to form a cyclohexene tetrasubstituted structure. The general reaction is also shown in Fig. 6.5.

4. From mass spectra, the existence of isomers for the different types of dimers can be deduced. Isomeric forms corresponded to those expected from the FAME involved.

## RADICAL REACTIONS

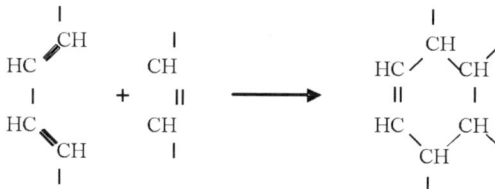

## DIELS ALDER REACTIONS

**Fig. 6.5.** Routes of formation of nonpolar dimers.

Even if conditions during thermal alteration in the absence of air are far different from those used in frying, techniques applied have been of great help in the identification and quantitation of nonpolar dimers in used frying fats. The levels found indicate that nonpolar dimers might be one of the most relevant groups among the alteration compounds (23).

*Polar Dimers*

The structure of polar dimers is still largely unknown. Difficulties are due to the heterogeneity in this group of compounds. First, different oxygenated functions are likely to be present in oxidized monomers before dimer formation, or generated by the oxidation of nonpolar dimers. Second, more than one functional group can be present in the same dimeric molecule. Last, the oxygen may or may not be involved in the dimeric linkage. Therefore, the large number of possible combinations results in a complex mixture that is difficult to separate. Under these circumstances, studies have paid more attention to defining the composition of alteration products than to the mechanisms involved in dimer formation.

The basic knowledge of polar dimers has been obtained by heating FAME,

TG, or fats and oils in the presence of air, or by thermal decomposition of FAME hydroperoxides.

Among the studies carried out, those giving information on the dimers found in fats and oils used in frying, or heated under simulated frying conditions are especially interesting. After separating fractions of different polarity by adsorption chromatography, identification techniques including mass spectrometry, nuclear magnetic resonance, and infrared spectroscopy, have been helpful in providing evidence of some of the dimeric structures formed. The following results are noteworthy:

1. Interestingly, the presence of dehydrodimers, bicyclic, tricyclic, and Diels Alder nonpolar dimers has been reported by different authors (24–26), supporting the importance of the allyl radical and formation of nonpolar conjugated dienes even in the presence of oxygen.
2. Among compounds including oxygenated functions, acyclic dimers with C-O-C linkages and tetrahydrofuran tetrasubstituted dimers have been isolated from low polarity fractions obtained after transesterification of heated soybean oil (25).
3. Structures found for polar dimers were mainly C-C linked dimers containing monohydroxy, dihydroxy, and keto groups (1,27).

The interesting series of papers by Christopoulou and Perkins (28–30) are good examples of a systematic study increasing our knowledge of dimeric structures. First, dimers from methyl stearate containing hydroxy and keto groups were synthesized to represent structures that may be formed during the thermal oxidation of fats (28). Second, mixtures of synthetic dimers were analyzed by GC, high-performance liquid chromatography (HPLC), and thin-layer chromatography (TLC). Different columns and conditions were used to define the most efficient systems for dimer analysis (29). Finally, dimers were isolated from used frying fats by size exclusion chromatography and further separated by GC and HPLC. Identification by gas chromatography/mass spectrometry provided evidence of the presence of monohydroxy, dihydroxy, and keto groups in the C-C linked dimers of methyl linoleate, together with the previously mentioned structures of nonpolar dimers (30).

In summary, due to the high number of non-oxygenated and oxygenated dimeric FAME found in fats and oils subjected to frying conditions, it is difficult to obtain more detailed analyses of these compounds. The difficulty increases even more when dealing with the underivatized TG.

*Higher Oligomers*
Definite structures for compounds with molecular weight higher than dimers have not been reported, either in methyl esters from frying fats or in model systems. This is not strange considering much more research remains to be done on structure elucidation and quantitation of simpler molecules, i.e., oxidized monomers and dimers, which are

intermediates in trimer and higher oligomer formation. Moreover, the potential number of different structures in trimer formation increases exponentially with respect to those compounds of lower molecular weight as many different dimeric structures may be combined with many different monomeric structures. Furthermore, the limitation of the techniques for isolation, separation, and identification increases. Nevertheless, some information has been reported on the chemical composition of the residual oligomeric fraction obtained after distillation of dimer methyl esters, solvent fractionation of polymeric material, or a combination of chromatographic techniques. In general, the results reported indicate that the polymers formed were essentially FAME dimers and trimers joined through C-C and C-O-C linkages (1,25,31). However, when drastic conditions were applied, even FAME pentamers were found in significant amounts (32,33).

Finally, it is important to remark that the information obtained from oxidized monomeric, dimeric, and oligomeric FAME has been of great help in isolating the fractions of interest without the interference from other modified, or unmodified, fatty acyls also attached to the glyceride molecule. Nevertheless, this approach implies a considerable loss of information on the original structures of oxidized monomeric, dimeric, and oligomeric TG generated during the thermal oxidation of fats.

# Analysis of Oxidized Monomers, Dimers, and Higher Oligomers

Improved chromatographic techniques have contributed significantly to qualitative and quantitative analyses resulting in the identification of new compounds formed during deep frying. Classical techniques, based on volatility differences, solvent fractionation, or urea adduction are described and commented on in the excellent review written by Artman (27). However, only the chromatographic techniques described in this chapter are applied at present.

Liquid chromatography is the most appropriate chromatographic technique, considering the main characteristics of the new compounds formed. First, excluding those minor nonpolar compounds originating through isomerization and cyclization reactions, new compounds possess higher polarity than their parent, intact TG. Second, the main groups of compounds formed, i.e., hydrolytic products, oxidized monomers, and dimers and oligomers, have different molecular weight ranges. Lastly, the new compounds have low volatility.

So far, adsorption chromatography has been widely used for determining total polar compounds. The standard method uses a classical silica column for a gravimetric determination (34,35). Limitations for polar compounds, around 25%, are established in the countries where frying oils are fully regulated. Some of these countries use additional criteria such as oxidized fatty acids insoluble in petroleum ether, smoke point, free fatty acids, and polymer content. However, it is necessary to obtain more precise information about the nature and quantity of the new compounds formed during frying, which may impair the nutritional value of the food as reported elsewhere in this book. For the separation of different groups of compounds, high-performance size-exclusion chromatography (HPSEC) is also of great utility, either for direct analysis of the fat or oil, or to concentrate fractions obtained from the oil or from FAME (36).

## Direct Analysis of Used Frying Oils: Quantitation of Dimers and Oligomers

Evaluation of dimers and higher oligomers by direct size exclusion chromatography is a simple and rapid analysis (37). Only a solution of the oil sample in a suitable solvent and short elution time (10–30 min) is required (38). Preliminary studies on the use of size-exclusion chromatography (39–41) indicated that a good separation of oligomers is possible with Sephadex LH-20 swelled in chloroform or ethanol, BioBeads in benzene or chloroform, or Styragel in tetrahydrofuran. Later, it was demonstrated that a good correlation existed between the amount of polymerized TG and polar components separated by column chromatography (42), indicating that the former is a reasonably good measurement of alteration in frying oils and fats. Further refinement of chromatographic matrices and organic solvents leading to optimal chromatographic conditions for obtaining quantitative data resulted in the column normally used now, which is approximately 30 cm × 0.8 cm i.d., with spherically shaped particles (5 or 10 μm) and a controlled pore size distribution. With this column, good resolution is achieved for sample concentrations between 30–50 mg/mL and loops of 10–20 μL (43).

Direct analysis of oils and fats by HPSEC is especially applicable for quality evaluation of used frying fats in which oligomers are the most representative group of compounds formed from thermoxidative alteration. Hence, the IUPAC Commission on Oils, Fats and Derivatives adopted the method, after two interlaboratory tests, for samples containing not less than 3% oligomers (35,44). The method proposes a single column of 30 cm × 0.77 cm i.d. packed with copolysterene divinyl benzene (5 μm particle size), tetrahydrofuran as the mobile phase, a refractive index detector, sample concentration of 50 mg/mL for an injection valve with a 10 μL loop, and a flow rate of 1 mL/min. Under these conditions, the analysis time is about 10 min and three peaks corresponding to higher oligomers, dimers, and monomers are resolved.

Results obtained from numerous samples have clearly shown that polymeric compounds constitute the major fraction among the different groups of alteration compounds formed during frying, normally accounting for more than 50%, and their levels have been found to correlate well with those of polar compounds (45–47).

Analysis of TG polymers as a quality control parameter for used frying fats is now used extensively (48–57). Recently, in this context, a new index called OSET (oxidative stability at elevated temperature) has been described for estimating the stabilizing activity of additives at simulated frying temperatures. The test consists of accelerating polymerization with acid-catalyzed silica gel. Oils or fats are heated for

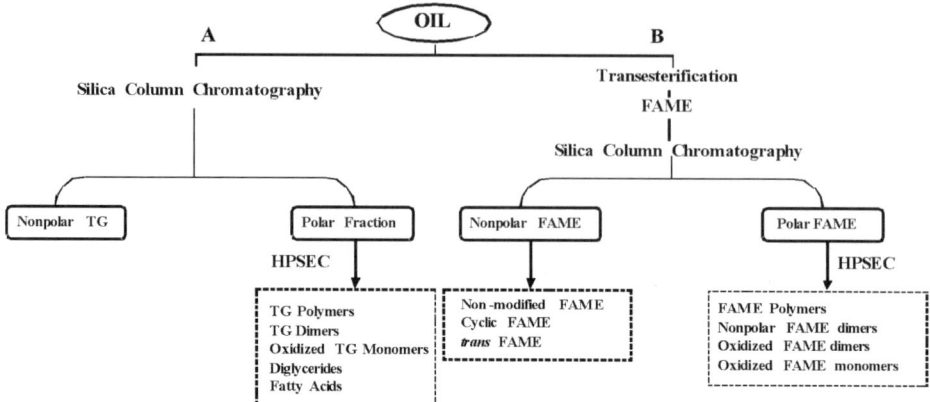

**Fig. 6.6.** Schematic representation of methodologies for quantitation of groups of compounds formed during frying. *Abbreviations*: TG, triglycerides; FAME, fatty acid methyl esters.

2 h at 170°C in the presence of the additive tested, and HPSEC is used to evaluate dimeric and polymeric TG contents (50).

Given the simplicity of the HPSEC technique, determination of polymers has been recently proposed as a good method to control the quality of used frying fats. A maximum content of 12% has been suggested for discarding used frying fats (58).

## Analysis of Oil Concentrated Fractions

As discussed above, quantitation of polar compounds by classical silica column is the basis of legislation in some European countries for quality control of used frying fats for human consumption. Further application of HPSEC to the isolated polar fractions is for quantitation of specific groups of altered compounds. Fig. 6.6 summarizes the analytical procedures starting with used frying oil or from its FAME.

When starting with used frying oil or fat (part A of the scheme in Fig. 6.6), the analytical procedure joins two IUPAC methods, that is, determination of polar compounds and determination of dimers and oligomers (36). With this combination, quantitative data on groups of compounds characteristic of the different types of degradation in used frying oils and fats are obtained. The procedure is briefly described below. Nonpolar and polar fractions are eluted with 150 mL of a mixture hexane:diethyl ether, 90:10, and 150 mL diethyl ether, respectively. After gravimetric determination of polar compounds, the fractions are further analyzed by HPSEC, using 100 and 500 Å columns with polysterene divinylbenzene highly cross-linked macroporous packing (particle size: 5 μm) connected in series; mobile phase: tetrahydrofuran (flow rate: 1 mL/min); detector: refractive index. Resolution in the range of lowest or highest MW can be improved by adding one styrene/divinylbenzene copolymer column of 50 Å (59) or 500 Å (60) molecular weight (MW), respectively, or using three-μm mixed-bed styrene/divinylbenzene copolymer columns (61). The methodology has

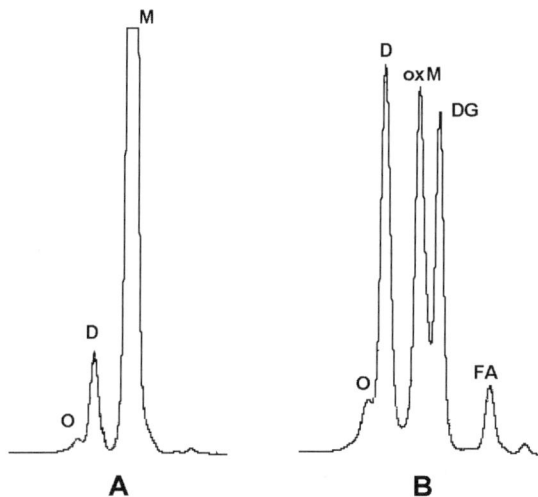

**Fig. 6.7.** High-performance size-exclusion chromatograms of total used frying oil (A) and its polar fraction (B). *Abbreviations*: O, oligomeric triglycerides; D, dimeric triglycerides; M, monomeric triglycerides; oxM, oxidized monomeric triglycerides; DG, diglycerides, FA, fatty acids.

been recently standardized by the IUPAC, with slight modifications (62), for analysis of used frying oils, and for virgin and refined oils.

An alternative technique intended to reduce the quantity of the sample and solvents, as well as shortening the analysis time, is based on the use of silica cartridges for the separation and monostearin as internal standard (38). Only 15 mL of elution solvents are required for each fraction. This modified procedure is useful for samples with a wide range of alteration products and especially for those of low levels of degradation. In the latter case, quantitation with an internal standard shows significantly lower errors compared to gravimetric determination, based on relative standard deviations.

The methodology based on silica column-HPSEC has proved to be an excellent alternative for the evaluation of used frying oils. In addition to providing both the determination of total polar compounds and polymerized TG, broader knowledge on the different groups of compounds formed is gained (8,20,36,63). Advantages offered by the combined technique are clearly reflected in Fig. 6.7, which shows the profile obtained by simply injecting the entire oil sample, and the enhanced possibilities for quantitation in the polar compound fraction. First, a substantial increase in the possibilities for quantitation of all the groups of alteration compounds is achieved because of the effect of concentration. Secondly, oxidized monomers, a measurement of oxidative degradation, can be determined independently, since the co-eluting major peak of nonoxidized TG is separated in the nonpolar fraction. Thirdly, concomitant evaluation of diglycerides as a marker of hydrolytic degradation is possible, otherwise overlapping with the abundant TG in direct HPSEC analysis of the whole sample. Finally, a substantial increase in sensitivity is achieved in the quantitation of polymerization compounds due to the effect of concentration, overcoming the limitation of the IUPAC Standard Method 2.508 to a minimum of 3% content for analyses of total samples by HPSEC (35). The groups of compounds quantitated can be differentiated between thermally oxidized compounds (oxidized monomers, dimers, and oligomers), associated with negative physiological effects, and hydrolytic products (diglycerides and fatty acids), naturally released from lipolysis in the gut before absorption.

In recent years, a plethora of studies on frying fats and oils based on the application of this methodology has been published, and results obtained have contributed to improved knowledge on important issues related to frying, such as performance of different oils (47,64–74), composition of oils absorbed by the fried food and lipid interchange between frying oil and food (66,75–77), relevance of hydrolysis among the reactions occurring during the frying process (47,64,65,75), the action of the main variables involved in the continuous and discontinuous frying processes (77,78), and the effect of oil replenishment during frying on oil quality (79–83).

Beside applications on used frying oils, the methodology has also been applied to thermally oxidized fats and oils (61,84–88), and food lipids and oils subjected to microwave heating (89–93).

## Analysis of FAME Concentrated Fractions

A combination of adsorption chromatography and HPSEC can be also applied to FAME for the analysis of oxidized monomers, dimers, and higher oligomers. In this case, the analytical procedure allows specific analysis of different groups of oxidized and polymerized fatty acyls included in TG molecules (37).

Briefly, FAME are obtained by transesterification of 1 g of sample with sodium methoxide and hydrochloric acid-methanol. FAME are quantitatively recovered and separated by silica column chromatography in two fractions, which are analyzed by HPSEC, under the same chromatographic conditions described previously. Depending on the objective of the application, two possibilities for the analytical scheme have been proposed. When using 150 mL hexane/diethyl ether, 95:5, the first fraction includes exclusively the most abundant nonpolar monomeric FAME. Then, elution with 150 mL diethyl ether yields a minor polar fraction including three groups of FAME (polymers, dimers, and oxidized monomers). This analytical approach is also outlined in Fig. 6.6 B. Alternatively, when using hexane/diethyl ether 88:12 for elution of the first fraction, the combined chromatographic analysis permits discrimination between nonpolar and oxidized FAME dimers. Thus, both nonoxidized FAME and the nonpolar or nonoxidized FAME dimers, linked by C-C bonds and lacking extra-oxygenated functions in their structure, are quantitated in the first fraction. FAME polymers, oxidized FAME dimers, and oxidized FAME monomers are determined in turn in the polar fraction. Therefore, global quantitation of the compounds eluted in the second fraction provides a measurement of the total oxidized fatty acyl groups included in TG molecules.

It is interesting to note that, as observed in Fig. 6.6 B, two groups of modified FAME eluted in the fraction of nonpolar FAME: *trans* FAME and cyclic FAME. These groups have similar polarity and molecular weight than those corresponding to nonmodified FAME and, consequently, they cannot be separated by HPSEC. However, their quantitative importance as compared to the total new FAME formed as well as to the group of nonmodified FAME is minimal. Concerning cyclic fatty acids, interesting information on their formation and analysis can be found in another chapter in this book. Formation of *trans* FAME has been less studied and the lack of interest might be due to the high levels of *trans* FAME present when stable, hydrogenated oils are used for frying. From nonhydrogenated vegetable oils, the levels of *trans* oleic and linoleic isomers formed during frying were very low, in the order of mg/kg (94,95). Thus it can be deduced that when present in significant amounts, *trans* fatty acids in fried foods come from the frying fat and are not formed due to the process conditions.

Application of HPSEC to the concentrated fractions obtained from both intact samples and FAME, through the silica column-HPSEC and transesterification-silica column-HPSEC procedures, respectively, has contributed to the ability to further study the composition of the polar fraction in used frying oils (23,96). Table 6.2 shows the results obtained by the methodologies included in Fig. 6.6 applied to samples of used frying fats and oils around the limit for rejection (21.1–27.6% polar compounds) collected by Food Inspection Services in Spain. As can be observed, values of total altered FAME from 8.1–11.3% were found and, among them, the major group was oxidized FAME monomers (about 30 mg/g oil). Also, results revealed some insight into the complexity of the structure of polymeric compounds, by comparing TG and FAME dimer and polymer values. In general, the low FAME-polymers-to-TG-polymers ratios in contrast to the FAME-dimers-to-TG-dimers ratios gave evidence of the considerable contribution of dimeric linkages to the structures of trimeric and higher oligomeric TG (23).

**TABLE 6.2**
**Quantitation of Polar Compounds and Polar FAME in Used Frying Oils and Fats Around the Level of Rejection for Human Consumption**

| Polar Compounds (TG) (wt% on oil) | | | | Polar FAME (wt % on FAME) | | | | |
|---|---|---|---|---|---|---|---|---|
| **Total** | Oligomers | Dimers | Oxidized monomers | **Total** | Oligomers | Oxidized dimers | Nonpolar dimers | Oxidized monomers |
| **23.1** | 4.0 | 9.2 | 6.8 | **8.1** | 0.8 | 2.0 | 2.1 | 3.2 |
| **25.5** | 4.7 | 9.0 | 8.8 | **10.4** | 1.1 | 3.1 | 2.8 | 3.4 |
| **26.4** | 5.5 | 8.9 | 6.4 | **10.8** | 0.7 | 3.8 | 2.9 | 3.4 |
| **27.6** | 6.5 | 10.9 | 7.2 | **11.3** | 1.1 | 3.8 | 2.7 | 3.7 |
| **27.6** | 3.7 | 7.2 | 9.4 | **8.7** | 0.5 | 1.6 | 2.8 | 3.8 |

## Analysis of Oxidized Monomers by Capillary Gas Chromatography

The next step forward in the analysis of the new compounds formed during frying is the detailed evaluation of the main constituents within the groups of monomers, dimers, and oligomers. In this respect, there is a growing interest in the oxidized monomers because of their possible nutritional implications due to their high absorbability and their presence in used frying fats at non-negligible levels (23). In recent years, nutritional aspects have emerged on the potential biological effects of oxidized lipids and there is increasing evidence that they may be detrimental to health, particularly in connection with the development of atherosclerosis, liver damage, and promotion of intestinal tumors (5).

Separation and identification of the main structures in FAME from used frying fats indicates that the fraction of oxidized monomers contains significant amounts of compounds with MW lower than that of the original fatty acids. Among these compounds are the short-chain n-oxo FAME, originally TG-bound aldehydes, resulting from hydroperoxide breakdown (11,97). The knowledge of the levels of the nonvolatile aldehyde derivatives in used frying fats and fried products is of great importance from the nutritional point of view since they remain attached to the TG and thus retained in the fried food ingested by the consumer. Some reports on 9-oxononanoic acid (98,99), the major esterified aldehyde in oxidized lipids, indicate that such structures could induce lipid peroxidation and affect hepatic metabolism.

Concerning oxidized compounds of molecular weight similar to that of the starting fatty acids, the main groups present corresponded to epoxyacids, ketoacids, and hydroxyacids (6,100).

*Qualitative Analysis*

In recent years, many interesting studies have been carried out to analyze the presence of oxidation compounds in intact lipid molecules, paying special attention to the aldehydes bound to the TG backbone, also known as core aldehydes. In this context, two different approaches have been used: either a combination of thin-layer chromatography (TLC) and high-performance liquid chromatography coupled with mass spectrometry (HPLC-MS) on dinitrophenyl hydrazone derivatives (101,102) or direct analysis of oxidized lipid molecules. (17,103–106) These studies are of enormeous interest due to the biological activity found for some specific structures (107,108) and have promoted the synthesis of complex standards for identification purposes and for nutritional and biological research (109–112). Normal phase (104) and reverse phase HPLC (105) have been applied to the difficult separation of oxidation compounds in oxidized fats and oils. Hundreds of different molecules have been identified, and among them hydroperoxides, epoxides, and aldehydes bound to TG stand out. In the case of fats and oils heated at frying temperature, the analysis is even more complex due to the rapid hydroperoxide decomposition by different radical reactions. In this regard, the studies are limited to the analysis of heated triolein and tristearin (113,114) where the complexity of the new molecules formed from a simple

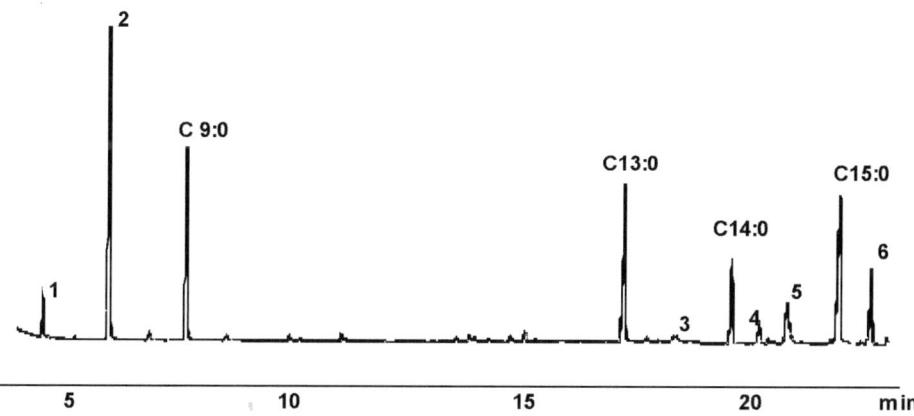

**Fig. 6.8.** Partial gas chromatogram showing separation of major short-chain oxidation compounds in used frying fats after base-catalyzed transesterification and diazomethane methylation. Conditions: HP Innowax capillary column (30 m × 0.25 mm i.d.). Temperature program: 90°C (2 min), 4°C /min, 240°C (25 min). Peak assignments: $C_{13:0}$, methyl tridecanoate (internal standard); $C_{14:0}$, methyl miristate, C15:0, methyl pentadecanoate (internal standard); 1, methyl heptanoate; 2, methyl octanoate; 3, methyl 8-oxooctanoate; 4, dimethyl octanodiate; 5, methyl 9-oxononanoate; 6, dimethyl nonanodiate.

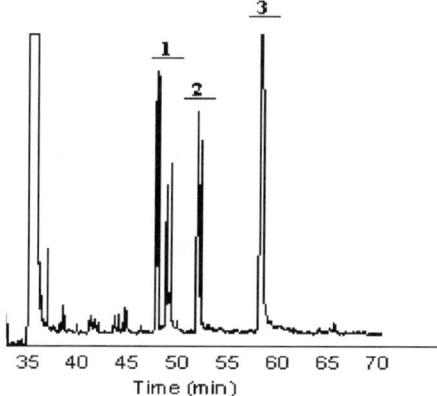

**Fig. 6.9.** Partial gas chromatogram of fatty acid methyl esters (FAME) of thermoxidized trilinolein after hydrogenation. Conditions: HP Innowax capillary column (30 m × 0.25 mm i.d.). Temperature program: 90°C (2 min), 4°C /min, 240°C (25 min). $C_{18:0}$, methyl stearate; 1, Epoxy FAME; 2, Keto FAME; 3, Hydroxy FAME.

TG gives a clear idea of the enormous efforts applied in this type of research. Analysis of triolein heated at high temperatures until reaching levels of polar compounds around 30% revealed the complex degradation of the samples as compared to those samples oxidized at low temperature. Apart from short-chain compound formation, epoxy, hydroperoxy, and keto functions in the TG molecules were detected (113). The number of molecules identified was very low and it was inferred that efficiency of separation techniques needs to be significantly improved for a better definition of the new compounds formed at high temperatures.

*Quantitative Analysis*
When the aim is to obtain quantitative data of oxidized compounds present in dietary lipids, a different approach is necessary in order to overcome the difficulties of separating hundreds or even thousands of different molecules present in heated fats and oils. It is important to remark that oxidized compounds can be present in hundreds of combinations depending on the other fatty acyl groups esterified in the TG molecule, which in turn may or may not contain one, two, or several oxygenated functions. Even more are present in the case of used frying fats where polymerization is one major reaction and hence oxidized fatty acyl groups are also present in dimeric or oligomeric TG (8).

Quantitative analysis could be obtained by converting fats and oils into FAME. Although the structure of the TG molecules is lost, the main advantage is that derivatization to FAME allows concentration of the compounds bearing any specific oxidized function. In this line, accordingly, the quantitation of methyl octanoate ($C_{8:0}$) in heated and used frying fats was the first example of an accurate determination of short-chain compounds formed during thermoxidation (115). Later, methyl heptanoate ($C_{7:0}$) formed from 8-hydroperoxide of oleic acid in monounsaturated frying oils was also quantitated. Interestingly, the ratio between both methyl esters was very useful in deducing the extent of degradation of the main fatty acids, oleic, and linoleic acids present in edible fats and oils. Also, it was found that the level of $C_{8:0} + C_{7:0}$ in used frying fats at the limit of rejection ranged from 1.8 to 2.4 mg/g (16). These results demonstrated that these compounds were not major compounds in frying fats and oils.

With regard to bound aldehydes, we have reported on qualitative analysis of aldehydic acids as FAME derivatives in used frying oils by gas/liquid chromatography mass spectrometry and the resulting need for more detailed studies on derivatization methods for quantitative purposes (97). Essential requirements of the derivatization technique selected had good repeatability and recovery of the compounds of interest, avoiding artifact formation. Conversion to the simplest volatile derivatives, methyl esters, is by far the most common procedure for gas chromatography analyses but chemical modifications in the case of oxidized structures have been reported (116).

Methylation procedures suitable for quantitation of major short-chain glycerol-bound oxidation compounds were assayed (15). To that end, methyl linoleate and trilinolein were subjected to thermoxidation conditions that closely simulate those for frying (117). Methyl linoleate was used as control, since it can be analyzed without further derivatization and represents the main fatty acid undergoing degradation in most vegetable oils at high temperatures. In this study, two base-catalyzed transesterification methods were selected as the most appropriate methods because of excellent recovery of the compounds of interest (15,18). Both methods permitted the accurate quantitation of major bound compounds in model TG systems. Not only short-chain FAME and aldehydic FAME were found. It was also deduced that the oxidation of aldehyde to acids and their quantitation was possible after a second methylation step with diazomethane (14,15). Fig. 6.8 shows the significant part of the chromatogram corresponding to used frying oil at the limit of rejection. The total amounts of this group of compounds was in the order of mg/g, confirming that all these bound compounds coming from hydroperoxide breakdown are only a minor part among those oxidized compounds formed.

Quantitation of the specific structures of oxidized compounds of molecular weight similar to that of the starting fatty acids remains a very difficult task even after fat derivatization to FAME. In Fig. 6.9, the representative part of a chromatogram shows the three main groups of compounds eluting after major fatty acids. The chromatogram corresponds to heated methyl linoleate after hydrogenation to reduce the complexity.

In recent years, quantitation of monoepoxy compounds in model TG systems, in thermoxidized oils and in used frying fats have been reported (18,118,119). Although quantitation in model systems was not difficult, when applied to heated fats and oils, the separation and accurate quantitation of the new compounds formed required more complex analyses and elimination of interference from other components naturally present in fats and oils (118).

Quantitation of epoxides in thermoxidized olive and sunflower oils, and in used frying oils demonstrated that, for similar levels of polar compounds, monounsaturated oils showed higher levels of monoepoxides than polyunsaturated oils. This was attributable to two concurrent facts. First, a lower tendency for monounsaturated oils to polymerize. Second, greater stability and thereby accumulation of the major monoepoxides formed in monounsaturated oils, i.e., monoepoxystearates, in contrast to susceptibility to further reactions of the most abundant monoepoxides found in polyunsaturated oils, i.e., monoepoxyoleates. It was also found that monoepoxides were major oxidized compounds, accounting for about 25% of the total oxidized monomers in actual used frying oils at the limit of rejection (119).

Less work has been done on the specific structures of keto and hydroxy FAME in used frying fats, although total amounts of oxoacids, estimated as 2,4 dinitrophenylhidrazones, and hydroxy and polyhydroxyacids, quantitated as pyruvic acid 2,6 dinitrophenylhidrazone esters, have been reported (120). Also, interesting information is available on different GC-MS techniques for quantitation of hydroxy compounds obtained from the reduction of hydroperoxides (121,122), which can be of great help in the near future to gain knowledge on the oxidized compounds present in used frying fats and oils.

## Final Remarks

1. Evaluation of all changes occurring in fats and oils during frying is a difficult task due to the complexity of the reactions involved and the great variety of compounds formed depending on the frying conditions. Still, efforts made during the last two decades to develop chromatographic techniques have enabled us to gain useful information on the rate of formation of new compounds in relation to the conditions of frying.
2. At present, determination of the main groups of compounds, namely, oxidized, polymerized, and hydrolytic compounds, by combination of adsorption and size-exclusion chromatographies, constitutes the best analytical tool for quality evaluation of used frying fats and oils as it allows evaluation of the relative importance of the three main pathways of degradation.
3. Studies are currently underway with the aim of going more deeply into the specific compounds formed at frying temperatures. However, much work remains to be done to clarify the specific structures of the new compounds formed during frying and to evaluate their nutritional implications at the levels normally found in used frying fats and oils and fried products.

## Acknowledgments

This work was supported in part by Ministerio de Educación y Ciencia (Project AGL 2004-00148) and Junta de Andalucía.

## References

1. Chang, S.S.; R.J. Peterson; and C. Ho. *J. Am. Oil Chem. Soc.* **1978**, *55*, 718.
2. Nawar, W.W. *Grasas Aceites* **1998**, *49*, 71.
3. Brewer, M.S.; J.D. Vega; and E.G. Perkins. *J. Food Lipids* **1999**, *6*, 47.
4. Gillat, P. In *Frying: Improving Quality*; J.B. Rossell, Ed.; CRC Press: Cambridge, U.K., 2001; pp. 266–336.
5. Dobarganes, M.C.; and G. Márquez-Ruiz. *Curr. Opin. Clin. Nutr. Metab. Care* **2003**, *6*, 157.
6. Capella, P. *Rev. Fr. Corps Gras.* **1989**, *36*, 313.
7. Frankel, E.N. *Prog. Lipid Res.* **1982**, *22*, 1.
8. Dobarganes, M.C. *OCL* **1998**, *5*, 41.
9. Lomanno, S.S.; and W.W. Nawar. *J. Food Sci.* **1982**, *47*, 744.
10. Scott, G. *Atmospheric Oxidation and Antioxidants*; Elsevier: Amsterdam, The Netherlands, 1965.
11. Kamal-Eldin, A.; and L.A. Appelqvist. *Grasas Aceites* **1996**, *47*, 342.
12. Frankel, E.N. *Lipid Oxidation*; The Oily Press: Dundee, Scotland, 1998.
13. Velasco, J.; S. Marmesat; G. Márquez-Ruiz; and M.C. Dobarganes. *Eur. J. Lipid Sci*

*Technol.* **2004,** *106,* 728.

14. Berdeaux, O.; J. Velasco; G. Márquez-Ruiz; and M.C. Dobarganes. *J. Am. Oil Chem. Soc.* **2002,** *79,* 279.
15. Berdeaux, O.; G. Márquez-Ruiz; and M.C. Dobarganes. *J. Chromatogr. A* **1999,** *863,* 171.
16. Marquez-Ruiz, G.; and M.C. Dobarganes. *J. Sci. Food Agric.* **1996,** *70,* 120.
17. Neff, W.E.; and W.C. Byrdwell. *J. Chromatogr. A* **1998,** *818,* 169.
18. Berdeaux, O.; G. Márquez-Ruiz; and M.C. Dobarganes. *Grasas Aceites* **1999,** *50,* 53.
19. Figge, K. *Chem. Phys. Lipids* **1971,** *6,* 164.
20. Figge, K. *Chem. Phys. Lipids* **1971,** *6,* 178.
21. Wheeler, D.H.; A. Milun; and F. Linn. *J. Am. Oil Chem. Soc.* **1970,** *47,* 242.
22. Sen Gupta, A.K. *Fette Seifen Anstrichm.* **1969,** *71,* 873.
23. Márquez-Ruiz, G.; M. Tasioula-Margari; and M.C. Dobarganes. *J. Am. Oil Chem. Soc.* **1995,** *72,* 1171.
24. Zeman, A.; and H. Scharmann. *Fette Seifen Anstrichm.* **1969,** *71,* 957.
25. Ottaviani, P.; J. Graille; P. Perfetti; and M. Naudet. *Chem. Phys. Lipids* **1979,** *24,* 57.
26. Christopoulou, C.N.; and E.G. Perkins. *J. Am. Oil Chem. Soc.* **1989,** *66,* 1338.
27. Artman, N.R. *Adv. Lipid Res.* **1969,** *7,* 254.
28. Christopoulou, C.N.; and E.G. Perkins. *J. Am. Oil Chem. Soc.* **1989,** *66,* 1344.
29. Christopoulou, C.N.; and E.G. Perkins. *J. Am. Oil Chem. Soc.* **1989,** *66,* 1353.
30. Christopoulou, C.N.; and E.G. Perkins. *J. Am. Oil Chem. Soc.* **1989,** *66,* 1360.
31. Williamson, L. *J. Appl. Chem.* **1953,** *3,* 301.
32. Perkins, E.G.; and F.A. Kummerow. *J. Am. Oil Chem. Soc.* **1959,** *36,* 371.
33. Firestone, D.; S. Nesheim; and W. Horwitz. *J. Assoc. Off. Agr. Chem.* **1959,** *44,* 466.
34. Waltking, A.E.; and M. Wessels. *J. Assoc. Off. Anal. Chem.* **1981,** *64,* 1329.
35. International Union of Pure and Applied Chemistry (IUPAC). *Standard Methods for the Analysis of Oils, Fats and Derivatives,* 7th ed.; Blackwell: Oxford, U.K., 1992.
36. Dobarganes, M.C.; M.C. Pérez-Camino; and G. Márquez-Ruiz. *Fat Sci. Technol.* **1988,** *90,* 308.
37. Márquez-Ruiz, G.; M.C. Pérez-Camino; and M.C. Dobarganes. *J. Chromatogr. A* **1990,** *514,* 37.
38. Márquez-Ruiz, G.; N. Jorge; M. Martín-Polvillo; and M.C. Dobarganes. *J. Chromatogr. A* **1996,** *749,* 55.
39. Aitzetmüller, K. *J. Chromatogr.* **1972,** *71,* 355.
40. Perkins, E.G.; R. Taubold; and A. Hsieh. *J. Am. Oil Chem. Soc.* **1973,** *50,* 223.
41. Unbehend, V.M.; H. Scharmann; H.J. Strauss; and G. Billek. *Fette Seifen Anstrichm.* **1973,** *75,* 689.
42. Schulte, E. *Fette Seifen Anstrichm.* **1982,** *84,* 178.
43. Dobarganes, M.C.; and G. Márquez-Ruiz. In *Advances in Lipid Methodology—Two;* W.W. Christie, Ed.; The Oily Press: Dundee, Scotland, 1993; pp. 113–137.
44. Wolff, J.P.; F.X. Mordret; and A. Dieffenbacher. *Pure Appl. Chem.* **1991,** *63,* 1163.
45. Gere, A. *Die Nahrung* **1982,** *26,* 923.
46. Perrin, J.L.; P. Perfetti; C. Dimitriades; and M. Naudet. *Rev. Fr. Corps Gras* **1985,** *32,* 151.
47. Masson, L.; P. Robert; N. Romero; M. Izaurieta; S. Valenzuela; J. Ortiz; and M.C. Dobarganes. *Grasas Aceites* **1997,** *48,* 273.

48. Lampi, A.-M.; L.H. Dimberg; and A. Kamal-Eldin. *J. Sci. Food Agric.* **1999,** *79,* 573–579.
49. Gertz, C. *Eur. J. Lipid Sci. Technol.* **1999,** *102,* 566.
50. Gertz, C.; S. Klostermann; and S.P. Kochhar. *Eur. J. Lipid Sci. Technol.* **2000,** *102,* 543.
51. Gertz, C.; and S.P. Kochhar. *inform* **2002,** *13,* 386.
52. Soheili, K.C.; W.E. Artz; and P. Tippayawat. *J. Am. Oil Chem. Soc.* **2002,** *79,* 287.
53. Soheili, K.C.; P. Tippayawat; and W.E. Artz. *J. Am. Oil Chem. Soc.* **2002,** *79,* 1197.
54. Neff, W.E.; K. Warner; and F. Eller. *J. Am. Oil Chem. Soc.* **2003,** *80,* 801.
55. Kiatsrichart, S.; M.S. Brewer; K.R. Cadwallader; and W.E. Artz. *J. Am. Oil Chem. Soc.* **2003,** *80,* 479.
56. El-Sayed, F.E.; and S.S.M. Allam. *J. Food Lipids* **2003,** *10,* 285.
57. Coscione, A.R.; P.C. Osidacz; S. Kiatsrichart; and W.E. Artz. *J. Food Lipids* **2004,** *11,* 57.
58. DGF (German Society for Fat Research). Proceedings of the 3rd International Symposium of Deep-Fat Frying—Final Recommendations, *Eur. J. Lipid Sci. Technol.* **2000,** *102,* 594.
59. Hopia, A.; V.I. Piironen; P.E. Koivistoinen; and L.E.-T. Hyvoenen. *J. Am. Oil Chem. Soc.* **1992,** *69,* 772.
60. Gomes, D. *J. Am. Oil Chem. Soc.* **1992,** *69,* 1219.
61. Abidi, S.L.; I.H. Kim; and K.A. Rennick. *J. Am. Oil Chem. Soc.* **1999,** *76,* 939.
62. Dobarganes, M.C.; J. Velasco; and A. Dieffenbacher. *Pure Appl. Chem.* **2000,** *72,* 1563–1575.
63. Dobarganes, M.C.; G. Márquez-Ruiz; O. Berdeaux; and J. Velasco. In *Frying of Foods*; D. Boskou and I. Elmadfa, Eds.; Technomic Publishing: Lancaster, PA, 1999; p. 143–161.
64. Dobarganes, M.C.; G. Márquez-Ruiz; and M.C. Pérez-Camino. *J. Agric. Food Chem.* **1993,** *41,* 678.
65. Arroyo, R.; C. Cuesta; J.M. Sánchez-Montero; and F. Sánchez-Muñiz. *Fat Sci. Technol.* **1995,** *95,* 292.
66. Sébédio, J.-L.; M.C. Dobarganes; G. Márquez; I. Wester; W.W. Christie; G. Dobson; F. Zwobada; J.M. Chardigny; T. Mairot; and R. Lahtinen. *Grasas Aceites* **1996,** *47,* 5.
67. Márquez-Ruiz, G.; R. Garcés; M. León-Camacho; and M. Mancha. *J. Am. Oil Chem. Soc.* **1999,** *76,* 1169.
68. Masson, L.; P. Robert; M. Izaurieta; N. Romero; and J. Ortiz. *Grasas Aceites* **1999,** *50,* 460.
69. Abidi, S.L.; and K. Warner. *J. Am. Oil Chem. Soc.* **1999,** *78,* 763.
70. Bastida, S.; and F.J. Sanchez-Muniz. *Food Sci. Technol. Int.* **2001,** *7,* 15.
71. Bastida, S.; and F.J. Sanchez-Muniz. *J. Am. Oil Chem. Soc.* **2001,** *79,* 447.
72. Masson, L.; P. Robert; M.C. Dobarganes; C. Urra; N. Romero; J. Ortiz; E. Goicoechea; P. Pérez; M. Salame; and R. Torres. *Grasas Aceites* **2001,** *53,* 190.
73. Abidi, S.L.; and K.A. Rennick. *J. Am. Oil Chem. Soc.* **2001,** *80,* 1057.
74. Houhoula, D.P.; V. Oreopoulou; and C. Tzia. *J. Sci. Food Agric.* **2001,** *83,* 314.
75. Pérez-Camino, M.C.; G. Márquez-Ruiz; M.V. Ruiz-Méndez; and M.C. Dobarganes. *J. Food Sci.* **1992,** *56,* 1644.
76. Pozo-Díaz, R.M.; T.A. Masoud-Musa; M.C. Pérez-Camino; and M.C. Dobarganes. *Grasas Aceites* **1995,** *46,* 85.

77. Jorge, N.; G. Márquez-Ruiz; M. Martín-Polvillo; M.V. Ruiz-Méndez; and M.C. Dobarganes. *Grasas Aceites* **1996,** *47*, 20.
78. Jorge, N.; G. Márquez-Ruiz; M. Martín-Polvillo; M.V. Ruiz-Méndez; and M.C. Dobarganes. *Grasas Aceites* **1996,** *47*, 14.
79. Cuesta, C.; F.J. Sánchez-Muñiz; C. Garrido-Polonio; S. López-Varela; and R. Arroyo. *J. Am. Oil Chem. Soc.* **1993,** *70*, 1069.
80. Romero, A.; C. Cuesta; and F.J. Sánchez-Muñiz. *Fett Wissens. Technol.* **1995,** *97*, 403.
81. Romero, A.; C. Cuesta; and F.J. Sánchez-Muñiz. *J. Am. Oil Chem. Soc.* **1998,** *75*, 161.
82. Cuesta, C.; and F.J. Sánchez-Muñiz. *Grasas Aceites* **1998,** *49*, 310.
83. Romero, A.; C. Cuesta; and F.J. Sánchez-Muñiz. *J. Agric. Food Chem.* **1999,** *47*, 1168.
84. Barrera-Arellano, D.; M.V. Ruiz-Méndez; G. Márquez-Ruiz; and M.C. Dobarganes. *J. Sci. Food Agric.* **1999,** *79*, 1923.
85. Barrera-Arellano, D.; M.V. Ruiz-Méndez; J. Velasco; G. Márquez-Ruiz; and M.C. Dobarganes. *J. Sci. Food Agric.* **2002,** *82*, 1696.
86. Verleyen, T.; A. Kamal-Eldin; M.C. Dobarganes; R. Verhe; K. Dewettinck; and A. Huyghebaert. *Lipids* **2001,** *36*, 719.
87. Verleyen, T.; A. Kamal-Eldin; R. Mozuraityte; R. Verhe; K. Dewettinck; A. Huyghebaert; and W. de Grey. *Eur. J. Lipid Sci. Technol.* **2001,** *10*, 228.
88. Kamal-Eldin, A.; J. Velasco; and M.C. Dobarganes. Oxidation of Mixtures of Triolein and Trilinolein at Elevated Temperatures. *Eur. J. Lipid Sci. Technol.* **2003,** *105*, 165–170.
89. García-Ayuso, L.E.; J. Velasco; M.C. Dobarganes; and M.D. Luque de Castro. *Chromatographia* **2000,** *52, 103.*
90. Caponio, F.; A. Pasqualone; M.T. Bilancia; D. Sacco; D. Delcuratolo; and T. Gomes. *Ind.-Aliment.* **2001,** *40*, 628.
91. Caponio, F.; A. Pasqualone; and T. Gomes. *Eur. Food Res.Technol.* **2002,** *215*, 114.
92. Luque-García, J.L.; J. Velasco; M.C. Dobarganes; and M.D. Luque de Castro. *Food Chem.* **2002,** *76*, 241.
93. Beatriz, M.; P.P. Oliveira; and M.A. Ferreira. *Grasas Aceites* **1994,** *45*, 113.
94. Gamel, T.H.; A. Kiritsakis; and C. Petrakis. *Grasas Aceites* **1999,** *50*, 421.
95. Priego-López, E.; J. Velasco; M.C. Dobarganes; G. Ramis-Ramos; and M.D. Luque de Castro. *Food Chem.* **2003,** *83, 143.*
96. Jorge, N.; L.A. Guaraldo-Goncalves; and M.C. Dobarganes. *Grasas Aceites* **1997,** *48*, 17.
97. Kamal-Eldin, A.; G. Márquez-Ruiz; M.C. Dobarganes; and L.A. Appelqvist. *J. Chromatogr. A* **1997,** *776*, 245.
98. Minamoto, S.; K. Kanazawa; H. Ashida; and M. Natake. *Biochem. Biophys. Acta* **1988,** *958*, 199.
99. Kanazawa, K.; and H. Ashida. *Arch. Biochem. Biophys.* **1991,** *288*, 71.
100. Gardner, D.R.; R.A. Sanders; D.E. Henry; D.H. Tallmadge; and H.W. Wharton. *J. Am. Oil Chem. Soc.* **1992,** *69*, 499.
101. Kamido, H.; A. Kuksis; L. Marai; and J.J. Myher. *Lipids* **1993,** *28*, 331.
102. Sjövall, O.; A. Kuksis; and H. Kallio. *J. Chromatogr. A* **2001,** *905,* 119.
103. Sjövall, O.; A. Kuksis; and H. Kallio. *Lipids* **2001,** *37,* 81.
104. Steenhorst-Slikkerveer, L.; A. Louter; H. Janssen; and C. Bauer-Plank. *J. Am. Oil Chem. Soc.* **2000,** *77,* 837.
105. Byrdwell, W. C.; and W.E. Neff. *J. Chromatogr. A* **2001,** *905,* 85.

106. Byrdwell, W. C.; and W.E. Neff. *Rapid Commun. Mass Spectrom.* **2002,** *16,* 300.
107. Gasser, H.; S. Hallstrom; H. Red; and G. Schlag. *Free Radical Res.* **1995,** *22,* 327.
108. Kamido, H.; A. Kuksis; L. Marai; and J.J. Myher. *J. Lipid. Res.* **1995,** *36,* 1876.
109. Kuksis, A. *inform* **2000,** *11,* 746.
110. Kamido, H.; A. Kuksis; L. Marai; J.J. Myher; and H. Pang. *Lipids* **1992,** *27,* 645.
111. Ravandi, A.; A. Kuksis; N. Shaikh; and G. Jackowski. *Lipids* **1997,** *32,* 989.
112. Sjövall, O.; A. Kuksis; L. Marai; and J.J. Myher. *Lipids* **1997,** *32,* 1211.
113. Byrdwell, W.C.; and W.E. Neff. *J. Chromatogr. A* **1999,** *852,* 417.
114. Byrdwell, W.C.; and W.E. Neff. *J. Am. Oil Chem. Soc.* **2004,** *81,* 13.
115. Peers, K.E.; and A.T. Swoboda. *J. Sci. Food Agric.* **1982,** *33,* 389.
116. Christie, W.W. In *Advances in Lipid Methodology*; W.W. Christie, Ed.; The Oily Press: Dundee, Scotland, 1998; Vol. 2, pp. 8–111.
117. Barrera-Arellano, D.; G. Márquez-Ruiz; and M.C. Dobarganes. *Grasas Aceites* **1997,** *48,* 231.
118. Velasco, J.; O. Berdeaux; G. Márquez-Ruiz; and M.C. Dobarganes. *J. Chromatog. A* **2002,** *982,* 145.
119. Velasco, J.; S. Marmesat; O. Berdeaux; G. Márquez-Ruiz; and M.C. Dobarganes. *J. Agric. Food Chem* **2004,** *52,* 4438.
120. Schwartz, D.P.; A.H. Rady; and S. Castañeda. *J. Am. Oil Chem. Soc.* **1994,** *71,* 441.
121. Thomas, D.W.; F.J.G.M. van Kuijk; E.A. Dratz; and R.J. Stephens. *Anal. Biochem.* **1991,** *198,* 104.
122. Wilson, R.; R. Smith; P. Wilson; M.J. Shepherd; and R.A. Riemersma. *Anal. Biochem.* **1997,** *248,* 76.

# Formation, Analysis, and Health Effects of Oxidized Sterols in Frying Fat

**Paresh C. Dutta[a], Roman Przybylski[b], Michael N.A. Eskin[c], and Professor emeritus Lars-Åke Appelqvist[a]**

[a]Department of Food Science, Swedish University of Agricultural Sciences, Uppsala, Sweden, [b]Department of Chemistry and Biochemistry, University of Lethbridge, Lethbridge, Alberta, Canada T1K3M4, [c]Department of Human Nutritional Sciences, University of Manitoba, Winnipeg, Manitoba, Canada

Sterols, minor compounds present in dietary fat, comprise a major portion of the unsaponifiable matter of most vegetable oils. They are mainly present as free sterols and esters of fatty acids, in addition to sterol glucosides and acetylated (1–3). Vegetable oil sterols are collectively known as plant sterols or phytosterols. Sterols vary with the origin of the fat, and are affected by food processing. Cholesterol is the main animal sterol, while β-sitosterol, campesterol, stigmasterol, brassicasterol, avenasterol, and stigmastenol are major plant sterols present in vegetable oils at much higher levels than cholesterol is in animal fats (4). Sterols share a similar chemical structure that undergoes oxidation in the presence of oxygen. The complete or partial structures of 80 primary and secondary compounds formed when cholesterol oxidizes have been identified (5,6). Cholesterol is absorbed by the human digestive system in significant amounts, while phytosterols have an absorption level of 5% of that of cholesterol (7). The difference in absorption affects the availability of plant sterols and cholesterol in terms of their nutritional implication and oxidation products. Recently, however, it was found that vegetarians absorb more phytosterols than do individuals fed a mixed diet. In addition, formula-fed infants have higher amounts of plant sterols in their blood than do adults (7). Recommendations to increase food intake of plant origin while limiting animal based foods have influenced the trends in use of frying oils. During recent decades, in European countries, vegetable oils namely refined rapeseed oil, partially hydrogenated rapeseed oil, palm oil/rapeseed oil or soybean oil blends, and palm olein or super olein are in use. Generally, in North America, the frying industry and fast-food restaurants employ frying fats and shortenings based on cottonseed oil, partially hydrogenated soybean oil, and/or rapeseed oil (8). In addition, the interest in fortifying food products (functional foods) with phytosterols has increased in recent times because of their cholesterol-lowering property in humans (9,10). These may affect biological properties of the cell, particularly when oxidized phytosterols are present among sterols absorbed with oxidized cholesterol (6,11–19).

# Chemistry of Sterols

## Sterol Structure and Consequences

The phytosterols present in vegetable oil are classified into three major groups: 4-desmethylsterols, cholestane series, and normal phytosterols (4). Examples are cholesterol (cholest-5-en-3β-ol), the main sterol component of animal fat and fish oil, and β-sitosterol (24α-methylcholest-5-en-3β-ol), a major phytosterol; 4-monomethylsterols; 4α-methyl-; cholestane series; gramisterol and obtusifoliol, minor sterols present in many vegetable oils; 4,4′-dimethylsterols; lanostane series; also known as triterpene alcohols; present in vegetable oil in minor amounts. Phytosterols are 28- and 29-carbon atom steroid alcohols, which are distinct from cholesterol, a 27-carbon atom sterol. The main phytosterols present in oil are campesterol, stigmasterol and β-sitosterol. These contain a double bond at the fifth carbon atom in ring B-$\Delta^5$ although some have a double bond at the seventh carbon atom in ring B-$\Delta^7$, for example, stigmastenol. The predominant 4-monomethylsterols in vegetable oil, such as obtusifoliol, gramisterol, and citrostadienol, are $\Delta^7$ and $\Delta^8$ sterols. The main phytosterols, campesterol, stigmasterol, β-sitosterol, and brassicasterol have the same backbone structure as cholesterol, differing only in side chains (Fig. 7.1).

β-sitosterol and campesterol have an ethyl and methyl group, respectively, attached to the twenty-fourth carbon atom ($C_{24}$) in the side chain. Brassicasterol, a major sterol in Brassica family plants, and stigmasterol both have double bonds at $C_{22}$ (Fig. 7.1). Sterols with double bonds at $C_5$ in the B ring have $C_7$ and $C_4$ as the weakest points in their structure. Abstraction of allylic hydrogen predominantly occurs at $C_7$. The attack at $C_4$ seldom occurs due to the adjacent hydroxy group at $C_3$ and tertiary $C_5$ (5). Epimeric hydroperoxides of cholesterol form once the chain reaction commences (20). The corresponding weak carbon atoms are also present in phytosterols, with some comparable products found (21,22). Cholesterol autoxidation is a well-established free radical process, sharing the same chemistry that occurs for the oxidation of unsaturated lipids (23). Sterols can be oxidized by ground-state oxygen ($^3O_2$), ozone ($O_3$), excited ground oxygen–singlet oxygen ($^1O_2$), hydroperoxide ($H_2O_2$), dioxygen cation ($O_2^+$), and hydroxyl radical (HO)$^{\cdot}$ though autoxidation is the most extensive process yielding a complicated mixture of oxidized sterols, only partially examined for cholesterol, while remaining incomplete for phytosterols.

Oxygen also attacks the tertiary carbons in the side chain, namely $C_{20}$ and $C_{25}$ in the cholesterol molecule, and $C_{24}$ in the major plant sterol molecules (Fig. 7.1). The additional double bond in the side chain might have both protective and stimulating effects on plant sterol oxidation. Stigmasterol shows the highest resistance to oxidation among major sterols, including cholesterol, while brassicasterol oxidizes at the fastest rate (24). This difference in oxidation rate may be due to the different configuration of the double bond in these sterols. $C_{25}$ is the next most reactive radical oxidation side to $C_7$ in cholesterol. Therefore, similar reactivity is expected for phytosterols.

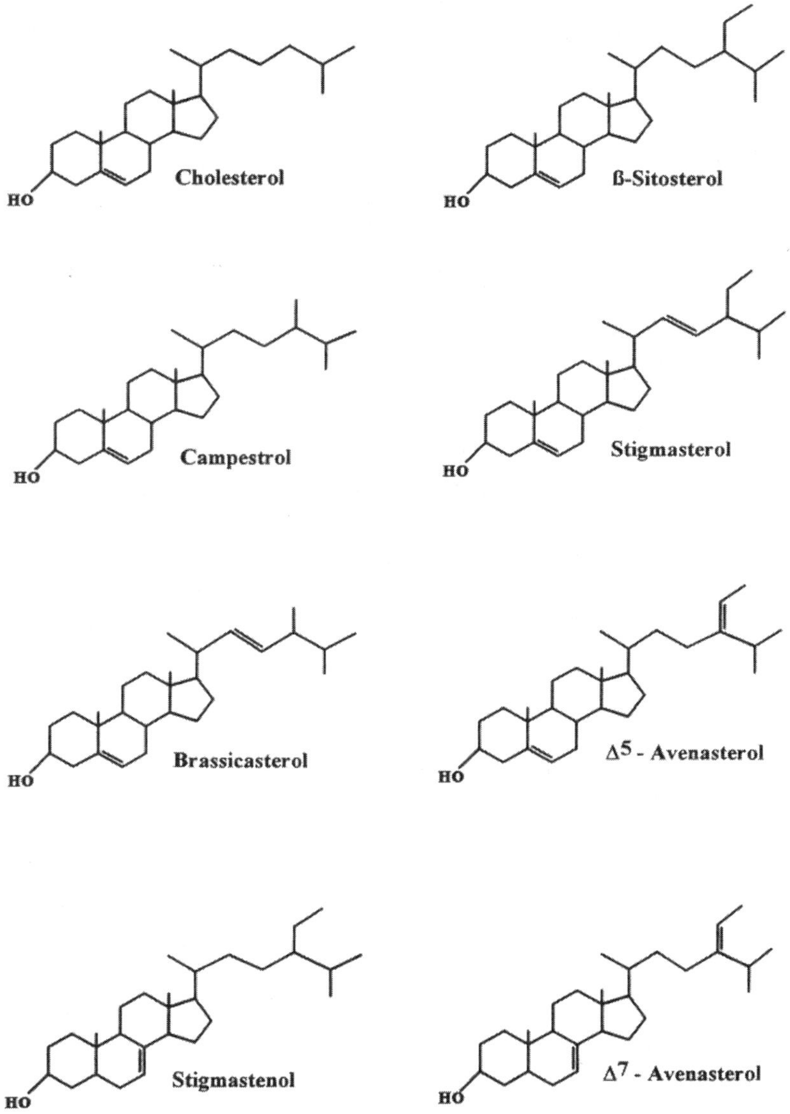

**Fig. 7.1.** Structure of common sterols present in vegetable oils.

Cholesterol oxidation in different media produces 20- and 25-hydroxycholesterols; similar products result for phytosterols (22). Sterols present as fatty acid esters oxidize in a manner similar to free sterols. Data on the rate of oxidation of esterified sterols remain controversial; when unsaturated fatty acids are present, there are trends that stimulate oxidation in aqueous dispersions, although they are saturated in the solid state (25–27).

### Sterol Nomenclature

Sterol structure is composed of two different units. The tetracyclic "nucleus" includes all carbon atoms in three 6- and one 5-carbon atom rings, marked with capital letters from left to right (28). To these rings, two methyl groups attach at the 10- and 13-carbon atom, and a side chain, usually 8-carbon atoms, at 17-carbon atom (Fig. 7.2). Within these rings, carbon atoms are numbered as shown in Fig. 7.2, further continued to methyl groups and side chains, respectively. Double bonds added to the structure are usually marked by Δ with the number of carbon atom(s) where they are present, together with configuration *trans* and *cis* (marked E and Z in European nomenclature). α and β placed after the number indicate the configuration of the group toward the plain of rings. In β-sitosterol, however, the Greek letter differentiates geometrical structure of this sterol (28). Nearly all phytosterol names can be formed as derivatives of 5α-cholestan-3β-ol (Fig. 7.2). See also Table 7.1.

# Sterol Oxidation

### Product Formation from and Conditions Affecting Cholesterol Oxidation

Cholesterol (cholest-5-en-3β-ol), a 27-carbon steroid alcohol mainly present in animal lipids oxidizes readily by autoxidation or enzymatic metabolism as well by photosensitized (photooxidation) processes (5,29,30). The molecular structures of

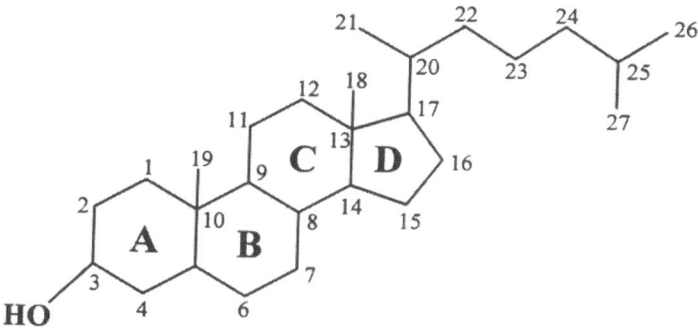

**Fig. 7.2.** Chemical structure of 5α-cholestan-3β-ol.

**TABLE 7.1**
**Trivial and Chemical names of Sterols**

| Trivial name | Chemical name | Additions to 5α-cholestan-3β-ol |
|---|---|---|
| Cholesterol | Cholest-5-en-3β-ol | $\Delta^5$ |
| Camperstrol | 24α-Methylcholest-5-en-3β-ol | $\Delta^5$, 24α-Methyl |
| Brassicasterol | 24β-Methylcholest-5,22-en-2β-ol | $\Delta^5$, $trans\ \Delta^{22}$, 24β-Methyl |
| β-sitosterol | 24α-Ethylcholest-5-en-3β-ol | $\Delta^5$, 24α-Ethyl |
| Stigmasterol | 24α-Ethylcholest-5,22-en-3β-ol | $\Delta^5$, $trans\ \Delta^{22}$, 24α-Ethyl |
| Stigmastenol | 24α-Ethylcholest-7-en-3β-ol | $\Delta^7$, 24α-Ethyl |
| $\Delta^5$-Avenasterol | 24-Ethylidenecholest-5-en-3β-ol | $\Delta^5$, $trans\ \Delta^{24}$, 24-Ethylidene |
| $\Delta^7$-Avenasterol | 24-Ethylidenecholest-7-en-3β-ol | $\Delta^7$, $trans\ \Delta^{24}$, 24-Ethylidene |
| Ergosterol | 24β-methylcholest-5,7-en-3β-ol | $\Delta^5$, $\Delta^7$, $trans\ \Delta^{22}$, 24β-Methyl |
| Lanosterol | 4,4,14α-Trimethylcholest-8,24-en-3β-ol | $\Delta^8$, $trans\ \Delta^{22}$, 4,4,14α-Trimethyl |
| Obtusifoliol | 4α,14α-Dimethyl-24-methylenecholest-8-ene-3β-ol | $\Delta^8$, 4α, 14α-Dimethyl, 24-Methylene |
| Gramisterol | 4α–Methyl-24-methylenecholest-7-ene-3β-ol | $\Delta^7$, 4α–Methyl, 24-Methylene |
| Citrostadienol | 4α-Methyl-24-ethylidenecholest-7-ene-3β-ol | $\Delta^7$, 4α-Methyl, $trans$-24-Ethylidene |
| Cycloatenol | 9,19-Cyclo-4,4,14α-trimethylcholest-24-ene-3β-ol | $\Delta^{24}$, 9,19-Cyclo,4,4,14α-Trimethyl |
| Cyclobranol | 9,19-Cyclo-4,4,14α,24-tetramethylcholest-24-ene-3β-ol | $\Delta^{24}$, 9,19-Cyclo,4,4,14α,24-Tetramethyl |
| Cyclosadol | 9,19-Cycle-4,4,14α,24-tetramethylcholest-24-ene-3β-ol | $\Delta^{24}$, 9,19-Cyclo,4,4,14α,-Tetramethyl |

**TABLE 7.2**

Nomenclature of Some Common Oxidation Products of Cholesterol, Sitosterol, Stigmasterol, Campesterol, and Brassicasterol

| Short name | Trivial name | Systematic name |
|---|---|---|
| **Cholesterol** | Cholesterol | Cholest-5-en-3β-ol |
| 7α-OH-cholesterol | 7α-hydroxycholesterol | Cholest-5-en-3β,7α-diol |
| 7β-OH- cholestero | 7β-hydroxycholesterol | Cholest-5-en-3β,7β-diol |
| 7-ketocholesterol | 7-ketocholesterol | Cholest-5-en-3β-ol-7-one |
| α-epoxycholesterol | Cholesterol -5α,6α-epoxide | 5α, 6α-epoxy-cholestan-3β-ol |
| β-epoxycholesterol | Cholesterol -5β,6β-epoxide | 5β,6β-epoxy-cholestan-3β-ol |
| Cholestanetriol | Cholestanetriol or 5α,6β-dihydroxycholesterol or 3β,5α,6β-trihydroxycholestane | Cholestan-3β,5α,6β-triol |
| **Sitosterol** | Sitosterol | (24R)-ethylcholest-5-en-3β-ol |
| 7α-OH-sitosterol | 7α-hydroxysitosterol | (24R)-ethylcholest-5-en-3β,7α-diol |
| 7β-OH-sitosterol | 7β-hydroxysitosterol | (24R)-ethylcholest-5-en-3β,7β-diol |
| 7-ketositosterol | 7-ketositosterol | (24R)-ethylcholest-5-en-3β-ol-7-one |
| α-epoxysitosterol | Sitosterol-5α,6α-epoxide | (24R)-5α,6α-epoxy-24-ethyl cholestan-3β-ol |
| β-epoxysitosterol | Sitosterol-5β,6β-epoxide | (24R)-5β,6β-epoxy-24-ethyl cholestan-3β-ol |
| Sitostanetriol | Sitostanetriol or 5α,6β-dihydroxysitosterol or 3β,5α,6β-trihydroxyethyl-cholestane | (24R)-ethylcholestan-3β,5α,6β-triol |
| **Stigmasterol** | Stigmasterol | Stigmasterol |
| 7α-OH-stigmasterol | 7α-hydroxystigmasterol | (24S)-ethylcholest-5,22-dien-3β,7α-diol |
| 7β-OH-stigmasterol | 7β-hydroxystigmasterol | (24S)-ethylcholest-5,22-dien-3β,7β-diol |
| 7-ketostigmasterol | 7-ketostigmasterol | (24S)-ethylcholest-5,22-dien-3β-ol-7-one |
| α-epoxystigmasterol | Stigmasterol-5α,6α-epoxide | (24S)-5α,6α-epoxy-24-ethyl cholest-22-en-3β-ol |
| β-epoxystigmasterol | Stigmasterol-5β,6β-epoxide | (24S)-5β,6β-epoxy-24-ethyl cholest-22-en-3β-ol |

**TABLE 7.2, CONT.**

**Nomenclature of Some Common Oxidation Products of Cholesterol, Sitosterol, Stigmasterol, Campesterol, and Brassicasterol**

| Short name | Trivial name | Systematic name |
|---|---|---|
| Stigmastentriol | Stigmastentriol or 5α,6β-dihydroxystigmasterol or 3β,5α,6β-trihydroxystig-mastene | (24S)-ethylcholest-22-en-3β,5α,6β-triol |
| **Campesterol** | Campesterol | (24R)-methylcholest-5-en-3β-ol |
| 7α-OH-campesterol | 7α-hydroxycampesterol | (24R)-methylcholest-5-en-3β,7α-diol |
| 7β-OH-campesterol | 7β-hydroxycampesterol | (24R)-methylcholest-5-en-3β,7β-diol |
| 7-ketocampesterol | 7-ketocampesterol | (24R)-methylcholest-5-en-3β-ol-7-one |
| α-epoxycampesterol | Campesterol-5α,6α-epoxide | (24R)-5α,6α-epoxy-24-methyl cholestan-3β-ol |
| β-epoxycampesterol | Campesterol-5β,6β-epoxide | (24R)-5β,6β-epoxy-24-methyl cholestan-3β-ol |
| Campestanetriol | Campestanetriol or 5α,6β-dihydroxycampesterol or 3β,5α,6β-trihydroxymeth-ylcholestane | (24R)-methylcholestan-3β,5α,6β-triol |
| **Brassicasterol** | Brassicasterol or Ergosta-5,22-dien-3-ol | (24S)-methylcholest-5,22-dien-3β-ol |
| 7α-OH-brassicasterol | 7α-hydroxybrassicasterol | (24S)-methylcholest-5,22-dien-3β,7α-diol |
| 7β-OH-brassicasterol | 7β-hydroxybrassicasterol | (24S)-methylcholest-5,22-dien-3β,7β-diol |
| 7-ketobrassicasterol | 7-ketobrassicasterol | (24S)-methylcholest-5,22-dien-3β-ol-7-one |
| α-epoxybrassicasterol | Brassicasterol-5α,6α-epoxide | (24S)-5α,6α-epoxy-24-meth ylcholest-22-en-3β-ol |
| β-epoxybrassicasterol | Brassicasterol-5β,6β-epoxide | (24S)-5β,6β-epoxy-24-meth ylcholest-22-en-3β-ol |
| Brassicastentriol | Brassicastentriol or 5α,6β-dihydroxybrassicasterol or 3β,5α,6β-trihydroxyer-gostene | (24S)-methylcholest-22-en-3β,5α,6β-triol |

phytosterols relate strictly to the structure of cholesterol (Fig. 7.1). Since the oxidation mechanism of phytosterols is limited, only a few published studies with phytosterols that discuss the mechanism of sterol oxidation are available. Until 1986, 66 different oxysterols were identified (6). Formation of different oxysterols by autoxidation, as documented by experiments, proceeds through a free radical chain reaction. Abstraction of hydrogen from allylic $C_7$ in the ring structure of cholesterol and tertiary carbons at the $C_{20}$ and $C_{25}$ positions (5,6,20,23,29,30) initiates the process. The initiation reaction, that is, the formation of free radicals, is not well understood. Some speculate, however, that some probable causative factors are nitrogen oxides, excited oxygen species or transition metal ions that act as initiation catalysts (31). The radicals thus formed react with oxygen to produce corresponding peroxyl radicals, which in turn stabilize by yielding different cholesterol hydroperoxides. The thermal decomposition of these hydroperoxides produces 7α-hydroxy-, 7β-hydroxy-, 7-keto-, 20-hydroxy-, and 25-hydroxycholesterol. Epimeric 7-hydroperoxides of cholesterol can also attack the $\Delta^5$ double bond of cholesterol, forming secondary oxidation products of cholesterol, such as epimeric epoxycholestanol. Both epoxides, in turn, convert to 5α-cholesta-3β,5,6β-triol, through epoxy ring opening by hydration. Formation of β-epoxide and its hydration to the triol is favored (6).

Cholesterol is present in association with other acyl lipids, in which unsaturated fatty acids readily oxidize, forming hydroperoxides by the same mechanism discussed previously (31). The different species of fatty acid hydroperoxides subsequently attack the cholesterol molecule and form various oxysterols (31). This mechanism is supported by the fact that cholesterol oxidation is strongly accelerated by the autoxidation of coexisting unsaturated triacylglycerols (32,33).

Concerning oxysterols in fat and oil used for frying and cooking, only eight components are generally reported (34,35). Many reviews and monographs on the content of oxysterols in food and food products have resulted in the past decade because of their health implications (36–39). Because cholesterol and other sterols structures are presented in Figs. 7.1 and 7.2, only the structures of the common oxycholesterol are shown here (Fig. 7.3). These structures should be familiar to students of cholesterol oxidation. Table 7.2 lists the nomenclature and abbreviations of common oxidation products of cholesterol and phytosterols.

Autoxidation of cholesterol is facilitated by many factors including temperature, light, oxygen, free radical initiators, metal ions, prooxidizing agents, and a shortage of antioxidants (40,41). The majority of cholesterol oxides in food products accumulate primarily via autoxidation during food preparation, although the formation of oxysterols by enzymes in food before processing cannot be excluded.

## Evidence of, Conditions of, and Products Formed from Phytosterol Oxidation

Phytosterols are mainly 28- and 29-carbon steroid alcohols. Major phytosterols in vegetable oil and fat are dominated by 4-desmethylsterols (4,42). These $C_{28}$ and $C_{29}$

steroid alcohols oxidize, forming homologs similar to known cholesterol oxides. Some oxidized phytosterols reported in the literature were found in plant materials, except for seed oil (5). Little work has been done on phytosterol oxides, particularly their presence in vegetable oil and different food products. The historical development of oxidation products of phytosterols is described elsewhere (5); only recent developments are discussed here. This is in sharp contrast to the extensive analytical work on different oxidation products of cholesterol and $C_{28}$ and $C_{29}$ phytosterols by Aringer and Nordström and later by other groups (43–48). These authors synthesized and isolated a large number of oxysterols by preparative thin-layer chromatography (TLC) and characterized them by gas chromatography (GC) and gas chromatography/mass spectrometry (GC/MS). Identification of different epimers of dihydroxy oxysterols, including 6β-hydroxy-, 7α- and 7β-hydroxy-, 15-hydroxy-, 25-hydroxy- and 26-hydroxy- derivatives of $C_{28}$ and $C_{29}$ steroids, resulted. Also, 3β-hydroxycholest-5-en-7-one of both $C_{28}$ and $C_{29}$ steroids, both epimers of 6-hydroxycholest-4-en-3-one of $C_{29}$ steroid, 7α-hydroxycholest-4-en-3-one of both $C_{28}$ and $C_{29}$ sterols, and 26-hydroxycholest-4-en-3-one of $C_{29}$ sterol were characterized. The authors identified the 5α-cholestane-3,12-dione of $C_{29}$ sterol, 7α-hydroxy-5β-cholestan-3-one of both $C_{28}$ and $C_{29}$ sterols, as well as 5β-cholestane-3,12 and 3,15-dione of $C_{29}$ sterol, in

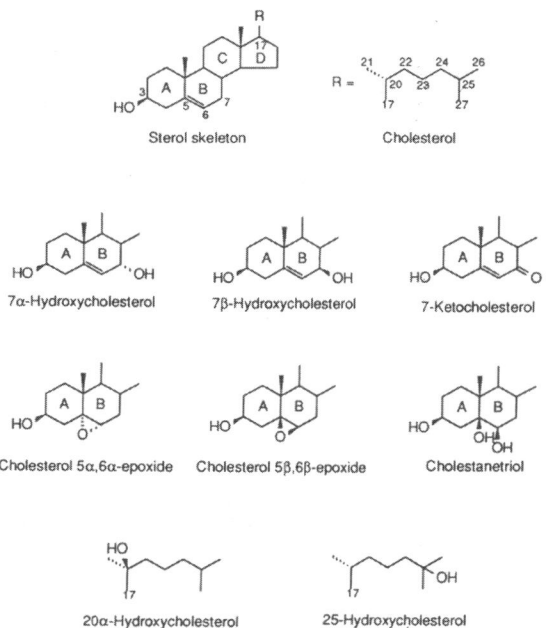

**Fig. 7.3.** Structures and trivial names of common oxidation products of cholesterol.

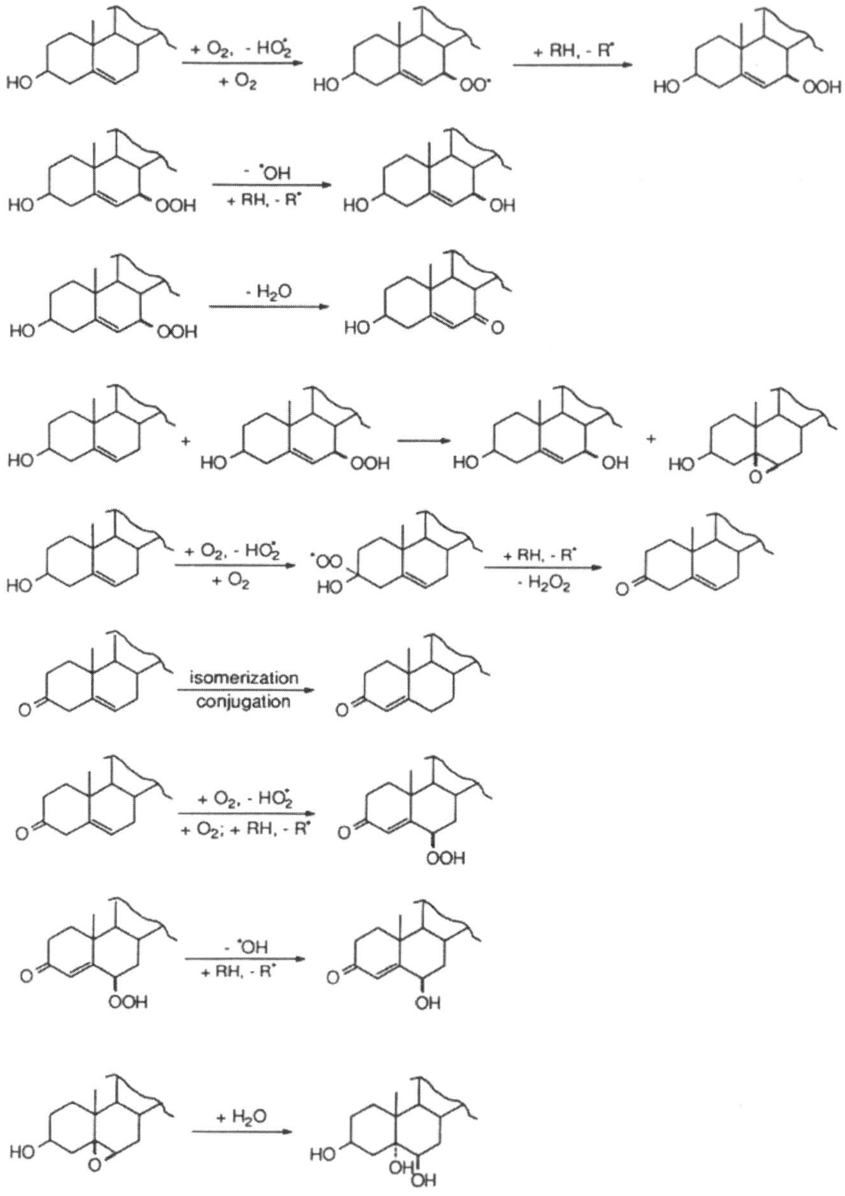

**Fig. 7.4.** Sequence of reactions leading to different products formed by A,B-ring transformations during autoxidation of D5 sterols. Yanishlieva et al. (51).

**Fig. 7.4, cont.** Sequence of reactions leading to different products formed by A,B-ring transformations during autoxidation of D5 sterols. Yanishlieva et al. (51).

addition to the characterization of both epimers of 5,6-epoxycholestan-3β-ol of both $C_{28}$ and $C_{29}$ sterol. Recently, additional MS data on some other phytosterol oxidation products were published (49,50).

The kinetics of pure sitosterol (stigmast-5-en-3β-ol) autoxidation at different temperatures and its esterified form were reported (26). This study showed that initially, esterified sitosterol had greater oxidizability compared to free sitosterol. In later stages, however, this difference in oxidizability was less pronounced. Also identified was 7-hydroperoxides, 7-hydroxy-, and 7-ketositosterol. Sitosterol was oxidized either by heating at 150°C for 1 h or at ambient temperature in air for 6 mo with oxidation products characterized by TLC, ultraviolet (UV), and direct MS (51). The authors proposed the sequence of reactions caused by autoxidation of A- and B-rings of sterols (Fig. 7.4).

At least 10 oxidation products of sitosterol were identified, including stigmasta-3,5-diene, stigmasta-3,5-diene-7-one, stigmast-4-en-3-one, stigmast-4-en-3,6-dione, stigmasta-5,24-diene-3β-ol, 6-hydroxystigmast-4-en-3-one, 3β-hydroxystigmast-5-en-7-one, stigmast-5-ene-3β,7β-diol, stigmast-5-ene-3α,7α-diol, 3β,5-dihydroxy-5α-stigmastan-6-one, and 5α-stigmastan-3β,5,6β-triol. Isolation and identification by TLC-UV and MS of some oxidation products of sitosterol stearate were reported (52). In this study, the authors showed fundamental differences between the oxidation of free and esterified sitosterol. For example, no oxidized products at the third carbon atom in ring A were found, and no displacement of the $\Delta^5$ double bond to the $\Delta^4$ position was observed in sitosterol stearate oxidized in air at 150°C as compared to free sitosterol. Other oxidation products identified were 5,6-epoxide- and 7-ketositosterol stearate.

Oxidation studies of 5% sitosterol in tristearin at 120°C for 7 h produced epimeric 7-hydroxysitosterol, as well as 5,6-epoxysitosterol, 5α-stigmastan-3β,5,6β-triol, and other oxidation products (51,53). From these studies, the authors observed that, in contrast to pure sitosterol, oxidation products formed in a tristearin medium yielded epimeric 5,6-epoxysitosterol and stigmast-4,6-dien-3-one. Formation of 6-hydroxystigmast-4-en-3-one and 5α-stigmastan-3β,5,6β-triol was considerably less, compared to that from pure sitosterol oxidation. Yanishlieva-Maslarova and Marinova-Tasheva (54) studied the effect of the unsaturation of lipid media on autoxidation of sitosterol. This study demonstrated that when sitosterol was added at a 5% level in different triacylglycerols with variable degrees of unsaturation, the percentage of changed sterol increased with increasing unsaturation of the medium. Formation of 5α-stigmastan-3β,5,6β-triol was observed only in sunflower oil, while triol was not observed in tristearin and lard containing 5% sitosterol, although epimeric 5,6-epoxysitosterol was found in both.

Daly et al. (55) characterized some oxidized products of pure sitosterol heated at 100°C. They found 7α-hydroxy-, 7β-hydroxysitosterol, and 7-ketositosterol by TLC and GC/MS, along with epimeric 5,6-epoxysitosterol, $\Delta^4$-sitosterol-3-6-dione, $\Delta^4$-sitosterol-3-one, and $\Delta^5$-sitosterol-3-one.

Studies on phytosterol oxidation focused mainly on sitosterol (5,55); however, oxidation of stigmasterol and $\Delta^5$-avenasterol were also reported. Gordon and Magos (56) isolated some oxidation products of $\Delta^5$-avenasterol produced by heat treatment at 180°C of pure triacylglycerol containing 0.1% $\Delta^5$-avenasterol. After isolation of unsaponifiables and enrichment by preparative TLC, products of oxidized $\Delta^5$-avenasterol, such as epimeric 7-hydroxy-, 7-ketoavenasterol, UV and GC/MS onfirmed cholesta-3,5-diene-7-one, epimeric 5,6-epoxide, cholest-5-en-3-one, cholest-4-en-3-one, and cholesta-4,6-diene-3-one.

Blekas and Boskou (57) isolated some oxidation products of stigmasterol by heating purified triacylglycerol containing 5% stigmasterol at 180°C. Using TLC, UV, and IR spectroscopy, the authors tentatively identified epimeric 7-hydroxystigmasterol, 5,6-epoxide, and stigmast-22-en-3,5,6-triol. Other oxidation products of stigmasterol were characterized by MS, including stigmasta-4,22-diene-3-one, stigmasta-3,5,22-triene-7-one, and stigmasta-3,5,22-triene.

In order to investigate side-chain autoxidation products, a mixture of the two phytosterols sitosterol and campesterol, oxidized for 72 h at 120°C in an air-ventilated oven (47). The following oxidation products were identified and characterized from the two phytosterols: 24-methylcholest-5-en-3β,24-diol, 24-methylcholest-5-en-3β,25-diol, 24-methylcholest-4-en-6α-ol-3-one, 4-methylcholest-4-en-6β-ol-3-one, 24-ethylcholest-5-en-3β,24-diol, 24-ethylcholest-5-en-3β, 25-diol, 24-ethylcholest-4-en-6α-ol-3-one, 24-ethylcholest-4-en-6β-ol-3-one (47). The authors further characterized three semipolar side-chain oxidation products of stigmasterol, 24-ethylcholest-5,22-dien-3β,25-diol, 24-ethylcholest-5,22-dien-3β,24-diol, and 24-ethyl-5,22-choladien-3β-ol-24-one, in addition to its common ring-structure oxidation compounds, by TLC, GC/MS, and NMR (48).

In order to study photooxidation of phytosterols, Bortolomeazzi et al. (49) produced 5-hydroperoxides of β-sitosterol, campesterol, stigmasterol, and brassicasterol of the respective sterols in the presence of hematoporphyrine as sensitizer in pyridine. The reduction of the hydroperoxides produced the corresponding 5α-hydroxy derivatives. The 7α- and 7β-hydroperoxides of the sterols were obtained by allowing an aliquot of the 5α-hydroperoxides to isomerize, which in turn epimerized to 7β-hydroperoxides. This study further supports that the phytosterols have the same behavior to photooxidation as cholesterol, and the different phytosterols photooxidized, at about the same rate.

Later in another study, Säynäjoki et al. (58) studied photooxidation of stigmasterol in the presence of methylene blue as a sensitizer. The results were similar to those of Bortolomeazzi, et al. (49). As soon as photooxidation started, 5α-OOH and 6α- and 6β-OOH formed, and the amount of 7α-OOH began to increase slightly more slowly than that of 5α-OOH. The authors suggested that this might be due to the rearrangement of 5α-OOH to 7α-OOH. The authors did not try to detect 7β-OOH knowing that hydroperoxides formed immediately after photooxidation allowing no time for rearrangement of 7α-OOH to 7β-OOH. More detailed discussion of the mechanism of sterol oxidation is available in a recent review (30).

Some comments on the possibility of oxidation of stanols, the saturated counterpart of sterols, are of interest because of their considerable use as functional food ingredients in many countries (59,60). Stanols do not contain a double bond at the $\Delta^5$ position (see Fig. 7.1), and they are considered more stable than their counterparts containing an olefinic unsaturation (5). Since cholesterol is autoxidized in the side chain producing a variety of hydroperoxides, the saturated counterpart, the stanol, might also generate similar products (59). In addition, the free stanols may also generate 3-keto compounds as in the case of free unsaturated sterols. The only known supporting report, published by Soupas et al. (60), showed that thermally induced oxidation of sitostanol in bulk, and nine different sitostanol oxides in purified oils, were formed, including epimers of 7- and 15-hydroxysitostanol and 6- and 7-ketositostanol. As expected, the oxidation of sitostanol in palm oil and rapeseed oil oxidized at 182°C for 2h, showed low, but quantifiable, amounts of 7α-OH-sitostanol, 6α-OH-sitostanol and 5β-OH-sitostanol.

# Appearance of Sterols in Frying Fat and Fried Food

During refining, fats and oils undergo various chemical and physical treatments, often at temperatures of 250°C and higher (61–63). Each refining step exerts different effects on the sterol content in refined vegetable oil, as has been comprehensively reviewed (4). Since then, new reports have been published, and are discussed in this chapter. Some general comments address quantitative and qualitative changes in the levels of different sterols, particularly concerning the oxidative changes of phytosterols during refining.

## Effect of Processing on Sterols in Refined Oil

During refining, the total content of sterols decreases, but generally not their relative proportions (4). Coconut oil selectively loses considerable amounts of $\Delta^5$-avenasterol from an initial 24–45% to 3–11% of total sterols after refining (64). Studies of olive and other vegetable oils established that, regardless of oil type, neutralization and deodorization processes were mainly responsible for the losses in sterol content (65).

Some vegetable oils, such as crude sunflower and corn oils, contain rather high amounts of sterol glucosides, removed exclusively during degumming (66–68). Studies on changes in sterols and steryl glycosides in sesame and rapeseed oils during manufacturing showed that the content of free sterols did not change in sesame oil (69). In rapeseed oil, however, 22% of the free sterols are removed, mainly, during deodorization. Dehydrated sterols also form in rapeseed oil during bleaching and deodorization, while degumming removes most of the sterol glycosides. In sesame oil, refining removes about 90% of the sterol glycosides (69).

Sterols, present as free sterols and sterol glycosides in crude vegetable oil, are also present as sterol esters of fatty acids at levels ranging from 25–80% of the total sterols (64,69). During bleaching, those sterol esters undergo deacylation. Some interesterification also occurs (70–76).

In addition to the previously mentioned effects on different sterols, sterols undergo oxidation with the formation of 7-hydroxy-, 7-ketosterols (73,77), and dehydroxylated sterols during alkali refining and bleaching. These are commonly known as sterenes, or sterol hydrocarbons (4,78,80). In early studies, many components reported as artifacts of sterol degradation products (4) might be misinterpretations because phytosterols are complex mixtures and each sterol has the potential to be oxidized. High performance liquid chromatography (HPLC) and GC separation techniques were insufficient to resolve these complex mixtures. In the past decade, analytical techniques to determine these compounds improved significantly. Parts of these components are removed during deodorization, (4); however, deodorization, which applies high temperatures of up to 250°C, also contributes to production of these compounds (4,79,80).

Results from studies on the effect of physical refining, also known as steam refining, on the content of unsaponifiables are presented in Table 7.3. (81). Crude soybean oil usually contained 0.39% sterols, which decreased to a level of 0.26–0.04% during physical refining performed at different temperatures and times. The most significant change for $\Delta^7$-stigmastenol was that crude oil contained 28 ppm of this sterol, increasing to 118 ppm following physical refining for 2 h at 300°C. This increase in stigmast-7-enol was attributed to isomerization of sitosterol, from $\Delta^5$ to $\Delta^7$ isomer. Meneghetti et al. (82) identified different transformation products of sterols, including saturated $C_{28}$ and $C_{29}$ sterols, in refined corn oil using GC/MS. However, *Arachis hipogaea* and *Soya hispida* oils, refined by alkali treatment did not contain any "detectable" transformation products of sterols (83).

In a study on the variation of sterols during the refining of olive and other vegetable oils, neutralization decreased the quantity of different sterols, but not their proportional composition (84). Bleaching, however, resulted in a decrease in quantity as well as a change in proportional composition. This was particularly evident for $\Delta^5$-avenasterol, which decreased the most, followed by terpenic di-alcohols, which reduced by one-half. The content of citrostadienol and 24-methylenecycloartanol also decreased significantly. Under mild bleaching, different transformation products from 24-methylencycloartanol resulted. In a more recent study on sterols and squalene degradation products in olive oil, bleaching earth exhibited stronger effects at higher concentrations and elevated temperatures (85). A level of 1% bleaching earth at 80°C for 1 h gave less than 5 ppm of the degradation products in olive oil compared to 2,408 ppm of degradation products generated at 160°C in the presence of 5% bleaching earth. The temperature of decolorization affected the concentration of stigmast-3,5-diene. At 80°C for 1 h with 3% bleaching earth, the concentration of stigmast-3,5-diene was 4% of the olefinic degradation products, whereas at 120 and 160°C, these concentrations were 8–11% and 13%, respectively (85). Deodorization generated more degradation products during the same period at higher temperatures while at the same time removing more of these products. Before deodorization, vegetable oils contained 1,206, 2,246, and 2,208 ppm of olefinic degradation products. After deodorization at 260°C, the amounts decreased to 148, 240, and 280 ppm, respectively (85).

Many refined, and some labeled "nonrefined," vegetable oils were investigated, including sunflower, safflower, corn, soybean, and rapeseed, as was cocoa butter. Sunflower oil "from the Swiss market" contained 16 ppm dehydroxylated sterols, unrefined safflower oil had 0.8–6.0 ppm, while unspecified safflower oil had 6.0 ppm of olefinic degradation products. Corn, soybean, and rapeseed oils also contained these components, although no quantitative data were presented. Cocoa butter heated with or without bleaching earth at different temperatures under laboratory conditions also showed different levels of these components. Very low levels of dehydroxylation products were present in cocoa butter subjected to heat treated at 110°C without bleaching earth. While bleached at 70°C in the presence of 1% bleaching earth, cocoa butter had 30 mg/kg 3,5-stigmastadiene, this amount increased at 90°C (85).

Production of dehydroxylated sterols (sterenes) at different stages of refining of maize (corn), rapeseed, soybean, sunflower, and palm oils was recently reported in detail (77,86). Sterenes were generated only during bleaching and deodorization. Deodorization contributed the highest amounts of these dehydroxylated sterols for all oils, with the exception of palm oil. None of these compounds were found in the crude oil or alkali neutralized oil. Table 7.4 shows the concentration of sterenes identified in different vegetable oils after bleaching and deodorization (78). Among the vegetable oils analyzed, maize (corn) oil contained the highest total amounts of sterenes, as it contains more total sterols compared to other oils. However, it cannot be generalized from this report that a high concentration of sterols yields more sterenes; after deodorization sunflower oil contained higher amounts of some of these components compared to rapeseed oil, the latter contained a higher total sterol content: 6 mg/kg vs. 2 mg/kg (78).

Virgin olive oil contained $n$-alkanes from $C_{14}$–$C_{35}$ and $n$-heptadecene, along with squalene (87). After refining $n$-alkanes, isomerization products of squalene, isoprenoidal polyolefins were found as compounds formed from degradation of hydroxy derivatives of squalene, and sterenes derived from sitosterol and 24-methylenecycloartenol. Both chemically and physically refined olive oil were examined. Physical refining generated much higher amounts, 21.53 ppm of stigmasta-3,5-diene, than chemical refining, which produced only 8.98 ppm. There were no quantitative data presented for sterenes formed from 24-methylenecycloartenol (87).

Deodorization of different vegetable oils generated detectable amounts of stigmast-3,5-diene at 180°C, which increased at 200°C (88). In studies on the effect of quantity and types of bleaching earth used, 1 and 3% Tonsil demonstrated no difference in 3,5-stigmastadiene formation at 50°C, whereas at 70°C, production of this component was much higher than at 50°C: about 0.3 and 10 ppm, respectively. Trisil, added at 3%, generated very small amounts of this component at both temperatures.

**TABLE 7.3**

**Individual Sterols in Crude Oil, Phosphoric Acid-Degummed and Bleached Oil and Oil Physically Refined Under Various Conditions (mg/kg)[a]**

| Soybean Oils | Total Sterols | Campesterols | Stigmasterol | β-Sitosterol | Δ⁷-Stigmastenol | Δ⁷-Avenasterol | Unkown |
|---|---|---|---|---|---|---|---|
| Crude | 3,900 | 764 (19.6) | 757 (19.4) | 2,312 (59.3) | 28 (0.7) | 39 (1.0) | 0.0 |
| PDB[b] | 3,200 | 601 (18.8) | 632 (19.9) | 1,890 (58.9) | 48 (1.5) | 29 (0.9) | 0.0 |
| PDBPHR[c] at: | | | | | | | |
| 240°C, 2h | 2,600 | 449 (17.2) | 472 (18.1) | 1,620 (62.5) | 35 (1.3) | 24 (0.9) | 0.0 |
| 250°C, 2h | 2,100 | 361 (17.2) | 399 (19.0) | 1,283 (61.5) | 40 (1.9) | 17 (0.8) | 0.0 |
| 260°C, 2h | 1,600 | 287 (17.9) | 285 (17.8) | 966 (60.4) | 35 (2.2) | 14 (0.9) | 13.0 (0.8) |
| 280°C, 2h | 900 | 155 (17.2) | 164 (18.3) | 509 (56.6) | 50 (5.5) | 17 (1.9) | 5.0 (0.5) |
| 300°C, 2h | 400 | 48 (12.2) | 49 (12.1) | 157 (39.4) | 118 (29.4) | 25 (6.2) | 3.0 (0.7) |
| 280°C, 1/2h | 1,200 | 205 (17.1) | 233 (18.6) | 730 (60.8) | 31 (2.6) | 11 (0.9) | 0.0 |
| 280°C, 1h | 100 | 162 (16.2) | 184 (18.4) | 607 (60.7) | 33 (3.3) | 14 (1.4) | 0.0 |
| 280°C, 3h | 600 | 89 (14.8) | 101 (16.9) | 323 (53.9) | 62 (10.4) | 11 (1.7) | 14.0 (2.3) |
| 300°C, 1/2h | 1,000 | 158 (15.8) | 168 (16.8) | 600 (60.0) | 53 (5.3) | 21 (2.1) | 0.0 |

*Source:* Adapted from Jawad et al. (81).

[a] Figures in parentheses correspond to percentage composition of sterols. Traces of cholesterol were found in crude and physically refined oil samples. The crude oil sample also contained a trace of Δ⁵-avenasterol.

[b] PDB = Phosphoric acid-degummed and bleached oil.

[c] PDBPHR = Phosphoric acid-degummed, bleached, and physically refined oil.

**TABLE 7.4**
Concentrations of Sterenes in Edible Oil (μg/g)

| Compound | Name | Maize B[a] | Maize DW[b] | Maize D[c] | Palm B | Palm D | Rapeseed B | Rapeseed D | Soybean B | Soybean D | Sunflower B | Sunflower D |
|---|---|---|---|---|---|---|---|---|---|---|---|---|
| B1 | 24-Ethylcholesta-3,5-diene | 1.8 | 2.0 | 42.0 | 3.2 | 2.9 | 1.4 | 4.9 | 0.7 | 2.6 | 1.0 | 10.0 |
| B2 | 24-Ethylcholesta-3,5,22-triene | 0.1 | 0.2 | 6.2 | 0.5 | 0.5 | 0.7 | 3.7 | 0.3 | 0.4 | 0.2 | 1.6 |
| B3 | 24-Methylcholesta-3,5-diene | 0.6 | 0.8 | 16.0 | 1.5 | 1.2 | 0.9 | 0.5 | 0.5 | 1.5 |  | 2.2 |
| B4 | 24-Methylcholesta-3,5-diene | 1.2 | 1.3 | 1.8 |  | 0.1 |  |  |  |  |  | 3.1 |
| B5 |  |  |  |  |  |  | 0.3 | 0.2 | 0.3 | 0.4 |  | 3.2 |
| B6 | 24-Methylcholesta-3,5,22-triene |  |  |  |  |  | 0.5 | 1.2 | 0.5 |  |  | 3.1 |
| B7 |  |  |  |  |  |  | 0.9 | 0.6 | 0.6 |  |  |  |
| D1 |  |  |  | 1.5 |  | 0.1 |  | 0.3 | 0.1 |  |  | TR[d] |
| D2 | 24-Ethylcholesta-2,5-diene |  |  | 15.0 |  |  | 0.4 |  | 0.5 |  | 0.3 | 2.4 |
| D3 | 24-Methylcholesta-2,5-diene |  |  | TR |  |  |  |  | 0.6 |  | 0.4 | TR |
| D4 | 24-Ethylcholest-2-ene |  |  | 1.1 |  |  |  |  |  |  |  |  |
| D5 |  |  |  | 2.1 |  |  |  |  |  |  |  |  |
| D6 | 24-Ethylcholesta-4,6-deine |  |  | 55.0 |  |  |  | TR |  | TR |  | 8.6 |
| D7 | 24-Ethylcholesta-4,6,22-triene |  |  | 4.6 |  |  |  |  |  |  |  |  |
| D8 | 24-Methylcholesta-4,6-diene |  |  | 17.3 |  |  |  |  |  |  |  | 1.4 |

Source: Mennie et al. (78).
[a]B = Bleached.
[b]DW = Dewaxed.
[c]D = Deodorized.
[d]TR = Trace.

**TABLE 7.5**
**Composition of Constituent Disteryl Ethers Isolated from Commercial Vegetable Oil and from a Table Margarine**

| Disteryl Ethers | Vegetable Oil[a] Found | (Calculated)[b] | Table Margarine[a] Found | (Calculated)[b] |
|---|---|---|---|---|
| | (%) | (%) | (%) | (%) |
| Dicholesterol ether | 3 | (<1) | –[c] | –[c] |
| Dibrassicasteryl ether | 3 | (1) | 5 | (1) |
| Dicampesteryl ether | 11 | (11) | 10 | (12) |
| Disitosteryl ether | 21 | (24) | 23 | (23) |
| Campesterylsitosteryl ether | 29 | (33) | 31 | (34) |
| Sitosterylbrassicasteryl ether | 13 | (9) | 12 | (8) |
| Campesterylbrassicasteryl ether | 9 | (6) | 11 | (6) |
| Others[d] | 11 | (16) | 8 | (16) |

Source: Schulte and Weber (89).
[a]Each contains about 1 μg disteryl ethers/g.
[b]Calculated for the formation of disteryl ethers from the sterol mixture.
[c]The disteryl ether fraction of margarine cotains 67% dicholesteryl ether, calculated as "100% from animal origin."
[d]Including small proportions of cholesterylcampesteryl ether, chlesterylbrassicasteryl ether, cholesterylsitosteryl ether, and traces of disteryl ethers containing stanyl moieties.

The formation of disteryl ethers, an artifact formed during the bleaching of commercial oil and table margarine, resulted (71,89). Figure 7.5 and Table 7.5 present the structural formula of disteryl ethers and their content in commercial vegetable oil and table margarine.

Gordon and Firman (90), in order to study the formation of stigmastadienes, stirred olive oil (100 g) with bleaching earth, using a mechanical stirrer, for 1 h at 80–120°C under nitrogen. The formation of stigmastadienes was much higher in olive oil bleached with Premiere bleaching earth (ranging from 3–45 μg/g) compared with Fulmont 237 (ranging from 1–10 μg/g) and Terrana 510 (ranging from < 1 to 5μg/g). The results showed that the bleaching earths had a major catalytic effect on stigmastadiene formation. The stigmastadiene content increased linearly with temperature. Also, the earth that was most active for pigment removal was also most active for sitosterol dehydration.

In a study under laboratory conditions, Bortolomeazzi et al. (91) demonstrated that steroidal hydrocarbons (sterenes) with three double bonds could partially originate from the corresponding hydroxysterols present in the crude oils. The authors later confirmed that the bleaching process caused a reduction of the hydroxyphytosterol with partial formation of steroidal hydrocarbons having three double bonds (steratrienes) in the ring system at the 2-, 4-, and 6-positions (92). In addition, results from the analysis of phytosterol oxidation products in the samples of crude sunflower oil, peanut oil, corn oil, lampante virgin olive oil, palm oil, coconut oil, and palm nut oil were reported. The concentrations of phytosterol oxidation products in the oils varied among different samples of the same type of oil, as well as between the different oils. It was demonstrated that sunflower and maize were the oils with the larger amount of total phytosterol oxidation products, with values in the range from 4.5–67.5 ppm and from 4.1–60.1 ppm, respectively. Considerably lower concentrations, between 2.7 and 9.6 ppm, were present in the peanut oil, and in the palm nut oil, 5.5 ppm. Among the lampante olive oils analyzed, only three samples of the six contained phytosterol oxidation products, and these were in the low concentration range, 1.5–2.5 ppm. In addition, the samples of palm and coconut oils were devoid of any detectable amounts of phytosterol oxidation products (92).

Further, the authors conducted refining processes in the laboratory with a sample of crude sunflower oil containing considerable amounts of various phytosterol oxidation products. After bleaching the crude sunflower oil with acidic earths, the recoveries of the hydroxysterols were very low (Table 7.6). In particular, at the level of 2% of earths, only ca. 6% of the initial 7-OH-sitosterol remained in the oil. However, in the case of the bleaching with neutral earths, recoveries of the hydroxysterols were higher, with average values of 29 and 14% at earth levels of 1 and 2%, respectively. In contrast, recoveries of 7-ketositosterol were similar for the different bleaching conditions with an average value of 76% (92).

The same sunflower oil sample was deodorized at 180°C, under vacuum, in the laboratory. No significant differences resulted between the trienes and dienes concentrations before and after deodorization. This indicates that heating at 180°C for 1 h was a condition not sufficient for the dehydration of either the hydroxysterols or sterols. Moreover, recoveries of the hydroxysterols were quantitative with no decomposition reactions. Table 7.6 shows similar recovery of 7-ketositosterol during the bleaching processes that was 80%. The authors concluded that dehydration of the hydroxysterols would occur at higher temperatures (92).

A recent investigation documented the content of sterol oxides in refined and partially hydrogenated vegetable oil (45). In this study, it was found that a blend of partially hydrogenated rapeseed and palm oils, refined regular sunflower oil, and high oleic sunflower oil contained 41, 40, and 47 ppm of total oxidized sterols, respectively (Table 7.7). The major oxysterols in the blend of partially hydrogenated rapeseed and palm oils were both epimers of epoxycampesterol and epoxysitosterol in the amounts of 11 and 17 ppm, respectively. Other oxidation products, identified as 7α-hydroxy-, 7β-hydroxy-, 7-keto-, and triol of both campesterol and sitosterol, ranged from 1–4 ppm in the blend. The major oxysterol in regular and high oleic sunflower oil was 7-ketositosterol detected at levels of 13 and 14 ppm, respectively. In addition, smaller amounts of other oxidized products were found, with levels of 5–8 ppm in these oils. The major difference in the sterol oxidation products in sunflower oil compared with the rapeseed oil/palm oil blend was marked in the higher contents of 7α-and 7β-

**TABLE 7.6**
**Recovery Percentages of Phytosterol Oxidation Products in a Sample of Sunflower Oil after Bleaching and Deodorization**

| POP | Bleaching | | | | Deodorization |
|---|---|---|---|---|---|
| | 1% acidic earths | 2% acidic earths | 1% neutral earths | 2% neutral earths | 180°C |
| 7α-OH-β-sitosterol | 11.8 | 5.8 | 30.4 | 10.6 | 99.6 |
| 7β-OH-β-sitosterol | 9.1 | nd[a] | 29.7 | 15.9 | 102.2 |
| 7α-OH-campesterol | 10.8 | nd | 29.5 | 11.9 | 99.9 |
| 7β-OH-campesterol | nd | nd | 27.6 | 15.5 | 110.1 |
| 7α-OH-stigmasterol | 7.5 | nd | 27.8 | 12.2 | 106.1 |
| 7β-OH-stigmasterol | nd | nd | 27.6 | 18.2 | 109.8 |
| 7-keto-β-sitosterol | 76.7 | 63.5 | 79.9 | 83.2 | 83.0 |

[a] = not determined
Adapted from Bortolomeazzi et al. (92).

**TABLE 7.7**
**Levels of Phytoterol Oxidation Products (mg/g)a in the Oil Samples of Rapeseed Oil/Palm Oil Blend, Sunflower Oil and High-Oleic Sunflower Oil[b]**

| Oil Sample | 7α-OH | 7β-OH | 7-Keto | Epoxy[c] | Triol | Total |
|---|---|---|---|---|---|---|
| RP | | | | | | |
| Sitosterol | 4.4 | 1.3 | 1.6 | 17.2 | 2.9 | 41.0 |
| Campesterol | 1.1 | nd | nd | 10.8 | 1.7 | |
| Stigmasterol | nd[d] | nd | nd | nd | nd | |
| SO | | | | | | |
| Sitosterol | 5.8 | 6.6 | 12.9 | 5.3 | 4.9 | 39.9 |
| Campesterol | 1.7 | 0.5 | nd | 0.9 | nd | |
| Stigmasterol | 0.7 | 0.6 | nd | nd | nd | |
| HOSO | | | | | | |
| Sitosterol | 7.7 | 5.4 | 14.1 | 7.8 | 5.7 | 46.7 |
| Campesterol | 2.8 | 1.0 | nd | 1.4 | nd | |
| Stigmasterol | 0.4 | 0.4 | nd | nd | nd | |

Source: Dutta (45).
[a] Means of duplicate analyses.
[b] Reproduced from reference 9.
[c] Include both 5α, 6α-epoxy- and 5β, 6β-epoxy-sterols.
[d] < 0.1 mg/g not detected.
RP, blend of partially hydrogenated rapeseed/palm oil; SO, sunflower oil; HOSO, high-oleic sunflower oil.

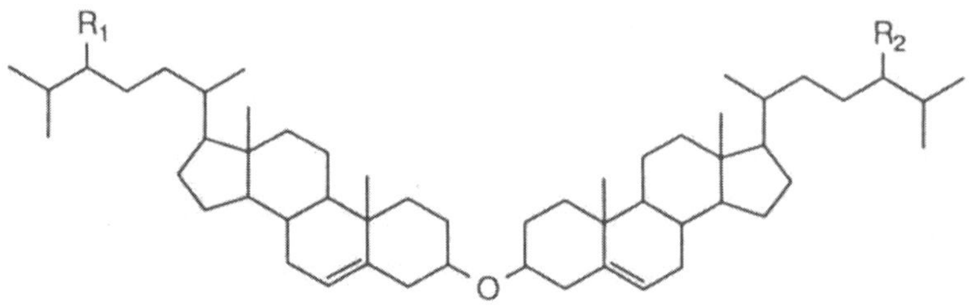

**Fig. 7.5.** Structure of various disteryl ethers. Schulte and Weber (89): $R_1 = R_2$ , uniform disteryl ethers; $R_1$, $R_2 = H$, dicholesteryl ether; $R_1$, $R_2 = CH_3$, dicampesteryl ether; $R_1 \neq R_2 = C_2H_5$, disitosteryl ether; $R_1$ ($R_2$, mixed disteryl ether; $R_1 = H$; $R_2 = CH_3$, cholesterylcampesteryl ether; $R_1 = H$, $R_2 = C_2H_5$, cholesterylsitosteryl ether; $R_1 = CH_3$, $R_2 = C_2H_5$, campesterylsitosteryl ether.

hydroxysitosterol, sitostanetriol, and 7-ketositosterol, and lower content of sterol-5,6 epoxides. A similar pattern resulted in the high-oleic sunflower oil, except that this oil contained more 7-ketositosterol and sterol-5,6-epoxides than sunflower oil. Small amounts of 7-hydroxystigmasterol were also detected, ranging from 0.4–0.8 µg/g in most of the oils (Table 7.7).

There are institutional regulations limiting the content of dehydrogenated sterols, particularly in virgin olive oil (88). The International Olive Oil Council (IOOC) limited stigmast-3,5-diene content at 0.15 ppm in virgin olive oil. No such regulations exist concerning refined oil, however. In addition, levels of other oxidized phytosterols in different refined oils are areas of concern. Levels of other oxidation products of phytosterols in different processed vegetable oils require further assessment.

## Effect of Hydrogenation on Sterols in Hydrogenated Fat

Hydrogenated fat and oil are required for many specialty food products (93). In a recent review, Edvardsson and Irandoust (94) described both industrial and laboratory scale hydrogenation reactors. Refined hydrogenated fat and oil also might contain different degradation products of sterols, generated not only during refining but also during hydrogenation at elevated temperatures (150–185°C). The degradation products of sterols in hydrogenated products concerning changes in the amount of sterols, transesterification of sterols, and generation of hydrogenated sterols, commonly known as stanols (4), have not been well studied. Due to the scarcity of recent work, some reports listed in the reviews are discussed here.

Kanematsu et al. (95) examined sterol contents during refining, including hydrogenation in several animal fats and vegetable oils. Sterol content decreased markedly during hydrogenation in this study. In a comprehensive study of different table margarines and hydrogenated vegetable oils (96), no stanols were found, leading to the assumption that the type of catalysts and conditions used to selectively hydrogenate polyunsaturated fatty acids might not be suitable for sterol hydrogenation. However, in a study on different vegetable oils and lard, Sugano et al. (97) demonstrated that during partial hydrogenation, total sterol content decreased, but a corresponding increase in stanols did not occur, whereas, after complete hydrogenation, an increase in amounts of stanols was observed. This study demonstrated the influence of different methods of hydrogenation. The authors also reported the presence of stanols in unhydrogenated fat and oil.

Modifications of desmethylsterols, 4α-methyl sterols, and 4,4′-dimethyl sterols during hydrogenation of edible vegetable oil were recently investigated (98–103). These studies showed slight changes in major desmethylsterol, while minor components such as $\Delta^5$ and $\Delta^7$ avenasterol decreased drastically and, to a small extent, so did the content of stigmasterol. In addition, the content of obtusifoliol, gramisterol, and citrostadienol decreased by 80% of their original levels. These decreases were attributed to double bonds in the side chains, which are reactive under hydrogenation. Cycloartenol and 24-methylenecycloartenol, the predominant 4,4′-dimethyl sterols (triterpene alcohols) in peanut, soybean, sunflower, and corn oils, undergo structural transformation during hydrogenation. For example, 91% of 24-methylenecycloartenol disappeared during hydrogenation. The authors concluded that during hydrogenation triterpene alcohols undergo different chemical processes, such as hydrogenation of double bonds, opening of the cyclopropane ring, and positional isomerization of the side chain double bonds. Fig. 7.6 shows a scheme for structural changes of 24-methylenecycloartenol (102). The principal components formed from 24-methylenecycloartenol are cyclobranol and cyclosadol; cycloartenol generated 9β,19-cyclo-5α-lanost-25-en-3β-ol. Schulte (86) reported that hydrogenated vegetable oil and margarine contain different amounts of steradienes, the dehydration products of sterols. One of these hydrogenated oils, however, showed diminished amounts of these products, from 2.3–0.9 ppm, after deep fat frying for a long period.

### Effect of Processing on Sterols in Lard and Tallow

After extraction, lard and tallow undergo more or less advanced refining processes, depending on the end use of the products (104). In contrast to vegetable oil, animal fat almost exclusively contains cholesterol as the main sterol, and thus is the major sterol in both lard and tallow. Content of cholesterol ranges from 0.37–0.42% in lard, 0.08–0.14% in beef tallow, and 0.23–0.31% in mutton tallow (42). Several recent reviews also deal with the content of degradation products of cholesterol in lard and tallow, along with different food products (36,37,39). However, these reports mainly concern cholesterol oxidation products in lard and tallow treated only in laboratory conditions.

Kanematsu et al. (95) studied different animal and vegetable fats and oils for content of sterols during refining. In crude beef tallow and pork fat, the sterol content was 1.94 mg/g and 1.03 mg/g, respectively, which decreased to 0.90 and 0.38 mg/g, respectively, after hydrogenation and deodorization. Sugano et al. (97) studied the effect of different methods of hydrogenation on cholesterol. They demonstrated that

**TABLE 7.8**
**Content of Cholesterol Oxides in Refined and Unrefined Lard**

| Company[a] | Lot | Quality | Cholesterol Oxides[b], ppm | | | | |
|---|---|---|---|---|---|---|---|
| | | | 5α,6α-epoxy | 7-keto | 7α-hydroxy | 20α-hydroxy | 25-hydroxy |
| A | 1 | Unrefined | 0.3 | TR[c] | TR | ND[d] | ND |
| A | 1 | Refined | 0.3 | TR | TR | ND | ND |
| A | 2 | Unrefined | ND | 0.2 | TR | TR | TR |
| A | 2 | Refined | ND | 0.2 | TR | TR | TR |
| B | 3 | Unrefined | TR | TR | ND | 0.3 | TR |
| B | 3 | Refined and deoderized[e] | TR | TR | TR | 0.2 | ND |
| B | 3 | Refined and deoderized[f] | TR | TR | TR | TR | ND |
| B | 4 | Unrefined | ND | 0.2 | TR | TR | TR |
| B | 5 | Unrefined | ND | 0.3 | TR | 0.3 | 0.2 |
| B | 6 | Unrefined | ND | 0.3 | TR | 0.3 | 0.2 |
| B | 7 | Unrefined | ND | 0.3 | TR | TR | TR |

Source: Nourooz-Zadeh and Appelqvist (35).
[a]A, Andelsflott plant, Göteborg, Sweden; B, Konvex plant Kävlinge, Sweden.
[b]The 5α,6α-epoxycholestanol, 7β-hydroxycholestertol, and cholestane-triol were not detected; detection limit 0.1 ppm.
[c]TR = Trace ( < 0.1 ppm).
[d]ND = Not detected.
[e]Refined, bleached with Tonsil LFF80 1% and deodorized in a pilot plant at the Margarine Company, Helsingborg, Sweden.
[f]Refined, bleached with Tonisil LFF80 1% and deodorized in a pilot plant at the Karlshamns AB, Karlshamn, Sweden.

almost all cholesterol converts to the saturated counterpart, cholestanol, after complete hydrogenation. The content of cholestanol in unhydrogenated lard was present only as 1.2% of cholesterol, which increased to 32% under partial hydrogenation, and to 96.4% following complete hydrogenation. Fedeli et al. (105) did not find any significant difference in proportions of different sterols in lard after processing. For instance, cholesterol was at 98.9% in lard from adipose tissue, whereas, after refining, it changed slightly to 98.5%, although the percentages of sterols in total unsaponifiables decreased from 69 to 66%. Nourooz-Zadeh and Appelqvist (35) in a comprehensive study on different refined and unrefined lard found traces or quantifiable amounts of oxidized cholesterols. Table 7.8 presents the content of different cholesterol oxides in different lard samples. Schulte (86) reported cholestadiene, the dehydration product of cholesterol, at a level of 17.3 ppm in commercial lard. Similarly, in margarine, the author identified cholestadiene in various amounts, probably generated during the refining of animal fat.

Recently, oxidation of cholesterol during the bleaching and deodorization process

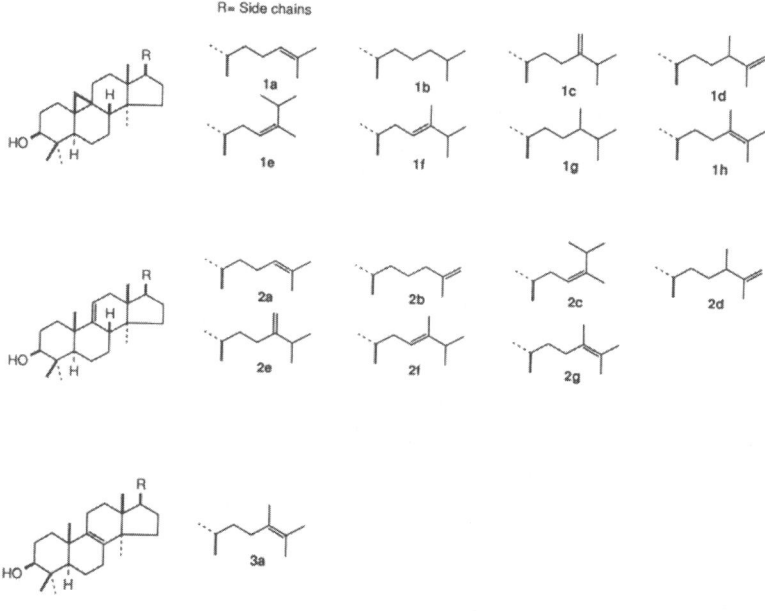

**Fig. 7.6.** Structural changes of 24-methylenecycloartanol during hydrogenation.: 1a, 5α-cycloart-24-en-3β-ol; 1b, 5α-cycloartan-3β-ol; 1c, 24-methylene-5α-cycloartan-3β-ol; 1d, [24s]-24-methyl-5α-cyclo-art-25-en-3β-ol; 1e, 24-methyl-5α-cycloart-Z-23-en-3β-ol; 1f, 24-methyl-5α-cycloart-E-23-en-3β-ol; 1g, 24-methyl-5α-cycloartan-3β-ol; 1h, 24-methyl-5α-cycloart-24-en-3β-ol; 2a, 5α-lanosta-9 (11), 24-dien-3β-ol; 2b, 5α-cycloart-25-en-3β-ol; 2c, 24-methyl-5α-lanosta-9 (11), Z-23-dien-3β-ol; 2d, 24-methyl-5α-lanosta-9 (11), 25-dien-3β-ol; 2e, 24-methylene-5α-lanost-9 (11)-en-3β-ol; 2f, 24-methyl-5α-lanosta-9 (11), E-23-dien-3β-ol; 2g, 24-methyl-5α-lanosta-9 (11), 24-dien-3β-ol; 3α, 24-methyl-5α-lanosta-8,24-dien-3β-ol. Strocchi and Marnaschio (102).

of tallow was studied (106). Only cholest-5-en-3β,7β-diol (7β-HC) and 5β, 6β-epoxy-5β-cholestan-3β-ol (β-CE) could be quantified in the samples of crude and refined tallow. The content of 7β-HC was slightly higher in the processed tallow (0.6–0.7 μg/g) compared with natural tallow (0.2 μg/g). However, the content of β-CE increased considerably during processing, ranging from 0.8 μg/g to 3.4 mg/g, compared with 0.6 μg/g in the natural tallow (Table 7.9). The authors observed that the content of β-CE was influenced by temperature, duration, and type of bleaching earth (106).

## Presence of Oxidation Products in Fried Food

The literature on fried food is not voluminous, as compared to that of other food. Some reports discussed here were part of earlier reviews (36–39); however, they are cited due to the scarcity of recent reports in this area. Lee et al. (107) used a qualitative method to analyze sterol epoxides in french fries, fried in unspecified fat and oil, probably mixed animal and vegetable fat and oil. Using HPLC, they reported that total epoxides were present at levels ranging from 0.02–0.34% of sterols. Lee et al. (108) reported cholesterol and sitosterol oxidation products in french fries and potato chips. In this study, sample chips were fried in cottonseed oil, while french fries were fried in mixtures of beef tallow and hydrogenated vegetable oil. Because the separation of different mixed oxysterols probably was not optimal, only a few common oxysterols could be analyzed. Chip samples, stored at 23°C for 150 days, did not produce detectable amounts of oxysterols. Conversely, chip samples stored at 40°C produced quantifiable amounts of sitosterol oxides after 95 days of storage. Levels of sitosterol β-epoxide, 7α-hydroxysitosterol, and 7β-hydroxysitosterol, in chips stored at 40°C for 95 days, were 6, 13, and 9 ppm in lipids, respectively. In french fry samples, both cholesterol and sitosterol oxides were detected. Samples obtained from five restaurants all had various levels of different sterol oxides, although in this study only four oxysterol components were reported (Table 7.10).

Zhang et al. (109) analyzed the content of oxysterols in french fries cooked in mixed animal/vegetable shortening from two local restaurants for 30 consecutive days. In this study, only cholesterol oxidation products were reported, although french fries were cooked in mixed animal and vegetable fat of unknown proportions. Separation and detection of different oxysterol products both from cholesterol and from phytosterols were indeed complicated. Results from the two restaurants varied qualitatively, with samples from restaurant A containing epoxy sterols as major components, and samples from restaurant B containing cholestanetriol as the major component. These differences were not explained, although the content of total sterol oxides in french fries from restaurants A and B ranged from 11–50 ppm and 11–39 ppm, respectively.

Park and Addis (110) determined cholesterol oxides in different foods, including french fries cooked in tallow, fried chicken meat, and fried chicken crust, along with other foods and food ingredients, using HPLC, saponification, or

**TABLE 7.9**
**Content of Two Cholesterol Oxidation Products (µg/g, mean of duplicate) in Natural-, Bleached-, and Deodorised Tallow**

| Sample No. | Bleaching earth type | Time (min) | Temper- ature (°C) | Steam (%) | Pres- sure (mbar) | Choles- terol | 7β-OH[a] µg/g tallow ± SEM | β-CE[b] |
|---|---|---|---|---|---|---|---|---|
| *Natural tallow* | | | | | | | | |
| 1 | – | – | – | – | – | 502 | 0.17 | 0.60 |
| | | | | | ± 3 | ± 0.03 | ± 0.15 | |
| *Bleached tallow* | | | | | | | | |
| 2 | 1% Optimum 215[c] | 30 | 100 | – | – | – | 0.69 | 2.65 |
| | | | | | | | ± 0.02 | ± 0.60 |
| 3 | 1% Optimum 215 | 60 | 80 | – | – | – | 0.69 | 3.37 |
| | | | | | | | ± 0.07 | ± 0.23 |
| 4 | 2% Optimim 215 | 30 | 80 | – | – | – | 0.49 | 1.48 |
| | | | | | | | ± 0.01 | ± 0.29 |
| 5 | 1% Standard 310 [d] | 30 | 100 | – | – | – | 0.60 | 1.98 |
| | | | | | | | ± 0.09 | ± 0.45 |
| 6 | 1% Ex 640[e] | 30 | 100 | – | – | – | 0.69 | 0.81 |
| | | | | | | | ± 0.08 | ± 0.12 |
| *Deodorised tallow* | | | | | | | | |
| 7[f] | | 45 | 230 | 1 | 2 | – | 0.60 | 2.09 |
| | | | | | | | ± 0.07 | ± 0.36 |

Source: Verleyen et al. (106).
[a] Cholest-5-en-3β,7β-diol.
[b] 5β, 6β-epoxy- 5β-cholestan-3β-ol.
[c] Very acid-activated bleaching earth.
[d] Standard acid-activated bleaching earth.
[e] Natural bleaching earth.
[f] Bleached according to condition of sample No. 2.

derivatization of oxysterols. Good recoveries were reported for 7-keto and epimeric 7-hydroxycholesterols, although co-elution of other components occasionally interfered with quantitation, particularly with epimeric 7-hydroxycholesterols. Large differences for 7β-hydroxycholesterol were evident; they ranged from 6.8–58.8 ppm in samples of french fries. 7α-Hydroxycholesterol was below detection levels, while 7-ketocholesterol was present at a level of 4.1 ppm. Fried chicken meat and fried chicken crust did not show any quantifiable amounts of the previously mentioned oxysterols.

**TABLE 7.10**
**Levels of Sterol Oxidation Products in French Fries (μg/g of lipids)**

| Restaurant | Cholestan-5α, 6α-epoxy-3β-ol | Cholestan-5β, 6β-epoxy-3β-ol | 5-Cholestan-3β,7β-diol | 5-Cholestan-3β,7α-diol |
|---|---|---|---|---|
| A | 19 | 25 | 39 | TR[a] |
|   | 10 | 27 | 14 | 2 |
| B | BD[b] | 3 | 11 | 7 |
|   | 9 | 23 | 44 | 21 |
| C | 17 | 18 | 27 | 8 |
|   | ND | 18 | 30 | 13 |
| D | TR | TR | 3 | 0 |
|   | ND | 6 | 0 | 0 |
| E | 6 | 9 | 81 | 21 |
|   | ND | 2 | 62 | 2 |

Source: Lee et al. (108).
[a]ND = Not detected due to co-eluting substances on high-performance liquid chromatography.
[b]TR = Trace on TLC but not quantifiable by high-performance liquid chromatography.

A new method to evaluate cholesterol oxides by TLC/flame ionization detection (FID) was compared with a GC method in tallow used for french fries (34). Although it was not a direct analysis of french fries, to some extent fat for frying may reflect products in french fries. In this study, it was reported that about 25% of cholesterol was lost after 60 h of frying. About 90% of the lost cholesterol was identified as triol, 90 ppm; epimeric 7-hydroxycholesterols, 40 ppm; $\Delta^{3,5,7}$-oxocholesterol, 20 ppm; epimeric epoxides, dominated by β-epoxide, 10 ppm; and 7-oxocholesterol, 10 ppm.

The presence of different oxides of cholesterol, sitosterol, and campesterol in different fried products, cooked in mixed animal and vegetable fat and oil from North America, were reported (111). Cholesterol oxides were found at different levels in all snack food samples analyzed. Both epimers of epoxysitosterol and epoxy-campesterol, along with both epimers of 7-hydroxysitosterol and 7α-hydroxycampesterol, were quantified in different amounts in these samples. Moreover, 7-ketositosterol and 7-ketocampesterol, along with sito- and campestatriol were quantified at different levels.

Temperature is one of the important factors facilitating sterol oxidation in food products in addition to processing and storage (112). The effect of cooking temperature in the generation of pre-cooked french fries is shown in Table 7.11 and Table 7.12. The contents of these oxysterols in samples of french fries cooked in a blend of rapeseed/palm oils contained the highest amounts of sterol oxides and epimers of epoxy sterols (22).

**TABLE 7.11**
**Levels of Sterol Oxides (ppm) in Lipids of French Fry Samples Fried in PO/RO Blend, SO, and HOSO[a]**

| Sample | Sterol | 7α-OH | 7β-OH | 7-Keto | Epoxy | Triol | Others[b] | Total |
|--------|--------|-------|-------|--------|-------|-------|-----------|-------|
| PO/RO | Sitosterol | 3.9 | 3.3 | 3.8 | 111.0 | 3.5 | 22.7 | 212 |
|  | Campesterol | 2.0 |  |  | 61.9 |  |  |  |
| SO | Sitosterol | 8.1 | 9.8 | 11.8 | 5.2 | 3.7 | 24.5 | 65 |
|  | Campesterol | 1.2 | 0.3 |  | 0.4 |  |  |  |
| HOSO | Sitosterol | 12.1 | 13.0 | 18.4 | 9.2 | 10.9 | 41.4 | 110 |
|  | Campesterol | 0.9 |  |  | 4.0 |  |  |  |

[a]PO/RO = Blend of hydrogenated palm oil/rapeseed oil; SO = sunflower oil; HOSO = high oleic sunflower oil, prefried frozen samples were prepared by heating at 250°C for 15 min (22).
[b]Others = Unidentified sterol oxides.

**TABLE 7.12**
**Levels of Phytosterol Oxidation Products (mg/g)[a] in the Lipids of French Fries Samples Fried in Rapeseed Oil/Palm Oil Blend, Sunflower Oil, and High-Oleic Sunflower Oil, Prepared for Consumption by Frying at 200°C for 15 Minutes[b]**

| Sample | 7α-OH | 7β-OH | 7-Keto | Epoxy[c] | Triol | Total |
|--------|-------|-------|--------|----------|-------|-------|
| RP |  |  |  |  |  |  |
| Sitosterol | 2.9 | 3.7 | 4.1 | 7.6 | 0.5 | 32.0 |
| Campesterol | 1.4 | 2.6 | 3.3 | 5.4 | 0.5 |  |
| SO |  |  |  |  |  |  |
| Sitosterol | 3.8 | 7.3 | 13.1 | 2.2 | 1.1 | 36.9 |
| Campesterol | 0.3 | 1.3 | 5.9 | 1.3 | 0.6 |  |
| HOSO |  |  |  |  |  |  |
| Sitosterol | 4.7 | 9.7 | 13.5 | 5.4 | 2.8 | 53.7 |
| Campesterol | 1.4 | 1.8 | 9.2 | 3.6 | 1.6 |  |

Source: Dutta (45).
[a] Means of duplicate analyses.
[b] Reproduced from reference 9.
[c] Include both 5α, 6α-epoxy- and 5β, 6β-epoxy-sterols
RP, lipids from French fries fried in a blend of palm/rapeseed oil; SO, lipids from French fries fried in sunflower oil; HOSO, lipids from French fries fryied in high oleic sunflower oil.

The same pre-cooked samples of french fries, fried in various vegetable oils at 200°C, were investigated later cooked at lower temperatures for the content of phytosterol oxidation products (POP) (45). The levels of total POP products in the lipids of french fries fried at 200°C in rapeseed oil/palm oil blend, sunflower oil, and high-oleic sunflower oil were 32, 37, and 54 µg/g, respectively. 7-Ketositosterol and 7-ketocampesterol were the dominating oxidation products in all the samples,

**TABLE 7.13**
**Levels of Phytosterol Oxidation Products (mg/g)[a] at 0, 10, and 25 Weeks of Storage in the Lipids of Potato Chips Fried in Palm Oil[b]**

| Sterol | Weeks | 7α-OH | 7β-OH | 7-Keto | Epoxy[c] | Triol | Total |
|---|---|---|---|---|---|---|---|
| Sitosterol | 0 | 0.6 | 0.8 | 0.9 | 0.9 | 0.2 | 5.0 |
| Campesterol | | 0.3 | 0.2 | 0.7 | 0.3 | 0.1 | |
| Sitosterol | 10 | 0.5 | 1.0 | 1.3 | 0.8 | 0.2 | 6.1 |
| Campesterol | | 0.3 | 0.3 | 1.0 | 0.6 | 0.1 | |
| Sitosterol | 25 | 1.2 | 1.4 | 2.0 | 1.9 | 0.9 | 8.6 |
| Campesterol | | 0.1 | 0.2 | 0.7 | 0.2 | nd[d] | |

Source: Dutta and Appelqvist (44).
[a] Means of duplicate analyses.
[b] Reproduced from reference 8.
[c] Include both 5α, 6α-epoxy- and 5β, 6b-epoxy-sterols.
[d] < 0.1 mg/g not detected.

**TABLE 7.14**
**Levels of Phytosterol Oxidation Products (μg/g)[a] at 0, 10, and 25 Weeks of Storage in the Lipids of Potato Chips Fried in Sunflower Oil[b]**

| Sterol | Weeks | 7α-OH | 7β-OH | 7-Keto | Epoxy[c] | Triol | Total |
|---|---|---|---|---|---|---|---|
| Sitosterol | 0 | 4.4 | 9.9 | 16.1 | 3.5 | 1.2 | 45.8 |
| Campesterol | | 0.9 | 1.5 | 4.7 | 2.4 | 1.2 | |
| Sitosterol | 10 | 4.9 | 10.4 | 16.5 | 4.4 | 1.2 | 49.4 |
| Campesterol | | 1.1 | 1.9 | 6.0 | 2.1 | 0.9 | |
| Sitosterol | 25 | 6.6 | 8.8 | 15.7 | 4.6 | 0.8 | 47.1 |
| Campesterol | | 0.6 | 1.5 | 5.3 | 2.6 | 0.6 | |

Source: Dutta and Appelqvist (44).
[a] Means of duplicate analyses.
[b] Reproduced from reference 8.
[c] Include both 5α, 6α-epoxy- and 5β, 6β-epoxy-sterols.

followed by epimers of sterol-5,6-epoxides (Table 7.12). In addition to epimers of 7-hydroxysterols originating from both campesterol and sitosterol, both campestanetriol and sitostanetriol were present in relatively low levels at 1 μg/g in a palm oil/rapeseed oil blend, 1.7 μg/g in common sunflower oil, and 4.4 μg/g in high-oleic sunflower oil. In general, the content of POP in french fries prepared in different vegetable oils reflected those of oils used for the frying operations but much lower in total POP than those cooked at 250°C (Table 7.11).

The formation of POP was studied in potato chips fried exclusively in the vegetable oils mentioned above (44). The chips were prepared under industrial conditions and vacuum-packed and were stored for 25 weeks at room temperature

**TABLE 7.15**
Levels of Phytosterol Oxidation Products ($\mu$g/g)[a] at 0, 10, and 25 Weeks of Storage in the Lipids of Potato Chips Fried in High-Sunflower Oil[b]

| Sterol | Weeks | 7$\alpha$-OH | 7$\beta$-OH | 7-Keto | Epoxy[c] | triol | Total |
|---|---|---|---|---|---|---|---|
| Sitosterol | 0 | 2.8 | 7.8 | 10.9 | 3.1 | 1.7 | 35.1 |
| Campesterol | | 1.0 | 1.7 | 3.3 | 2.0 | 0.8 | |
| Sitosterol | 10 | 2.7 | 11.3 | 14.2 | 4.2 | 2.3 | 54.8 |
| Campesterol | | 1.3 | 2.4 | 12.2 | 2.9 | 1.3 | |
| Sitosterol | 25 | 5.3 | 11.5 | 18.1 | 6.3 | 3.6 | 58.5 |
| Campesterol | | 0.7 | 1.5 | 6.2 | 3.5 | 1.8 | |

Source: Dutta and Appelqvist (44).
[a] Means of duplicate analyses.
[b] Reproduced from reference 8.
[c] Include both 5$\alpha$, 6$\alpha$-epoxy- and 5$\beta$, 6$\beta$-epoxy-sterols.

(Tables 7.13– 7.15). The levels of total POPs in the lipids of chips fried in palm oil increased from 6 $\mu$g/g at 10 weeks to 9 $\mu$g/g at 25 weeks of storage (Table 7.13). The contribution from different sitosterol oxidation products was almost at equal levels from 0.6 to 0.9 $\mu$g/g, except for the sitostanetriol, which was at a considerably lower level, i.e., 0.2 $\mu$g/g in the lipids. The content of POP in chips fried in sunflower oil was virtually unchanged during storage of up to 25 weeks (Table 7.14). 7-Ketositosterol was present in the highest amount, about 16 $\mu$g/g, and remained almost entirely unchanged during storage. The content of POP in the chips fried in high-oleic sunflower oil was 55 $\mu$g/g after 10 weeks of storage. This amount increased to about 59 $\mu$g/g after 25 weeks of storage (Table 7.15). The proportions of different oxidation products of campesterol and sitosterol in the chips prepared in high-oleic sunflower oil were similar, also dominated by 7-ketocampesterol and 7-ketositosterol. Results from this study show that storage at 25°C up to 25 weeks did not increase the amounts of POP to a great extent, except for chips prepared in high-oleic sunflower oil, where the content of sterol oxidation products tended to increase after 10 weeks of storage (44).

Published data on the levels of POP in food products prepared in vegetable oils are very limited and so is the harmonized method to analyze them. The authors anticipate that in the future more research will be conducted in a neglected but important area that concerns public health and quality of life.

# Analytical Methodologies for Sterol Oxidation Products

The purpose of this section is to review techniques used for the analysis of sterol oxidation products in food and biological matrices. Presently, there is not an official method available to determine oxidized sterols in food products. Reference methods for the analysis of plant sterols in fats and oils, as part of unsaponifiables, are as follows: ISO 6799 (113), ISO 12228 (114), AOCS Ch 6-91 (115), IUPAC method 2.401 (116) and method 2.403 (117). Analysis of oxidized sterols is complicated by the fact that phytosterols are present in the free and conjugated form with fatty acids, phenolic acids, and sugars in food matrices. Furthermore, the minute amounts and very complex mixture of sterols and their oxidation products necessitates a multi-step analytical procedure, including: a) efficient extraction of all lipids from analyzed matrix; b) purification to eliminate impurities, ballast components, and enrich sterol oxides; c) isolation of sterols and oxidation components; d) detection and quantification of sterol oxidation products; e) confirmation of chemical structure and identity of oxidation products.

## *Extraction*

Sterols and their oxidation products are minor components in food lipid fractions, often including a complex mixture of many components. Knowing the history of the sample prior to the isolation of the lipid components is critical. For instance, it was observed that levels of 7-ketocholesterol in fresh human plasma increased from 24–60 ng/mL to 50–300 ng/mL when serum was frozen for up to 6 months (118). Dzeletovic et al. (119) observed that thawing plasma samples more than once can significantly increase the amounts of epoxy and triol derivatives of cholesterol. Butter oil usually contains approximately 30 μg/g each of 5α- and 5β-epimers of cholesterol, but when it was stored for a year at −20°C, the amount of these oxidized cholesterols increased to 90 and 60 μg/g, respectively (120).

Sterols and their oxidation products must be isolated from sample matrices in as pure and unchanged form as possible, primarily through extraction of all lipids. Exhaustive extraction of lipids is crucial in measuring sterols and their polar derivatives. The polarity of sterol oxides ranges widely from the almost nonpolar, 7-ketocholesterol, to the polar sterol hydroperoxides. Nonpolar solvents, such as hexane, quantitatively extract free and esterified sterols with fatty acids (121). However, some cholesterol oxides, namely 25-hydroxycholesterol and 5α-cholestane-3β,5,6β-triol, are not freely soluble in nonpolar solvents, such as petroleum ether and hexane (122). Moreau et al. (123) compared how hexane, methylene chloride, ethanol, and isopropanol extract polar sterol conjugate. Each solvent extracted > 95% of sterol esters and free sterols, the nonpolar derivatives. Polar fractions of sterol derivates extracted only partially, where efficiency increased with polarity of solvent applied. Most lipid analysts use a mixture of chloroform-methanol (1:2, v/v), with endogenous water in the sample, as a ternary solvent system to extract lipids from animal, plant, and bacterial tissue (124,125). With these solvents, a two-phase system is formed when equal volumes of water and chloroform are added to the extract of lipids. A separated bottom layer contains lipids in chloroform while the upper layer contains the contaminants.

The ratio between solvent volume and sample amount is also important and affects completeness of extraction. The minimum ratio of sample to extracting solvent is usually specified as 1:20 (wt/vol). Polar solvents showed the best extraction efficiency for different food matrices (124,125). Currently, there is an interest in the application of an isopropanol/hexane mixture (2:3, v/v) for lipids extraction, mainly because of its lower toxicity. However, lower extraction efficiency results when these solvents are used, compared to the previous system (126,127). Lipid extracts from different matrices tend to contain appreciable amounts of non-lipid contaminants, such as sugars, amino acids, salts, and others. These compounds must be removed before the lipids are analyzed due to the interference, possible interaction, and/or role as catalysts to initiate lipid deterioration (128). Each sample or matrix requires modification of the extraction procedure and/or solvent system in order to accommodate differences in characteristics, and to achieve a proficient isolation of all lipids. Detailed descriptions of extraction procedures for specific matrices are available (127,129,130).

Additional free sterols can be liberated from their more complex derivatives by acid and/or alkaline hydrolysis before extraction; however, the portion of freed sterols is usually small (131). When samples are already in the form of purified lipids, such as animal fats and vegetable oils, extraction is not required.

## Enrichment of Sterols and Their Oxidation Products

Total lipid extracts are crude preparations, with triacylglycerides (TG) and/or phospholipids (PL) as major components. Sterols are the minor components often present at 0.05–10% of the total lipids extract, while in plant origin lipids, different derivatives of sterols can be expected. To efficiently separate this complex mixture of sterol derivatives, particularly when oxidation products are analyzed, very efficient separation techniques are required. It is easier to analyze animal-origin lipids because they contain only one sterol, cholesterol. More than 30 different oxidation derivatives of this sterol have been identified (6).

## Saponification

The first step often used to separate TG and PL from sterols is saponification. The basic principle of saponification is to hydrolyze ester bonds of TG, PL, and sterol esters in an alkaline media, such as KOH in methanol. It is particularly important for phytosterols, often present in an esterified form in vegetable oil. This is the case for rapeseed/canola oil, both low and high erucic acid, where more than 50% of the all sterols are present in the form of esters with fatty acids (132). In samples containing only free sterols, saponification can be omitted.

Hot saponification, recommended by all official methods, is a concentration step used before separation and quantification by chromatographic methods (133,134). Tsai et al. (135) reported that 75% of $5\alpha,6\alpha$-epoxycholestanol disappeared after hot saponification when using the official procedure. Moreover, formation of epimeric forms of 7-hydroxycholesterol and 7-ketocholesterol as artifacts were reported (6,136). These official procedures were developed to analyze unsaponifiables, mainly sterols and tocopherols in fats and oils, where sterols are more stable than their oxidation products. Despite the damaging effect of conventional hot saponification on the integrity of some sterol oxides, this procedure is often used in the preparation of samples for sterol oxides analysis.

To prevent artifact formation or the decomposition of existing oxides, cold saponification was proposed (137). Park and Addis (138) reported that only negligible amounts of artifacts were formed using the cold saponification technique, but require longer time to saponify. Recovery of 7-ketocholesterol was close to 100% when added to a food sample and using the cold saponification technique for isolation (138,139). Removing oxygen from solvents, containers, and solutions, protection from light exposure, and performing saponification at ambient temperatures further eliminated artifact formation when standards were assessed (111). The application of all these precautions made this procedure laborious, so other, simpler ways to protect sterols and oxysterols from decomposition or formation of artifacts were developed. Nourooz-Zadeh (140) replaced chemical saponification with enzymatic hydrolysis of sterol esters to eliminate harsh conditions of chemical hydrolysis. This process is as slow and laborious as cold saponification.

Klatt et al. (141) proposed direct saponification in a sample and extraction of released sterols. This procedure, hot saponification, is a modification of an official AOAC method, but was only tested on intact sterols. Oxysterols assessment will need further developmental work (142,143).

The problems with efficient isolation of intact oxysterols do not end with saponification; extraction of saponifiables also should be evaluated for adequacy of the solvent applied. Neutral sterols and nonpolar oxides, such as 7-ketocholesterols, are extracted efficiently with nonpolar solvents including petroleum ether and hexane (144,145). Commonly reported sterol oxides are more polar than cholesterol, and some are not freely soluble in nonpolar solvents. Tsai (146) reported poor recovery of cholesterol 5α-epoxide after hot saponification, citing structural alteration of this component as responsible for it. The use of petroleum ether as a solvent to recover unsaponifiables can be responsible for the poor recovery of oxides. In the official methods cited above, hexane, diethyl ether, or a mixture of both, are recommended for extraction after saponification. Plant sterols are very soluble in a mixture of these solvents; however, oxides are poorly extracted by hexane. Diethyl ether also causes poor separation of layers. Additionally, relatively good solubility of this solvent in water further complicates separation of layers, a condition that is directly responsible for loss of sterols and their derivatives. Common cholesterol oxides extracted with diethyl ether showed higher than 80% recovery, while 5α-cholestane-3β,5,6β-triol recovery was below 60% (138). This clearly indicates the necessity of a polar solvent that also offers the best solubility for sterol components. Chloroform is a unique solvent for lipids and for extracting sterols; it offers easier separation of layers and thus better efficiency (131).

## Chromatographic Methods

Since sterols and their oxidation derivatives are more polar than TG, but less than PL, chromatographic techniques can be applied to separate those groups of components and concentrate them. Thin-layer chromatography (TLC) is frequently used to separate these fractions from other lipids. Usually, it is run in a preparative mode with a suitable solvent mixture to achieve separation. Visualized spots are then scraped, extracted with organic solvent, concentrated, and analyzed by gas chromatography (GC) or high performance liquid chromatography (HPLC) (140,144–150). The TLC technique is often used for the separation of free and esterified sterols from TG and PL by developing with 10% acetone in chloroform. Under these conditions, oxidized cholesterols are separated from PL, and the latter remains at the origin due to its insolubility in acetone. Oxide spots are scraped off and components extracted with 80% dichloromethane in methanol (151). The applicability of TLC as a substitute for saponification was demonstrated for the evaluation of oxides in biological samples (152). Cholesterol oxides were concentrated as individual fractions by TLC and quantified by GC/MS. This procedure is much milder than saponification, but it is labor-intensive and prone to error due to the multistep extraction required to fully recover components from TLC spots. This separation technique is also used to separate sterol fractions from oxidized sterols after saponification, the previous must be removed due to their high concentration and interference during GC/MS analysis (139).

Column chromatography with ion-exchange packing was also used. Nourooz-Zadeh and Appelqvist (153) utilized trimethylaminohydroxypropyl-Lipidex packing to separate sterol fractions. This method again proved laborious this time because the packing has a very low capacity, limiting the amount of sample to be separated. This necessitates using columns with large diameters to achieve a fraction size sufficient to analyze by GC or HPLC, and a large volume of solvents to elute fractions.

Silicic acid has long been used for separation of lipid classes. A disposable pipet with 1g silicic acid was used to separate oxide fractions from egg yolk lipids by eluting with 20 mL hexane/ethyl acetate (5:3, v/v) (135). Park and Addis and De Vore (110, 154) modified this procedure by incorporating two hexane/ethyl acetate washings to remove TG and cholesterol while using acetone to elute oxides free of PL. A further improvement in the isolation of the oxide fractions was provided by introduction of prefabricated silica cartridges (Sep-Pak); however, silica activity, controlled by the amount of water, is limited and varies with manufacturers. Amount of water directly affects performance and separation of lipid components, and each batch of these columns require testing for separations (155,156). Both TLC and classical column chromatography require large volumes of solvents, are time consuming, and cannot be used where humidity is not controlled due to its influence on silica activity (157).

Recently, several effective methods applying solid phase extraction (SPE) were developed to isolate purified sterols and their derivatives. Solid-phase extraction is very effective for the separation of esterified sterols and their oxidation products into individual fractions (158–164). Using a small volume of solvents and cartridges packed with neutral alumina, aminopropyl, and octadecyl modified silica, fast and efficient separation is accomplished without artifact formation and with recovery higher than 95% (Table 7.16).

Kou and Holmes (158) used a $C_{18}$ Sep-Pak column to pre-purify cold saponified cholesterol derivatives, which were further analyzed by HPLC. Use of coated silica improved separation and recovery due to the controlled interaction with coating molecules. Bonded silica prevented bonding of highly polar material to the packing, a common problem with conventional silica (157). Sevanian and McLeod (159) applied diol-columns with a mobile phase of toluene/ethyl acetate (3:2, v/v) to isolate oxidized cholesterol from liposomes. This isolation resulted in 99% recovery. When this isolation procedure was combined with enzymatic saponification, a very effective tool was developed to enrich oxidized cholesterol without changes before GC analysis (140).

Differently sized molecules were also used to separate sterols from TG and PL. Human plasma lipid extracts were separated on gel chromatography with Sephadex LH-20 prior to quantifying vitamin D by HPLC. This component has a structure similar to 25-hydroxycholesterol, which was successfully separated under these conditions. Limitations of Sephadex are: low sample capacity requiring multiple separations to achieve a sufficient amount of material for further analysis, and use of large volumes of solvents. Removing excess of solvent to concentrate analytes may be a source of artifacts formation (35,153,165).

HPLC on normal and reversed-phase columns was used to isolate normal

**TABLE 7.16**
**Methods Used for Isolation and Separation of Sterol Oxides**

| Sample | SPE Mode | Solvent | Saponification | Type of sterols | Reference |
|---|---|---|---|---|---|
| Edible Oils and Fats | Neutral Alumina | DEE/HEX | Yes | FS; SE | 160 |
| Oils | Silica | Acetone | Yes | FS; SE; OS | 46 |
| Wheat | $C_{18}$ | $CHCl_3/CH_3OH$ | Yes | FS; SE; SD | 161 |
| Milk | Amino | Acetone | Yes | OS | 162 |
| Oils | Silica | Acetone | Yes | OS | 60 |
| Oils | Silica | DEE/$CH_3OH$ | Yes | OS | 91 |
| Cheese, Pork[a] | Amino | Acetone | Yes | OS | 163 |
| Dry Milk, Butter | Amino, Silica | HEX/EAAcetone | No | OS | 164 |

Abbreviations: DEE = Diethyl ether; HEX = hexane; FS = free sterols; SE = esterified sterols; SD = sterol derivatives, other than SE; OS = oxidized sterols.
[a] Lipids were transesterified with sodium methoxide.

and oxidized sterols (148,166,167). A normal-phase column with silica was used in earlier analyses to obtain purified fractions of oxidized cholesterols, which were further analyzed by GC, and/or to separate individual components directly (136). Improved separation of the oxidized sterols fraction was achieved on a column with aminopropyl packing run in normal mode where silica was deactivated (140,166). Separation of all individual oxidized cholesterol derivatives on one HPLC column under isocratic conditions is difficult to achieve due to a complicated mixture of components requiring very high-resolution packing (159,168). Application of a complex gradient elution system, with nonlinear programming and a high-efficiency column, permitted separation of ten major cholesterol oxidation products, including some epimeric forms (169). Sevanian and McLeod (159) achieved similar separation efficiency using isocratic conditions, 3-μm particle size and standard length of column. These two separation techniques can be used for biological samples; however, their application to the assessment of oxysterols in food products requires further testing due to a complex mixture of various components.

The spectrophotometric detector in ultraviolet (UV) regions is the most commonly used detector in HPLC because of its relatively high sensitivity and specificity. Most sterol oxides are not strong UV absorbers, unlike their parent components, and require detection below 210 nm. This not only limits specificity but also restricts the solvents used for the mobile phase because of increased absorption. UV detector and the similar version—photodiode detector—are able to detect cholesterol oxides at levels as low as 40 ng (110,158,170). Oxidized sterols such as 7-ketosterol with conjugated double bonds, enables detection at 235–240 nm range to the nanogram level (110,155). Absorbance for 7-ketocholesterol was reported to be twice as high as that of dihydroxy derivatives and 25-hydroxycholesterol (171). Quantitative analysis requires a linear response of the detector within quantitation range for all components analyzed. The UV detector does not respond equally to all components because its signal depends on the type of chromophore present in a molecule. Thus, it is important to know the response factor and to calibrate for each component analyzed.

Sensitivity of the spectrophotometric detector is improved when sterol oxidation products are derivatized with components having a strong chromophore, such as benzoic acid. Derivatization is usually performed prior to separation on HPLC, and the detection limit can be lowered by one order of magnitude (172). Sensitivity improves further by derivatization to anthronyl fluorescent components. Yokohama et al. (165) developed a highly selective reagent that quantitatively reacts only with the hydroxyl group in position 3 of the cholestane molecule. Using this derivative, the detection limit with the fluorescence detector was reduced to the range of 10 pg.

Neither 5,6-epoxides nor triols have sufficient absorption even below 210 nm and are usually detected with other detectors, such as lower sensitivity refractive index (168). This detector can be used to quantify cholesterol epoxides and has a detection limit at about 5 μg. When a higher-efficiency column was applied with less band broadening, the detection level decreased tenfold to 0.5 μg (159,167). Note that the performance of another HPLC detector, wire transported FID, has been tested for the detection and quantification of oxides. A good linear response was found for ten common cholesterol oxides over the range of 0.39–100 µg per component, with the lower value representing the detection limit (169).

Hydroperoxides of cholesterol were also detected using an electrochemical detector after HPLC separation, with a detection limit in the range of 25 pmol (173). This particular detector can only detect peroxides because it measures change in electrical potential, and other oxidation products cannot be measured.

Recently, the evaporative light scattering detector has been applied for sterol component detection after HPLC separation. With this detector, ng amounts of sterol derivatives were detected (174). HPLC is more often used to separate different types of sterol derivatives, which are endogenous components present in food products and raw materials. The main reason for the lack of HPLC methods for oxidized sterols is its limited separation power, which is lower than GC capillary columns, and the inability to separate complex mixtures. HPLC is often used as a preliminary separation method to collect fractions of components where separation is based on differences in polarity or affinity of a group of components to packing. Applying these principles, different sterol groups of compounds can be isolated/purified for further analysis on GC, HPLC, and MS (175).

A very useful method for identification of complex sterol mixtures, including oxysterols, involves capillary HPLC with MS as the detector (6,176). This detection system provides information about the molecular structure of components analyzed and their purity, although MS for HPLC is still very expensive and not readily available. New developments in MS software allowing deconvolution of peaks, isolation of individual components in chromatographic non-separated mixture, provides an additional tool in the identification of components. Development of an ion trap, time-of-flight and cyclotron ion separation detectors for MS may provide a reasonably priced spectrometer for use with HPLC and GC and offer new capabilities for the separation and identification of complex mixtures. Particularly promising is the development in cyclotron resonance mass spectrometry where prior separation of components is not required as in other techniques. The ability to identify components is due to extremely high resolution of this type of spectrometer. Wu et al. (177) were able to separate and identify about 2000 components in oils using "direct injection" of oil sample into this mass spectrometer. The cyclotron spectrometer has resolution at 2 ppm meaning six decimal places of molecular mass. This identification method will allow fast and accurate analysis of lipids, based on identification and quantification without prior separation and manipulation of the sample. Furthermore, if some separation techniques are applied prior to MS, such as separation of lipid classes by HPLC, even further enhanced identification and quantification potentially exists (177). The main limitation of this technique today is the cost of instrumentation and availability. Currently, only two companies offer a market version of the mass spectrometer.

## Gas Chromatography

Gas chromatography (GC) is used to separate and quantify cholesterol oxides in biological and food samples more often than any other technique. Oxysterols are detected mainly with a flame ionization detector (FID) because of its universality and ability to detect components at the ng-pg level. The most effective separation can be achieved using capillary columns with nonpolar, e.g., 5% phenyl and 95% dimethyl polysiloxane (Rtx-5m), and polar, 14% cyanopropylphenyl and 86% dimethyl polysiloxane (Rtx-1701) phases (137,139,140,147,153,178). Separation of some oxysterols on polar and nonpolar phases in Fig. 7.7 and 7.8 are presented (179).

Separation of the same oxidation products on two different polarity columns showed small changes in retention time and pattern of components, where on polar columns the elution order changed for 19-hydroxycholesterol, cholesteroltriol, and 5α-epoxycholesterol. Furthermore, different co-elution of oxysterols on both columns was observed (Fig. 7.7 and 7.8).

A phase film thickness of 0.10–0.25 μm, with hydrogen as the carrier gas, are preferable to protect sterols from decomposition and maintain as low an elution temperature as possible (180). To obtain better resolution and higher capacity, a thicker film layer of phase is typically applied. However, this requires higher separation temperatures and with it is expected higher phase bleeding and the potential of component thermal decomposition. Thicker phase layer results in longer analysis time but with better separation of oxides and higher column capacity. Also, changes in the elution order of components have to be expected, compared to standard columns (139). Capillary columns, particularly with phases containing phenyl, successfully separate most cholesterol oxidation products (178). Phytosterols usually contain a mixture of four main sterols. Hence, there is a possibility of many more oxidation products, which in turn dramatically increase the possibility of overlapping due to the set separation potential of the column (Fig. 7.7 and 7.8).

Oxysterols are generally analyzed as trimethylsilyl (TMS) derivatives due to their higher volatility and stability during separation at temperatures higher than 250°C (180–182). For TMS derivatization, different reagents are used, including N,O-bis-trimethylsilylacetamide (BSA), N-trimethylsilyl imidazole (TMSI), and bis-trimethylsilyl- trifluoroacetamide (BSTFA). Trimethylchlorosilane (TMCS) is usually used as a catalyst, and derivatization occurs in pyridine under anhydrous conditions to prevent ether bond hydrolysis. In oxysterols, hindered hydroxy groups are possible and enough aggressive derivatization agent has to be applied. Underivatized sterol were also separated by GC, but thermal alteration of analyzed compounds can occur during the chromatographic run, causing problems with separation of epimers and other thermally labile components (178,181,183).

A split injection mode is frequently used to introduce a sample onto a capillary column resulting in problems with discrimination, to which components with high boiling points are particularly prone. Use of split injection requires that the temperature of the injector be about 10–20°C higher than the highest temperature of the column, usually in the range of 270–340°C or higher. This can cause "thermal shock" and thermal alteration to analytes (184). Maeker and Unruh (136) used an on-column injection technique, which eliminates discrimination so that all injected components are transferred equally onto the column. This method, however, suffers from tailing of the solvent peak and potential thermal degradation of compounds due to applied high temperature and long exposure time. A programming temperature vaporizing injector (PTV), a modified on-column injection system, where the sample is injected at ambient temperature and injector temperature is increased rapidly, injects thermally sensitive components with minimal "thermal shock" (111). This injector transfers analytes into the capillary column under the gentlest conditions possible and eliminates the problems of both split and on-column injection systems. It also concentrates the sample, effectively increasing the sensitivity of the GC detector by forming and eluting narrow bands of components.

Quantification of components by GC is usually done with FID, the most commonly utilized detector with a wide range of linear response. The relative response factors (RRF) and linearity of detector response have to be established for each component as part of quantitative analysis protocol (179). Despite the critical role of the RRF in quantification, how calibration was performed and what factors were used for calculations are rarely described in the literature. In most cases, a RRF equal to 1.0 is applied for all analyzed components. However, actual response factors for

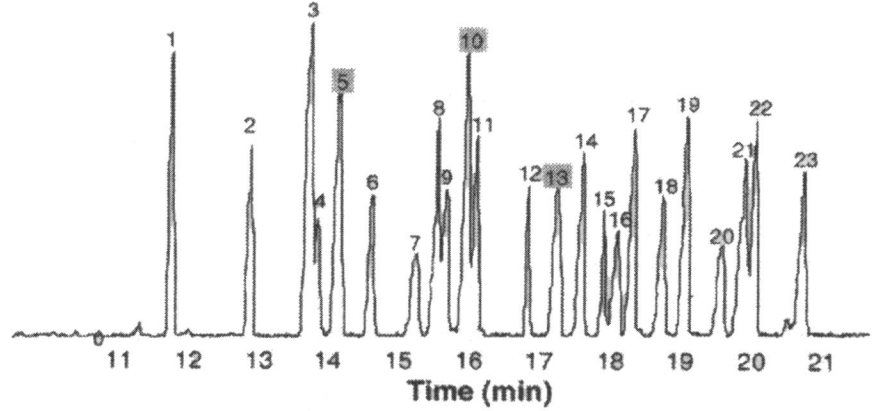

**Fig. 7.7.** Separation of major oxidation products formed from cholesterol, campesterol, stigmasterol and β-sitosterol on the 50% methyl and 50% phenyl polysiloxane (DB-17, 30m × 0.25mm, 0.15μm; column temperature: 90°C–1min; 30°C/min to 270°C–1min and at 3°C/min to 300°C – 12 min).
1–5α cholestane; 2 – 7α-hydroxycholesterol; 3 – 19-hydroxycholesterol; 4 - 7α-hydroxycampesterol; 5–7α–hydroxystigmasterol + 7β-hydroxycholesterol; 7β-hydroxystigmasterol; 9–5β-epoxycholesterol; 10 - 5α-epoxycholesterol + cholesteroltriol; 11 - 7β-hydroxysitosterol; 12–5β-epoxycampesterol; 13–5α-epoxycampesterol + 5β-epoxystigmasterol; 14–5α-epoxystigmasterol; 15–5β-epoxysitosterol; 16–7-ketocholesterol; 17–5α-epoxysitosterol; 18-campesteroltriol; 19-stigmasteroltriol; 20–7-keto-campesterol; 21-sitosteroltriol; 22–7-ketostigmasterol; 23– 7-ketositosterol. Adapted from Apprich and Ulberth (179).

systems used in the analysis are published (139,153,178,185) (Table 7.17). As can be observed from Table 7.17, RRF lower and higher than 1.0 can be expected and applied for proper quantification of oxysterols. Also, RRF is dependent on the type of internal standard used.

Quantitative GC demands that all analytes produce peaks with reproducible areas at the concentration analyzed. When analytes undergo thermal decomposition and/or irreversible adsorption in the system, calibration curves may not pass through the origin and/or cannot be linear. These effects are corrected by deactivation of the analytical system, namely, the injector insert, column, and detector passage, by silylation of the system or by using silanized parts (184).

Variability in analytical procedure can be compensated for by incorporating an internal standard with a similar chemical character as the analytes. The internal standard should go through all preparation steps and be separated from all analyzed components. For best results, the standard peak should appear between analyzed components. Often, a sterol component not present in the samples being analyzed can be used as a standard, mainly α-cholestane and nonpolar sterol; some polar derivatives of cholesterol, 19-hydroxycholesterol, are often used when oxides of phytosterols are analyzed. Polar internal standards were also proposed such as hydroxy derivatives of cholesterol for assessment of plant sterols (121,138,139,175).

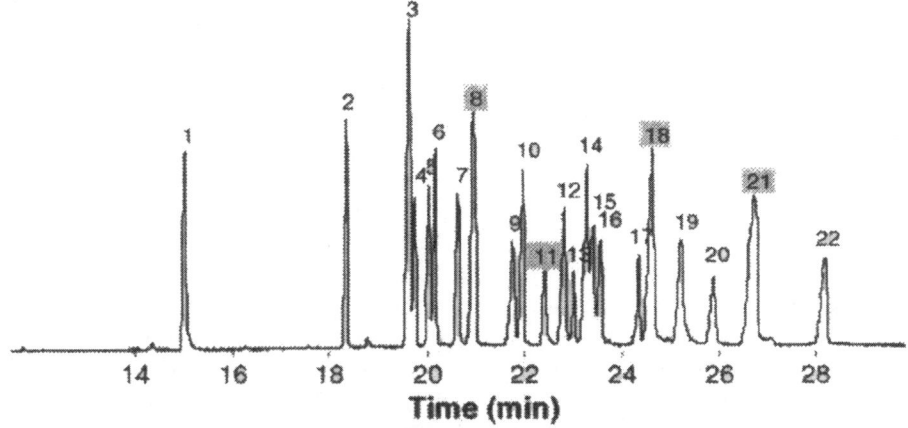

**Fig. 7.8.** Separation of cholesterol, campesterol, stigmasterol and β-sitosterol major oxidation products on 95% dimethyl and 5% diphenyl polysiloxane (HP-5, 30 m × 0.25 mm; 0.25 μm, temperature program: 90°C–1min; 30°C/min to 270°C–1 min and 3°C/min to 300°C–12 min). 1–5α-cholestane; 2–7α-hydroxycholesterol; 3–19-hydroxycholesterol; 4–7α-hydroxycampesterol; 5–7α-hydroxystigmasterol; 6–7β-hydroxycholesterol; 7–5β-epoxycholesterol; 8–5α-epoxycholesterol + 7α-hydroxysitosterol; 9–7β-hydroxycampesterol; 10–7β-hydroxystigmasterol; 11–cholesteroltriol + 5β-epoxycampesterol; 12–5α-epoxycampesterol; 13–5β-epoxystigmasterol; 14–7β-hydroxysitosterol; 15–5α-epoxystigmasterol; 16–7-ketocholesterol; 17–5β-epoxysitosterol; 18–5α-epoxysitosterol + campesteroltriol; 19–stigmasteroltriol; 20–7-ketocampesterol; 21–7-ketostigmasterol + sitosteroltriol; 22–7-ketositosterol. Adapted from Apprich and Ulberth (179).

**TABLE 7.17**
**Response Factors for Trimethylsilyl Ethers of Sterol Oxidation Products**

| Sterol Derivatives | RRF | RSD (%) |
|---|---|---|
| 7α-hydroxy cholesterol | 1.037 | 0.8 |
| 7β-hydroxy cholesterol | 0.966 | 0.7 |
| 7α-hydroxy campesterol | 1.110 | 1.8 |
| 7β-hydroxy campesterol | 1.090 | 1.1 |
| 7-ketocampesterol | 1.258 | 3.7 |
| 7-ketostigmasterol | 1.261 | 2.5 |
| 7-ketositosterol | 1.282 | 1.3 |
| Campesterol-triol | 1.205 | 1.7 |
| Stigmasterol-triol | 1.198 | 3.3 |
| Sitosterol-triol | 1.206 | 1.1 |
| α-epoxycampesterol | 1.137 | 2.8 |
| α-epoxystigmasterol | 1.137 | 2.8 |
| α-epoxysitosterol | 1.139 | 2.6 |

Adapted from Apprich and Ulberth (179).

## Mass Spectrometry

Generally, "identification" in chromatographic methods is based on the comparison of retention data between compounds analyzed and authentic standards, if available. Re-running actual samples with added standard(s) improves identification confidence when the peak of the evaluated component is augmented. It is also possible that more than one compound has the same or very similar retention time and multiple components may appear as one peak. Consequently, confirmation of the chromatographically separated peaks must be performed by other methods, such as mass spectrometry (MS) and/or infrared spectrometry (IR). Capillary GC with MS as a detector has emerged as the most powerful identification and separation technique available today. Similarly, MS as a detector for HPLC is a very powerful tool. In this separation method, oxides are separated under relatively mild conditions and without derivatization; however, the separation efficiency of HPLC columns is much lower than capillary GC and good separation rarely occurs. Mass spectrometry (MS), with electron impact (EI) or chemical ionization (CI), can provide fragmentation spectra, chemical structure, and identity of the separated molecules of sterol oxides (186,187). Similar modes of MS are applied for GC as discussed for HPLC. Using high resolution MS provides data about elemental structures of molecules and makes it easier to define the chemical formula. Molecular fragmentation data are more often available for silylated sterols and oxysterols than for underivatized components (43, 137,166). Typical diagnostic fragments for oxidized sterols using EI in Table 7.18 are discussed (43,46,49,138,152,178,179,188–194).

**TABLE 7.18**
**Some Diagnostic Ions for Trimethylsilyl Derivatives of Oxidized Sterols**

| | | Molecular Mass of Ions (daltons) | | | |
|---|---|---|---|---|---|
| Oxide | Typical Ions | Cholesterol | Stigmasterol | Sitosterol | Campesterol |
| 7α(β)- Hydroxy | (M -TMS)+ | 456 | 482 | 484 | 470 |
| 7-Keto | M+; (M – TMS)+ | 472, 367 | 498, 408 | 500, 410 | 486, 396 |
| 5α(β) - Epoxy | M+; (M – TMS)+ | 474, 384 | 500, 410 | 502, 412 | 488, 398 |
| Triol | (M – TMS)+, (M – A ring)+ | 547, 404 | 572, 429 | 574, 431 | 560, 417 |

Abbreviations: M+ = molecular ion; TMS = trimethyl silyl group. Adapted from references: 46, 49, 179, 194.

Dutta and Appelqvist (22) and Bortolomeazzi et al. (49) analyzed phytosterol oxidation product standards by MS and obtained mass spectra having fragmentation patterns similar to respective cholesterol derivatives. The ions, however, had different molecular masses due to the different structures of the parent molecules. For quantification, selected ion MS or mass fragmentography is often used to confirm identity and to quantify oxysterols in complex mixtures. This approach offers much lower detection limits than full spectrum MS (43,46,195,196). The MS method discussed above is related to the detection of positive ions formed during fragmentation of analyzed molecules. Negative ion MS and selective negative ion spectrometry were also used to quantify and identify oxidized sterols; however, this technique is not popular and requires more work to establish fragmentation of oxysterols (43).

Recently, a combination of HPLC, GC, and MS was used to analyze free and esterified sterols in vegetable oil. Sterol fractions, and their esters, were directly transferred from a liquid chromatograph to a gas chromatograph for separation on a capillary column, allowing the quantification of each free and esterified sterol using MS as a detector (180,197,198). This combination of the best separation and detection techniques is extremely powerful but requires further developments for the evaluation of oxysterol.

# Health Implications of Cholesterol- and Phytosterol-Oxidized Products

The formation of cholesterol oxidized products (COP) in many processed foods is of increasing concern. Recent attention has focused on the oxidized products of phytosterols as well. This section will discuss the health implications of phytosterol-oxidized products in addition to those reported for cholesterol-oxidized products.

## Cholesterol-Oxidized Products

Cholesterol-oxidized products (COP) formed in many processed foods, including deep fat frying, raises serious health concerns. These compounds, also referred to as oxysterols, have a hydroxyl- or a keto-group on the cholesterol molecule and include 7-OH, 7-keto, 19-OH-, 22-OH, and 25-OH cholesterol. In addition to being mutagenic, cytotoxic, and carcinogenic, COP also inhibit cholesterol biosynthesis, causing arterial damage (16,199–203). As a result, increased attention has focused on the need to eliminate or prevent the formation of COP in the human diet.

## Cholesterol Biosynthesis

The ability of COP to inhibit cholesterol biosynthesis is attributed to their effect on 3-hydroxy-3-methyl glutaryl CoA reductase (HMG-CoA reductase). The latter is a key enzyme regulating biosynthesis of cholesterol. A study by Peng et al. (204) found that 25-hydroxycholesterol was the most potent inhibitor of this enzyme, followed by cholestanetriol. Earlier work by Brown et al. (205) showed that suppression of cholesterol biosynthesis by 25-hydroxycholesterol retarded the growth of cultured cells in human fibroblasts. The overall effect of COP may be to decrease cholesterol in the cell membrane, resulting in altered membrane function or disruption and premature cell death, which in turn may lead to necrosis, abnormal cell proliferation, and the formation of atheromas (206).

Osada and co-workers (207) found a significant reduction in food intake, body weight gain, and relative liver weights in young (4 wk) and adult (8 mo) rats fed diets containing 0.5% oxidized cholesterol. In addition, the activity of two key enzymes involved in cholesterol biosynthesis and catabolism, HMG-CoA reductase, and cholesterol 7αhydroxylase in liver microsomes, were both lowered, particularly in adult rats. As expected, the level of cholesterol was reduced in both serum and liver. However, in sharp contrast was 6-desaturase, a key enzyme involved in the metabolism of linoleic acid to arachidonic acid, increased significantly in adult rats fed oxidized cholesterol. Thus, oxidized cholesterol could influence eicosanoid production from arachidonic acid.

## Atherosclerosis

Deposition of cholesterol in arteries and aorta, resulting in atherosclerotic plaque from increased cholesterol consumption, was first established by Anitschow (208). This model of plaque formation was accepted for many years. Schwenk and co-workers (209), however, reported that oxidation products of cholesterol were much more atherogenic. Nevertheless, the role of cholesterol in plaque formation was still the accepted model until Imai et al. (14,210) demonstrated the primary role that oxygenated sterols played in arterial wall injury and lesion development. Subsequent feeding, venous infusion, and in vitro studies confirmed that cholesterol alone showed little atherogenic or cytotoxic effects compared to cholesterol oxidation products

(211–213). Other researchers showed that feeding oxidized cholesterol, cholestanetriol, to pigeons increased the incidence of atherosclerotic plaque and aortic accumulation of calcium, while intravenous injection of 25-hydroxycholesterol induced lesions on aortic surfaces of rabbits (201,214).

Vine and co-workers (215) also confirmed the pro-atherogenic properties of dietary oxysterols compared to unoxidized cholesterol by their incorporation into the plasma triglyceride-rich lipoproteins in rabbits, increasing aortic cholesterol concentration.

The possible role of arachidonate metabolism in the development of atherosclerosis has attracted considerable attention (216). Production of $PGI_2$ by endothelial cells decreases platelet adhesion to the blood vessel wall surface and subsequent platelet aggregation, exerting a protective effect against atherosclerosis. Alteration of arachidonate metabolism was reported as a result of damage to vessels (217). Many factors contribute to dysfunction or injury to endothelial cells, including cholesterol oxides. Using human umbilical vein endothelial cell cultures, Peng and co-workers (218) showed that inhibition of $PGI_2$ production by 5,6-epoxide and cholestane-3,5,6-triol reached a maximum after 12 h, while inhibition by 25-hydroxycholesterol and 7-ketocholesterol reached a maximum after 24 h. These differences were attributed to the direct membrane sterol replacement of cholestane-3,5,6-triol and 7-ketocholesterol on inhibition of cholesterol biosynthesis. Because $PGI_2$ functions as a vasodilator and inhibitor of platelet aggregation, any decrease in its production is expected to enhance platelet and leukocyte adhesion. In the same study, it was noted that platelet adhesion was enhanced in the presence of endothelia exposed to COP. These and other studies all pointed to the role of cholesterol oxidation products as powerful atherogenic agents (37,206,219,220).

## Cytotoxic and Mutagenic Properties of COP

The cytotoxic properties of many COP are widely recognized (11,221–223). Their formation in membranes occurs during lipid peroxidation and includes 7-keto, 7-hydroxy-, and 25-hydroxy-cholesterols, together with epoxides. Sevanian and Peterson (224) studied the cytotoxicity and mutagenicity of COP in bovine membrane preparations. Among the oxidized products identified were 5,-6-epoxides and 7-ketocholesterol. These were found to increase under conditions of oxidant stress or antioxidant deficiency. The weak electrophilic nature of cholesterol epoxides was thought to account for their mutagenic activity; however, these epoxides can be converted by cholesterol hydrolase to cytotoxic cholestanetriol. Carboni and co-workers (225) and Hrelia et al. (226) both showed that supplementation of cultured cardiomyocytes with cholesterol-5,6-epoxide reduced cellular protein levels without affecting viability, and that cholestanetriol was detected in the lipid fraction of the supplemented cells. Because of the extreme toxicity of cholestanetriol, Carboni and co-workers (227) examined its in vitro effects on cultured rat cardiomyocytes. These researchers also found a significant reduction in cell protein content with no effect on cell viability; theirs was the first report on triol toxicity of cardiomyocytes.

A model using 25-hydroxycholesterol (25-OHC), over a concentration range of 1–10 μM, was found to induce apoptosis in monocytemacrophage (228,229) and in lymphanoid cell lines (230,231). Another model using 7-ketocholesterol induced apoptosis in vascular endothelial and smooth muscle cells (232). Rusinol and co-workers (233) showed that the cytotoxic agent in oxidized LDL was an oxysterol which enhanced reflux of calcium. Further work by Panini et al. (234) reported that either oxidized LDL or 25-OHC activated cytosolic phospholipase $A_2$ ($cPLA_2$). The latter, an enzyme in signal transduction pathways activated by calcium, was previously found to induce apoptosis by the tumor necrosis factor (235,236). Panini and co-workers (234) showed that oxidized LDL or 25-OHC activated cPLA2 in both mouse macrophages and fibroblasts.

### Oxidized Phytosterol

The structures of phytosterols, as discussed earlier in this chapter, are analogous to cholesterol and therefore would be expected to oxidize in a similar manner. Little information, however, is available on the potential toxicity of phytosterols relative to COP. Using cultured macrophage-derived cell lines Adcox and co-workers (237) compared the relative toxicities of a 50%/40% mixture of β-sitosterol and campesterol and their oxides with COP. Similar patterns of toxicity were observed with respect to LDL leakage, cell viability, and mitochondrial dehydrogenase activity. Cell damage was greatest in the presence of 5α,6α-cholesterol epoxides and other cholesterol oxides followed by β-sitosterol/campesterol oxides, cholesterol and β-sitosterol. Thus the ability of phytosterols to oxidize during frying has the potential to cause cellular damage. A recent study by Lea et al. (238), however, showed that a phytosterol concentrate containing 30% phytosterol oxides did not exhibit genotoxic potential or any evidence of subchronic toxicity when fed to rats over 90 consecutive days. The no-observed-effect-level or NOEL established for phytosterol oxides by Lea and co-workers (238) was 128 mg/kg/day for male rats and 144 mg/kg/day for female rats. Based on this data, phytosterol oxides were not considered to be an obvious concern to human health.

## Summary

Oxidation products of cholesterol are of considerable interest because of their possible effects on human health. As more vegetable oil is used for cooking, it is necessary to consider the occurrence and effect of plant sterol oxides. This became particularly apparent when it was observed that vegetarians absorbed more phytosterols than did those on non-vegetarian diets. This implies that the effects could be greater than previously perceived. Work with oxidation products of plant sterols has expanded due to the availability of analytical methods. Separation of very complex mixtures of parent phytosterols and their oxidized products requires techniques with very high resolution. In addition, the presence of very low concentrations of these compounds, and their lability, requires new, more efficient and milder isolation techniques. Such improvements have permitted a greater examination of phytosterol oxides in foods that should lead to more comprehensive studies of their absorption and metabolism.

## References

1. Kiribuchi, T.; T. Mizunaga; and S. Funanashi. *Agric. Biol. Chem.* **1966,** *30,* 770.
2. Lepage, M. *J. Lipid Res.* **1964,** *5,* 587.
3. Swern, D. In *Bailey's Industrial Oil and Fat Products,* 4th ed.; John Wiley & Sons: New York, 1979; Vol. 1, p. 53.
4. Kochar, S.P. *Prog. Lipid Res.* **1983,** *22,* 161.
5. Smith, L.L. In *Cholesterol Autoxidation*; Plenum Press: New York, 1981; p. 49.
6. Smith, L.L. *Chem. Phys. Lipids* **1987,** *44,* 87.
7. Vuoristo, M.; T.A. Miettinen. *Am. J. Clin. Nutr.* **1994,** *59,* 1325.
8. Kochhar, S.P. In *Frying—Improving Quality*; Woodhead Publishing: Cambridge, U.K., 2001; p. 87.
9. Normén, L.; J. Froblich; and E. Trautwein. In *Phytosterols as Functional Food Components and Nutraceuticals*; P.C. Dutta, Ed.; CRC Press: Boca Raton, FL, 2004; p. 243.
10. Moreau, R.A. *Ibid.*, p. 317.
11. Sevanian, A.; A.R. Peterson, *Proc. Natl. Acad. Sci.* **1984,** *81,* 4198.
12. Raaphorst, G.P.; E.I. Azzam; R. Lanlois; and J.E. van Lier. *Biochem. Pharm.* **1987,** *36,* 2369.
13. Kandusch, A.; H. W. Chen; and H. J. Heiniger. *Science* **1978,** *201,* 498.
14. Imai, H.; N.T. Werthessen; V. Subramanyam; P. W. LeQuesne; A.H. Soloway; and M. Kanisawa. *Science* **1980,** *207,* 651.
15. Peng, S.K.; C.B. Taylor. *World Rev. Nutr. Diet.* **1984,** *44,* 117.
16. Addis, P.B. *Food Chem. Toxicol.* **1986,** *24,* 1021.
17. Grandgirard, A. In *Cholesterol and Phytosterol Oxidation Products: Analysis, Occurrence, and Biological Effects*; F.P. Guardiola, P.C. Dutta, F. Codony, and G.P. Savage, Eds.; AOCS Press: Champaign, IL, 2002; p. 375.
18. Ratnayake, W.M.; E. Vavasour. In *Phytosterols as Functional Food Components and Nutraceuticals*; P.C. Dutta, Ed.; CRC Press: Boca Raton, FL, 2004; p. 365.
19. Dean, L.O.; L.C. Boyd. *Ibid.*, p. 419.
20. Maeker, G. *J. Am. Oil Chem. Soc.* **1987,** *64,* 388.
21. Nourooz-Zadeh, J.; L.-Å. Appelqvist. *J. Am. Oil Chem. Soc.* **1992,** *69,* 288.
22. Dutta, P.C.; L.-Å. Appelqvist. *inform* **1995,** *6,* 500.
23. Smith, L.L. In *Biological Effects of Cholesterol Oxides*; S.-K. Peng and R.J. Morin, Eds.; CRC Press: Boca Raton, FL, 1992; p. 7.
24. Przybylski, R.; W. Li. *inform* **1995,** *6,* 499.
25. Korahani, V.; J. Bascoul; and A. Crastes de Paulet. *Lipids* **1982,** *17,* 703.
26. Yanishlieva, N.; H. Schiller; and E. Marinova. *Riv. Ital. Delle Sost. Grasse* **1980,** *57,* 572.
27. Bascoul, J.; N. Donergue; and A.C. de Paulet. *J. Steroid Biochem.* **1983,** *19,* 1779.
28. Nes, W.R. *Adv. Lipid Res.* **1983,** *15,* 233.
29. Nourooz-Zadeh, J. Cholesterol Oxides in Food: Analytical Methods, Levels in Some Swedish Foods and Food Safety Aspects; Ph.D. Thesis, Swedish University of

Agricultural Sciences, Uppsala, Sweden, 1988; p. 9.

30. Lercker, G.; M.T. Rodriguez-Estrada. In *Cholesterol and Phytosterol Oxidation Products: Analysis, Occurrence, and Biological Effects*; F. Guardiola, P.C. Dutta, F. Codony, and G.P. Savage, Eds.; AOCS Press: Champaign, IL, 2002; p. 1.

31. Breuer, O. Oxysterols: Analysis, Occurrence and Biological Effects; Ph.D. Thesis, Huddinge University Hospital, Stockholm, Sweden, 1995; p. 8.

32. Nawar, W.W.; S.K. Kim; Y.J. Li; and M. Vajdi. *J. Am. Oil. Chem. Soc.* **1991,** *68*, 496.

33. Li, N.; T. Ohshima; K.-I. Shozen; H. Ushio; and C. Koizumi. *J. Am. Oil Chem. Soc.* **1991,** *71*, 623.

34. Bascoul, J.; N. Domergue; M. Olle; and A.C. de Paulet. *Lipids* **1986,** *21*, 383.

35. Nourooz-Zadeh, J.; L.-Å. Appelqvist. *J. Am. Oil Chem. Soc.* **1989,** *66*, 586.

36. Finocchiaro, E.T.; T. Richardson. *J. Food Prot.* **1983,** *46*, 917.

37. Kumar, N.; O.P. Singh. *J. Sci. Food Agric.* **1991,** *55*, 497.

38. Park, P.S.W.; P.B. Addis. In *Biological Effects of Cholesterol Oxides*; S.-K. Peng and R.J. Morin, Eds.; CRC Press: Boca Baton, FL, 1992; p. 71.

39. Bösinger, S.; W. Luf; and E. Brandl. *Int. Dairy J.* **1993,** *3*, 1.

40. Sarantinos, J.; K. O'Dea; and A.J. Sinclair. *Food Austral.* **1993,** *54*, 485.

41. Lai, S.-M.; J.I. Gray; D.J. Buckley; and P.M. Kelly. *J. Agric. Food Chem.* **1995,** *43*, 1127.

42. Padley, F.B. In *The Lipid Handbook*; F.D. Gunstone, J.L. Harwood, and F.B. Padley, Eds.; Chapman and Hall: Cambridge, U.K., 1986; p. 104.

43. Aringer, L.; L. Nordström. *Biomed. Mass Spectrom.* **1981,** *8*, 183.

44. Dutta, P.C.; L.-Å. Appelqvist. *J. Am. Oil Chem. Soc.* **1997,** *74*, 647.

45. Dutta, P.C. *J. Am. Oil Chem. Soc.* **1997,** *74*, 659.

46. Lampi, A.-M.; L. Juntunen; J. Toivo; and V. Piironen. *J. Chromatogr. B* **1997,** *777*, 83.

47. Johnsson, L.; P.C. Dutta. *J. Am. Oil Chem. Soc.* **2003,** *80*, 767.

48. Johnsson, L.; R.E. Andersson; and P.C. Dutta. *J. Am. Oil Chem. Soc.* **2003,** *80*, 777.

49. Bortolomeazzi, R.; M. De-Zan; L. Pizzale; and L.S. Conte. *J. Agric. Food Chem.* **1999,** *47*, 3069.

50. Dutta, P.C. In *Cholesterol and Phytosterol Oxidation Products: Analysis, Occurrence, and Biological Effects*; F. Guardiola, P.C. Dutta, F. Codony, and G.P. Savage, Eds.; AOCS Press: Champaign, IL, 2002; p.235.

51. Yanishlieva, N.; E. Marinova. *Riv. Ital. Delle Sost. Grasse* **1980,** *57*, 477.

52. Yanishlieva-Maslarova, N.; H. Schille; and A. Seher. *Fette Seife Anstrichm.* **1980,** *84*, 308.

53. Yanishlieva, N.; H. Schiller. *J. Sci. Food Agric.* **1980,** *35*, 219.

54. Yanishlieva-Maslarova, N.; E.M. Marinova-Tasheva. *Grasa Aceites* **1980,** *37*, 343.

55. Daly, G.G.; E.T. Finocchiaro; and T. Richardson. *J. Agric. Food Chem.* **1983,** *31*, 46.

56. Gordon, M.H.; P. Magos. *Food Chem.* **1984,** *14*, 295.

57. Blekas, G.; D. Boskou. *Food Chem.* **1989,** *33*, 301.

58. Säynäjoki, S.; S. Sundberg; L. Soupas; A.-M. Lampi; and V. Piironen. *Food Chem.* **2003,** *80*, 415.

59. Dutta, P.C. In *Phytosterols as Functional Food Components and Nutraceuticals*; P.C. Dutta, Ed.; CRC Press: Boca Raton, FL, 2004; p. 397.

60. Soupas, L.; L. Juntunen; S. Saynajoki; A.-M. Lampi; and V. Piironen. *J. Am. Oil Chem. Soc.* **2003,** *81*, 135.

61. Young, F.V.K.; C. Poot; E. Biernoth; N. Krog; L.A. O'Neill; and N.G.J. Davidson. In

*The Lipid Handbook*; F.D. Gunstone, J.L. Harwood, and F.B. Padley, Eds.; Chapman and Hall: Cambridge, U.K., 1986; p. 181.

62. Carr, R. In *Oil Crops of the World*; G. Röbbelen, R.K. Downey, and A. Ashri, Eds.; McGraw-Hill: New York, 1989; p. 226.
63. Smouse, T.H. In *Methods to Assess Quality and Stability of Oils and Fat-Containing Foods*; K. Warner and N.A.M. Eskin, Eds.; AOCS Press: Champaign, IL, 1995; p. 17.
64. Touche, J.; M. Derbesy; J. Castang; M. Olle; and J. Estienne. *J. Ann. Falsif. Expert. Chim.* **1977,** *70*, 263.
65. Serani, A.; D. Piacenti. *Riv. Ital. Delle Sost. Grasse* **1992,** *69*, 311.
66. Popov, A.; Z. Milkova; and M. Marekov. *Oli Grassi Deriv.* **1992,** *11*, 83.
67. Popov, A.; T.S. Milkova; and N. Marekov. *Nahrung* **1992,** *19*, 547.
68. Prokhorova, L.T.; N.N. Frolova; and E.I. Gorshkova. *Maslob-Zhir. Promist. No.* **1979,** *5*, 18.
69. Murui, T.; A. Fukushima. *Yukagaku* **1979,** *42*, 442.
70. Homberg, E. *Fette Seifen Anstrichm.* **1974,** *76*, 433.
71. Homberg, E. *Fette Seifen Anstrichm.* **1974,** *77*, 8.
72. Kaufmann, H.P.; Y. Hamza. *Fette Seifen Anstrichm.* **1970,** *72*, 432.
73. Kaufmann, H.P.; E. Vennekel; and Y. Hamza. *Fette Seifen Anstrichm.* **1970,** *72*, 242.
74. Niewiadomski, H. *Nahrung* 1975, *19*, 525.
75. Tiscornia, E.; G.C. Bertini. *La Riv. Ital. Delle Sost. Grasse* **1975,** *52*, 101.
76. Johansson, A. *J. Am. Oil Chem. Soc.* **1979,** *56*, 886.
77. Niewiadomski, H.; J. Sawicki. *Zesz. Probl. Postepow Nauk Roln.* **1965,** *53*, 178.
78. Mennie, D.; C.F. Moffat; and A.S. McGill. *J. High Res. Chromatogr.* **1994,** *17*, 831.
79. Niewiadomski, H.; J. Sawicki. *Vitalst. Zivilisationskr.* **1968,** *13*, 239.
80. Niewiadomski, H.; J. Sawicki. *Zesz. Probl. Postepow Nauk Roln.* **1968,** *80*, 253.
81. Jawad, I.M.; S.P. Kochhar; and B.J.F. Hudson. *Lebensm. Wiss. Technol. 17*, 155.
82. Meneghetti, O.; G. Amelotti; and A. Griffini. *La Riv. Ital. Delle Sost. Grasse* **1984,** *64*, 461.
83. Zunin, P.; A. Bocca; and E. Tiscornia. *Riv. Ital. Delle Sost. Grasse* **1989,** *66*, 133.
84. Leone, A.M.; V. Liuzzi; E. La Notte; and M. Santoro. *Riv. Ital. Delle Sost. Grasse* **1984,** *61*, 69.
85. Grob, K.; A. Artho; and C. Mariani. *Fat Sci. Technol.* **1992,** *94*, 394.
86. Schulte, E. *Fat Sci. Technol.* **1994,** *96*, 124.
87. Lanzon, A.; T. Albi; A. Cert; and J. Gracian. *J. Am. Oil Chem. Soc.* **1994,** *71*, 285.
88. Grob, K.; M. Bronz. *Fat Sci. Technol.* **1994,** *96*, 252.
89. Schulte, E.; N. Weber. *Lipids* **1987,** *22*, 1049.
90. Gordon, M.H.; C. Firman. *J. Sci. Food Agric.* **2001,** *81*, 1530.
91. Bortolomeazzi, R.; M. De-Zan; L. Pizzale; and L.S. Conte. *J. Agric. Food Chem.* **2000,** *48*, 1101.
92. Bortolomeazzi, R.; F. Cordaro; L. Pizzale; and L.S. Conte. *J. Agric. Food Chem.* **2003,** *51*, 2394.
93. Grothues, B.G.M. *J. Am. Oil Chem. Soc.* **1985,** *62*, 390.
94. Edvardsson, J.; S. Irandoust. *J. Am. Oil Chem. Soc.* **1994,** *71*, 235.
95. Kanematsu, H.; T. Maruyama; I. Niiya; M. Imamura; and T. Matsumoto. *Yukagaku* **1973,** *22*, 814.
96. Parodi, P.W. *J. Am. Oil Chem. Soc.* **1975,** *52*, 345.

97. Sugano, M.; K. Imaizumi; H. Taniguchi; and H. Kubota. *Sci. Bull. Fac. Agric. Kyushu Univ.* **1977,** *32*, 21.
98. Strocchi, A.; G. Guerzoni. *Riv. Della Soc. Ital. Sci. Aliment.* **1988,** *14*, 125.
99. Strocchi, A. *Riv. Ital. Delle Sost. Grasse* **1987,** *64*, 401.
100. Strocchi, A. *Riv. Della Soc. Ital. Sci. Aliment.* **1988,** *17*, 41.
101. Strocchi, A. *Rev. Fr. Corps Gras* **1988,** *35*, 163.
102. Strocchi, A.; G. Marascio. *Fat Sci. Technol.* **1993,** *95*, 293.
103. Strocchi, A.; G. Marascio. *Riv. Ital. Delle Sost. Grasse* **1993,** *70*, 7.
104. Griffiths, W.A. In *Proceedings of the Lipidforum Symposium*; R. Marcuse, Ed.; Kållered, Sweden, 1986; p. 124.
105. Fedeli, E.; A. Gasparoli; and M. Cardillo. *Riv. Ital. Delle Sost. Grasse* **1981,** *58*, 1.
106. Verleyen, T.; P.C. Dutta; R. Verhe; K. Dewettinck; A. Huyghebaert; and W. de Greyt. *Food Chem.* **2003,** *83*, 185.
107. Lee, K.; A.M. Herian; and T. Richardson. *J. Food Protect.* **1983,** *47*, 340.
108. Lee, K.; A.M. Herian; and N. A. Higley. *J. Food Protect.* **1985,** *48*, 158.
109. Zhang, W. B.; P.B. Addis; and T.P. Krick. *J. Food Sci.* **1991,** *56*, 716.
110. Park, S.W.; P.B. Addis. *J. Food Sci.* **1985,** *50*, 1437.
111. Przybylski, R.; N.A.M. Eskin. In *Proceedings of the Eighth International Rapeseed Congress*; D.I. McGregor, Ed.; Saskatoon, Saskatchewan, Canada, 1991; p. 888.
112. Rudzinska, M.; J. Korczak; A. Gramza; P.C. Dutta; and E. Wasowicz. *J. AOAC Int.* **2004,** *87*, 499.
113. International Organisation for Standardization (ISO). ISO 6799. ISO: Geneva Switzerland, 1991.
114. International Organisation for Standardization (ISO). ISO 12228. ISO: Geneva Switzerland, 1999.
115. American Oil Chemists' Society. Method Ch 6-91. AOCS Press: Champaign, IL, 1997.
116. International Union of Pure and Applied Chemistry (IUPAC). IUPAC 2.401. IUPAC: Research Triangle Park, NC, 1998.
117. International Union of Pure and Applied Chemistry (IUPAC). IUPAC 2.403. IUPAC: Research Triangle Park, NC, 1999.
118. Björkhem, I. *Anal. Biochem.* **1986,** *154,* 497.
119. Dzeletovic, S.; O. Breuer; E. Lund; and U. Diczfalusy. *Anal. Biochem.* **1995,** *225*, 73.
120. Fischer, K.H.; G. Laskawy; and W. Grosch. *Z. Lebensm. Unters. Forsch.* **1985,** *181*, 14.
121. Piironen, V.; D.G. Lindsay; T.A. Miettinen; J. Toivo; and A.-M. Lampi. *J. Sci. Food Agric.* **2000,** *80*, 939.
122. Park, S.W.; P.B. Addis. *J. Food Sci.* **1987,** *52*, 1500.
123. Moreau, R.A.; M.J. Powell; and K.B. Hicks. *J. Agric. Food Chem.* **1996,** *44*, 2149–2154.
124. Bligh, E.G.; W.J. Dyer. *Can. J. Biochem. Physiol.* **1958,** *37*, 911.
125. Folch, J.; M. Lees; and G.H. Sloane Stanley. *J. Biol. Chem.* **1957,** *726*, 497.
126. Radin, N.S. *Methods Enzymol.* **1981,** *72*, 5.
127. Christie W.W. In *Gas Chromatography and Lipids*; The Oily Press: Ayr, U.K., 1992; p. 28.
128. Christie W.W. In *High-Performance Liquid Chromatography and Lipids*; Pergamon Press: Oxford, U.K., 1987; p. 26.
129. Christie W.W. In *Lipid Analysis*; Pergamon Press: Oxford, U.K., 1982; p. 22.

130. Kates, M. In *Techniques of Lipidology*; Elsevier: Amsterdam, The Netherlands, 1986; p. 64.
131. Toivo, J.; A.-M. Lampi; S. Aalto; and V. Piironen. *Food Chem.* **2000**, *68*, 239.
132. Appelqvist, L-Å.; A.K. Kornfeldt; and J.E. Wennerholm. *Phytochemistry* **1981**, *20*, 207.
133. American Oil Chemists' Society, *Official Methods and Recommended Practices*, AOCS Press: Champaign, IL, 1994.
134. Association of Official Analytical Chemists (AOAC), *Official Methods of Analysis*, 14th ed.; AOAC: Washington, DC, 1984.
135. Tsai, L.S.; K. Ijichi; C.A. Hudson; and J.J. Meehan. *Lipids* **1980**, *15*, 124.
136. Maeker, G.; J. Unruh. *J. Am. Oil Chem. Soc.* **1986**, *63*, 767.
137. Bergstrom, S.; O. Wintersteiner. *J. Biol. Chem.* **1941**, *141*, 597.
138. Park, S.W.; P.B. Addis. *J. Agric. Food Chem.* **1986**, *34*, 653.
139. Pie, J.E.; K. Spahis; and C. Seillan. *J. Agric. Food Chem.* **1990**, *38*, 973.
140. Nourooz-Zadeh, J. *J. Agric. Food Chem.* **1990**, *38*, 1667.
141. Klatt, L.V.; B.A. Mitchell; and R.L. Smith. *J. AOAC Int.* **1995**, *78*, 75–79.
142. Kovacs, M.I.P. *J. Cereal Sci.* **1990**, *11*, 291.
143. Jekel, A.A.; H.A. Vaessen; and R.C. Schothorst. *Fresenius J. Anal. Chem.* **1998**, *360*, 595.
144. Gerhardt, K.O.; C.W. Gehrke. *J. Chromatogr.* **1977**, *135*, 341.
145. Flanagan, V.P.; A. Ferretti; D.P. Schwartz; and J.M. Ruth. *J. Lipid Res.* **1975**, *16*, 97.
146. Tsai, K.S. Quantification of Cholesterol Oxidation Products by Capillary Gas Chromatography. M.S. Thesis, University of Minnesota, Minneapolis–St. Paul, 1984.
147. Maeker, G.; F.J. Bunick. *J. Am. Oil Chem. Soc.* **1986**, *63*, 771.
148. Addis, P.B.; H.A. Emanuel; S.D. Bergmann; and J.H. Zavoral. *Free Radical Biol. Med.* **1989**, *7*, 179.
149. Toschi, T.G.; M.F. Caboni. *Ital. J. Food Sci.* **1992**, *4*, 223.
150. Sugino, K.; J. Terao; H. Murakami; and S. Matsushita. *J. Agric. Food Chem.* **1986**, *34*, 36.
151. Gruenke, L.D.; J.C. Craig; N.L. Petrakis; and M.B. Lyon. *Biomed. Environ. Mass Spectrom.* **1987**, *14*, 335.
152. Björkhem, I.; O. Breuer; B. Angelin; and S.H. Wikstrom. *J. Lipid Res.* **1988**, *29*, 1031.
153. Nourooz-Zadeh, J.; L.-Å. Appelqvist. *J. Food Sci.* **1987**, *52*, 57.
154. De Vore, V.R. *J. Biol. Chem.* **1988**, *233*, 311.
155. Morgan, J.N.; D.J. Armstrong. *J. Food Sci.* **1989**, *54*, 427.
156. Schulte, E. *Eur. J. Lipid Sci. Technol.* **2004**, *106*, 772–776.
157. Geiss, F. In *Fundamentals of Thin Layer Chromatography*; Huthig Verlag: Heidelberg, Germany, 1987; p. 226.
158. Kou, I.L.; R.P. Holmes. *J. Chromatogr.* **1985**, *330*, 339.
159. Sevanian, A.; L.L. McLeod. *Lipids* **1987**, *22*, 627.
160. Phillips, K.M.; D.M. Ruggio; J.I. Toivo; M.A. Swank; and A.H. Simpkins. *J. Food Comp. Anal.* **2002**, *15*, 123.
161. Rosenberg, R.; N.L. Ruibal-Mendieta; G. Petitjean; P. Cani; D.L. Delacroix; N.M. Delzenne; M. Meurens; J. Quetin-Leclercq; and J.L. Habib-Jiwan. *J. Cereal Sci.* **2003**, *38*, 189.
162. Rose-Sallin, C.; A.C. Hugget; J.O. Bosset; R. Tabacchi; and L.B. Fay. *J. Agric. Food Chem.* **1995**, *43*, 935.

163. Schmarr, H.G.; H.B. Gross; and T. Shibamoto. *J. Agric. Food Chem.* **1996,** *44*, 512.
164. Ulberth, F.; D. Rossler. *J. Agric. Food Chem.* **1998,** *46*, 2634.
165. Yokohama, H.; I. Ohtsuka; H. Shiojiri; K. Katayama; and S. Ishikawa. *Anal. Biochem.* **1998,** *157*, 186.
166. Missler, S.R.; B.A. Wasilchuk; and C. Merrit. *J. Food Sci.* **1985,** *50*, 595.
167. Tsai, L.S.; C.A. Hudson. *J. Food Sci.* **1985,** *50*, 229.
168. Smith, L.L. *J. Liquid Chromatogr.* **1985,** *16*, 1731.
169. Maeker, G.; E.H. Nungesser; and M.I. Zulak. *J. Agric. Food Chem.* **1988,** *36*, 61.
170. Baggio, S.R.; A.M.R. Miguel; and N. Bragagnolo. *Food Chem.* **2005,** *89*, 475–484.
171. Csallany, A.S.; S.E. Kindom; P.B. Addis; and J.H. Lee. *Lipids* **1989,** *24*, 645.
172. Fillion, L.; J.A. Zee; and C. Gosselin. *J. Chromatogr.* **1991,** *547*, 105.
173. Korytowski, W.; G.J. Bachowski; and A.W. Girotti. *Anal. Biochem.* **1991,** *197*, 149.
174. Breinholder, P.; L. Mosca; and W. Lindner. *J. Chrom. B* **2002,** *777*, 67.
175. Moreau, R.A.; B.D. Whitaker; and K.B. Hicks. *Prog. Lipid Res.* 2002, *41*, 457.
176. Sevanian, A.; R. Seraglia; P. Traldi; P. Rossato; F. Ursini; and H. Hodis. *Free Radical Biol. Med.* **1994,** *17*, 397.
177. Wu, Z.; R.P. Rodgers; and A.G. Marshall. *J. Agric. Food Chem.* **1994,** *52*, 5322.
178. Park, S.W.; P.B. Addis. *Anal. Biochem.* **1985,** *149*, 275.
179. Apprich, S.; F. Ulberth. *J. Chrom. A* **2004,** *1055*, 169–176.
180. Grob, K. In *On-Line Coupled LC-GC*; Huthig Verlag: Heidelberg, Germany, 1990; p. 406.
181. Shen, C.S.J.; A.J. Sheppard. *J. Am. Oil Chem. Soc.* **1986,** *63*, 472.
182. Brooks, C.J.W.; W.J. Cole; T.D.V. Lawrie; J. MacLachlan; J.H. Borthwick; and G.M. Barrett. *J. Steroid Biochem.* **1983,** *19*, 189.
183. Van Lier, J.E.; L.L. Smith. *Anal. Biochem.* **1968,** *24*, 419.
184. Grob, K.; S. Rennhard. *J. High Res. Chromatogr.* **1968,** *3*, 627.
185. Van de Boenkamp, P.; T.G. Kosmeijer-Schull; and M.B. Katan. *Lipids* **1988,** *23*, 1079.
186. Lin, Y.Y.; C.E. Low; and L.L. Smith. *J. Steroid Biochem.* **1981,** *14*, 563.
187. Lin, Y.Y.; L.L. Smith. *Mass Spectrom. Rev.* **1984,** *3*, 319.
188. Lin, Y.Y.; L.L. Smith. *Biomed. Mass Spectrom.* **1978,** *5*, 604.
189. Tsai, L.S.; C.A. Hudson. *J. Food Sci.* **1978,** *49*, 1245.
190. Aringer, L.; P. Eneroth. *J. Lipid Res.* **1974,** *15*, 389.
191. Park, S.W.; P.B. Addis. *J. Am. Oil Chem. Soc.* **1989,** *66*, 1632.
192. Brooks, C.J.; W. Henderson; and G. Steel. *Biochim. Biophys. Acta* **1973,** *296*, 431.
193. Ansari, G.A.S.; L.L. Smith. *Methods Enzymol.* **1973,** *233*, 332.
194. Johannes, C.; R.L. Lorenz. *Anal. Biochem.* **2004,** *325*, 107.
195. Sanghvi, A.; E. Grassi; C. Bartman; R. Lester; M.G. Kienle; and G. Giovanni. *J. Lipid Res.* **1981,** *22*, 720.
196. Sanghvi, A.; M.G. Kienle; and G. Galli. *Anal. Biochem.* **1981,** *85*, 430.
197. Kamm, W.; F. Dionisi; L.B. Fay; C. Hischenhuber; H.G. Schmarr; and K.H. Engel. *J. Chrom. A* **2001,** *918*, 341.
198. Grob, K.; M. Lanfranchi; and C. Mariani. *J. Am. Oil Chem. Soc.* **1990,** *67*, 626.
199. Baranowski, A.; C.W.M. Adams; O.B. Bayliss High; and D.B. Bowyer. *Atherosclerosis* **1982,** *41*, 255.
200. Kendall, C.W.; M. Koo; E. Sokoloff; and A.V. Rao. *Cancer Lett.* **1982,** *66*, 241.
201. Peng, S.K.; C.B. Taylor; J.C. Hill; and R.J. Morin. *Atherosclerosis* **1985,** *54*, 121.

202. Peterson, A.R.; H. Peterson; C.P. Spears; J.E. Trosko; and A. Sevanian. *Mutat. Res.* **1985,** *203,* 355.
203. Seillan, C.; C. Dubuquoy. *Prostaglandins Leukot. Essent. Fatty Acids* **1990,** *36,* 11.
204. Peng, S.K.; P. Tham; C.B. Taylor; and B. Mikkelson. *J. Clin. Nutr.* **1979,** *32,* 1033.
205. Brown, M.S.; S.E. Danna; and J.L. Goldstein. *J. Biol. Chem.* **1975,** *250,* 4027.
206. Hubbard, R.W.; Y. Ono; and A. Sanchez. *Prog. Food Nutr. Sci.* **1975,** *13,* 17.
207. Osada, K.; T. Kodama; L. Cui; Y. Ito; and M. Sugano. *Biosci. Biotechnol. Biochem.* **1994,** *58,* 1062.
208. Anitschkow, N. *Beitr. Path. Anat. Allg. Path.* **1913,** *56,* 379.
209. Schwenk, E.; D.F. Stevens; and R. Altschul. *Proc. Soc. Exp. Biol. Med.* **1959,** *102,* 42.
210. Imai, H.; N.T. Werthessen; C.B. Taylor; and K.T. Lee. *Arch. Pathol. Lab. Med.* **1976,** *100,* 565.
211. Toda, T.; D. Leszcynski; and F. Kummerow. *Paroi Arterielle.* **1981,** *7,* 167.
212. Imai, H.; H. Nakamura. *Fed. Proc.* **1981,** *39,* 772a.
213. Taylor, C.B.; S.K. Peng; N.T. Werthessen; P. Tham; and K.T. Lee. *Am. J. Clin. Nutr.* **1979,** *32,* 40.
214. Jacobson, M.S.; M.G. Price; A.E. Shamoo; and F.P. Heald. *Artheriosclerosis* **1985,** *57,* 209.
215. Vine, D.F.; J.C.L. Mamo; L.J. Beilin; T. Mori; and K.D. Croft. *J. Lipid Res.* **1998,** *39,* 1995.
216. Moncada, S. *Ibid.* **1982,** *2,* 193.
217. Majerus, P.W. *J. Clin. Invest.* **1983,** *72,* 1521.
218. Peng, S.-K.; B. Hu; A.Y. Peng; and R.J. Morin. *Artery* **1993,** *20,* 122.
219. Kubow, S., *Nutr. Rev.* **1993,** *51,* 33.
220. Smith, L.L.; B.H. Johnson. *Free Radical Biol. Med.* **1989,** *7,* 285.
221. Black, H.S.; J.T. Chan. *Oncology* **1976,** *33,* 119.
222. Kelsey, M.I.; R.J. Pienta. *Cancer Lett.* **1979,** *6,* 143.
223. Peng, S.K.; B.C. Taylor; P. Tham; N.T. Werthessen; and B. Mikkelson. *Arch. Path. Lab. Med.* **1978,** *102,* 57.
224. Sevanian, A.; A.R. Peterson. *Food Chem. Toxicol.* **1986,** *24,* 1102.
225. Carboni, M.F.; P.L. Bordoni; C.A. Rossi; and S. Hrelia. In *Proceedings of the Fifteenth International Symposium on Capillary Chromatography*; P. Sandra and D. Devos, Eds.; Huethig Verlag: Heidelberg, Germany, 1993; Vol. 1, p. 701.
226. Hrelia, S.; A. Bordoni; M.F. Carboni; G. Lercker; P. Capella; E. Turchetto; and P.L. Biagi. *Biochem. Biol. Int.* **1994,** *32,* 565.
227. Carboni, M.F.; S. Hrelia; A. Bordoni; G. Lercker; P. Capella; E. Turchetto; and P.L. Biagi. *J. Agric. Food Chem.* **1994,** *42,* 2367.
228. Aupiex, K.; D. Weltin; J.E. Mejia; M. Christ; J. Marchal; J.-M. Freyssinet; and P. Bischoff. *Immunobiology* **1994,** *194,* 415.
229. Harada, K.; S. Ishibashi; T. Miyashita; J.-L. Osuga; H. Yagyu; K. Ohashi; Y. Yazaki; and N. Yamada. *FEBS Lett.* **1994,** *411,* 211.
230. Bansal, N.; A. Houle; and G. Melynkovych. *FASEB Lett.* **1994,** *5,* 211.
231. Christ, M.; B. Luu; J.E. Mejia; I. Moosbrugger; and P.Z. Bischoff. *Immunology* **1993,** *78,* 455.
232. Lizard, G.; S. Monier; C. Cordelet; L. Gesquierre; V. Deckert; S. Gueldry; L. Lagrost; and P. Gambert. *Atherscler. Thromb. Vasc. Biol.* **1999,** *19,* 1190.

233. Rusinol, A.E.; L. Yang; D. Thewke; S.R. Panini; M.F. Kramer; and M. S. Sinensky. *J. Biol. Chem.* **2000,** *275*, 7296.

234. Panini, S.R.; L. Yang; A.E. Rusinol; M.S. Sinensky; J.V. Bonventre; and C.C. Leslie. *J. Lipid Res.* **2001,** *42*, 1678.

235. Enari, M.; H. Hug; M. Hayakawa; F. Ito; Y. Nishimura; and S. Nagata. *Eur. J. Biochem.* **1996,** *236*, 533.

236. Wu, Y.L.; X.R. Jiang; D.M. Lillington; P.D. Allen; A.C. Newland; and S.M. Kelsey. *Cancer Res.* **1998,** *58*, 633.

237. Adcox, C.; L. Boid; L. Oehrl; J. Allen; and G. Fenner. *J. Agric. Food Chem.* **2001,** *49*, 2090.

238. Lea, L.J.; P.A. Hepburn; A.M. Wolfreys; and P. Baldrick. *Food Chem. Toxicol.* **2004,** *42*, 771.

# PART 3

# Nutrition

# Role of Fat in the Diet

**Bruce E. McDonald and Michael N.A. Eskin**

*University of Manitoba, Department of Human Nutritional Sciences, Winnipeg, Manitoba, Canada R3T 2N2*

Dietary fat plays several important nutritional roles. As mentioned in the previous chapter, it is a major source of energy for the body, in spite of expressed concerns about levels of intake in developed countries. In the breast-fed infant, for example, it provides 45–50% of the dietary energy. Fat provides appreciably more energy, on an equal weight basis, than carbohydrate or protein; fat provides 9 kcal/g of metabolizable energy compared to an average energy level of 4 kcal/g for carbohydrate and protein.

Dietary fat also is the source of essential fatty acids (EFA) and thus must be present in the diet. The EFA derive from two families of fatty acids, namely those of the n-6 and those of the n-3 families of unsaturated fatty acids. The latter are often referred to as omega-6 and omega-3 fatty acids, respectively, in the popular press. Linoleic acid (LA; 18:2 n-6), which accounts for more than 50% of the fatty acids in many vegetable oils (e.g., corn, soybean, sunflower, cottonseed, and safflower), is the main n-6 fatty acid in the diet. LA accounts for between 85 and 90% of the n-6 fatty acids in the diet with the balance coming from arachidonate acid (AA; 20:4 n-6) and γ-linolenic acid (18:3 n-6). Borage oil, evening primrose oil, and black currant (seed) oil are relatively rich sources of γ-linolenic acid, but it is also present in small quantities in a number of vegetable oils. α-Linolenic acid (18:3 n-3), by contrast, is a member of the n-3 family of fatty acids. Like LA, α-linolenic acid (LNA; 18:3 n-3) is the main n-3 fatty acid in the diet, although canola oil and soybean oil are the only common vegetable oils that contain any appreciable quantity of LNA (approx. 11 and 8%, respectively). Other common vegetable oils contain less than 1% LNA. Fatty fish (e.g., herring, mackerel, salmon, and trout) and fish oils also are rich sources of n-3 fatty acids but in this case, the principal n-3 fatty acids are the long-chain homologs of LNA, namely eicosapentaenoic acid (EPA; 22:5 n-3) and docosahexaenoic acid (DHA; 22:6 n-3). The body is capable of converting (elongating and further desaturating) LA and LNA to the long-chain polyunsaturated isomers (n-6 and n-3 LC-PUFA, respectively). However, humans are relatively poor converters of LNA to n-3 LC-PUFA, although the latter varies with the LA-to-LNA ratio because LA and LNA are elongated and desaturated by the same enzyme pathway. Since diets usually contain appreciably higher levels of LA than LNA, the effect is primarily on the conversion of LNA to the n-3 LC-PUFA (viz., EPA and DHA). EFA are important constituents of biological membranes, which surround cells and subcellular particles.

They are present in biological membranes primarily as constituents of phospholipids, namely phosphatidylethanolamine, phosphatidylcholine, and phosphatidylserine. DHA is an important constituent of the membrane lipids of nerve tissue and the retina and is thought to play an important role in neural development and visual function. DHA can account for 50% or more of the total fatty acids in these tissues. EFA also serve as the precursors of a variety of biologically active compounds referred to collectively as eicosanoids (e.g., prostaglandins, thromboxanes, and leukotrienes), which are important regulators of a host of physiological reactions. The Panel on Macronutrients (1) gave the Dietary Reference Intakes (DRI) for LA and LNA as the AI (Adequate Intake) because there was insufficient data to establish Estimated Average Requirements for these fatty acids and thus to specify a RDA (Recommended Dietary Allowance). The AI for adults are based on the highest median intakes of LA and LNA by men and women in the United States. The AI for LA is 14–17 g/day for men and 11–12 g/day for women, while the AI for LNA is 1.6 g/day for men and 1.1 g/day for women, of which 10% can be EPA and DHA. However, the recommendations of the Panel on Macronutrients have been challenged (2), primarily on the basis of the relatively high levels of LA in the U.S. diet and the accompanying high ratio of LA-to-LNA (viz., 9-10:1 for men and 10-11:1 for women) perpetuated by the AI for n-6 and n-3 PUFA. Nonetheless, the recently updated Dietary Guidelines for Americans (3) recommend an n-6 PUFA intake of between 5 and 10% of energy and an LNA intake between 0.6 and 1.2% of energy.

In addition to its role as an important source of energy and the source of EFA, dietary fat serves as a carrier of the fat-soluble vitamins (vitamins A, D, E, and K) and carotenoids, aids in their absorption, and as in the case of vitamin E, is an important source of the vitamin. In addition to vitamin E, vegetable oils are a rich source of phytosterols (plant sterols), which have been found to lower blood cholesterol levels. Margarines and shortenings containing phytosterols and their hydrogenated derivatives (phytostanols) are known to lower serum cholesterol levels, primarily by interfering with the absorption of dietary cholesterol and the reabsorption of biliary cholesterol (4). Fat also plays an important role in the texture and palatability of foods that accounts for the popularity of fried foods.

Much of the interest in dietary fat over the past quarter century, however, stems from its implication in the etiology of chronic diseases, in particular cardiovascular disease (CVD) but also other chronic diseases such as cancer, diabetes, and hypertension. Both the level and the type of fat in the diet have been of concern in relation to CVD. Current dietary recommendations, however, focus on the type of fat in the diet, that is, the fatty acids that increase CVD risk factors, namely saturated and *trans* fatty acids. Recommendations with respect to level of fat in the diet have eased somewhat over the past few years. The current recommendations of the NCEP (5) call for a range in fat intake of 25–35% of total energy compared to earlier recommendations that emphasized intakes of less than 30% of energy. Similarly, the DRI for Macronutrients (1) set the acceptable AMDR (Acceptable Macronutrient Distribution Range) for total fat at 20–35% of total energy. The adverse effect of saturated fat on serum total cholesterol and low-density lipoprotein (LDL) cholesterol is well established (6,7). In addition, epidemiological studies found a higher incidence of coronary heart disease in populations that consumed high levels of saturated fat (8). Hence, the NCEP (5) has lowered the recommendation for total saturated fat in the diet from the earlier recommendation of 10% or less of total energy to 7% or less of total energy. There also is the recognition that the long-chain saturated fatty acids (viz., 12:0, 14:0, 16:0, and 18:0) are not equal in their effect on serum cholesterol and lipoprotein levels (6,7,9). Myristic acid (14:0) causes the greatest increase in total and LDL cholesterol levels whereas stearic acid (18:0) has little effect on these risk factors for CVD. By contrast, both monounsaturated fatty acid (MUFA) and polyunsaturated fatty acid (PUFA) have an ability to decrease total and LDL cholesterol levels when they replace carbohydrate or saturated fat in the diet.

Although there has been an emphasis on the amount of saturated fat in the diet, the decrease over the period 1971–2000 was disappointingly slow, from an intake of 13–13.5% of total energy to approximately 11% of total energy (10). A portion of this apparent decrease was the result of an increase in total energy intake, in essence due to an increase in carbohydrate intake during the 30-year period. In fact, the absolute intake of saturated fat by women actually increased slightly (from 22.3 g to 22.9 g per day). By contrast, the absolute intake of saturated fat by men decreased by about 5 g per day (36.7–31.7 g) during this period.

Concern about the adverse effect of saturated fat led to the replacement of animal fats in deep frying by partially hydrogenated vegetable oils. However, the finding (11) that *trans* fatty acids not only produced an increase in LDL level, the major atherogenic lipoprotein, but a decrease in HDL level, which protects against atherosclerosis, raised questions with this change from saturated fat to partially hydrogenated vegetable oils. Confirmation of the adverse effect of *trans* fats on CVD risk factors (12,13) led to the mandatory declaration of *trans* fatty acids in the nutrition label of conventional foods and supplements in the United States (effective January 1, 2006) and Canada (effective December 12, 2005). The regulatory agencies in both countries define *trans* fatty acids as those with one or more isolated double bonds in a *trans* configuration, thereby excluding conjugated fatty acids (e.g., conjugated linoleic acid, CLA) present in milk fat and adipose fat of ruminant animals. These developments created a serious conundrum for deep frying operations and the food manufacturing and food processing industry generally since both saturated fat and *trans* fat have adverse effects on CVD risk factors. Hence, industry has turned to high-oleic, low-linolenic acid vegetable oils, such as mid- and high-oleic acid sunflower oil, high-oleic acid canola (rapeseed) oil, and low-linolenic acid soybean oil for deep frying. These oils represent a new generation of frying oils that eliminate the need to hydrogenate, thus avoiding the formation of *trans* fatty acids, while maintaining the oxidative stability required for deep frying operations and the foods produced by these processes (14). The development of these cultivars occurred very quickly. For example, hybrids of mid-oleic sunflower (commercially trademarked as Nu-Sun™), which first became available in 1998, accounted for an estimated 65% of the oil-type sunflower grown in the United States in 2004 (15). In addition to their acceptable oxidative stability, the switch to high-oleic acid oils was aided by research that found diets enriched with MUFA (namely, oleic acid) were equally as effective as those enriched with PUFA (namely, linoleic acid) in lowering total and LDL cholesterol levels when each replaced saturated fatty acids in the diet (11,16–18).

Concern about the level of dietary fat in the diet and its proposed association with the prevalence of obesity in the United States and other industrialized countries has been a topic of debate for several years. In the United States, obesity increased from 14.5% in 1971 to 30.9% in 2000 (10). However, an analysis of the trends in dietary intake revealed that the main change during this period was an increase in mean energy intake of approximately 170 kcals/day (from 2450–2618) for men and 335 kcals/day (from 1542–1877) for women. The primary contributor to the increase in total energy intake was carbohydrate for both men and women. In fact, fat intake as a percent of total energy intake decreased from 36.9–32.8% for men and from 36.1–32.8% for women. However, absolute fat intake increased approximately 7 g for women during the 30-year period. By contrast, absolute fat intake decreased about 5 g for men. These data suggest that the increased prevalence of obesity is not directly connected to dietary fat intake but rather to increased total energy intake and other factors such as a sedentary lifestyle. In an effort to address the broader question embodied in the imbalance between energy intake and energy expenditure by Americans and the accompanying obesity epidemic, the U.S. Department of Agriculture (USDA) undertook an extensive revision of the Food Guide Pyramid. The original Pyramid was developed to help Americans select healthful diets. The new food guidance system, dubbed MyPyramid (19) that is designed to help individuals make healthy food and physical activity choices, translates the principles of the 2005 Dietary Guidelines for Americans (3) into an action plan for a healthier lifestyle. One of the feature components of MyPyramid is a web-based system of interactive tools that provides an individual with specific, consumer-friendly guidance in support of behavioral change. MyPyramid, launched in April 2005, offers the consumer a personalized approach to healthy food choices and a reminder of the importance of daily physical activity. The overall aim of the revision of the original Pyramid was to improve the Guide's effectiveness in inspiring individuals to make healthy food choices. The revision also attempts to ensure that the USDA food guide system reflects the latest nutritional science. Actions leading up to MyPyramid include the publication of new nutritional standards (Dietary Reference Intakes) by the National Academy of Sciences' Institute of Medicine 2002 (1) and the revision and update of the Dietary Guidelines for Americans (3).

Frying is an economical and widely accepted process for producing products with unique flavors and textures. The introduction of high-oleic frying oils as alternatives to hydrogenated oils will ensure that these products will be healthier with respect to both *trans* and saturated fatty acids. In addition, the enhanced oxidative stability expected from these new frying oils will extend the frying period by minimizing the degree of oil degradation. Accompanying the development of these oils must be good manufacturing practices to ensure that deep fryers and oils are properly maintained and cleaned so that fried food products sold to the consumer are both safe and wholesome.

## References

1. Institutes of Medicine (IOM). *Dietary Reference Intakes: Energy, Carbohydrate, Fiber, Fat, Fatty Acids, Cholesterol, Protein, and Amino Acids*; National Academies Press: Washington, DC, 2002.
2. Lands, W.E.M. *inform* **2002,** *13*, 876.
3. www.health.gov/dietaryguidelines/dga2005/report/
4. Yankah, V.V.; and P.J.H. Jones. *inform* **2001,** *12*, 899.
5. National Cholesterol Education Program. Third Report of the National Cholesterol Education Program (NCEP) on Detection, Evaluation, and Treatment of High Blood Cholesterol in Adults (Adult Treatment Panel III). *Circulation* **2002,** *106*, 3143–3421.
6. Kris-Etherton, P.M.; and S. Yu. *Am. J. Clin. Nutr.* **1997,** *65*, 1628S.
7. Mensink, R.P.; P.L. Zock; A.D. Kester; and M.B. Katan. *Am. J. Clin. Nutr.* **2003,** *77*, 1146.
8. Keys, A.; A. Menotti; M.J. Karvonen; C. Aravanis; H. Blackburn; R. Buzina; B.S. Djordjevic; A.S. Dontas; F. Fidanza; M.H. Keys, et al. The Diet and 15-Year Death Rate in the Seven Countries Study. *Am. J. Epidemiol.* **1986,** *124*, 903–915.
9. Müller, H.; B. Kirkhus; and J.I. Pederson. *Lipids* **2001,** *35*, 783.
10. Trends in Intake of Energy and Macronutrients—United States, 1971–2000; www.cdc.gov/mmwr/preview/mmwrhtml/mm5304a3.htm
11. Mensink, R.P.; and M.B. Katan. *New Eng. J. Med.* **1989,** *321*, 436.
12. Lichtenstein, A.H.; L.M. Ausman; S.M. Jalbert; and E.J. Schaefer. *New Engl. J. Med.* **1999,** *340*, 1933.
13. Judd, J.T.; D.J. Baer; B.A. Clevidence; P. Kris-Etherton; F.A. Muesing; and M. Iwane. *Lipids* **1999,** *37*, 123.
14. Sakurai, H.; T. Yoshihashi; H.T.T. Nguyen; and J. Pokorny. *Czech J. Food Sci.* **1999,** *21*, 145.
15. Kleingartner, L. *inform* **1999,** *15*, 748.
16. Chan, J.K.; V.M. Bruce; and B.E. McDonald. *Am. J. Clin. Nutr.* **1991,** *53*, 1230.
17. Wardlaw, G.M.; J.T. Snook; M.C. Lin; M. Puangco; and J.S. Kwon. *Am. J. Clin. Nutr.* **1991,** *54*, 104.
18. Lichtenstein, A.H.; L.M. Ausman; W. Carrasco; J.L. Jenner; L.J. Gualtieri; B.R. Goldin; J.M. Ordovas; and E.J. Schaefer. Effects of Canola, Corn, and Olive Oils on Fasting and Postprandial Plasma Lipoproteins in Humans as Part of a National Cholesterol Education Program Step 2 Diet. *Arterioscler. Thromb. Vasc. Biol.* **1993,** *13*, 1533–1542.
19. www.mypyramid.gov

# 9

# Nutritional and Physiological Effects of Used Frying Oils and Fats

**Gloria Márquez-Ruiz and M. Carmen Dobarganes**

*Instituto de la Grasa (C.S.I.C.), Avda. Padre García Tejero, 4, 41012 Sevilla, Spain*

The nutritional and physiological effects of frying fats and oils have been subjects of intensive investigations since the 1950s. Comprehensive reviews are available on general or specific aspects of this topic (1–17). However, whether used frying fats are detrimental to human health remains one of the most difficult questions to answer in light of the multifaceted views requiring coordinated, multidisciplinary studies to prove.

During frying, fats and oils are heated to high temperatures while exposed to air, resulting in a complex series of reactions that generates a wide array of compounds. From the nutritional point of view, the nonvolatile degradation products of used frying fats and oils are most relevant since they remain in the oil, are retained in the food, and are subsequently ingested. Such nonvolatile products include polymeric triglycerides, monomeric triglycerides containing oxidized or cyclic fatty acyls, and breakdown products. Except for monomeric triglycerides containing cyclic fatty acyls, all nonvolatile compounds formed during frying are determined globally by adsorption chromatography. The standard method, polar compound determination uses a classical silica column for a gravimetric determination (18). Limits for polar compounds, around 25%, are established in most of the countries where frying oils are fully regulated (19). The degree of alteration, and types of compounds formed, depend principally on the heating conditions (temperature, time, surface-to-oil volume ratio, and continuous or discontinuous heating), unsaturation of the oil, and impact of prooxidants and antioxidants.

The assessment of metabolic and toxicological effects of used frying fats offers significant challenges. First, the multitude of reaction products makes the analytical evaluation of oils and fats tested very difficult. Second, animal trials involve those drawbacks associated with formulation of balanced diets, that is, selection of the degree of deterioration, appropriate feeding levels of oil, duration of feeding, and finally, extrapolation of data obtained. Also, one of the problems most often encountered in nutritional studies is the distinction between the oxidized compounds coming from the diet and those produced in vivo, the latter being closely related to the efficiency of dietary antioxidants and enzymatic defense systems. Apart from animal studies, in recent years the number of research groups specializing in the use of cell cultures and human studies has increased considerably.

Among the very different experimental approaches used for selection of samples, a high number of studies test whole oils in experimental diets, or in vitro experiments, or fractions isolated from them. Alternatively, other studies are based on model systems, using either synthesized compounds or fractions obtained from methyl linoleate, or linoleic acid subjected to thermoxidative conditions. Some researchers used extremely abusive conditions in an attempt to generate sufficient amounts of degradation products for their metabolic studies. It is important to keep in mind, however, that compounds formed in this manner may not be representative of those encountered in oils and fats during normal culinary practices.

According to the substrates used in the studies, this chapter is divided into three major sections: i) heated oils and fats, ii) used frying oils and fats, and iii) model systems and specific compounds.

# Heated Oils and Fats

This section discusses studies that start with oils and fats heated at high temperatures without food.

According to the different conditions applied during heating, especially available oxygen, that greatly influence the type of compounds formed, studies were arranged in the following sequence: i) highly oxidized oils and fats subjected to forced aeration, ii) oils and fats heated in the absence of air, and iii) oils and fats thermally oxidized in the presence of air under conditions that mimic those of frying. Although experiments on fats and oils highly oxidized or heated in the absence of air could seem irrelevant to used frying fats, results under such conditions are very valuable for comprehending the reactions occurring during frying and the effects of abusive conditions.

## Highly Oxidized Oils and Fats (Forced Aeration)

The main type of compounds generated by heating with forced aeration are dimers and polymers containing numerous oxygenated functions.

Johnson et al. (20,21) heated different fats and oils to 200°C while aerating at the rate of 100 mL/min and correlated the level of linoleate, and the effect of oxidative damage, to growth inhibition in rats. The rate of absorption of oxidized fats was slower, as was the rate of hydrolysis by pancreatic lipase in vitro when linoleate was increased. Toxicity was attributed to the polymeric fraction (nondistillable residue from the nonurea-adduct of the fatty acids). This fraction was isolated from corn oil held at 200°C for 48 h and fed to rats at 12% of the diet, resulting in 100% mortality within 5 days (22). Similar experiments were later carried out by Ohfuji et al. (23–25). Soybean oils held at 185°C for 90 h during aeration were fed to rats at 20% of the diet for 4 weeks. Thermally oxidized oils were fractionated by silica column chromatography. The fraction eluted with diethyl ether proved to be toxic.

The effect of aeration on the toxicity of whole oils during heating has been reported in animal studies (26–29), but extreme effects were only observed in rats fed highly polyunsaturated fish oil (30,31). In this context, a distillation residue from herring oil held at 185°C during aeration caused 100% mortality among test rats.

Eder and co-workers published a large number of papers during recent years on the metabolic effects of oils heated under various forced aeration temperature and time conditions. Normally, a mixture of lard and sunflower oil was heated at 110°C or 150°C for 6 days, or 190°C for 24 h, or else soybean or sunflower oils were heated from 110–130°C for 22–48 h, respectively, with continuous aeration. In all cases, a high level of oxidative deterioration was observed, as shown by a marked decrease in unsaturated fatty acids and the complete loss of tocopherols. To avoid the potential nutritional effects of the latter changes, both fatty acid composition and vitamin E were adjusted to equal levels in the control diet. Still, it was found that the thermally oxidized lard–sunflower oil mixture reduced thyroid hormone concentrations in miniature pigs (32) and rats (33), affected the desaturation of linoleic and α-linolenic acids in rats (34), hepatic lipogenesis in rats (35), and reduced antioxidant status of erythrocytes in rats and pigs (36). However, it did not adversely affect the lipoprotein profile in the rat model (37). In turn, the thermally oxidized sunflower oil altered thyroid hormone status (38) while only vitamin E deficiency, and not ingestion of thermally oxidized soybean oil, accounted for the stimulated synthesis of hepatic metallothionein isoforms (39).

In other experiments, although the researchers used lower temperatures (50–55°C), the overall conditions also achieved high levels of oxidative deterioration. Thus, the sunflower oil–linseed oil mixture held at 50°C for 16 days, and sunflower oil–lard mixture heated at 55°C for 49 days gave rise to peroxide values of about 800 mEq $O_2$/kg oil and 40% polar compounds, and 1500 mEq $O_2$/kg oil and 53% polar compounds, respectively. Diets containing such oxidized oils influenced the antioxidant status of the fetus when fed to pregnant rats (40), did not affect lipogenic enzymes in mammary glands (41), increased formation of oxysterols in liver (42), altered gene expression of hepatic enzymes (43), affected the selenium status of rats (44), and the morphology and function of the thyroid gland (45).

## Oils and Fats Heated in Absence of Air

Research on oils and fats heated to high temperatures in the absence of oxygen facilitated the ability to obtain very specific compounds, the most notable being cyclic monomers, nonpolar dimers, and polymers.

The early works of Crampton et al. (46–48) in this area are most significant. Inclusion of linseed oil heated under a carbon dioxide atmosphere for 12 h at 275°C fed to rats at 20% of the diet caused weight loss and high mortality. They learned that toxicity lay in the distillable, nonurea-adductable fraction containing cyclic fatty acid monomers. The polymeric fraction caused diarrhea but did not appear to be inherently toxic. These findings were confirmed and extended by many studies using polyunsaturated oils (49–54) and by specific works on cyclic monomers (55–61).

It should be noted, though, that the harmful effects observed are associated with highly unsaturated oils heated over 200°C that favor the formation of cyclic monomers from the intramolecular cyclization of the $C_{18}$ polyunsaturated fatty acids. Cyclic monomers are cause for nutritional concern and constitute the topic of another chapter in this book.

## *Thermally Oxidized Oils and Fats*

Simulation of frying conditions by heating at 180–220°C in the presence of air is, so far, the most common protocol in many of the studies aimed at better understanding the physiological effects of used frying oils and fats, since representative degradation compounds under more controlled conditions can be obtained. However, contrasting results can arise from different experimental approaches. In some cases, for example, the nutritional adequacy of the diet, in terms of the levels of protein, essential fatty acids as linoleic acid, and antioxidants, determined the magnitude of the effects observed.

In general, most studies have not shown chronic toxicity of oils and fats heated at temperatures around 200°C when fed to rats at levels of up to 20% of the diet, although growth retardation and liver enlargement or modifications in hepatic enzymes were usually reported (62–74).

Conversely, other studies reported serious detrimental effects (75–81). Generally, deleterious effects were attributed to the fraction containing cyclic monomers. In this context, probably the most alarming results were obtained by Shue and co-workers (77), who administered 1 mL nonurea-adduct methyl esters to rats by gastric intubation, resulting in 100% mortality in test animals after 8 days. Mortality decreased drastically with balanced diets, however, indicating that dietary nutrients other than heated fat can exert an important influence. Alexander et al. (80,81) fed rats a fraction isolated by vacuum distillation, 1 Torr and temperatures between 150–180°C, from thermally oxidized low-erucic acid rapeseed oil, corn oil, olive oil, and lard. They reported extensive damage from histological evaluations of heart, liver, and kidney tissues.

Dietary protein levels have been shown to play a role in toxic effects of thermally oxidized oils and fats. Witting et al. (82) observed a tendency for lack of growth in rats fed a diet containing heated corn oil and 10% casein. At 30% or more protein in the diet, however, only a mild effect was detected. Also along this line are the studies of Hemans (83) and Govind Rao et al. (84) on the effect of dietary protein and vitamin components. Weanling rats were fed heated corn oil (190°C for 132 h) with protein levels of 8, 10, 20, and 30% of the diet in two separate phases by varying vitamin levels by a factor of 100. Animals fed the heated oil at the low vitamin level gained significantly less weight than those fed fresh corn oil. At a 10% protein level, the effect of heated corn oil on growth was more apparent due to the stress of a low-protein diet. At higher protein levels, the differences in body weight were not significant. The authors concluded that the growth response of rats was apparently due to an interrelationship among the levels of nonvolatile oxidation products, vitamin levels, and protein levels in the total diet (83). Further, the biochemical oxidation effect of lipids, followed by the conversion of $^{14}C$-acetate to expelled $^{14}CO_2$, was studied. In the groups fed 10% protein, one-half of the administered radioactivity was recovered 50 min after acetate injection in rats fed fresh oil, whereas 90 min were required to reach the same level in rats fed heated oil. Lower percentages were recovered in the 20 and 30% protein groups. Around 60% more lipids were deposited in livers of rats fed heated versus fresh oil, and recoveries of total expired $^{14}CO_2$ were lower whenever heated fat was fed. These findings may indicate a preferential utilization of acetate for lipid synthesis in the liver, resulting in accumulation of hepatic lipids that were mainly unsaturated (84). When Andia and Street (85) investigated hepatic microsomal enzyme activities associated with the hepatomegaly commonly observed upon feeding thermally oxidized oils and nonurea-adduct-forming fatty acids, they found that ingestion of corn oil held at 180°C for 24 h increased activities of both methyltransferase and the mixed function oxidase system. These effects were influenced by dietary protein levels.

In general, animal studies on chronic feeding of thermally oxidized oils showed evidence of oxidative stress in the liver, as indicated by the larger induction of hepatic microsomal activities important to metabolizing lipid oxidation products. This includes cytochromes P-450, UDP-glucuronyl transferase, glutathione S-transferase, and superoxide dismutase (85–87). In addition to finding changes in the antioxidative defense system of rats fed 15% soybean oil held at 180°C for 48 h, Corcos Benedetti et al. (87) reported that circulating and stored vitamin A were reduced in rats, with effects appearing earlier in young animals. Serum and liver vitamin E status were suggested as the most sensitive indicators for the antioxidative capability of the organism.

Absorbability of heated fats is one of the most essential ways for obtaining evidence of metabolic interference. Perkins and co-workers (88) approached the evaluation of absorption through the collection of lymph from the abdominal duct of rats. Corn oil was transesterified with 1-$^{14}C$-linoleic acid and subjected to thermal oxidation. The nonurea-adductable fraction of methyl esters was administered to rats via gastric intubation, and lymph was collected for 48 h. Absorption was retarded in comparison to the labeled corn oil used for the control. The maximum absorption peak of nonurea-adduct occurred at 24 h, and the global absorption was around 40%. Lymph lipids were analyzed by thin-layer chromatography in which 103 compounds were isolated, among which 25 monomers were identified as cyclic monomers, long-chain aldehydes and ketones, hydroxy esters, and keto esters. This study was followed by others evaluating the absorption of specific compounds. In this context, the work of Combe et al. (89,90) is also worth commenting on. Lymphatic absorption of nonvolatile oxidation products formed in thermoxidized fats and synthetic triglycerides containing a definite aromatic or alicyclic fatty acid was studied. Results showed that 4% of total polymeric acids, 53% of total oxidized monomeric acids, and 96% of total cyclic monomeric acids were recovered in lymphatic lipids. The different fractions were evaluated by photodensitometry, followed by separation of methyl esters by thin-layer chromatography. Although accurate quantitative determination could not be achieved, valuable data on compounds other than oxidized monomers were provided.

Quite a different approach to determining absorbability is measuring digestibility coefficients. Overall, reports on digestibility indicate significantly lower values for thermoxidized oils compared to fresh oils, and this fact was generally attributed to the presence of the nondistillable nonurea-adductable fraction, that is, the polymeric fraction (91–93).

In studies performed in our laboratory, we applied a combined chromatographic analysis to gain insight into the digestibilities of monomeric, dimeric, and polymeric fatty acyl groups included in thermoxidized oils. Rats were fed diets supplemented with fresh olive oil, olive oil held at 180°C for 150 h, and a 1:1 mixture of nonheated/heated olive oil at 6, 12, and 20% of the diet (94). After 24 days, fecal lipids were extracted, derivatized to methyl esters, and analyzed by a combination of adsorption and high-performance size-exclusion chromatographies (95). The methodology enabled separation and quantitation of five fatty acid methyl ester groups in dietary oils and fecal lipids: nonpolar monomers, oxidized monomers, nonpolar dimers, oxidized dimers, and polymers. We found oil digestibility decreased with increasing degree of alteration, regardless of the dietary oil level. As expected, digestibility of nonpolar (nonoxidized) fatty acids was generally very high. Nevertheless, their digestibility was found to decrease significantly as the extent of alteration of dietary oil increased, independently of the dietary oil content (96). Further experiments showed a marked reduction in the in vitro hydrolysis rate of thermoxidized oils due to the difficulties involved in the pancreatic lipase action on high-molecular weight glyceridic molecules (97). In the last years, Sánchez-Muñiz and co-workers have continued to investigate these aspects, and have greatly contributed to the knowledge of in vitro enzymatic hydrolysis kinetics in thermally oxidized, and used, frying oils and fats (98–101).

In vitro results obtained in our laboratory were confirmed through quantitation of free and esterified levels of nonaltered fatty acids in fecal fats (102). From these results, the specific decrease in digestibility of nonaltered fatty acids as the global oil alteration level is increased may be attributed to impaired lipolysis of polymeric triglycerides, including, in part, intact fatty acyl groups. Digestibility of oxidized fatty acid monomers was considerably high, averaging 76.6%. Among polymeric fatty acids, nonpolar dimers had the lowest digestibility (10.9% on average), whereas oxidized dimers and polymers possessed higher apparent absorbability than expected, ranging from 22.7–49.6% (96). The low digestibility of nonpolar dimers, especially when compared to polar dimers, along with the presence of high free levels in nonabsorbed lipids, specifically suggested difficulties during the absorption process

(102). Analyses of adipose tissue lipids did not show appreciable amounts of alteration compounds, but the profile of fatty acid composition of these triglycerides reflected clear differences in digestibility coefficients, which, in turn, are dependent on dietary content of thermally oxidized oil (103).

Interestingly, excretion of unsaponifiable components was markedly higher in rats on diets containing thermoxidized oils versus nonheated oils, even after correcting for endogenous contribution (104). This suggested that oil alteration may influence endogenous lipid metabolism. In view of these results, major neutral sterols and nonpolar fatty acids originating from microflora metabolism were quantitated in fecal lipids, as representative compounds of the major sources of endogenous lipid products. Fecal endogenous sterols, particularly cholesterol, were significantly higher when diets contained oil, and excretion increased as the dietary oil alteration was greater. Similar results were obtained for endogenous fatty acids. Increases in fecal sterols, dependent on oil alteration, could be explained by the impairment in triglyceride hydrolysis and the subsequent effect on cholesterol micellar solubilization. Additionally, high concentrations of poorly digestible lipids could have led to intestinal microbial modifications (105). Under extreme conditions, Hochgraf et al. also reported greater cholesterol excretion in rats fed diets with 10% lipid model preparation containing 60% highly oxidized linoleic acid (106,107).

A different approach was later used by Sánchez-Muñiz and co-workers to study digestibility, based on true digestibility measurements by means of esophageal probes (108,109). After 4 h experiments in Wistar rats, true digestibility coefficients of thermoxidized palm oleins, measured by lipid disappearance from intestinal lumen, were significantly lower than those of the non-used palm olein. Analyses of the luminal remaining lipids showed that the true digestibility values of triglyceride polymers and triglyceride dimers decreased as the oil alteration was greater and, also, that hydrolysis of non-oxidized triglycerides was negatively affected by the presence of large amounts of thermally oxidized compounds.

As to the effect of dietary thermally oxidized oils in cardiovascular disease events, feeding trials with rabbits have shown a higher incidence of atherosclerosis when diets contained thermoxidized polyunsaturated oils (110). Naruszewicz et al. (111) observed that ingestion of thermally oxidized soybean oil by humans caused certain changes in the metabolism of chylomicrons and, particularly, a rise in their assimilation by macrophages. Oral administration of 100 g thermoxidized oil (heated for 7 h at 220°C) in a test meal increased levels of thiobarbituric acid-reactive substances (TBARS) in chylomicrons. Furthermore, incubation of chylomicrons with mouse macrophages caused a tenfold increase in the accumulation of cholesterol ester mass in the cells than did incubation with chylomicrons from subjects fed fresh oil.

## Used Frying Oils and Fats

The greatest concern about the nutritional effects of used frying fats was expressed over intermittent or discontinuous frying because the highest degradation levels were found under these conditions. Conversely, there is less opportunity for significant alteration in commercial frying operations using continuous frying because of the high turnover and constant protection of the surface oil by steam (water) from the food; or in pan-frying, where oils are heated only for short periods of time and rarely reused.

A survey of the literature on nutritional effects of used frying oils and fats shows discrepancies among studies since, as with studies on oils thermally oxidized in the absence of food, experimental conditions vary widely. In general, information on the used frying oils tested is based exclusively on the duration of heating, temperature selected, and products fried. Therefore, in most cases insufficient analytical criteria are provided to establish valid relationships between degradation compounds in frying fats and effects observed.

With a few exceptions, papers describing works with used frying fats are less alarming than others dealing with artificially abused fats. In general over the years, no evidence of toxicity in terms of biological and histological changes, and only slight effects on growth rate and liver enlargement, have been shown (1,64,66,112–120). Similarly mild effects, if any, were found in those studies of the most practical value using oils and fats from actual commercial frying operations (121–123).

Long-term studies were conducted by Nolen et al. (124) to assess realistic, practical hazards of consuming used frying fats. Cottonseed oil, partially hydrogenated soybean oil, and lard were used for intermittent frying of different foods, with replacement of frying oil or fat only when foaming prevented further use, a point reached after 49, 60, and 116 h, respectively. Rats were fed used frying oils and fats at 15% in the diet for 2 years. Used oils and fats were less absorbable than fresh oils and fats and hence gave slower growth rates. Otherwise, no differences were found in clinical, metabolic, or pathological criteria. Only the distillable nonurea-adductable fractions were toxic when fed by stomach tube to Weanling rats in large quantities. The authors stated that oils and fats under actual frying conditions may cause the formation of toxic substances, but the level and degree of toxicity are too low to have any practical dietary significance. As in the rats, experiments with dogs showed some limitation of weight gain due to the lower coefficient of absorbability of used oil or fat (125).

Probably the most elaborate long-term feeding trials using oils heated under conditions of good commercial practice are the series of studies by Lang and co-workers, which took 10 years to complete. Soybean oils and hardened groundnut oil, heated for 96 h, with and without frying food, at 175°C, were fed to three generations of rats at 10% in the diet without any adverse effects observed. Paradoxically, an appreciable reduction in the incidence of both benign and malignant tumors was noted. The authors attributed these findings to a reduction of carcinogenic polycyclic aromatic hydrocarbons in the heated fats and the induction of microsomal enzymes (115,126–129).

Interestingly, some researchers have approached the evaluation of used frying oils and fats by comparatively testing the same fats heated in the laboratory. Thus

Poling et al. (123) did not find differences between samples of frying oils and fats from restaurants, bakeries, potato chip and doughnut fryers, and similar oils and fats laboratory-heated at 182°C for 120 h. Only slightly and significantly increased liver weights were observed for frying and thermoxidized samples, respectively, as compared to fresh samples. Alexander et al. (130) found that corn oil, peanut oil, and hydrogenated soybean oil from pressure deep-fat fryers producing fried chicken (at 120°C up to 40 cycles over 2 days) resulted in less deterioration than did laboratory-heated oils (at 180°C for 72 h). This was shown by chemical analyses, coefficients of digestibility, and metabolizable energy studies. Extensive histological examinations, however, revealed damage to the thymus and liver when laboratory-heated corn oil or peanut oil and commercial pressure deep-fried peanut oil were fed to the rats. Testes and epididymes were damaged by deep-fried peanut oil and laboratory-heated peanut oil (131). Also, soybean oil thermally oxidized at 200°C was compared with used frying soybean oil administered to rats at 15% of the diet for 28 days, and only the laboratory-heated oil depressed weight gain and feed consumption (132).

In a series of works, the effects of peanut, sesame, and coconut oils heated in an oven at 180°C for 72 h with intermittent stirring were compared with those of the same oils used for frying potato chips for 1.5 h continuously. Rats fed 20% oils in the diet for 5 months did not show any change of erythrocyte fatty acid composition, haematological and histological status (133), nor did they present any deleterious effect on growth, plasma, and tissue lipid profile (134), or changes in digestibility coefficients (135). When testing peanut, rice bran, and palm oil heated and used in frying under the same conditions, results indicated no appreciable damage on hepatic antioxidant enzymes or decreased digestibility (136).

Of special interest among investigations on used frying oils are the extensive studies reported by Billek and co-workers (112). They developed the first analytical method for global quantitation of all degradation compounds present in used frying oils and fats, based on the separation of nonpolar and polar fractions by silica column chromatography (137). Analyzing more than 400 used frying oils, it was concluded that over 30% polar compounds was unacceptable according to sensory and analytical criteria. The authors indicated that oils and fats heated under usual household or commercial practice contained approximately 10–20% polar material. Sunflower oil used in industrial frying of fish and about to be discarded was fractionated by column chromatography. Fresh oil, frying oil, and both nonpolar and polar fractions of used frying oil were fed to rats for 18 months at 20% in the diet. Only diets containing 20% polar compounds caused a small reduction in growth and increased liver and kidney weights. Changes of clinical or histological parameters were not found, except for higher activities of serum glutamic puruvic transaminase and glutamic oxaloacetate transaminase. Frying fats were concluded to be safe, considering that animals recovered rapidly with a normal diet and that the amount of alteration compounds ingested was excessive, with a daily intake of about 1.2 kg polar material if extrapolated to human consumption (112).

Digestibility generally decreases when frying fats are consumed (25,138) although some authors have not found important changes (66,139). More recent studies added information on digestion and absorption of specific groups of polar compounds formed during frying: oxidized triglyceride monomers, triglyceride dimers, and triglyceride polymers. In our laboratory, we demonstrated the difficulties of hydrolysis for polymeric compounds in used frying oils from restaurants, and fried food outlets, collected by Food Inspection Services, containing 7.5–61.4% polar compounds (1.8–47.6% dimers plus polymers). Among polar compounds, oxidized triglyceride monomers were extensively hydrolyzed but dimers and higher oligomers gave low hydrolysis values (11–42%). Furthermore, the hydrolysis rate of nonoxidized, intact triglycerides was markedly affected, e.g., from 95% in a nonheated oil to 52% in the most altered used frying oil analyzed. Even those samples around the limit of rejection showed as much as 20% of nonhydrolyzed nonoxidized triglycerides (140). True digestibility measurements by means of esophageal probes (141) also showed that hydrolysis of nonoxidized triglycerides was negatively affected by the presence of large amounts of polar compounds.

Decreased hydrolysis and absorption of a fraction of used frying oils may have an influence on other nutrients throughout the gastrointestinal tract, but these aspects have often been ignored. In this context, Pérez-Granados et al. reported that inclusion of olive oil, sunflower oil, and palm olein used for potato frying until reaching 25% polar compounds used in rat diets at 8% enhanced magnesium absorption, although magnesium retention was not affected due to increased urinary excretion. Magnesium absorption occurs largely in the colon, and the authors suggested undigested polar compounds may bind magnesium and form some soluble products more easily absorbed, even when in a form that can't be utilized, and hence later excreted in the urine (142).

In studies on lipid and lipoprotein metabolism intended to gain insight into the potential effects of frying fats on cardiovascular disease, contradictory results were obtained (6,143–150). In general, implications of dietary oxidized lipids in lipoprotein metabolism, and their link to the development of atherosclerosis, were largely investigated in humans during recent years (151–156). However, many key questions in this line of evidence remain unanswered (157,158), and one of them is the role of vitamin E in the prevention of cardiovascular disease.

Some specific aspects examined on the effects of used frying oils on cardiovascular disease and associated risk factors are discussed below. In relation to the important process of atherome formation, Giani et al. (144) studied the effects of used frying oils on prostaglandin synthesis. Rats were fed diets containing 10% unsaturated lipids, which had been used for frying potato sticks during ten cycles of 9 min at 180°C, and sufficient levels of vitamin E. Although growth, plasma lipids, and fatty acid profiles of plasma, liver, and heart lipids were unchanged, results showed an increased thromboxane:prostacyclin ratio principally due to higher platelet thromboxane $A_2$. Oxidative stress was observed, since supplementation with pharmacological doses of vitamin E (300 mg/kg diet) neutralized these adverse effects.

In recent studies, it has been investigated whether vegetable oils used in short-term frying (210°C, 8 h) and those obtained in a fast food restaurant after use for a week, could affect postprandial endothelial function in humans, as tested by consuming milkshakes with 60 g of used frying olive or sunflower oils. Postprandial changes were observed only in the case of samples obtained from the fast food restaurant (159,160), which also gave rise to a significant decrease in the activity of serum paraoxonase, an esterase enzyme associated with HDL that hydrolyzes oxidized phospholipids (161). When effects on postprandial serum paraoxonase activity were tested in diabetic patients, a significant increase of activity was observed only in the case of women consuming the olive oil, and the authors suggested that the use of olive oil versus polyunsaturated oils in frying may have a positive effect (162). The latest data obtained by this research group suggest that susceptibility to oxidation of lipoproteins in low antioxidant environments may be increased in the postprandial period following meals rich in polyunsaturated oils, whether heated or unheated (163). Unfortunately, information provided on the oxidized and frying oils tested was too scarce to establish valid relationships between oxidation compounds and the effects observed. As to cardiovascular disease risk factors, Soriguer et al. examined the effect of used frying oils on the risk of hypertension in an extensive cross-sectional study. Oil samples were taken from the kitchens of a random subset of 538 participants and 10% of the oils collected contained over 20% polar compounds. A strong association was found between consumption of such oils and the risk of hypertension, even after inclusion in the models of variables influencing hypertension, such as age, sex, and obesity (164).

Studies on the effect of used frying oils on hepatic metabolism in animals have been numerous. Izaki et al. (165) showed that liver and serum tocopherol levels decreased and hepatic TBARS increased according to the level of deterioration. Interesting correlations were made among several biochemical criteria related to oxidative stress and various deterioration indices of the oils. Liver TBARS were well correlated with oxidation parameters in the used frying oils, such as petroleum ether-insoluble oxidized fatty acids ($r = 0.9191$), polar fraction separated by column chromatography ($r = 0.9056$), glyceride dimer fraction ($r = 0.9023$), and carbonyl value ($r = 0.8647$). It is well known that normal defense mechanisms against in vivo lipid peroxidation include antioxidants (mainly vitamin E) and detoxification enzymes, such as superoxide dismutase, catalase, and glutathione peroxidase, which act in concert as cellular free radical and peroxide scavengers. Huang et al. (166) reported that rat hepatic microsomal enzymes were induced and cytochrome P-450 content increased by using soybean oil for frying potato chips at 200°C for 24 h (four 6 h periods) at 15% in the diet, and that the level of induction was significantly influenced by dietary protein level. These findings agree with other studies already commented on in that deleterious effects of thermally oxidized oils are minimized when rats consume a high-protein diet concomitantly. Further studies in guinea pigs using the same procedure to prepare the oil reported that it contained as much as 54.5% polar compounds, and the results obtained showed evidence of the induction of a hepatic xenobiotic metabolizing enzyme system (167,168) and upregulation of the expression of the hepatic peroxisome proliferator-activated receptor at target genes in rats (169), not finding sex-related differences in the response level (170). In other studies focused on the effect of used frying soybean oil on tocopherol retention and depletion, authors found significantly lower $\alpha$-tocopherol concentrations in rat plasma, liver, kidney, muscle, brain, epididymal fat, and lung for diets containing 50 ppm but effects were alleviated by increasing the content tenfold (171). Further, through a depletion-repletion and radioisotope tracer study, the authors suggested that $\alpha$-tocopherol catabolism and turnover was accentuated (172). Vitamin C supplementation also seemed to restore the impaired vitamin E status of guinea pigs fed the oil (173).

Perkins and Lamboni evaluated partially hydrogenated soybean oils used for frying variable foodstuffs in a fast food restaurant. Oil samples taken out after 4 and 7 days of frying, showing 6.4 and 9.9% triglyceride polymers content, respectively, were fed to rats at 15% of the diet. In the case of the 7-day–used oil, rats showed larger amounts of cytochromes $P_{450}$ and $b_5$, greater activity of NADPH-cytochrome $P_{450}$ reductase, and lower activities of carnitine palmitoyltransferase-I, isocitrate dehydrogenase, and glucose 6-phosphate dehydrogenase (174). After treatment of the used frying oil with magnesium silicate and filtering, the amount of polymers decreased to 4% and positive effects were observed in terms of levels of cytochromes $P_{450}$ and $b_5$ (175). More recently, it has been reported that olive and sunflower oils used for frying affected mitochondrial respiratory chain components and induced oxidative stress in liver microsomes in rats, even at levels of polar compounds lower than 7% of used frying oil, included at only 8% of the diet (176,177).

For many years there has been speculation that used frying fats might be carcinogenic, but most studies failed to detect mutagens in frying oils or a carcinogenic effect in feeding trials of normally used frying oils (1,178–180). The significance of exposure of the digestive tract to various lipid oxidation products known to act as mutagens, promoters, and carcinogenic substances is not clear, however. Studies on model compounds showed that some specific oxidation products are potentially carcinogenic (181–183). Also, volatile lipid oxidation products were found to be mutagenic through inhalation of vapor during frying (184).

Assays on mutagenesis are usually conducted following the Ames test, which uses various histidine-dependent *Salmonella typhimurium* strains as indicator microorganisms. Under the influence of mutagenic substances, some bacteria revert back to the histidine nondependent wild type. There is relatively good agreement between data from this test and data from carcinogenic experiments using animals (179). In this area, contributions of two research groups, namely, those of Taylor and Hageman, deserve special mention.

Taylor et al. (185) found a general lack of mutagens in deep-fried foods obtained at the retail level. Additionally, they carried out repetitive frying experiments using french fried potatoes and fish fillets. The authors concluded that deep-fried foods possessed low levels of mutagenic activity, and only under severe frying conditions were such levels appreciable. It was also pointed out, however, that effects could be much greater in small-volume commercial establishments using abusive frying conditions (186).

Hageman et al. (187) did not completely exclude a more specific role of frying fats in mutagen formation. Deep-frying fats collected from restaurants were fractionated into nonpolar and polar compounds by column chromatography. Polar compounds reached up to 44% in some samples. Mutagenic activity in *Salmonella* assays was found predominantly in polar fractions and was positively correlated with levels of TBARS. Short-term feeding trials showed that consumption of frying oils could cause increased cell proliferation in the gastrointestinal tract and induce oxidative stress, but results did not indicate any effect related to colon tumorigenesis (188). The authors suggested that compounds other than oxidized and polymeric products could be partially responsible for the mutagenic effects observed, such as heterocyclic amines and other pyrolysates, which may form in foods during deep frying and migrate to frying fats. Other researchers found higher mutagenicity in meat samples when butter or margarine was used as the frying fat. These authors suggested that the frying fat, merely as an effective heat conductor, played an essential role (189).

## Model Systems and Specific Compounds

To guarantee the safety of used frying fats, detailed information is necessary about the action of specific alteration compounds at levels higher than those found in the average human diet. In fact, much of the information available is derived from studies on model systems or individual compounds. A major handicap for researchers focused on individual compounds formed during frying is that few specific compounds have been separated and identified so far. Alternatively, some investigators make use of model compounds, usually methyl linoleate or linoleic acid, subjected to thermal oxidation under controlled conditions.

In the following sections, nutritional studies on model systems and specific compounds are discussed, except for cyclic monomers and sterol oxides, which are subjects of other chapters. Primary oxidation products, i.e., hydroperoxides, are first discussed, followed by three groups of secondary compounds arranged by molecular weight, that is, low-molecular-weight compounds coming from breakdown reactions, oxidized compounds of molecular weight similar to parent molecules, and high-molecular weight compounds.

## Hydroperoxides

Although the occurrence of hydroperoxides in used frying fats is limited due to their instability at high temperatures (they are practically absent above 150°C), it should be noted that cooling and storage of fried foods favor the formation of hydroperoxides. Additional interest comes from conversion of hydroperoxides in vivo into more stable oxidized compounds which are similar to those found in used frying oils and fats.

Triglyceride monohydroperoxides appear to be hydrolyzed by pancreatic lipase at almost the same degree as their original triglyceride (190,191). However, previous reactions in the stomach seem to play an important role. Thus, Kanazawa et al. have reported that trilinolein hydroperoxides were hydrolyzed in the rat stomach, and linoleic acid hydroperoxide administered intragastrically was converted into secondary oxidation compounds, including hydroxyls, epoxyketones, hexanal, and 9-oxononanoic acid, in a time-dependent manner (192,193). Hence, it now seems clear why the first studies on fatty acid hydroperoxides showed that they were very toxic to experimental animals when administered intravenously (194,195), but not when given orally. Other authors reported that gastrointestinal glutathione peroxidase plays an important role in the mucosal transport and conversion of hydroperoxides to less reactive hydroxy or aldehydic compounds (196–200). Muller et al. investigated the metabolic fate of radiolabeled 13-hydroperoxyoctaecadienoic acid, 13-hydroxyoctadecanoic acid, and linoleic acid in Caco-2 cell monolayers as a model of the intestinal epithelium, and found that radioactivity was recovered for both hydroperoxy and hydroxy compounds in diglycerides, phospholipids, and cholesterol esterified with oxidized fatty acids. They concluded that food-borne hydroperoxy fatty acids are reduced by the gastrointestinal glutathione peroxidase (201).

Paradoxically, it was suggested that human gastric fluid, which contains absorbed oxygen and a low pH, may be an excellent medium for enhancing oxidation of lipids in the presence of catalysts found in foods. Thus, in vitro experiments using muscle tissues, metmyoglobin or iron ions, incubated in simulated or human gastric fluid, showed enhanced hydroperoxide formation in samples of linoleic acid or soybean oil (202). In a recent series of studies, this research group has also determined the effect of pH, estimated the oxygen content in the stomach, and investigated the interaction of selected prooxidants and antioxidants in gastric fluids (203–205).

When methyl linoleate hydroperoxides were administered to animals, deleterious effects, such as encephalomacia in chicks (206) and creatinuria and erythrocyte hemolysis in rabbits (207), were observed and attributed to vitamin E and essential fatty acid deficiencies. Likewise, damaging effects detected in culture cells were partially counteracted by adding vitamin E (208,209). Antioxidants also inhibited tissue lipid peroxidation induced by short- and long-term feeding of methyl linoleate hydroperoxide to rats, as measured by chemiluminescence intensity and TBARS in the liver, lung, and heart (210,211).

With regard to the extensively studied role of oxidized lipids in atherogenesis, experimental evidence implicating lipid hydroperoxides in all three phases of the atherosclerotic process (i.e., initiation of endothelial injury, accumulation of plaque, and thrombosis) have been reported (212–214). The contribution of hydroperoxides from the dietary fat is necessarily very limited, however, due to its low absorption in the form of intact peroxide. Otherwise, they are most likely formed endogenously under particular dietary circumstances involving impaired antioxidant status (215, 216).

A high number of studies focused on the apparently important role of lipid hydroperoxides in mediating cellular and molecular events in degenerative intestinal disorders, with particular emphasis on colorectal cancer. It has been reported that the gut immune system constitutes a possible line of defense, as indicated by both in vitro experiments using 13-hydroperoxylinoleic acid on human colonic lamina propria lymphocites and in vivo assays (217,218).

Lipid hydroperoxides were reported to stimulate cell proliferation and DNA synthesis and to induce ornithine decarboxylase activity, indicating tumor-promoting effects (219–221). Studies on human colonic Caco-2 cells have shown that subtoxic levels of hydroperoxides disrupted intestinal redox homeostasis, which contributes to apoptosis (222). Complex metabolic effects from chronic exposure to subtoxic levels of hydroperoxides have been reported in rats (223–225) as well as effects on the expression of vascular endothelial growth factor in human colorectal tumor cells (226).

In an interesting paper, Bull and Bronstein (227) demonstrated the production of relatively large amounts of unsaturated carbonyl-containing fatty acids during metabolism of 13-hydroperoxyoctadecadienoic acid by rat colonic tissue. The major product, 13-oxo-9Z,11E-octadecadienoic acid, when instilled intrarectally, stimulated colonic mucosa DNA synthesis and induced ornithine decarboxylase activity in vivo. The interaction of lipid hydroperoxides and their secondary products with DNA was also investigated in vitro by evaluating fluorescence formed in the presence of metals and reducing agents. Hydroperoxy epoxides of linoleate, dihydroperoxides of linoleate, and linolenate were among the most active products (182,183). Other authors also described the reactions of lipid hydroperoxides and aldehydes which lead to DNA strand breakage (228–230). In an excellent review, Kanazawa et al. discussed the role of dietary lipid-derived radicals and nonradical oxidized lipids in carcinogenesis promotion. They stressed the fact that dietary components other than lipids may have a great influence on the apparent adverse effects of the latter, such as consumption of red meat and the metabolic activity of the microflora in the colon (230). Thus, a hemoglobin-iron-rich diet containing safflower oil was found to lead to increased incidence of colon cancer in rats, attributable to the generation of peroxyl radicals from dietary or membrane lipids of intestinal epithelial cells (231).

## Low-Molecular Weight Compounds

A number of papers were published by Kanazawa and co-workers on the absorption, utilization, and tissue distribution of hydroperoxides and their secondary products

(SP), particularly low-molecular weight compounds, by follow-up of radioactivity in rats administered different fractions of [14]C-labeled methyl linoleate or linoleic acid autoxidized at low temperatures. Comparatively, low-molecular weight compounds were more easily absorbed and distributed in rat tissues (232–234). Linoleic acid oxidized at 37°C for 7 days was fractionated by silica gel column. The SP obtained were separated using Sephadex LH-20 gel filtration to components of molecular weight higher (polymer-rich fraction) and lower (aldehyde-rich fraction) than that of hydroperoxides. The total SP fractions comprised about 36% mixed polymers, 26% epoxyhydroperoxides, and 38% low-molecular-weight compounds. In rats fed SP as a whole, 45% of the administered radioactivity was excreted through feces, and 55% was incorporated into the body, of which 50% was discharged by urination and 25% was exhaled. Accumulation in tissues and organs led to an elevation of serum transaminase activities and slight hypertrophy of the liver. Further studies showed significant incorporation of radioactive SP into hepatic mitochondrial and microsomal lipids. Although most of the radioactivity in mitochondria was in the neutral lipid fraction, lipid peroxidation was induced, as indicated by increases of lipid peroxide levels and detoxifying enzyme activities (235). Kanazawa et al. (236) also examined toxicity on intestinal mucosa by measuring activities of sucrase, maltase, and alkaline phosphatase, which can reflect the representative integrity and function of the brush border membrane. Both hydroperoxides and their SP, and especially hydroperoxides, decreased enzyme activities in jejunum at 6 h after exposure, then increased them in jejunum and ileum at 15 h.

The authors suggested that 9-oxononanoic acid may be one of the toxic products in the low-molecular-weight SP fraction. This compound was reported to induce endogenous hepatic lipid peroxidation (237) and to affect hepatic metabolism by decreasing the de novo synthesis of fatty acids (238). Also, the inactivation of some hepatic enzymes was examined in vitro, demonstrating that targets of low-molecular weight SP were mitochondrial glucose-6-phosphate dehydrogenase, CoA, and NAD-dependent aldehyde dehydrogenase, the latter being strongly inactivated by 9-oxononanoate (239). 9-Oxononanoate is the major short-chain glycerol-bound compound generated from the predominant fatty acyls in vegetable oil triglycerides. Thus, it has been recently reported that levels of 9-oxononanoate, determined as methyl ester, in used frying oils and fats with polar compound levels close to 25%, ranged from 0.56–1.61 mg/g of oil (240).

The main mechanism for aldehydes formation from lipid hydroperoxides is homolytic scission of the C–C bonds on either side of the alkoxy radical (241). This cleavage results in two types of aldehydes in the triglyceride molecule: aliphatic aldehydes derived from the methyl terminus of the fatty acid chain, and aldehydes still bound to the parent triglyceride. Biological effects of aldehydes released from the triglyceride molecule, such as 4-hydroxynonenal or 4-hydroxyhexenal, have been studied extensively since the 1990s (242–244). In contrast, glycerol-bound aldehydes, as it is 9-oxononanoate, have rarely been investigated, even though their nutritional importance is greater because they are part of nonvolatile triglycerides present in the diet.

The cytotoxic and mutagenic effects observed for 4-hydroxynonenal and 4-hydroxyhexenal are attributable to their high biological reactivity with proteins and DNA (242,243,245–249). Haynes et al. evaluated the role of each functional group of 4-hydroxy-2-nonenal in its cytotoxicity using analogous compounds and reported that aldehyde and adjacent double bond had the greatest effect, and that increasing chain length yielded increasing toxicity (250). In recent years, interest in 4-hydroxy-2-nonenal has increased so rapidly that it has probably achieved the status of the most studied oxidized structure among the multitude of compounds that can be formed during oxidation or frying. Most recent investigations focus on recognition of genes and pathways for regulating cell-signaling events. There are numerous reports on cell-signaling mechanisms. Citing only some examples, 4-hydroxy-2-nonenal has been suggested to cause cycle arrest and apoptosis, and induce or inhibit proliferation, hence showing seemingly opposite effects (251,252), and being associated with most degenerative diseases, including cancer (253–255), Alzheimer (256), and sclerosis (257). However, the endogenous/exogenous origin of this compound (258), its occurrence in foods, and the amounts required to exert deleterious effects, remain unknown. Only recently, 4-hydroxy-2-nonenal was detected in thermally oxidized soybean oil held at 180°C for 8 h during aeration, showing a maximum concentration as low as 42.5 mg/g of oil (259,260). Potato strips fried in the thermoxidized oil showed similar concentrations of 4-hydroxy-2-nonenal in oils extracted from them (261).

Malonaldehyde (MDA), formed from lipids containing polyunsaturated fatty acids with three or more methylene-interrupted double bonds, is a very reactive compound for cross-linking reactions with proteins and DNA because of its bifunctionality (262,263). However, the mutagenicity of MDA has been questioned (264). The TBA reagent, traditionally used to analyze MDA in foods and biological samples, was criticized since the reaction is not specific and the amounts measured can be overestimated by at least twofold. Most important, it seems likely that dietary MDA is of minor relevance when compared to that formed in vivo. Malonaldehyde usually binds to lysine residues of food proteins, from which it can be released and volatilized during heating. It seems that free MDA is not generated in the course of digestion (262), but one of the DNA adducts formed by malondialdehyde has been detected in human tissues (265).

## Oxidized Monomers

Oxidized monomers refer here to final stable products of molecular weight similar to their parent primary oxidation compounds (hydroperoxides), bearing oxygen-containing functional groups such as aldehyde, ketone, hydroxide, and epoxide. In the case of triglycerides, the major constituents of oils and fats used for frying (over 95%), oxidized triglyceride monomers include triglycerides with oxygenated functions present in one or more fatty acyl chains; hence it is easy to imagine the variety of compounds formed.

Globally, oxidized triglyceride monomers constitute the most abundant group of oxidation compounds found in fried products and foods in general (7). Quantitation of total polar compounds and their distribution in used frying fats and oils around the limit of rejection (25% polar compounds) has shown that the amount of oxidized triglyceride monomers was considerable, ranging from 5.9–9.4% expressed on fat or oil weight (266). The interest in oxidized monomers for nutritional studies is also supported by the high digestibility coefficients of oxidized fatty acyl groups. As already stated, this can be expected, considering that they have greater polarity and do not differ much in molecular weight when compared to unchanged fatty acids.

Recently, a novel analytical approach based on HPLC-electrospray ionization-MS was used to obtain molecular-level information for oxidized triglyceride structures present in small intestinal mucosa, adipose tissue, and lipoproteins of pigs fed diets with oxidized sunflower oils (peroxide value close to 200 mEq $O_2$/kg). The authors did not find triglyceride hydroperoxides in tissues or lipoproteins but they detected secondary oxidation products, such as triglyceride hydroxides, ketones, and epoxides (267,268).

Hydroxy fatty acyl groups are not only present as relatively stable compounds in triglycerides of oxidized and used frying fats, but are also largely formed from the conversion of hydroperoxy fatty acids before absorption. Studies on ricinoleic acid, a naturally occurring hydroxy fatty acid, indicated it is efficiently absorbed from the intestinal tract and deposited in the adipose tissue of rats (269–271), although no evidence was obtained for the incorporation of [14]C-hydroxy acids into liver lipids in their intact form (272). Recently, it has been reported that hydroperoxylinoleic acid and hydroxylinoleic acid, obtained enzymatically from linoleic acid, are efficiently absorbed in vitro by Caco-2 intestinal cells (273), and enhance solubilization and intestinal absorption of cholesterol (274). Only in mice fed cholesterol did 13-hydroxylinoleic acid increase atherogenicity of the plasma cholesterol profile (275). The authors concluded that, in the absence of hypercholesterolemia, oxidized lipids might be favorable since incubation of Caco-2 cells with oxidized linoleic acid led to increased apo A-I secretion and steady-state mRNA expression, whose level is negatively correlated with atherosclerotic cardiovascular disease (276).

In connection with mechanisms responsible for intestinal tumor promotion by dietary fat, Bull et al. (220) investigated the action of hydroxy oleate and stearate, ricinoleic acid, and the a,b-unsaturated ketone derived from hydroxyoctadecenoic acid, after intrarectal instillation to rats. It was concluded that the minimal requirement for stimulation of colon cell proliferation was the presence of an oxidized functionality adjacent to a carbon–carbon double bond. Furthermore, an a,b-unsaturated ketone was suggested as a likely candidate for the ultimate active form. This specific oxidized monomer was also associated with the induction of chick nutritional encephalopathy (277). Production of 2,4-dienone containing fatty acids was later demonstrated during the metabolism of hydroperoxy fatty acids in rat colonic homogenates (227). Other experiments described enzymatic activity that catalyzed the conversion of 13-hydroxyoctadecadienoic acid into 2,4-dienone containing fatty acids. The enzyme activity is widely distributed, with the highest activity in the colon and liver (221, 278).

Absorption of dietary hydroxy and epoxy fatty acids incorporated in triglycerides was recently reported in humans, through an excellent approach based on labeled fatty acids included in triglycerides and analysis by gas liquid chromatography-mass spectrometry (279,280). Hydrolysis of epoxy compounds to render the corresponding diols may also occur before absorption, under the acidic conditions in gastric media (281,282). Wilson et al. reported differences between monoepoxy and diepoxy fatty acids in the rate of absorption, i.e., although their early rate of absorption appeared similar, diepoxy fatty acids were absorbed much less as time went on (279). Specifically, monoepoxy compounds have been recently found to be one of the major oxidized groups formed at frying temperatures (283,284), ranging from 4–10 mg/g oil in real used frying oils at the limit of rejection (25% polar compounds). Therefore, overall results obtained so far stress the relevance of analytical and nutritional studies on epoxides.

Further, cytotoxic effects of monoepoxy linoleate or leukotoxin and its corresponding diol, leukotoxindiol, were reported (285–288), as well as their associations with severe pathological conditions such as adult respiratory distress syndrome, severe burns, and multiple organ failure (289). However, although there is evidence of epoxide formation in vivo through cytochrome P450s (290), physiological levels of epoxides in humans and significance of dietary epoxides are unknown.

Clearly, more research is needed to gain insight into the metabolic and toxicological significance of the numerous oxidized monomeric structures that may be generated from frying (291,292). For this purpose, it is essential to increase our knowledge of compound structures and to determine their occurrence under actual frying conditions (7,284).

## High-Molecular Weight Compounds: Dimers and Polymers

Polymerization reactions are accelerated by the high temperatures used in frying and involve radical and nonradical reactions. As a result, dimers and polymers of complex structure, mainly with C–C and C–O–C linkages, are generated (293–296). Refined oils already have 1–2% of dimeric triglycerides before culinary use, and used frying oils and fats at the limit of rejection (25% polar compounds) contain as much as 12% of dimeric plus polymeric triglycerides (297).

Contradictory results were obtained in early nutritional studies using fractions enriched in polymers, principally nondistillable residues or nondistillable nonurea-adductable fractions, obtained from oils heated under very different alteration conditions. In general, only when rats were fed abnormally high levels of polymer concentrates (12–20% of the diet), adverse effects, such as depressed growth and high mortality rate, were found. The toxicity could have been caused in large part by malabsorption of essential food constituents as a result of diarrhea.

One of the first studies on model systems was performed by Bottino (298), who tested the residual polymeric fraction obtained after distillation of methyl linoleate held at 300°C for 10 h in the absence of air. Significant amounts administered to rats by stomach tube gave rise to abnormal growth curves, marked diarrhea, and loss of weight and hair, but toxicity disappeared after the addition of fresh fat to the diet. Using quite different alteration conditions, Michael et al. (299) heated methyl linoleate in the presence of air for 200 h at 200°C. Monomeric, dimeric, and polymeric fractions were concentrated by molecular distillation. Dimers were further fractionated by silica column chromatography into polar and nonpolar fractions. Heated methyl linoleate, polymeric material, and nonpolar dimers were innocuous, whereas polar dimers were toxic when administered to Weanling male rats by stomach tube at 0.5 mL/day for three consecutive days.

In feeding experiments aimed at testing acute and chronic toxicity of dimeric triglycerides from refined oils, mice were fed a fraction containing 20% nonpolar dimeric triglycerides at 15% in the diet for one year; extensive examination of the animals showed no toxic effects. A high percentage of nonpolar dimeric fatty acids were excreted with feces. The acute $LD_{50}$ (24 h and 7 days) was higher than 20 mL/kg (300).

Measurements of absorption are of primary importance in evaluating biological activity of dimers and polymers, but unfortunately the literature is confusing in this regard. Some authors showed apparent digestibility of dimers between 30 and 70% (24,49,298), but others questioned such high values (91).

Studies on nonpolar dimers, either cyclic or noncyclic, indicated very low lymph recoveries. Hsieh and Perkins (301) found in lymph cannulation studies, approximately 0.4% of the labeled cyclic nonpolar dimeric fatty acid methyl esters was absorbed within 12 h. Similarly, values as low as 1% were later found for lymphatic absorption of nonpolar dimers (90). These levels seemed to be underestimated, however, as parallel experiments with rats fed labeled dimers showed recoveries of around 3% radioactivity in urine and $CO_2$. Furthermore, approximately 80% of the radioactivity was recovered in the gastrointestinal tract (ca. 40%) and feces after 48 h (300). Likewise, later studies on noncyclic nonpolar dimers showed an average of 85% of radioactivity in the gastrointestinal tract and feces, plus about 5% recovered as $CO_2$ following gastric intubation (302). Rats fed either noncyclic or cyclic nonpolar dimers up to a maximum realistic level of 0.75% in the diet did not exhibit significant differences in growth rate, feed efficiency, and liver size from those fed normal diets (301, 302).

In contrast, polar dimers and polar polymers appeared to be comparatively better absorbed than nonpolar dimers in lymph cannulation studies (90). Feeding trials in our laboratory using thermoxidized labeled linoleic acid at 1% in diets showed higher digestibility values for oxidized dimers and polymers than expected, as evaluated by a combination of chromatographic techniques and radioactivity measurements (303). This could be due, in part, to depolymerization reactions occurring under the strongly acidic conditions in the stomach, as suggested by the presence of nonaltered labeled fatty acids in feces, which were absent in diets.

Although complexity of polymers is a major handicap for nutritional studies, two key points support further analytical and nutritional research. First, polymers constitute a major fraction in used frying oils and fats, and second, their low absorption does not necessarily mean lack of health risk but involves increased levels of nondigested, nonabsorbed fat throughout the gastrointestinal tract that might potentially affect epythelial cells and microflora metabolism. In connection with this subject, it has been reported that both unabsorbed fats and bile acids secreted in response to a high fat intake might injure the intestinal mucose by their detergent activity, and metabolites of bile acids formed by intestinal bacteria (secondary bile acids) act as tumor promoters (304). Also, an interesting aspect is the potential contribution of intestinal flora to the production of mutagens from the oxidation of fecal lipids and the effect of vitamin E as a chemopreventive agent (305,306).

## Concluding Remarks

From the literature reviewed, there is general agreement that some compounds found in used frying oils and fats can impair their nutritional value or be potentially harmful. Therefore, of primary importance is improvement of the quality control of used frying fats and fried products, particularly in the case of fried foods prepared through discontinuous frying processes in which the used frying oil can reach a considerable degradation level.

Overall, a great part of the nutritional studies reported on the effects of used frying oils and fats lacks the analytical data necessary to establish valid relationships between the alteration compounds present and the extensively evaluated effects on molecular targets, metabolic pathways, and chronic diseases. An interesting aspect approached by other research groups is the thorough investigation of the metabolic effects of specific compounds formed during frying but so far focusing only on the few compounds that can be easily synthesized and characterized.

In order to avoid confounding and alarming results, and to define clearly the physiological role of dietary used frying oils and fats, a crucial research assignment is the development of methodologies directed toward defining the chemical structure and actual levels of nonvolatile compounds generated during frying, those present in the food chain and in the diet. Improvements in analytical evaluation are also essential to increase knowledge on bioavailability and thus achieve an accurate risk assessment of the degradation compounds ingested. Finally, a specific aspect that requires further exploration is the fate of poorly absorbed compounds present in used frying oils and fats, in terms of their interactions with other nutrients, with gastrointestinal mucose and fluids, and with microflora metabolism.

# Acknowledgments

This work was supported in part by Ministerio de Educación y Ciencia (Project AGL 2004-00148) and Junta de Andalucía.

## References

1. Artman, N.R. *Adv. Lipid Res.* **1969**, *7*, 245.
2. Billek, G. *Fat Sci. Technol.* **1992**, *94*, 161.
3. Billek G. *Eur. J. Lipid Sci. Technol.* **2000**, *102*, 587.
4. Causeret, J. *Cahier Nutr. Diet* **1982**, *17*, 19.
5. Clark, W.L.; and G.W. Serbia. *Food Technol.* **1991**, *45*, 84.
6. Cuesta, C.; F.J. Sánchez-Muñiz; and G. Varela. In *Frying of Food*; G. Varela, A.E. Bender, and I.D. Morton, Eds.; Ellis Horwood: Chichester, U.K., 1988; pp. 112–128.
7. Dobarganes, M.C.; and G. Márquez-Ruiz. *Curr. Opin. Clin. Nutr. Metabol. Care* **1991**, *6*, 157.
8. Grandgirard, A. *Ann. Nutr. Aliment.* **1980**, *34*, 377.
9. Grandgirard, A. In *Aspects Nutritionnels des Constituants des Aliments. Influences des Technologies*; A. Bernard and H. Carlier, Eds.; Technique et Documentation: Lavoisier, Paris, 1992; pp. 49–67.
10. Lee, D.M. *Survey of the Literature on the Heating of Fats and Oils*; The British Food Manufacturing Industries Research Association: Leatherhead, Surrey, U.K., 1973.
11. Mahungu. S.M.; W.E. Artz; and E.G. Perkins. In *Frying of Food*; D. Boskou and I. Elmadfa, Eds.; Technomic Publishing: Lancaster, PA, 1999; pp. 25–46.
12. Márquez-Ruiz, G.; M.C. Pérez-Camino; and M.C. Dobarganes. *Grasas Aceites* **1990**, *41*, 432.
13. Perkins, E.G. *Rev. Fr. Corps Gras* **1976**, *23*, 313.
14. Potteau, B.; and J. Causeret. *Rev. Fr. Corps Gras* **1976**, *18*, 591.
15. Ruiz-Gutiérrez, V. *Grasas Aceites* **1987**, *38*, 326.
16. Saguy I.S.; and D. Dana. *J. Food Eng.* **2003**, *56*, 143.
17. Viola, P.; and A. Bianchi. In *Frying of Food*; G. Varela, A.E. Bender, and I.D. Morton, Eds.; Ellis Horwood: Chichester, U.K., 1988; pp. 129–138.
18. International Union of Pure and Applied Chemistry (IUPAC). Standard Method 2.507. In *Standard Methods for the Analysis of Oils, Fats and Derivatives*, 7th ed.; IUPAC, Ed.; Blackwell: Oxford, U.K., 1992.
19. Firestone D. In *Deep Frying: Chemistry, Nutrition and Practical Applications*; E.G. Perkins and M.D. Erickson, Eds.; AOCS Press: Champaign, IL, 1996; pp. 323–334.
20. Johnson, O.C.; E.G. Perkins; M. Sugai; and F.A. Kummerow. *J. Am. Oil Chem. Soc.* **1957**, *34*, 594.
21. Johnson, O.C.; T. Sakuragi; and F.A. Kummerow. *J. Am. Oil Chem. Soc.* **1957**, *33*, 433.
22. Perkins, E.G.; and F.A. Kummerow. *J. Nutr.* **1957**, *68*, 101.
23. Ohfuji, T.; and T. Kaneda. *Lipids* **1973**, *8*, 353.
24. Ohfuji, T.; S. Iwamoto; and T. Kaneda. *J. Japan Oil Chem. Soc.* **1970**, *19*, 887.
25. Ohfuji, T.; K. Sakurai; and T. Kaneda. *J. Japan Oil Chem. Soc.* **1972**, *21*, 68.
26. Kieckebusch, W.; K. Jahr; G. Czok; W. Griem; K.H. Baessler; D.C.H. Hammar; and K. Lang. *Fette Seifen Anstrichm.* **1972**, *64*, 1154.
27. Simko, V.; A. Bucko; J. Babala; and R. Ondreicka. *Nutr. Dieta* **1963**, *6*, 91.

28. Nwanguma, B.C.; A.C. Achebe; L.U.S. Ezeanyika; and L.C. Eze. *Food Chem. Toxicol.* **1999,** *37,* 413.

29. Ammouche, A.; F. Rouaki; A. Bitam; and M.M. Bellal. *Ann. Nutr. Metab.* **1999,** *46,* 268.

30. Raulin, J.; and J. Petit. *Arch. Sci. Physiol.* **1999,** *16,* 77.

31. Raulin, J.; and T. Terroine. *Arch. Sci. Physiol.* **1999,** *16,* 89.

32. Eder, K. *J. Anim. Physiol. Anim. Nutr.* **1999,** *82,* 271.

33. Eder, K.; P. Skufca; and C. Brandsch. *J. Nutr.* **1999,** *132,* 1275.

34. Eder, K. *Lipids* **1999,** *34,* 717.

35. Eder, K.; A. Suelzle; P. Skufca; C. Brandsch; and F. Hirche. *Lipids* **2003,** *38,* 31.

36. Keller, U.; C. Brandsch; and K. Eder. *J. Anim. Physiol. Anim. Nutr.* **2004,** *88,* 59.

37. Eder, K.; U. Keller; F. Hirche; and C. Brandsch. *J. Nutr.* **2004,** *133,* 2830.

38. Eder, K.; and G.I. Stangl. *J. Nutr.* **2000,** *130,* 116.

39. Stangl, G.I.; K. Nostelbascher; K. Eder; and M. Kirchgessner. *Eur. J. Nutr.* **2000,** *39,* 112.

40. Brandsch, C.; and K. Eder. *Br. J. Nutr.* **2004,** *92,* 267.

41. Brandsch, C.; N. Nass; and K. Eder. *J. Nutr.* **2004,** *134,* 631.

42. Keller, U.; C. Brandsch; and K. Eder. *Eur. J. Nutr.* **2004,** *43,* 353.

43. Sulzle, A.; F. Hirche; and K. Eder. *J. Nutr.* **2004,** *134,* 1375.

44. Skufca, P.; K. Schafer; C. Brandsch; O. Simon; and K. Eder. *Trace Elem. Electrolytes* **2003,** *20,* 45.

45. Skufca, P.; C. Brandsch; F. Hirche; and K. Eder. *Annals Nutr. Metab.* **2003,** *47,* 207.

46. Crampton, E.W.; R.H. Common; F.A. Farmer; F.M. Berryhill; and L. Wiseblatt. *J. Nutr.* **1951,** *44,* 177.

47. Crampton, E.W.; R.H. Common; F.A. Farmer; A.F. Wells; and D. Crawford. *J. Nutr.* **1953,** *49,* 333.

48. Crampton, E.W.; R.H. Common; E.T. Pritchard; and F.A. Farmer. *J. Nutr.* **1956,** *60,* 13.

49. Friedman, L.; W. Horwitz; G.M. Shue; and D. Firestone. *J. Nutr.* **1961,** *73,* 85.

50. Gabriel, H.G.; J.C. Alexander; and V.E. Valli. *Nutr. Rep. Int.* **1979,** *20,* 411.

51. Grandgirard, A. *Ann. Biol. Anim., Bioch. Biophys.* **1978,** *18,* 287.

52. Potteau, B. *Nutr. Aliment.* **1976,** *30,* 67.

53. Potteau, B. *Nutr. Ailment.* **1976,** *30,* 89.

54. Potteau, B.; and A. Grandgirard. *Ann. Biol. Anim. Biochem. Biophys.* **1974,** *14,* 855.

55. Iwaoka, W.T.; and E.G. Perkins. *J. Am. Oil Chem. Soc.* **1978,** *55,* 734.

56. Le Queré, J.L.; E. Semon; B. Lanher; and J.-L. Sébédio. *Lipids* **1989,** *24,* 347.

57. Rojo, J.A.; and E.G. Perkins. *J. Am. Oil Chem. Soc.* **1987,** *64,* 414.

58. Sébédio, J.L.; and A. Grandgirard. *Prog. Lipid Res.* **1989,** *28,* 303.

59. Sébédio, J.L.; J.L. Le Quere; E. Semon; O. Morin; J.M. Vatele; and A. Grandgirard. *J. Am. Oil Chem. Soc.* **1987,** *64,* 1324.

60. Sébédio, J.L.; J.L. Le Quere; O. Morin; J.M. Vatele; and A. Grandgirard. *J. Am. Oil Chem. Soc.* **1989,** *66,* 704.

61. Sébédio, J.L.; J. Prevost; and A. Grandgirard. *J. Am. Oil Chem. Soc.* **1987,** *64,* 1026.

62. Alfin-Slater, R.B.; S. Auerbach; and L. Aftergood. *J. Am. Oil Chem. Soc.* **1959,** *36,* 638.

63. Coquet, B.; D. Guyot; X. Fouillet; and J.L. Rouaud. *Rev. Fr. Corps Gras* **1977,** *24,* 483.

64. Lanteaume, M.T.; P. Ramel; P. Acker; A.M. Le Clerc; and C. Wirth. *Rev. Fr. Corps Gras*

**1968**, *15*, 71.

65. Lanteaume, M.T.; P. Ramel; A.M. Le Clerc; and J. Rannaud. *Rev. Fr. Corps Gras* **1966**, *13*, 603.

66. Le Floch, E.; P. Acker; P. Ramel; M.T. Lanteaume; and A.M. Le Clerc. *Ann. Nutr. Aliment.* **1968**, *22*, 249.

67. Miller, K.W.; and P.H. Long. *Food Chem. Toxicol.* **1990**, *28*, 307.

68. Osim, E.E.; D.U. Owu; E.U. Isong; and I.B. Umoh. *Discovery Innovat.* **1994**, *6*, 389.

69. Owu, D.U.; E.E. Osim; and P.E. Ebong. *Acta Tropica* **1998**, *69*, 65.

70. Poling, C.E.; E. Eagle; E.E. Rice; A.M.A. Durand; and M. Fisher. *Lipids* **1970**, *5*, 128.

71. Ramel, P.; M.T. Lanteaume; A.M. Le Clerc; and J. Rannaud. *Rev. Fr. Corps Gras* **1967**, *14*, 505.

72. Rodríguez-Consuegra, M.A.; V. Ruiz-Gutiérrez; P. Sanz; R. Gutiérrez; and M. Repetto. *Rev. Toxicol.* **1985**, *2*, 149.

73. Ruiz-Gutiérrez, V.; P. Sanz; C. Domínguez; R. Gutiérrez; and M. Repetto. *Grasas Aceites* **1985**, *36*, 120.

74. Sanz, P.; V. Ruiz-Gutiérrez; M.C. Rodríguez-Vicente; P. Villar; R. Gutiérrez; and M. Repetto. *Rev. Toxicol.* **1984**, *1*, 135.

75. Binet, L.; and G. Wellers. *Ann. Nutr. Aliment.* **1966**, *20*, 25.

76. Indart, A.; M. Viana; M.C. Grootveld; C.J.L. Silwood; I. Sánchez-Vera; and B. Bonet. *Free Radical Res.* **2002**, *36*, 1051.

77. Kantorowitz, B.; and S. Yannai. *Nutr. Rep. Intl.* **1974**, *9*, 331.

78. Shue, G.M.; C.D. Douglass; D. Firestone; L. Friedman; and J.S. Sage. *J. Nutr.* **1968**, *94*, 171.

79. Yang, C.M.; C.W.C. Kendall; D. Stamp; A. Medline; M.C. Archer; and W.R. Bruce. *Nutr. Cancer Int. J.* **1998**, *30*, 69.

80. Alexander, J.C. *J. Am. Oil Chem. Soc.* **1978**, *55*, 711.

81. Gabriel, H.G.; J.C. Alexander; and V.E. Valli. *Lipids* **1978**, *13*, 49.

82. Witting, L.A.; T. Nishida; O.C. Johnson; and F.A. Kummerow. *J. Am. Oil Chem. Soc.* **1957**, *34*, 421.

83. Hemans, C.; F. Kummerow; and E.G. Perkins. *J. Nutr.* **1973**, *103*, 1665.

84. Govind Rao, M.K.; C. Hemans; and E.G. Perkins. *Lipids* **1973**, *8*, 342.

85. Andia, A.M.G.; and J.C. Street. *J. Agr. Food Chem.* **1975**, *23*, 173.

86. Ashwin, J.L.; P.G. Harris; and J.C. Alexander. *Nutr. Res.* **1991**, *11*, 79.

87. Corcos Benedetti, P.; M. Di Felice; V. Gentili; B. Tagliamonte; and G. Tomassi. *Ann. Nutr. Metab.* **1990**, *34*, 221.

88. Perkins, E.G.; S.M. Vachha; and F.A. Kummerow. *J. Nutr.* **1970**, *100*, 725.

89. Combe, N.; M.J. Constantin; and B. Entressangles. *Rev. Fr. Corps Gras* **1978**, *25*, 27.

90. Combe, N.; M.J. Constantin; and B. Entressangles. *Lipids* **1978**, *16*, 9.

91. Kajimoto, G.; and K. Mukai. *J. Japan Oil Chem. Soc.* **1970**, *19*, 66.

92. Potteau, B.; A. Grandgirard; M. Lhuissier; and J. Causeret. *Bibl. Nutr. Dieta* **1977**, *25*, 122.

93. Potteau, B.; M. Lhuissier; J. Le Clerc; F. Custot; R. Mezonnet; and R. Cluzan. *Rev. Fr. Corps Gras* **1970**, *17*, 143.

94. Márquez-Ruiz, G.; M.C. Pérez-Camino; V. Ruiz-Gutiérrez; and M.C. Dobarganes. *Grasas Aceites* **1991**, *42*, 32.

95. Márquez-Ruiz, G.; M.C. Pérez-Camino; and M.C. Dobarganes. *J. Chromatogr.* **1990**,

*514*, 37.
96. Márquez-Ruiz, G.; M.C. Pérez-Camino; and M.C. Dobarganes. *J. Am. Oil Chem. Soc.* **1992,** *69*, 930.
97. Márquez-Ruiz, G.; M.C. Pérez-Camino; and M.C. Dobarganes. *Fat Sci. Technol.* **1992,** *94*, 307.
98. Arroyo, R.; F.J. Sánchez-Muñiz; C. Cuesta; F.J. Burguillo; and J.M. Sánchez-Montero. *Lipids* **1996,** *31*, 1133.
99. Arroyo, R.; F.J. Sánchez-Muñiz; C. Cuesta; J.V. Sinisterra; and J.M. Sánchez-Montero. *J. Am. Oil Chem. Soc.* **1997,** *74*, 1509.
100. Sánchez-Muñiz, F.J.; R. Arroyo; J.M. Sánchez-Montero; and C. Cuesta. *Food Sci. Technol. Int.* **2000,** *6*, 449.
101. Sánchez-Muñiz, F.J.; and J.M. Sánchez-Montero. In *Frying of Food*; D. Boskou and I. Elmadfa, Eds.; Technomic Publishing: Lancaster, PA, 1999; pp. 105–141.
102. Márquez-Ruiz, G.; M.C. Pérez-Camino; and M.C. Dobarganes. *Ann. Nutr. Metab.* **1993,** *37*, 121.
103. Salgado, A.; G. Márquez-Ruiz; and M.C. Dobarganes. *Grasas Aceites* **1992,** *43*, 87.
104. Márquez-Ruiz, G.; M.C. Pérez-Camino; V. Ruiz-Gutiérrez; and M.C. Dobarganes. *Grasas Aceites* **1992,** *43*, 198.
105. Márquez-Ruiz, G.; and M.C. Dobarganes. *Scand. J. Gastroenterol.* **1992,** *27*, 1069.
106. Hochgraf, E.; S. Mokady; and E. Cogan. *J. Nutr.* **1997,** *127*, 681.
107. Hochgraf, E.; S. Mokady; and E. Cogan. *J. Nutr. Biochem.* **2000,** *11*, 176.
108. Sánchez-Muniz, F.J.; S. Bastida; and M.J. Gónzalez-Muñoz. *Lipids* **1999,** *34*, 1187.
109. González-Muñoz, M.J.; S. Bastida; and F.J. Sánchez-Muñiz. *J. Sci. Food Agric.* **2003,** *83*, 413.
110. Kritchevsky, D.; and S.A. Tepper. *J. Atheroscl. Res.* **1967,** *7*, 647.
111. Naruszewicz, M.; E. Wozny; E. Mirkiewicz; G. Nowicka; and W.B. Szostak. *Atherosclerosis* **1987,** *66*, 45.
112. Billek, G. In *The Role of Fats in Human Nutrition*; F.B. Padley and J. Podmore, Eds.; Ellis Horwood: Chichester, U.K., 1985; pp. 163–171.
113. Causeret, J.; B. Potteau; and A. Grandgirard. *Ann. Nutr. Aliment.* **1978,** *60*, 483.
114. Homrowski, S. *J. Nutr. Diet.* **1969,** *6*, 22.
115. Lang, K. *Fette Seifen Anstrichm.* **1973,** *75*, 73.
116. López-Varela, S.; F.J. Sánchez-Muñiz; and C. Cuesta. *Food Chem. Toxicol.* **1995,** *33*, 181.
117. López-Varela, S.; F.J. Sánchez-Muñiz; A.M. Pérez-Granados; and C. Cuesta. *J. Physiol. Biochem.* **1998,** *54*, 23.
118. Perry, M.N.; and A.M. Campbell. *J. Am. Diet. Assoc.* **1968,** *53*, 575.
119. Ramel, P.; M.T. Lanteaume; A.M. Le Clerc; J. Rannaud; and E. Morel. *Rev. Fr. Corps Gras* **1965,** *12*, 517.
120. Rodríguez, A.; C. Cuesta; F.J. Sánchez-Muñiz; and G. Varela. *Grasas Aceites* **1965,** *35*, 22.
121. Keane, K.W.; G.A. Jacobson; and G.H. Krieger. *J. Nutr.* **1959,** *68*, 57.
122. Nolen, G.A. *J. Am. Oil Chem. Soc.* **1972,** *49*, 688.
123. Poling, C.E.; W.D. Warner; P.E. Mone; and E.E. Rice. *J. Nutr.* **1960,** *72*, 109.
124. Nolen, G.A.; J.C. Alexander; and N.R. Artman. *J. Nutr.* **1967,** *93*, 337.
125. Nolen, G.A. *J. Nutr.* **1973,** *103*, 1248.

126. Kracht, K.; K. Lang; and J. Henschel. *Z. Ernahrungswiss.* **1974,** *13,* 132.
127. Lang, K. *Fette Seifen Anstrichm.* **1974,** *76,* 145.
128. Lang, K. *Z. Ernahrungswiss.* **1978,** *(Suppl. 21).*
129. Lang, K.; E.H. von Jan; and J. Henschel. *Z. Ernahrungswiss.* **1978,** *9,* 363.
130. Alexander, J.C.; B.E. Chanin; and E.T. Moran. *J. Food Sci.* **1983,** *48,* 1289.
131. Alexander, J.C.; V.E. Valli; and B.E. Chanin. *J. Toxicol. Environ. Health* **1983,** *21,* 295.
132. Chanin, B.E.; V.E. Valli; and J.C. Alexander. *Nutr. Res.* **1988,** *8,* 921.
133. Narasimhamurthy, K.; S. Vishwanatha; and P.L. Raina. *Nutr. Res.* **1998,** *18,* 1245.
134. Narasimhamurthy, K.; and P.L. Raina. *Mol. Cell. Biochem.* **1999,** *195,* 143.
135. Narasimhamurthy, K.; and P.L. Raina. *Eur. Food Res. Technol.* **2000,** *210,* 402.
136. Purushothama, S.; H.D. Ramachandran; K. Narasimhamurthy; and P.L. Raina. *Mol. Cell. Biochem.* **2003,** *247,* 95.
137. Billek, G.; G. Guhr; and J. Waibel. *J. Am. Oil Chem. Soc.* **1978,** *55,* 728.
138. Alexander, J.C. *Lipids* **1978,** *13,* 254.
139. Varela, G.; O. Moreiras-Varela; B. Ruiz-Rozo; and R. Conde. *J. Sci. Food Agric.* **1986,** *37,* 487.
140. Márquez-Ruiz, G.; G. Guevel; and M.C. Dobarganes. *J. Am. Oil Chem. Soc.* **1998,** *75,* 119.
141. González-Muñoz, M.J.; S. Bastida; and F.J. Sánchez-Muniz. *J. Agric. Food Chem.* **1998,** *46,* 5188.
142. Pérez-Granados, A.M.; M.P. Vaquero; and M.P. Navarro. *J. Sci. Food Agric.* **1999,** *79,* 699.
143. Garrido-Polonio, M.C.; M.C. García-Linares; M.T. García-Arias; S. López-Varela; M.C. García-Fernández; A.H.M. Terpstra; and F.J. Sánchez-Muñiz. *Br. J. Nutr.* **2004,** *92,* 257.
144. Giani, E.; J. Masi; and C. Galli. *Lipids* **1985,** *20,* 439.
145. Guillaumin, R.; B. Coquet; and J.D. Rouand. *Ann. Nutr. Aliment.* **1978,** *32,* 467.
146. Jethmalani, S.M.; G. Viswanathan; C. Bandyopadhyay; and J.M. Noronha. *Ind. J. Exp. Biol.* **1989,** *27,* 1052.
147. Lu, Y.F.; and Y.C. Lo. *Nutr. Res.* **1995,** *15,* 1783.
148. Liu, J.F.; and C.J. Huang. *J. Nutr.* **1995,** *125,* 3071.
149. Sánchez-Muñiz, F.J.; C. Cuesta; A. Rodríguez; and G. Varela. *Rev. Esp. Fisiol.* **1986,** *42,* 105.
150. Sánchez-Muñiz, F.J.; S. López-Varela; M.C. Garrido-Polonio; and C. Cuesta. *J. Sci. Food Agric.* **1998,** *76,* 364.
151. Riemersma, R.A.; R. Wilson; J.A. Payne; and M.J. Shepherd. *Free Radical Res.* **2003,** *37,* 341.
152. Staprans, I.; D.A. Hardman; X.M. Pan; and K.R. Feingold. *Diabetes Care* **1999,** *22,* 300.
153. Staprans, I.; J.H. Rapp; X.M. Pan; J.H. Rapp; K.Y. Kim; and K.R. Feingold. *Artheriosc. Thrombosis* **1994,** *14,* 1900.
154. Ursini, F.; and A. Sevanian. *Biol. Chem.* **2002,** *383,* 599.
155. Ursini, F.; A. Zamburlini; G. Cazzolato; M. Maiorino; G. Bittolo; and A. Sevanian. *Free Radical Biol. Med.* **1998,** *25,* 250.
156. Wilson, R.; K. Lyall; E.M. Millar; L. Smyth; C. Pearson; and R.A. Riemersma. *Thromb. Haemost.* **2003,** *89,* 654–659.

157. Cohn, J. *Curr. Opin. Lipidol.* **2002,** *13,* 19.
158. Stanley, J. *Lipid Technol.* **2002** (*May*), 59.
159. Williams, M.J.A.; W.H.F. Sutherland; M.P. McCormick; S.A. De Jong; R.J. Walker; and G.T. Wilkins. *J. Am. Coll. Cardiol.* **1999,** *33,* 1050.
160. Williams, M.J.A.; W.H.F. Sutherland; M.P. McCormick; D. Yeoman; S.A. De Jong; and R.J. Walker. *Nutr. Metab. Cardiovasc. Dis.* **2001,** *11,* 147.
161. Sutherland, W.H.F.; R.J. Walker; S.A. De Jong; A.M. van Rij; V. Phillips; and H.L. Walker. *Artheriosc. Thromb. Vasc. Biol.* **1999,** *19,* 1340.
162. Wallace, A.J.; W.H.F. Sutherland; J.I. Mann; and S.M. Willliams. *Eur. J. Clin. Nutr.* **2001,** *55,* 951.
163. Sutherland, W.H.F.; S.A. De Jong; R.J. Walker; M.J.A. Williams; C.M. Skeaff; A. Duncan; and M. Harper. *Atherosclerosis* **2002,** *160,* 195.
164. Soriguer, F.; G. Rojo-Martínez; M.C. Dobarganes; J.M. García-Almeida; I. Esteva; M. Beltrán; M.S. Ruiz-De Adana; F. Tinahones; J.M. Gómez-Zumaquero; E. García-Fuentes; and S. González-Romero. *Am. J. Clin. Nutr.* **2003,** *78,* 1092.
165. Izaki, Y.; S. Yoshikawa; and M. Uchiyama. *Lipids* **1984,** *19,* 324.
166. Huang, C.-J.; N.-S. Cheung; and V.-R. Lu. *J. Am. Oil Chem. Soc.* **1988,** *65,* 1796.
167. Liu, J.F.; and F.C. Chang. *J. Nutr. Sci. Vitaminol.* **2000,** *46,* 240.
168. Liu, J.F.; Y.W. Lee; and F.C. Chang. *J. Nutr. Sci. Vitaminol.* **2000,** *46,* 137.
169. Chao, P.M.; C.Y. Chao; F.J. Lin; and C.J. Huang. *J. Nutr.* **2001,** *131,* 3166.
170. Chao, P.M.; S.C. Hsu; F.J. Lin; Y.J. Li; and C.J. Huang. *Lipids* **2004,** *39,* 233.
171. Liu, J.F.; and C.J. Huang. *J. Nutr.* **1995,** *125,* 3071.
172. Liu, J.F.; and C.J. Huang. *J. Nutr.* **1996,** *126,* 2227.
173. Liu, J.F.; and Y.W. Lee. *J. Nutr.* **1997,** *128,* 116.
174. Lamboni, C.; and E.G. Perkins. *Lipids* **1996,** *31,* 955.
175. Perkins, E.G.; and C. Lamboni. *Lipids* **1998,** *33,* 683.
176. Battino, M.; J. Quiles; J.R. Huertas; M.C. Ramírez-Tortosa; M. Cassinello; M. Mañas; M. López-Frías; and J. Mataix. *J. Bioenerg. Biomembr.* **2002,** *34,* 127.
177. Quiles, J.L.; J.R. Huertas; M. Battino; M.C. Ramírez-Tortosa; M. Cassinello; J. Mataix; M. López-Frías; and M. Mañas. *Br. J. Nutr.* **2002,** *88,* 57.
178. Kalpagam, P.; and C. Rukmini. *J. Oil Technol. Assoc. Ind.* **1987,** *19,* 15.
179. Scheutwinkel-Reich, M.; G. Ingerowski; and H.-J. Stan. *Lipids* **1980,** *15,* 849.
180. Van Gastel, A.; R. Mathur; V.V. Roy; and C. Rukmini. *Food Chem. Toxicol.* **1984,** *22,* 403.
181. Earles, S.M.; J.C. Bronstein; D.L. Winner; and A.W. Bull. *Biochim. Biophys. Acta* **1991,** *1081,* 174.
182. Fujimoto, K.; W.E. Neff; and E.N. Frankel. *Biochim. Biophys. Acta* **1984,** *795,* 100.
183. MacGregor, J.T.; R.E. Wilson; W.E. Neff; and E.N. Frankel. *Food Chem. Toxicol.* **1985,** *23,* 1041.
184. Gao, Y.T.; W.J. Blot; W. Zheng; A.G. Ershow; C.W. Hsu; L.I. Levin; R. Zhang; and J.F. Fraumeni. *Int. J. Cancer* **1987,** *40,* 604.
185. Taylor, S.L.; C.M. Berg; N.H. Shoptaugh; and E. Traisman. *J. Am. Oil Chem. Soc.* **1983,** *60,* 576.
186. Taylor, S.L.; C.M. Berg; N.H. Shoptaugh; and V.N. Scott. *Food Chem. Toxicol.* **1982,** *20,* 209.
187. Hageman, G.; R. Kikken; F. Ten Hoor; and J. Kleinjans. *Mutat. Res.* **1988,** *204,* 593.

188. Hageman, G.; H. Verhagen; B. Schutte; and J. Kleinjans. *Food Chem. Toxicol.* **1991,** *29,* 689.
189. Barrington, P.J.; R.S.U. Baker; A.S. Truswell; A.M. Bonin; A.J. Ryan; and A.P. Paulin. *Food Chem. Toxicol.* **1990,** *28,* 141.
190. Matsushita, S. *J. Agric. Food Chem.* **1975,** *23,* 150.
191. Miyashita, K.; T. Takagi; and E.N. Frankel. *Biochim. Biophys. Acta* **1990,** *1045,* 233.
192. Kanazawa, K.; and H. Ashida. *Biochim. Biophys. Acta* **1998,** *1393,* 336.
193. Kanazawa, K.; and H. Ashida. *Biochim. Biophys. Acta* **1998,** *1393,* 349.
194. Cortesi, R.; and O.S. Privett. *Lipids* **1972,** *7,* 715.
195. Findlay, G.M.; H.H. Draper; and J.G. Bergan. *Lipids* **1970,** *5,* 970.
196. Aw, T.Y. *Free Radical Res.* **1998,** *28,* 637.
197. Aw, T.Y. *Toxicol. Applied Pharmacol.* **2005,** *204,* 320.
198. Kaneda, T.; and T. Miyazawa. *World Rev. Nutr. Diet.* **1987,** *50,* 186.
199. Kowalski, D.P.; R.M. Feely; and D.P. Jones. *J. Nutr.* **1990,** *120,* 1115.
200. Wingler, K.; C. Muller; K. Schmehl; S. Florian; and R. Brigelius-Flohe. *Gastroenterology* **2000,** *119,* 420.
201. Muller, C.; R. Friedrichs; K. Wingler; and R. Brigelius-Flohe. *Biol. Chem.* **2002,** *383,* 637.
202. Kanner, J.; and T. Lapidot. *Free Radical Biol. Med.* **2001,** *31,* 1388.
203. Gorelik, S.; T. Lapidot; I. Shaham; R. Granit; M. Ligumsky; R. Kohen; and J. Kanner. *J. Agric. Food Chem.* **2005,** *53,* 3397.
204. Lapidot, T.; R. Granit; and J. Kanner. *J. Agric. Food Chem.* **2005,** *53,* 3383.
205. Lapidot, T.; R. Granit; and J. Kanner. *J. Agric. Food Chem.* **2005,** *53,* 3391.
206. Nishida, T.H.; H. Tsuchiyama; M. Inoue; and F.A. Kummerow. *Proc. Soc. Exp. Biol. Med.* **1960,** *105,* 308.
207. Kokatnur, M.G.; J.G. Bergan; and H.H. Draper. *Proc. Soc. Exp. Biol. Med.* **1966,** *123,* 254.
208. Hennig, B.; and C.K. Chow. *Adv. Free Radical Biol. Med.* **1988,** *4,* 99.
209. Hennig, B.; C. Enoch; and C.K. Chow. *Nutr. Res.* **1987,** *7,* 1253.
210. Miyazawa, T.; A. Nagaoka; and T. Kaneda. *Agric. Biol. Chem.* **1983,** *47,* 1333.
211. Miyazawa, T.; C. Sato; and T. Kaneda. *Agric. Biol. Chem.* **1983,** *47,* 1577.
212. Chisolm, G.M.; and D. Steinberg. *Free Radical Biol. Med.* **2000,** *28,* 1815.
213. Jessup, W.; L. Krithairides; and R. Stocker. *Biochem. Soc. Trans.* **2004,** *32,* 134.
214. Stocker, R.; and J.F. Keany. *Physiol. Rev.* **2004,** *84,* 1381.
215. Duthie, G.G.; K.W.J. Wahle; and W.P.T. James. *Nutr. Res. Rev.* **1989,** *2,* 51.
216. Yagi, K. *Chem. Phys. Lipids* **1987,** *45,* 337.
217. Elitsur, Y.; A.W. Bull; and G.D. Luk. *Dig. Dis. Sci.* **1990,** *35,* 212.
218. Gualde, N.; H. Chable-Rabinovuitch; C. Motta; J. Durand; J.L. Beneytout; and M. Rigaud. *Biochim. Biophys. Acta* **1983,** *750,* 429.
219. Bull, A.W.; N.D. Nigro; W.A. Golembieski; J.D. Crissman; and L.J. Marnett. *Cancer Res.* **1984,** *44,* 4924.
220. Bull, A.W.; N.D. Nigro; and L.J. Marnett. *Cancer Res.* **1988,** *48,* 1771.
221. Earles, S.M.; J.C. Bronstein; D.L. Winner; and A.W. Bull. *Biochim. Biophys. Acta* **1991,** *1081,* 174.
222. Wang, T.G.; Y. Gotoh; M.H. Jennings; C.A. Rhoads; and T.Y. Aw. *FASEB J.* **2000,** *14,* 1567.

223. Tsunada, S.; R. Iwakiri; T. Noda; K. Fujimoto; J. Fuseler; C.A. Rhoads; and T.Y. Aw. *Digest. Dis. Sci.* **2003,** *48,* 210.

224. Tsunada, S.; R. Iwakiri; K. Fujimoto; and T.Y. Aw. *Digest. Dis. Sci.* **2003,** *48,* 2333.

225. Tsunada, S.; R. Iwakiri; T. Noda; K. Fujimoto; J. Fuseler; C.A. Rhoads; and T.Y. Aw. *Gastroenterology* **2003,** *124,* A261.

226. Jurek, D.; N. Udilova; A. Jozkowicz; H. Nohl; B. Marian; and R. Shulte-Hermann. *Pharmacology* **2004,** *72,* 151.

227. Bull, A.W.; and J.C. Bronstein. *Carcinogenesis* **1990,** *11,* 1699.

228. Kanazawa, A.; T. Sawa; T. Akaik; and H. Maeda. *Cancer Letters* **2000,** *156,* 51.

229. Kanazawa, A.; T. Sawa; T. Akaik; and H. Maeda. *Eur. J. Lipid Sci. Technol.* **2002,** *104,* 439.

230. Yang, M.H.; and K.M. Schaich. *Free Radical Biol. Med.* **1996,** *20,* 225.

231. Sawa, T.; T. Akaike; K. Kida; Y. Fukushima; K. Takagi; and H. Maeda. *Cancer Epidemiol. Biomarkers Prev.* **1998,** *7,* 1007.

232. Kanazawa, K.; E. Kanazawa; and M. Natake. *Lipids* **1985,** *20,* 412.

233. Oorada, M.; T. Miyazawa; K. Fujimoto; E. Ito; K. Terao; and T. Kaneda. *Agric. Biol. Chem.* **1988,** *52,* 2101.

234. Oorada, M.; T. Miyazawa; and T. Kaneda. *Lipids* **1986,** *21,* 150.

235. Kanazawa, K.; H. Ashida; S. Minamoto; and M. Natake. *Biochim. Biophys. Acta* **1986,** *879,* 36.

236. Kanazawa, K.; H. Ashida; S. Minamoto; G. Danno; and M. Natake. *J. Nutr. Sci. Vitaminol.* **1988,** *34,* 363.

237. Minamoto, S.; K. Kanazawa; H. Ashida; G. Danno; and M. Natake. *Agric. Biol. Chem.* **1985,** *49,* 2747.

238. Minamoto, S.; K. Kanazawa; H. Ashida; and M. Natake. *Biochim. Biophys. Acta* **1988,** *958,* 199.

239. Kanazawa, K.; and H. Ashida. *Arch. Biochem. Biophys.* **1991,** *288,* 71.

240. Velasco, J.; S. Marmesat; G. Márquez-Ruiz; and M.C. Dobarganes. *J. Agric. Food Chem,* **2005,** *53,* 4006.

241. Frankel, E.N. *Prog. Lipid Res.* **1982,** *22,* 1.

242. Esterbauer, H.; H. Zollner; and R.J. Schaur. In *Membrane Lipid Oxidation;* C. Vigo-Pelfrey, Ed.; CRC Press: Boca Raton, FL, 1990; *Vol. 1,* pp. 239–268.

243. Esterbauer, H.; R.J. Schaur; and H. Zollner. *Free Radical Biol. Med.* **1991,** *11,* 81.

244. KamalEldin, A.; and L.A. Appelqvist. *Grasas Aceites* **1996,** *47,* 342.

245. Carbone, D.L.; J.A. Doorn; and D.R. Petersen. *Free Radical Biol. Med.* **2004,** *37,* 1430.

246. Chung, F.L.; R.G. Nath; J. Ocando; A. Nishikawa; and L. Zhang. *Cancer Res.* **2000,** *60,* 1507.

247. Hu, W.; Z. Zhaohui; J. Eveleigh; G. Iyer; J. Pan; S. Amin; F.L. Chung; and M.S. Tang. *Carcinogenesis* **2002,** *23,* 1781.

248. Kurtz, A.J.; and R.S. Lloyd. *J. Biol. Chem.* **2003,** *278,* 5970.

249. Oe, T.; J.S. Arora; S.H. Lee; and I.A. Blair. *J. Biol. Chem.* **2003,** *278,* 42098–42105.

250. Haynes, R.L.; L. Szweda; K. Pickin; M.E. Welker; and A.J. Townsend. *Mol. Pharmacol.* **2000,** *58,* 788.

251. Laurora, S.; E. Tamagno; F. Briatore; P. Bardini; S. Pizzimenti; C. Toaldo; P. Reffo; P. Costelli; M.U. Dianzani; O. Danni; and G. Barrera. *Free Radical Biol. Med.* **2005,** *38,* 215.

252. Sharma, R.; D. Brown; S. Awasthi; Y. Yang; A. Sharma; B. Patrick; M.K. Saini; S.P. Singh; P. Zimmiak; S.V. Singh; and Y.C. Awasthi. *Eur. J. Biochem.* **2004**, *271*, 1690.

253. Chung, F.L.; R.G. Nath; J. Ocando; A. Nishikawa; and L. Zhang. *Cancer Res.* **2000**, *60*, 1507.

254. Feng, Z.; W. Hu; and M.S. Tang. *Biochem.* **2004**, *101*, 8598.

255. West, J.D.; C.A. Ji; S.T. Duncan; V. Amarnath; C. Schneider; C.J. Rizzo; A.R. Brash; and L.J. Marnett. *Chem. Res. Toxicol.* **2004**, *17*, 453.

256. Liu, Q.; M.A. Smith; J. Avila; J. DeBernardis; M. Kansal; A. Takeda; X.W. Zhu; A. Nunomura; K. Honda; P.I. Moreira; et al. *Free Radical Biol. Med.* **2005**, *38*, 746–754.

257. Perluigi, M.; H.F. Poon; K. Hensley; W.M. Pierce; J.B. Klein; V. Calabrese; C. De Marco; and D.A. Butterfield. *Free Radical Biol. Med.* **2005**, *38*, 960.

258. Uchida, K. *Progr. Lipid Res.* **2003**, *42*, 318.

259. Seppanen, C.M.; and A.S. Csallany. *J. Am. Oil Chem. Soc.* **2001**, *78*, 1253.

260. Seppanen, C.M.; and A.S. Csallany. *J. Am. Oil Chem. Soc.* **2002**, *79*, 1033.

261. Seppanen, C.M.; and A.S. Csallany. *J. Am. Oil Chem. Soc.* **2004**, *81*, 1137.

262. Draper, H.H.; and M. Hadley. *Xenobiotica* **1990**, *20*, 901.

263. Summerfield, F.W.; and A.L. Tappel. *Mutat. Res.* **1984**, *126*, 113.

264. Marnett, L.J.; and M.A. Tuttle. *Cancer Res.* **1980**, *40*, 276.

265. Marnett, L.J. *Mutation Res.* **1999**, *424*, 83.

266. Márquez-Ruiz, G.; M. Tasioula-Margari; and M.C. Dobarganes. *J. Am. Oil Chem. Soc.* **1995**, *72*, 1171.

267. Suomela, J.P.; M. Ahotupa; and H. Kallio. *Lipids* **2005**, *40*, 349.

268. Suomela, J.P.; M. Ahotupa; O. Sjovall; J.P. Kurvinen; and H. Kallio. *Lipids* **2004**, *39*, 639.

269. Govind Rao, M.K.; N. Risser; and E.G. Perkins. *Proc. Soc. Exp. Biol. Med.* **1969**, *131*, 1369.

270. Kieckebusch, W.; W. Griem; G. Czok; K.H. Baessler; E. Degkwitz; E. Schaeffner; and K. Lang. *Z. Ernaehrungswiss.* **1963**, *4*, 26.

271. Perkins, E.G.; J.G. Endres; and F.A. Kummerow. *J. Nutr.* **1961**, *73*, 291.

272. Reber, R.J.; and H.H. Draper. *Lipids* **1970**, *5*, 983.

273. Penumetcha, M.; N. Khan-Merchant; and S. Parthasarathy. *J. Lipid Res.* **1970**, *41*, 1473.

274. Penumetcha, M.; N. Khan-Merchant; and S. Parthasarathy. *J. Lipid Res.* **2002**, *43*, 895.

275. Khan-Merchant, N.; M. Penumetcha; O. Meilhac; and S. Parthasarathy. *J. Nutr.* **2002**, *132*, 3256.

276. Rong, R.; S. Ramachandran; M. Penumetcha; N. Khan; and S. Parthasarathy. *J. Lipid Res.* **2002**, *43*, 557.

277. Budowski, P.; I. Bartov; Y. Dror; and E.N. Frankel. *Lipids* **2002**, *14*, 768.

278. Bull, A.W.; S.M. Earles; and J.C. Bronstein. *Prostaglandins* **1991**, *41*, 43.

279. Wilson, R.; C.E. Fernie; C.M. Scrimgeour; K. Lyall; L. Smyth; and R.A. Riemersma. *Eur. J. Clin. Invest.* **2002**, *32*, 79.

280. Wilson, R.; K. Lyall; L. Smyth; C.E. Fernie; and R.A. Riemersma. *Free Radical Biol. Med.* **2002**, *32*, 162.

281. Zaher, F.A.; and S.M. Elshami. *Grasas Aceites* **1990**, *41*, 361.

282. Piazza, G.J.; A. Nunez; and T.A. Foglia. *Lipids* **2003**, *38*, 255.

283. Velasco, J.; S. Marmesat; O. Berdeaux; G. Márquez-Ruiz; and M.C. Dobarganes. *J. Agric.*

*Food Chem.* **2004,** *52,* 4438.

284. Velasco, J.; S. Marmesat; G. Márquez-Ruiz; and M.C. Dobarganes. *Eur. J. Lipid Sci. Technol.* **2004,** *106,* 728.
285. Mitchell, L.A.; J.H. Moran; and D.F. Grant. *Toxicology Letters* **2002,** *126,* 187.
286. Moran, J.H.; T. Mont; T.L. Hendrickson; L.A. Mitchell; and D.F. Grant. *Chem. Res. Toxicol.* **2001,** *14,* 431.
287. Moran, J.H.; G. Nowak; and D.F. Grant. *Toxicol. Appl. Pharmacol.* **2001,** *172,* 150.
288. Slim, R.; B.D. Hammock; M. Toborek; L.W. Robertson; J.W. Newman; C.H.P. Morisseau; B.A. Watkins; V. Saraswathi; and B. Hennig. *Toxicol. Appl. Pharmacol.* **2001,** *171,* 184.
289. Kosaka, K.; K. Suzuki; M. Hayakawa; S. Sugiyama; and T. Ozawa. *Mol. Cell. Biochem.* **1994,** *139,* 141–148.
290. Moran, J.H.; L.A. Mitchell; J.A. Bradbury; W. Qu; D.C. Zeldin; R.G. Schnellman; and D.F. Grant. *Toxicol. Appl. Pharmacol.* **2000,** *168,* 268.
291. Artman, N.R.; and J.C. Alexander. *J. Am. Oil Chem. Soc.* **1968,** *45,* 643.
292. Artman, N.R.; and D.E. Smith. *J. Am. Oil Chem. Soc.* **1968,** *49,* 318.
293. Christopoulou, C.N.; and E.G. Perkins. *J. Am. Oil Chem. Soc.* **1989,** *66,* 1360.
294. Dobarganes, M.C. *OCL* **1998,** *5,* 41.
295. Dobarganes, M.C.; and Márquez-Ruiz. In *Deep Frying: Chemistry, Nutrition and Practical Applications*; E.G. Perkins, and M.D. Erickson, Eds.; AOCS Press: Champaign, IL, 1996; pp. 89–111.
296. Ottaviani, P.; J. Graille; P. Perfetti; and M. Naudet. *Chem. Phys. Lipids* **1979,** *24,* 57.
297. DGF (German Society for Fat Research). *Proceedings of the 3rd International Symposium of Deep-Fat Frying—Final Recommendations, Eur. J. Lipid Sci. Technol.* **1979,** *102,* 594.
298. Bottino, N.R. *J. Am. Oil Chem. Soc.* **1962,** *39,* 25.
299. Michael, W.R.; J.C. Alexander; and N.R. Artman. *Lipids* **1966,** *1,* 353.
300. Strauss, H.-J.; H. Piater; and W. Sterner. *Fette Seifen Anstrichm.* **1982,** *84,* 199.
301. Hsieh, A.; and E.G. Perkins. *Lipids* **1976,** *11,* 763.
302. Perkins, E.G.; and R. Taubold. *J. Am. Oil Chem. Soc.* **1978,** *55,* 632.
303. Márquez-Ruiz, G.; and M.C. Dobarganes. *J. Chromatog. B* **1978,** *675,* 1.
304. Vonk, R.J.; M. Kalivianakis; D.M. Minich; C.M.A. Bijleveld; and H.J. Verkade. *Scand. J. Gastroenterol.* **1997,** *32,* 65.
305. Campbell, S.; W.L. Stone; S. Whaley; and K. Krishnan. *Crit. Rev. Oncology Hematology* **2003,** *47,* 249.
306. Stone, W.L.; and A.M. Papas. *J. Nat. Cancer Inst.* **1997,** *89,* 1006.

# ·•10•·

# Physiological Effects of *trans* and Cyclic Fatty Acids

**Jean-Louis Sébédio[a], Jean-Michel Chardigny[b], and Corinne Malpuech-Brugère[c]**

*[a]Human Nutrition Unit, Mass spectrometry platform UMR INRA- Université D'auvegrne, Centre de Clermont-Theix, 63122 Saint Genes Champanelle, France; [b]Univ. Clermont 1, UMR1019, UFR Nedecene, Cemont-Ferrand, F-63001 France; [c]CRNH Auverand, Clermont-Ferrand, F-63001 France*

## Introduction

Frying is a complex process that involves lipids, water, temperature, and oxygen. Consequently, many reactions are taking place such as oxidation, hydrolysis, polymerization, cyclization, and isomerization. The reactions of cyclization and isomerization are due to the temperature used to fry foods, and the products formed are dependent not only on the temperature that is utilized, but also the nature, and more specifically, the degree of unsaturation of the fatty acids (oleic versus linoleic versus linolenic acids, for example). Sébédio and Juaneda reported structures of the major *trans* fatty acids and cyclic fatty acid monomers in Chapter 5. This chapter will deal with the possible physiological effects of these components.

## *trans* Fatty Acids

### Intestinal Absorption

Several *trans* isomers of linoleic, and to a larger extent α-linolenic acids, are present in the human diet. The question of their intestinal absorption, which may influence their bioavailability, is of primary importance. Using rats as an experimental model, Trus et al. (1) demonstrated that the major *trans* isomers of α-linolenic acid, 18:3 9*cis*,12*cis*,15*trans*, 18:3 9*trans*,12*cis*,15*cis* and 18:3 9*trans*,12*cis*,15*trans* from heated linseed oil (240°C, 10 h), were well absorbed (>98.8%). No differences between oil (triacylglycerols), free fatty acids, or esters resulted, suggesting that these dietary forms are suitable for experimental studies. Moreover, it means that *trans* isomers of α-linolenic acid are absorbed in the intestinal tract as well as α-linolenic acid itself.

The content in *trans* isomers of linoleic acid in food is lower than α-linolenic acid isomers, particularly for di-*trans* isomers. However, it was demonstrated by Ono and

Fredrickson (2) that linolelaidic acid (i.e., 18:2 9*trans*,12*trans*) is absorbed to a lesser extent than its *cis* counterpart, linoleic acid. On the other hand, the extent of linoleic acid mono-*trans* isomers absorption in animal models has not been assessed. Taken altogether, these data suggest *trans* isomers of essential fatty acids are made available for all metabolic pathways like their *cis* analog.

## Incorporation in Biological Tissues

### 18:3 Isomers

Piconneaux reported that *trans* 18:3 isomers are incorporated in all liver lipid classes of rats fed heated linseed oil, a dietary model used extensively to assess the metabolic fate of *trans* 18:3 (3). Similar data were reported in other tissues, including brain and retina (4,5,6). Interesting data on acylation in cardiolipids was reported by Wolff et al. (7). They showed that 9*cis*,12*cis*,15*trans*-18:3 is preferentially incorporated in the 1(1′) position of liver mitochondrial cardiolipids, as what is observed for linoleic acid. It was concluded that the acylation fate of the *trans* double bond was that of a saturated bond.

Similarly, Bretillon et al. reported the same pattern in liver phosphatidylcholine (8). They demonstrated that the 15*trans* isomer of $C_{18:3}$ behaves like linoleic acid and is acylated at the sn2 position for about 90% (Fig. 10.1). At the same time, linoleic acid was acylated at the sn2 position for 95%, whereas less than 80% of α-linolenic acid was acylated at this position. Taken together, these data indicate that the *trans*

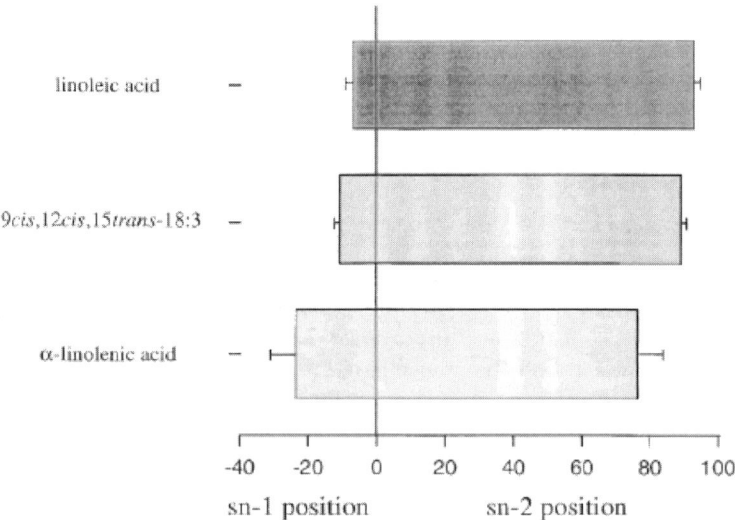

**Fig. 10.1.** $^{14}$C repartition between sn-1 and sn-2 positions of phosphatidylcholine of isolated rat livers infused with [1-$^{14}$C]-linoleic acid, α-linolenic acid and 9*cis*,12*cis*,15*trans*-18:3. Data are means ± SEM of 3 independent determinations.

double bond is considered a saturated bond for acylation in tissue lipids.

In human tissues, Adlof and Emken (9) mentioned these isomers did not occur. Perhaps analytical procedures at the time of the study were not powerful enough to detect these minor fatty acids. More recent data shows that human milk (10,11), as well as plasma and blood cells (12,13), contain *trans* isomers of 18:3. However, these data are a few years old now. Unfortunately, to our knowledge, no recent data are available for samples collected.

### 18:2 Isomers

In 1991, Beyers and Emken reported on the incorporation of deuterium-labeled mono-*trans* isomers of linoleic acid fed fat-free diets (14). They showed that after 4 days of intake, the 9*trans* 18:2 isomer was the most incorporated (5.57 mg/g wet weight in the mouse liver). The 12*trans* isomer content was 1.41 mg/g wet weight, which was not significantly different from the control (9*cis*,12*cis*-18:2).

Berdeaux et al. (15) also showed that both mono-*trans* isomers were incorporated in liver, heart, testes, and adipose tissue lipids of rats fed 4 weeks with *trans* 18:2 isomers from heated sunflower oil. The 9*trans*,12*cis* isomer was also better incorporated in all tissues, which may be related to its low bioconversion (see below).

In humans, linoleic acid mono-*trans* isomers were reported in milk and adipose tissue. In the latter, the 9*cis*,12*trans* isomer was incorporated more than the 9*trans*,12*cis*. Koletzko also reported the occurrence of linolelaidic acid in human milk as the major 18:2 *trans* isomer (16), but the analytical procedures should probably be re-examined, as this isomer has never been reported elsewhere. The 9*cis*,13*trans* 18:2 isomer, resulting from catalytic hydrogenation of oils and fats as demonstrated by Ratnayake et al. (17), may be co-eluted (18).

A French study called "The Aquitaine study" dealt with *trans* fatty acids in tissue lipids from parturient women (18). In adipose tissue, the major *trans* isomers were 18:1 fatty acids, which result from catalytic hydrogenation of oils and biohydrogenation in ruminants. However, some 18:2 *trans* isomers were also detected, with the 9*cis*,12*trans* isomer as the major one again (0.13 ± 0.06 % of total fatty acids, n = 90). On the other hand, cholesteryl esters from umbilical plasma contained *trans* isomers of linoleic acid, with the 9*trans*,12*cis* isomer being the major one (19).

# Bioconversion

### 18:3 Isomers

Some $C_{18:3}$ isomers are bioconverted into desaturated and/or elongated metabolites. Besides *trans* isomers of 18:3 (see above), metabolites with 20 and 22 carbon chain lengths were also identified and reported to occur in tissues. In the liver from rats fed heated linseed oil, the major *trans* n-3 fatty acid is 17-*trans*-EPA. 19 *trans* DHA was also identified (20).

These long-chain *trans* fatty acids are described in numerous tissues of animals, including heart, testes, kidneys, platelets, and plasma (6). Also, particular attention was paid to brain structures and retina, which are generally DHA-rich tissues. 19 *trans* DHA is reported in myelin and synaptosomes of rat brain (5), whereas specific brain structure (cortex, hippocampus, striatum, cerebellum) was more recently studied in depth in young (newborn piglet) (21) and aged animals (aging rat) (22). We reported that during brain development, *trans* n-3 fatty acids are incorporated to a small extent, representing less than 0.03% of total fatty acids in the frontal cortex or the hippocampus (23). This is lower than what was observed in liver and plasma (21). In the retina, the major *trans* fatty acid incorporated is 19 *trans* DHA, but it never reached 1% of total phospholipids fatty acids (24), representing only 2.6% of total $C_{22:6}$ fatty acids after 12 months of diet.

$C_{20}$ and $C_{22}$ metabolites of the 18:3 9*cis*,12*cis*,15*trans* were also reported in human platelets and plasma (12,13). The 17 *trans* isomer of eicosapentaenoic acid (EPA), and in some cases the 19 *trans* isomer of docosahexaenoic acid (DHA), were detected in these human samples, as previously described in liver from rats fed heated linseed oil (20). These findings suggest that humans, like rats, can desaturate and elongate the 15 *trans* isomer of linolenic acid. Animal studies showed these long-chain *trans* n-3 polyunsaturated fatty acids (PUFA) are incorporated in all tissues studied, including liver (20), heart, testes, and kidney (6).

Using an isolated perfused liver model, we also showed that the *trans*9,*cis*12,*cis*15-18:3 isomer was better converted via elongation to *trans* 20:3 than desaturated and further elongated (8). This was a confirmation of the low $\Delta^6$ desaturation of *trans*9*cis*12*cis*15-18:3 observed in vitro with liver microsomes (25). This was also in agreement with previous data indicating that a *cis* double bond at the $\Delta^9$ position is necessary for a fatty acid to be a substrate for the $\Delta^6$ desaturase (25).

The biosynthesis of *trans* 22:6 showed involvement of $C_{24}$ PUFA, as reported for DHA synthesis with elongations and $\Delta^6$ desaturation rather than a $\Delta^4$ desaturase (26). In cultured bovine endothelial cells, we identified *trans* $C_{24:5}$ and *trans* $C_{24:6}$ as intermediates between 17*trans*-20:5 and 19*trans*-22:6 (27).

## 18:2 Isomers

Data from studies by Beyers and Emken (14), as well as older ones (28), were previously reviewed in the first edition of this book. Briefly, both mono- and *trans*-isomers of linoleic acid (9*cis*,12*trans* and 9*trans*,12*cis*) are desaturated and elongated into *trans* isomers of arachidonic acid (mainly 20:4 5*cis*, 8*cis*,11*cis*, 14*trans* and, to a lesser extent, $C_{20:4}$ 5*cis*, 8*cis*, 11*trans*, 14*cis*). The 18:2 9*trans*,12*cis*, however, was preferentially elongated and then desaturated at the $\Delta^5$ position, leading to a "dead end" product (20:3 5*cis*, 11*trans*, 14*cis*). This pathway was not demonstrated for the 18:2 9*cis*,12*trans* isomer, whereas it was reported for all *cis* $C_{18:2}$ fatty acids (linoleic acid). These data demonstrate that the position of the *trans* double bond is of particular importance for the metabolic fate of these fatty acids.

The study of Beyers and Emken (14) illustrated possible metabolic routes for geometric isomers of linoleic acid. It is important to note that the dietary *trans* isomers were the only source of dietary lipids and that the level was high (2% by weight of diet). Recently, we fed rats a purified fraction of *trans* isomers of linoleic acid using much lower levels of *trans* fatty acids (0.5% by weight of dietary lipids) included in an equilibrated diet. Under these conditions, the 18:2 9*cis*,12*trans* isomer was transformed into 20:4 14*trans*, 5 times more than the 18:2 9*trans*, 12*cis* isomer into 20:4 11*trans*. This was illustrated by the difference in the 18:2 9*cis*,12*trans*/18:2 9*trans*,12*cis* ratio in the diet (0.81) and in liver phospholipids (0.16) (15).

New data were obtained from the isolated perfused liver model (8). As reported for 18:3 isomers (see above), the *trans*-9 isomer was preferentially elongated into "dead end" product, whereas the *trans*-12 isomer was a good substrate for the $\Delta^6$ desaturase (Table 10.1), as reported earlier in mice (14) or using liver microsomes (29).

## Oxidative Metabolism

One of the major metabolic pathways for essential fatty acids is β-oxidation (30). We performed some studies to assess this aspect for *trans* isomers of 18:2 and 18:3. In rats force-fed with a bolus of $[1\text{-}^{14}C]$-radio-labeled fatty acids (31), we observed that as their *cis* counterpart, *trans* isomers were extensively used for β-oxidation. $^{14}CO_2$ production ranged between 60 and 70% of the dose ingested. No differences were observed between α-linolenic acid and its $\Delta^9$- and $\Delta^{15}$-*trans* isomers. On the other hand, the 12 *trans*-18:2 isomer was used more for β-oxidation than the all *cis* or the 9*trans* isomer.

The extent of β-oxidation of *trans* PUFA in men was assessed using synthesized $^{13}C$ labeled fatty acids (32). Ten healthy males were involved in this study. After an overnight fasting period, they were fed a bolus (0.87–1.40 g in 30 g olive oil) of linoleic, α-linolenic, *trans*12-18:2, or *trans*15-18:3. The $^{13}CO_2$ production was assessed in breath tests for 8 h after the bolus intake. The data (Fig. 10.2) showed that, similar to the rat data (see above), the *trans*12-18:2 isomer was more β-oxidized than linoleic acid, whereas no difference was seen between α-linolenic acid and its 15*trans* isomer.

## CHD Risk

The only study dealing with these aspects is the Transline study (13). Eighty-eight healthy subjects were randomly assigned to a 6-week experimental period during which they had to consume, or not, 0.6 % of their energy intake as *trans* n-3 fatty acids from deodorized canola oil. At the end of the experimental period, both the LDL:HDL and total HDL:cholesterol ratios were significantly increased after the *trans* fatty acid intake, suggesting an increased risk of CHD of 8%.

**A**

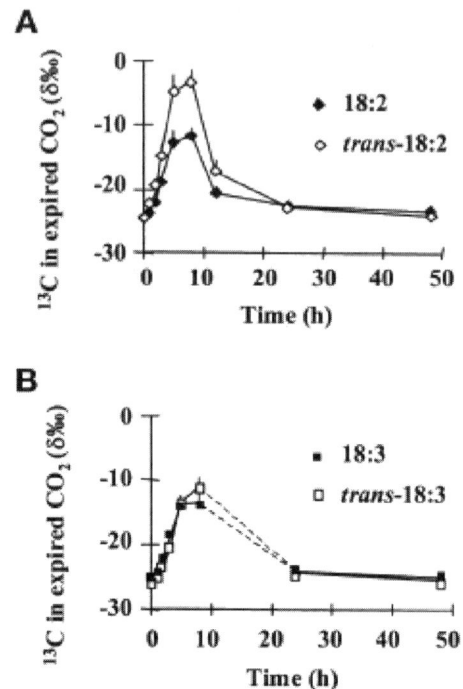

**B**

**Fig. 10.2.** $^{13}CO_2$ enrichment (d ‰ versus PBD) after ingestion of linoleic acid or 9*cis*,12*trans*-18:2 (A) or after ingestion of α-linolenic acid or 9*cis*,12*cis*,15*trans*-18:3 (B) by 10 healthy men.

## Eicosanoid Synthesis, Platelet Function, and Endothelial Cell Metabolism

Due to the structural analogy between the 17*trans* isomer of EPA and arachidonic acid and to the role of the latter in eicosanoid synthesis, we carried out several studies on the impact of *trans* EPA isomers in platelet and endothelial cells eicosanoid metabolism. In isolated platelets, it was suggested that *trans* n-3 fatty acids alter platelet function. The 17-*trans* isomer of EPA and the 19-*trans* isomer of DHA were particularly considered. These preliminary data were previously reviewed in the first edition of this book.

More recently, Loï et al. (33) have considered the effects of various EPA isomers resulting from different *trans* 18:3 isomers (34). Rat platelets were incubated in the presence of arachidonic acid as a triggering agent. Increasing concentrations of 11*trans*, 11,17di-*trans* and 17*trans* isomers of EPA were tested. All the *trans* isomers of EPA were antiaggregant, with those having a *trans* double bond at the $\Delta^{11}$ position being more potent (Table 10.2). These effects on platelet aggregation were dose-dependent and correlated to a decrease in thromboxane A2 production. HHT, another cycloxygenase metabolite, was depressed as well, whereas 12-HETE

**TABLE 10.1**
Radiolabelled Fatty Acids Formed From 100 nmol of [1-$^{14}$C] Linoleic Acid or From its *trans* Isomers in Total Lipids of Post Perfused Rat Livers. Results are Expressed as nmol ± SEM of 3 Independent Determinations

| Substrates fatty acids | Substrates fatty acids | | |
|---|---|---|---|
| | Linoleic acid | 9*cis*,12*trans*-18:2 | 9*trans*,12*cis*-18:2 |
| 16:0 | 5.1 ± 0.88 | 2.9 ± 0.64 | 2.9 ± 1.05 |
| 18:3 n-6 + 20:4 n-6 | 0.4 ± 0.08[a] | 6.5 ± 1.27[b] | 0.9 ± 0.57[a] |
| 20:2 n-6 | 0.9 ± 0.10[a] | 1.5 ± 0.17[a] | 4.5 ± 0.92[b] |

[a,b] Values are significantly different (P<0.05)

**TABLE 10.2**
Percentage of Inhibition of Platelet Aggregation After 4 Min Incubation

| PUFA[1] | 20:5 / 20:4 ratio | | | |
|---|---|---|---|---|
| | 0.25[a] | 0.5[b] | 1[c] | 5[d] |
| EPA[e] | 16.4 ± 3.2 | 24.8 ± 4.5 | 26.7 ± 6.6 | 77.1 ± 6.7 |
| 20:5 Δ11t[f] | 17.4 ± 2.4 | 31.7 ± 5.8 | 50.1 ± 5.7 | 89.5 ±3.0 |
| 20:5 Δ17t[e] | 9.1 ± 1.2 | 21.3 ± 6.8 | 32.8 ± 7.4 | 79.9 ± 8.0 |
| 20:5 Δ11t,17t[f] | 9.0 ± 2.9 | 31.24 ± 6.2 | 52.3 ± 8.4 | 93.1 ± 1.9 |

Platelets (108 platelets per mL) were stimulated with AA (5 μM) in the presence of increasing concentrations of 20:5 fatty acids. Aggregation obtained with 5 μM of AA represented the maximum platelet aggregation. It was assigned the values of 100 % of aggregation and 0 % of inhibition (results are expressed as means ± SEM of 10 independent determinations).

[1]FA with different superscripts have different effects (*P* < 0.0001). Values in columns corresponding to 20:5/20:4 ratios having different superscripts are significantly different (*P* < 0.0001).

production, through the 12-lipoxygenase pathway, was increased. We also studied the metabolism of radio labeled 17*trans*-EPA by platelets. This fatty acid behaved like arachidonic acid, with conversion into metabolites that may be *trans* isomers of thromboxane A3 and 12-HEPE.

The Transline study (see above) was the opportunity to assess the effects of *trans* isomers of α-linolenic acid in dietary food on platelet function in human volunteers. Neither the response to collagen nor U46619, the TxA2 analog, was modified after a 6-week intake of *trans* n-3 fatty acids (0.6% of energy) (35). Circulating eicosanoids were not measured in this study.

The influence of *trans* isomers of EPA on the eicosanoid synthesis from endothelial cells was also studied. These cells typically convert $C_{20}$ PUFA into prostacyclin (PGI2) as a powerful vasodilatator and antiaggregant agent. Bovine aortic endothelial cells were grown on media containing arachidonic acid alone, or with the addition of EPA, or its *trans* isomers (11*trans*, 17*trans* and 11,17 di-*trans*) (36). Each 20:5 fatty

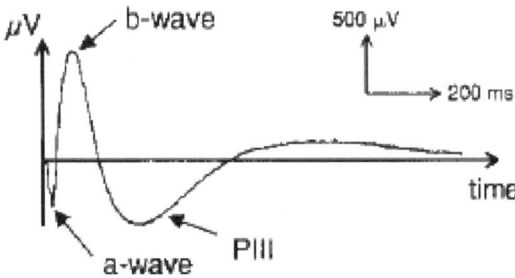

**Fig. 10.3.** Typical response of the rat retina to a light stimulus: the ERG comprises a negative a wave, a strong positive b-wave and a final slow negative variation of potential which corresponds to the end of the $P_{III}$ process.

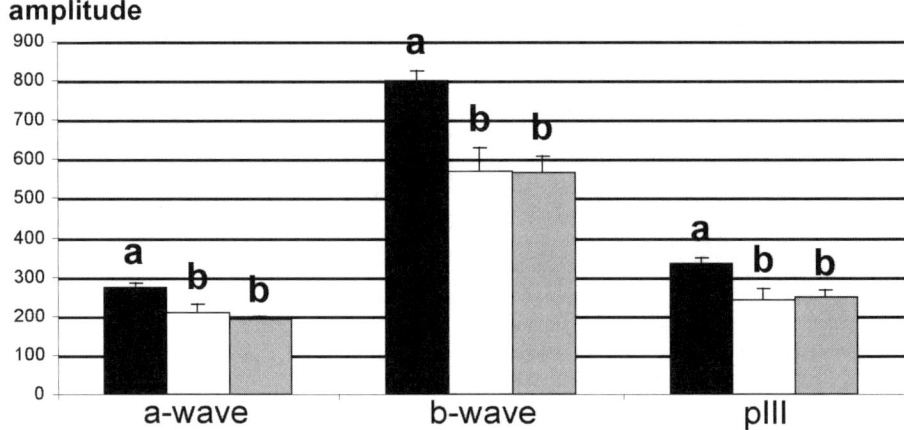

**Fig. 10.4.** Amplitude (µV) of the a-, b- and PIII-waves of the electroretinogram of rats fed for 21 months a diet containing no *trans* fatty acids and 2.0% (by weight) of α-linolenic acid (black bars), 1.3% of α-linolenic acid and 0.7% of *trans*-18:3 (white bars) or 2.0% of α-linolenic acid and 0.7% of *trans*-18:3 (grey bars)

acid significantly inhibited the PGI2 production from arachidonic acid by cultured endothelial cells. The inhibition ranged between 62 and 72% according to the isomers and was greater than the inhibitory effect of EPA (–49% PGI2 release). As reported above, in platelets the 17*trans* EPA isomer was bioconverted into eicosanoids, the major one being supposed to be a *trans* isomer of PGI3, as suggested by HPLC analyses.

# Retinal Function

Due to the importance of n-3 fatty acids in neural and retinal function, some studies assessed the impact of *trans* n-3 fatty acids in retinal function evaluated as the response to light stimuli. Early studies were carried in vitro using the isolated perfused

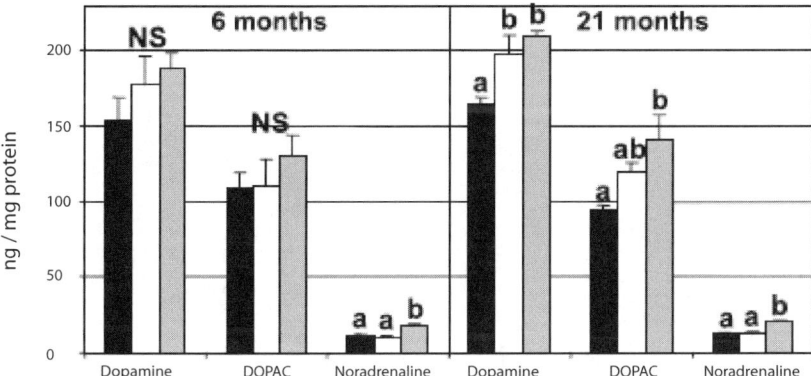

**Fig. 10.5.** Contents of endogenous monoaminergic neurotransmitters in the striatum of rats fed for 6 or 21 months with diets containing different levels of *cis* and *trans* isomers of 18:3 n-3. (Black bars: 2.0% *cis* 18:3 n-3, white bars: 1.3% *cis* 18:3 n-3 plus 0.7% of *trans* 18:3 n-3, grey bars: 2.0% *cis* 18:3 n-3 plus 0.7% of *trans* 18:3 n-3). Data are means ± SD of six independent determinations. Bars having different superscripts are statistically different ($P<0.05$).

retina model (37). It was suggested that dietary *trans* isomers of α-linolenic acid have deleterious effects on the survival of the retina as well as on the amplitude of the *b*-wave of the electroretinogram, a parameter considered a marker of retinal function.

More recently, the effects of a low intake of *trans* isomers of α-linolenic acid in aging rats were also studied (Fig. 10.3) (24). *Trans* n-3 intake (0.7% of total fatty acids) induced an incorporation of the 19*trans* isomer of DHA of up to 0.8% of total fatty acids after 12 months of the diet (24), and 1.2% after 21 months (Acar et al., unpublished).

Despite this low incorporation, which is associated with a decrease in the DHA content (–9% after 12 months, –20% after 21 months), the amplitude of the 3 waves of the electroretinogram was significantly altered after *trans* n-3 intake for 12 months (24) or 21 months (Fig. 10.4).

## Neurotransmission

As n-3 fatty acids are reported to alter the dopaminergic neurotransmission (38), studies were carried out to assess the effect of *trans* n-3 isomers. Both piglets and aged rats were used. Diets contained either 1.3% of *cis* linolenic acid and 0.7% of *trans* 18:3 isomers or 2.0% *cis* linolenic acid and the same amount (0.7%) of *trans* 18:3 isomers. In both animal models, dopamine concentration increased in frontal cortex and striatum (Fig. 10.5). On the other hand, dietary *trans* 18:3 isomers may contribute to a decrease of the dopamine concentration in the hippocampus, with some differences in the amplitude of the effects considering the diet. Taken all together, these data indicate that *trans* n-3 fatty acids may interfere with brain neurotransmission, in

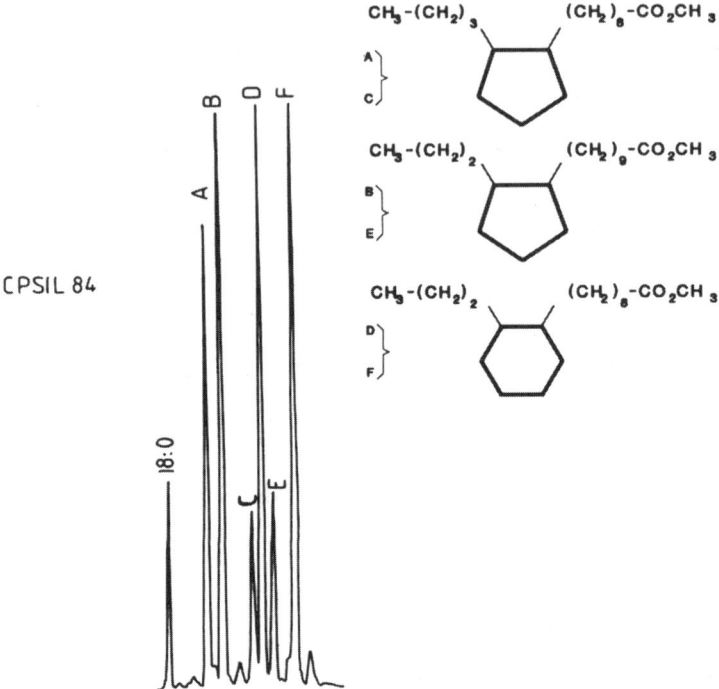

**Fig. 10.6.** Gas–liquid chromatographic analysis (CPSIL 84) of hydrogenated CFAM fraction isolated from heated linseed oil.

relationship with cognitive functions. Consequently, their intake has to be as low as possible.

## Cyclic Fatty Acid Monomers

Many nutritional studies used either heated oil containing many compounds (polymers, oxidized compounds, CFAM, *trans* fatty acid isomers, and so forth) or partially enriched fractions of cyclic fatty acids. Few reports deal with pure synthesized or isolated CFAM fractions. Therefore, it is very difficult in some cases to relate physiological effects observed to the CFAM only. Therefore, this chapter only describes studies using pure characterized cyclic monomer fractions. For that purpose, methods to isolate CFAM from heated oils were developed. They were a combination of column chromatography, urea inclusion, and preparative high-performance liquid chromatography (HPLC) (39,40). Generally, CFAM formed from linolenic acid were isolated from heated linseed oil. CFAM formed from linolenic acid (Fig. 10.6) are basically a mixture (1:1) of disubstituted 5-carbon ring isomers, while those arising from linoleic acid (heated sunflower oil; Fig. 10.7) are mainly 5-carbon ring isomers (39,41–46,58). So far, most work dealing with the biological effects of these molecules

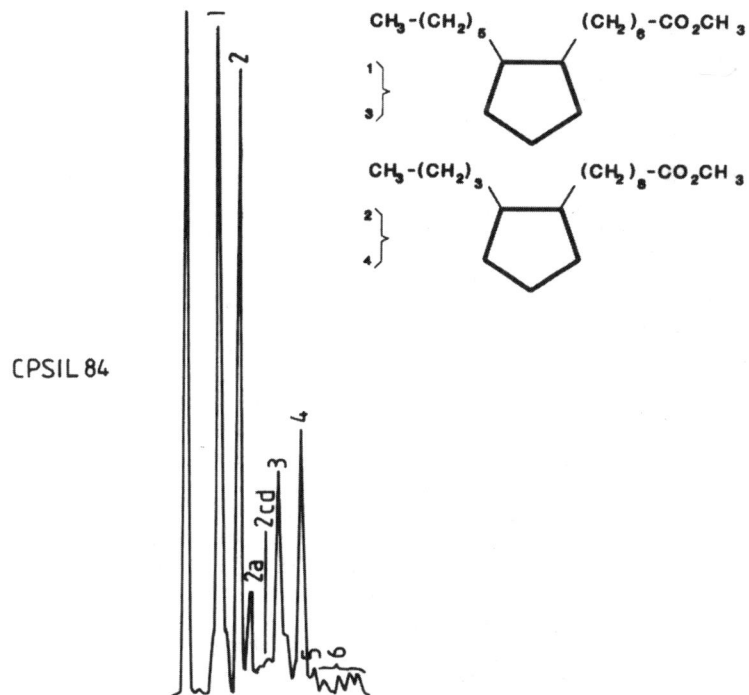

**Fig. 10.7.** Gas–liquid chromatographic analysis (CPSIL 84) of hydrogenated CFAM fraction isolated from heated sunflower oil.

has been performed on rats.

## Metabolism and Incorporation in Biological Tissues

Earlier studies (42) demonstrated that CFAM that displayed high intestinal absorption. More recently, Martin et al. (45) demonstrated the effects of CFAM upon intestinal metabolism are greatly influenced by their positioning within the triacylglycerol, and that the structure of CFAM influences their lymphatic recovery only when they are absorbed as free fatty acids. For example, the $C_5$ ring isomers are better recovered in the lymph relative to the $C_6$ ones when absorbed as free fatty acids; that was not the case when these were absorbed as 2-monoacyl-*sn*-glycerol. One explanation for these results is the existence of a selective binding to the Fatty Acid Binding Protein (FABP).

In vitro studies (47) using hepatic sub-cellular fractions showed a lower oxygen consumption and a lower activity of Carnitine Palmitoyl Transferase compared to 16:0 and 18:2n-6 for the mitochondrial oxidation. For peroxisomal oxidation CFAM formed from 18:2 showed the same kinetic parameters as 18:2n-6 and 16:0 (control) while a lower Acyl-CoA Oxidase (ACO) activity of CFAM isolated from 18:3 than

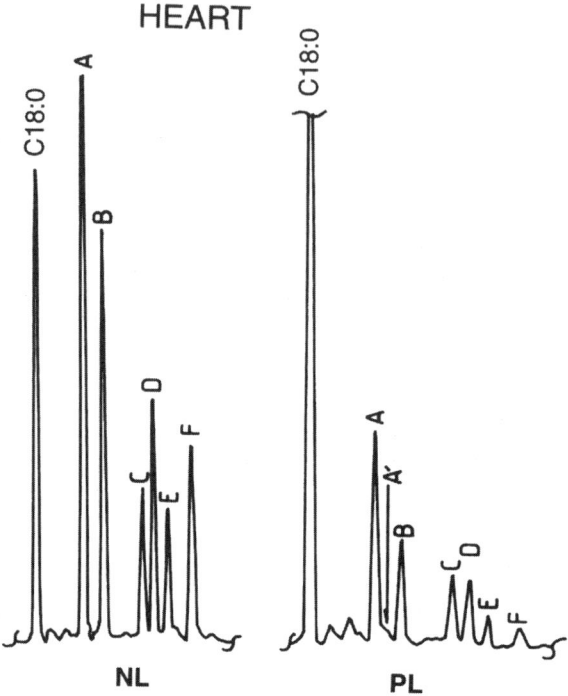

**Fig. 10.8.** Gas–liquid chromatographic analysis (CPSIL 84) of heart neutral (NL) and phospholipids (PL) of rats that received a diet containing cyclic fatty acid monomers isolated from heated linseed oil (see Fig. 6 for peak identifications).

**TABLE 10.3**
**Incorporation of Cyclic Fatty Acid Monomers (CFAM) Isolated from Heated Linseed and Sunflower Oils in Heart and Liver Lipids of Rats (average value of 8 rats)**

|  |  | From heated linseed oil | From heated sunflower oil |
|---|---|---|---|
| Liver | NL[a] | 1500.5 ± 196 | 1724.3 ± 211 |
|  | PL[b] | 734.0 ± 142 | 319.8 ± 103 |
| Heart | NL | 46.6 ± 20 | 21.4 ± 12 |
|  | PL | 35.7 ± 8 | 18.4 ± 12 |

[a]NL = nonphosphorus lipids
[b]PL = phospholipids

**TABLE 10.4**
Proportions (%) of CFAM (Mean ± SEM of 8 Rats) in PL and NL of Heart and Liver of Rats
Fed a CFAM Fraction Isolated from Heated Linseed Oil (Fig. 10.7.)

| Components Figs. 10.6 and 10.8 | Isolated | Heart | | Liver | |
|---|---|---|---|---|---|
| | CFAM Fraction | PL | NL | PL | NL |
| A | 16.41 | 45.88 ± 1.36 | 29.77 ± 1.52 | 23.51 ± 1.34 | 30.23 ± 0.95 |
| B | 21.20 | 18.10 ± 0.82 | 21.96 ± 1.79 | 18.64 ± 1.04 | 19.86 ± 1.25 |
| C | 7.23 | 12.46 ± 0.44 | 9.41 ± 0.89 | 12.91 ±2.25 | 12.70 ± 0.34 |
| D | 22.40 | 12.15 ± 0.78 | 16.19 ± 1.00 | 30.30 ± 1.53 | 11.31 ±1.21 |
| E | 7.92 | 6.90 ± 0.26 | 8.28 ± 0.48 | 8.69 ± 0.83 | 10.64 ± 0.98 |
| F | 24.84 | 4.51 ± 0.62 | 14.39 ± 2.24 | 5.95 ± 0.88 | 15.26 ± 1.23 |
| ΣC5 | 52.76 | 83.34 | 69.42 | 63.75 | 73.43 |
| ΣC6 | 47.24 | 16.66 | 30.58 | 36.25 | 26.57 |

Abbreviations as in Table 10.3.

for the other fatty acids was demonstrated. CFAM isolated from 18:2 and CFAM from 18:3 were poor substrates of Acyl-CoA Synthase (ACS) and consequently, the activation of these CFAM into their CoA derivatives may be the limiting step in their metabolism.

An in vivo study (48) using rats force-fed with 1-[$^{14}$C] labeled CFAM isolated from a heated sunflower oil, and heated linseed oil, showed both types of CFAM can be used for energy production. Furthermore, no significant differences in $CO_2$ production were found 24 h after the administration of CFAM and their respective precursors. These data demonstrate that, at least for the first β-oxidation cycle, CFAM oxidize in a similar way to both essential fatty acids. For the CFAM isolated from 18:2, the larger part of the radioactivity recovered in the lower portion of the gastrointestinal tract and in the urine suggests these cyclic fatty acids might be poorly metabolized compared to their corresponding precursor, linoleic acid.

CFAM can further incorporate in all types of rat tissue, both in non-phosphorus (neutral) lipids (NL) and in phospholipids (PL). For example, in adult male rats, recent studies carried out in our laboratory showed incorporation into tissue lipids that may be selective. Incorporation of CFAM (Table 10.5) is significantly higher (about twice) in NL, compared to PL, especially in the liver. The same pattern resulted for the heart, but no significant differences between NL and PL were observed. It should also be noted that quantities incorporated in the liver are at least 20 times higher compared to those found in the heart. Furthermore, for PL, the incorporation of CFAM isolated from linseed oil is twice as much as from CFAM isolated from heated sunflower oil.

The CFAM composition of adipose tissue reflected the original CFAM mixture extracted from the oil (Fig. 10.6), while differences existed between the oil and PL and NL fractions of the heart (Fig. 10.8), consistent with a relative increase of components

**TABLE 10.5**
**Proportions of CFAM in PL and NL of Liver from Rats Fed a CFAM Fraction Isolated from Heated Sunflower Oil**

| Components | CFAM fraction | PL | NL |
|---|---|---|---|
| 1 | 38.52 | 30.85 ± 3.34 | 53.38 ± 1.23 |
| 2 | 31.34 | 34.90 ± 3.69 | 18.31 ± 1.72 |
| 3 | 12.99 | 15.46 ± 1.98 | 18.07 ± 1.39 |
| 4 | 17.15 | 18.79 ± 1.22 | 10.24 ± 1.08 |

Abbreviations as in Table 10.3.

having a 5-carbon ring (compounds A, B, C, and E). Similar observations were made for liver lipids (Table 10.4).

The size of the ring is not the only factor influencing incorporation of CFAM. For example, other factors, such as *cis* versus *trans* configuration (ring substitution), or propyl versus butyl substitution on the ring, also contribute to the observed differences. Moreover, a large decrease of component F (*cis* isomer) was observed for PL compared to NL, while component D accumulated in liver PL (Table 10.4).

Similar results were reported by Ribot et al. (49) when looking at incorporation of CFAM isolated from heated linseed oil in rat heart cell cultures. There was also preferential incorporation of the 5-carbon ring isomers, especially those with a *trans* configuration (components A and B).

Selective incorporation was also observed when feeding animals with the CFAM fraction isolated from heated sunflower oil (Table 10.5). In NL, components 2 and 4 were the least incorporated. This was also the case for component 1 in PL.

Currently, it is very difficult to explain these selective incorporations, as all CFAM analyses were conducted after hydrogenation of the total fatty acid methyl esters. This method is typically used because of its convenience. In fact, these molecules are very complex. CFAM isolated from linseed oil are usually a mixture of $C_{18}$ diunsaturated *cis/trans* disubstituted $C_5$- and $C_6$-ring isomers which, upon hydrogenation, give a mixture of six peaks (A, B, C, D, E, and F; Fig. 10.6). CFAM isolated from sunflower oil are mainly a mixture of $C_{18}$ monounsaturated *cis/trans* disubstituted $C_5$-ring isomers, which upon hydrogenation lead to a mixture of four major peaks (1, 2, 3, and 4; Fig. 10.7). In this case, many factors affect selective incorporation (geometry of the double bond, size of the ring, configuration at the ring, size of the substitutes, and so forth). Methods enabling isolation and quantification of CFAM without hydrogenation need developing to better understand the selective incorporations of CFAM in biological tissues. Silver nitrate HPLC might be one technique (24,30).

# Biological Effects

As previously mentioned, many studies were done with heated oils containing cyclic monomers, and, in most cases, the physiological effects observed could not be related

**TABLE 10.6**
**Percentage of Administered Radioactivity Recovered After 2 Days in Urine and $CO_2$ Formation from Rats Fed Various Levels of Cyclic Fatty Acids**

| % CFAM in diet | Substance administered | % Radioactivity in urine | % Radioactivity in $CO_2$ |
|---|---|---|---|
| 0 | Methyl linoleate | 1.1 | 35.8 |
| 0 | Cyclic fatty acids | 40.5 | 13.4 |
| 0.0075 | Cyclic fatty acids | 37.3 | 14.4 |
| 0.0225 | Cyclic fatty acids | 40.9 | 15.1 |
| 0.1500 | Cyclic fatty acids | 42.3 | 13.1 |

Source: Iwaoka, W.T., and E.G. Perkins, *J. Am. Oil Chem. Soc. 55*:734–738 (1978).

to the presence of cyclic monomers. These studies are reviewed by Artman (50), Perkins (51), and Dobarganes in this book. In this chapter, we outline the major studies carried out with isolated CFAM fractions.

Evidence that CFAM could be harmful was shown by Crampton et al. in the 1950s (52,53), using elementary measurements such as growth and mortality. Quantities of cyclic monomers in the diets, however, were very high, constituting the weak point of the work. Some early studies were also performed by Iwaoka and Perkins (54), who prepared CFAM fractions by cyclization of methyl linolenate in the presence of NaOH and ethylene glycol for 1 h at 240°C under nitrogen. In that case, cyclohexadienic compounds with 10% aromatic components were obtained. These cyclic monomers were given at levels ranging from 75–1,500 ppm in diets with different protein levels (8, 10, and 15%). After several weeks on this diet, rats of different groups did not differ significantly in weight gain, feed efficiency, or organ weight (at similar protein levels). On the contrary, the lipid content of livers of rats fed 1,500 ppm of cyclic monomers rose significantly. This effect was very clear in rats with low protein levels (8 and 10%). In another experiment, Iwaoka and Perkins fed rats the same diets with 15% protein until they reached 200 g. They were then intubated with 0.2 mL uniformly labeled cyclic monomers. Control rats received the same quantity of labeled linoleate. Only a low quantity of radioactivity was detected as $CO_2$ during 48 h (14% compared to 36% in controls). Urine contained 40% of the radioactivity compared to 1% in controls (Table 10.6). The amount of cyclic monomers previously ingested had no effect on percentages of radioactivity detected in urine and $CO_2$.

Of interest in the work of Perkins and Iwaoka (54) was the increase of radioactivity of urine in rats. This suggests that a detoxification mechanism exists and is quite consistent with the preferential incorporation of 5-carbon ring cyclic monomers in tissue lipids observed in many experiments (see Tables 10.3 and 10.4). Iwaoka and Perkins explained that rapid excretion accounts for the very low harmfulness of 6-carbon ring cyclic monomers. This point is actually only hypothetical and requires

**TABLE 10.7**
**Reproduction Study in Rats Using Isolated Cyclic Monomers: Some Characteristics of Experimental Diet**

| Experimental groups | A | B | C | D | E |
|---|---|---|---|---|---|
| CFAM (% of diet) | 1 | 0.1 | 0.01 | – | – |
| Methyl esters of fresh linseed oil | – | 0.9 | 0.99 | 1 | 1 |
| Soybean oil (% of diet) | 9 | 9 | 9 | 9 | 9 |
| Food intake (g) | | | *ad libitum* | | Restricted to food intake of group A |

Source as in Table 11.2.

further study by comparing synthetic $C_5$ and $C_6$ carbon ring cyclic labeled fatty acids in order to look at their excretion in urine. Some previous results (55) demonstrated that compounds resulting from detoxification, such as glucuronides, appeared in urine of rats after ingestion of heated oil containing cyclic monomers. Furthermore, an induction of hepatic drug-metabolizing enzymes was observed in rats after ingestion of isolated CFAM formed from linolenic acid (56).

Using an acute mortality test on mice, Saito and Kaneda (57) tried to determine the most toxic fraction isolated from linseed oil heated under drastic conditions. Four components in this fraction were cyclic monomers. The first two had a disubstituted cyclohexanic structure with two identified ethylenic bonds on a lateral chain; another had a cyclohexyl-1,2-diylidene structure. The last one presented a disubstituted cyclohexenic structure with an ethylenic bond on one of the chains. The most recent studies on the structure of cyclic monomers (39,41,43,44,46,58,59), however, did not find all compounds described by Saito and Kaneda (60) in oil heated under the same conditions.

Many reproduction studies were carried out to study the potential toxicity of heated oil (61–63). Little work, however, was done on the subject with isolated CFAM fraction as described by Grandgirard and Sébédio (64). In their experiment, female Wistar rats received a purified diet, including 18% casein and 10% lipids, just after mating. They were divided into five experimental groups (Table 10.7). Groups A, B, and C received 1, 0.1, and 0.01% cyclic monomers respectively, and D and E were control groups. Groups B, C, D, and E also received methyl esters of fresh linseed oil, so that the total amount of methyl esters reached 1% in all groups. The remaining lipids were soybean oil to fit essential fatty acid demands. All groups were fed ad libitum, except for group E, in which food intake was restricted to that of group A.

A decrease in food intake of group A (and of group E, which was pair-fed) was observed compared to the three other groups. The number of pups per litter at birth did not vary among groups. Pup weight was significantly different, however. When calculations were made using double covariance with number per litter and quantity of diet eaten, we observed that pup weight of groups C, D, and E were not different

**TABLE 10.8**
**Reproduction Study in Rats Using Cyclic Monomers Isolated From Heated Linseed Oil and Given as Triacylglycerols**

| | CFAM (% in diet) | | | |
|---|---|---|---|---|
| | Control | 0.3 | 0.6 | 0.9 |
| Food Intake[1] of female (g/day) | 36.9$^a$ | 33.6$^a$ | 34.2$^a$ | 24.6$^b$ |
| Body weight gain of female during gestation (g) | 166$^a$ | 107$^b$ | 104$^b$ | 35$^c$ |
| Average number of fetus | 12.4$^a$ | 15.4$^a$ | 14.6$^a$ | 15.9$^a$ |
| Mean fetus weight At day 20 (g) | 3.28$^a$ | 2.49$^a$ | 2.90$^a$ | 2.84$^a$ |
| Liver weight of Fetus at day 20 (mg) | 247$^a$ | 157$^b$ | 179$^b$ | 172$^b$ |
| % CFAM in liver of fetus | / | 0.35$^a$ | 0.42$^a$ | 0.40$^a$ |
| C5/C6 | / | 2.0$^a$ | 1.8$^{ab}$ | 1.4$^b$ |

[1]Values having different superscripts are significantly different ($P<0.05$)

and that only the two highest levels of cyclic monomers affected pup weight at birth. This means that the lower weight at birth for pups of group A is partly due not only to the spontaneous decrease of food intake of their mothers but also to a specific effect of cyclic monomers. The most spectacular effect observed concerned pup mortality during the first days after birth. In group A, mortality of pups reached 98% during the first three days, and the few survivals did not live longer than five days. This effect had already been observed by Farmer et al. (61) and Potteau (62) with diets containing equivalent quantities of cyclic monomers present in heated oil. Mortality during the first three days also seemed higher in group B than in other groups, but this difference was not significant, owing to the variation between litters and also to the high mortality in controls. A complementary experiment (Grandgirard, personal communication) revealed that the following factors might have induced this high mortality in controls:

1. The diet was purposely poor in protein (18% casein) so that a high level would not reduce a potentially harmful effect, as is the case, sometimes, in nutritional experiments.
2. Ingestion of compounds as methyl esters also had an effect. It was assumed that animals could eat 1% of methyl esters without any problem, but it was not the case in this experiment.
3. The slow-growing breed of rat chosen revealed that it was unable to have sufficient food intake to supply energy necessary to feed a litter with many pups.

To verify these results, another reproduction experiment where the amount of protein in the diet increased (21%) and CFAM incorporated as ethyl esters to avoid toxicity of methanol. In that case, high mortality (81%) was found in the group of pups fed heated linseed oil (about 1% CFAM), whereas the same mortality (3%)

**TABLE 10.9**
Lipid Composition (%, $n = 5$) of Liver of Fetus From Rats Fed or Not CFAM Isolated From a Heated Linseed Oil. (85)

| PL | Chol | TAG | Control[1] |
|---|---|---|---|
| | $77.9 \pm 191^a$ | $6.1 \pm 0.38$ | $16.0 \pm 1.58^a$ |
| 0.3 | $57.4 \pm 6.33^b$ | $3.8 \pm 0.71$ | $38.8 \pm 6.93^b$ |
| 0.6 | $53.6 \pm 4.88^b$ | $5.4 \pm 0.78$ | $41.0 \pm 5.63^b$ |
| 0.9 | $62.3 \pm 1.52^b$ | $5.8 \pm 1.51$ | $31.9 \pm 0.74^b$ |

[1]Values having different superscripts in a row are significantly different ($P<0.05$).

was found in control group and groups receiving increasing amounts of CFAM (0.2, 0.4, 0.6, and 1%) (65). For the group receiving 1% CFAM as ethyl esters, however, the quantity of CFAM found in the livers was about four times lower than what was found in the livers of pups fed the heated oil containing about 1% CFAM. These results indicate ethyl esters of CFAM are not well absorbed and their toxicity depends on the form fed to rats (methyl esters, ethyl esters, triglycerides). The same selectivity of incorporation of CFAM (preferential incorporation of $C_5$-ring isomers) resulted for animals fed heated oil or purified CFAM fractions, however.

In a recent study, Joffre (66) also carried out a reproduction study with increasing amounts of CFAM in the diet (0.3, 0.6, and 0.9%) in order to determine if these molecules could have an effect during the gestation period by inducing cellular lesions in the liver of 20 day-old fetuses. Feeding female rats with CFAM as triacylglycerols resulted in a reduction of food intake for females fed the highest quantity of CFAM, which resulted in a lower body weight gain (Table 10.8) while no differences were observed for the average number of fetus and the mean fetus weight at day 20. The liver weight of fetus from rats fed CFAM was significantly lower than that of the control. Increasing the quantity of CFAM in the diet from 0.3 – 0.9% did not result in any changes in the CFAM content of liver of fetus but a modification in the CFAM profile was evident with a higher $C_5/C_6$ ratio at 0.3% CFAM than at 0.9% (Table 10.8). Furthermore, an accumulation of triglycerides also resulted in the liver of the fetus in all the groups fed CFAM (Table 10.9). Morphological studies also demonstrated that CFAM formed from linolenic acid induced accumulation of glycogen and changes in mitochondria ultrastructure. It would then be interesting to determine if these morphological changes would induce any changes in the function of the mitochondria for fatty acid oxidation. If so these changes could explain why at birth mortality could be high as at birth the pup is changing from a diet low in lipids to a diet rich in lipids (maternal milk).

Feeding studies with an isolated CFAM fraction were carried out by Lamboni et al. (67) to determine the effect of cyclic monomers on liver enzyme activity. CFAM fraction was isolated from a batch of partially hydrogenated soybean oil (PHSBO) commercially used for 7 days (group CFA) for frying fish, french fries, and chicken. Results were compared to two other groups of rats; one was fed total PHSBO after frying (group 7 DH) and the other was fed PHSBO before frying (NH). While an increased level of liver protein resulted in both experimental groups (7 DH and CFA) compared to the control (PHSBO), several enzyme activities changed. For example, activity of the NADPH-cytochrome P450 reductase was significantly increased, as already shown by Siess et al. (56), suggesting active detoxification by the liver of toxic compounds at high CFAM levels. The difference between the two experimental groups, however, also suggested that other components from used frying fat have an effect in conjunction with CFAM. In addition, activities of carnitine palmitoyl transferase-1 and isocitrate dehydrogenase significantly decreased when compared to the control. Significant depressed activity of glucose 6-phosphate dehydrogenase resulted for animals of the experimental groups. An increase of total liver lipids and protein and a decrease in the glycogen content in the liver was also observed by Lamboni et al. (68). This was later confirmed by Martin et al. (69). They also showed indirect evidence that CFAM have a peroxisome proliferators–like effect considering an elevation of some characteristic enzyme activities such as peroxisomal Acyl-CoA oxidase, and the microsomal ω-and (ω-1)-laurate hydroxylase. No differences occured in the CPT-1 activity, while a reduction of the stearoyl-CoA desaturase activity resulted. Recently, Bretillon et al. (70) demonstrated using wild type and PPARα null mice that PPARα is not the exclusive mediator of the effects of CFAM in lipid metabolism of mice.

Heart cell cultures have also been used for observing biological effects of CFAM arising from either n-3 or n-6 fatty acids (49,71). The 5-carbon ring compounds accumulated much more in cellular lipids (phospholipids as well as triglycerides) than in the 6-carbon ring compounds. Incorporation of CFAM isolated from sunflower oil (CFAMS) in cellular phospholipids affected the spontaneous beating rate of cardiomyocytes. This effect was not observed with CFAM isolated from linseed oil (CFAML). Activation of the α- and β-adrenergic systems was obtained by the addition of isoproterenol and phenylephrine, respectively. Neither the biochemical response (cAMP production) nor the physiological response (chronotropic effect) to the β-adrenergic agonist was altered by either CFAML or CFAMS, suggesting that the β-adrenergic system is not affected by the incorporation of CFAM in cellular lipids. Conversely, both biochemical (inositide phosphate production) and physiological (chronotropic effect) responses to the α-adrenergic agonist significantly decreased in cardiomyocytes containing CFAMS, but not CFAML. This differentiated sensitivity between the α- and β-adrenergic systems toward CFAM suggests that specific incorporation of CFAMS into phosphatidyl-inositols may be responsible for the observed effect, rather than a more general "membrane effect."

## Conclusion

*Trans* fatty acids and cyclic fatty acid monomers can be used as "normal" fatty acids for energy production and incorporate in membrane phospholipids. Consequently, they may affect fatty acid metabolism, and, in some cases, some important physiological functions in both animals and humans. Frying temperature is of primary importance and 200°C seems to be a critical point for producing these isomeric and cyclic species. However, data have shown that normal conditions would not likely produce such a quantity of these molecules as to fall into the conditions allowing these effects.

## References

1.  Trus, M.; A. Grandgirard; and J.L. Sébédio. Utilisation digestive des isomères trans de l'acide linolénique chez le rat. *Reprod. Nutr. Dev.* **1991,** *31,* 294.

2.  Ono, K.; and D.S. Fredickson. The Metabolism of 14C-labelled cis and trans Isomers of Octadecenoic and Octadecadienoic Acids. *J. Cell Biol.* **1964,** *26,* 725–735.

3.  Piconneaux, A. Etude de la Désaturation et de l'Elongation *in vivo* d'Isomères Géométriques de l'Acide Linolénique. Burgundy, Dijon, 1987.

4.  Chardigny, J.M.; A. Bron; J.L. Sébédio; P. Juanéda; and A. Grandgirard. Kinetics of Incorporation of n-3 trans Fatty Acids in the Rat Retina. *Nutr Res.* **1994,** *14,* 909–917.

5.  Grandgirard, A.; J.M. Bourre; P. Homayoun; O. Dumont; M. Piciotti; and J.L. Sébédio. Incorporation of trans Long-Chain n-3 Polyunsaturated Fatty Acids in Rat Brain Structures and Retina. *Lipids* **1994,** *29* (4), 251–258.

6.  Grandgirard, A.; A. Piconneaux; J.L. Sébédio; and F. Julliard. Trans Isomers of Long-Chain n-3 Polyunsaturated Fatty Acids in Tissue Lipid Classes of Rats Fed with Heated Linseed Oil. *Reprod. Nutr. Dev.* **1998,** *38* (1), 17–29.

7.  Wolff, R.L.; N.A. Combe; B. Entressangles; J.L. Sébédio; and A. Grandgirard. Preferential Incorporation of Dietary cis-9,cis-12,trans-15 18:3 Acid into Rat Cardiolipins. *Biochim. Biophys. Acta* **1993,** *1168,* 285–291.

8.  Bretillon, L.; J.M. Chardigny; J.P. Noel; and J.L. Sébédio. Desaturation and Chain Elongation of [1-14C]Mono-trans Isomers of Linoleic and α-Linolenic Acids in Perfused Rat Liver. *J. Lipid Res.* **1998,** *39* (11), 2228–2236.

9.  Adlof, R.O.; and E.A. Emken. Distribution of Hexadecenoic, Octadecenoic and Octadecadienoic Acid Isomers in Human Tissue Lipids. *Lipids* **1986,** *21* (9), 543–547.

10. Chardigny, J.M.; R.L. Wolff; E. Mager; J.L. Sébédio; L. Martine; and P. Juanéda. Trans Mono- and Poly-unsaturated Fatty Acids in Human Milk. *Eur. J. Clin. Nutr.* **1995,** *49,* 523–531.

11. Chen, Z.Y.; G. Pelletier; R. Hollywood; and W.M. Ratnayake. Trans Fatty Acid Isomers in Canadian Human Milk. *Lipids* **1995,** *30* (1), 15–21.

12. Chardigny, J.M.; J.L. Sébédio; P. Juanéda; J.M. Vatèle; and A. Grandgirard. Occurence of n-3 trans Polyunsaturated Fatty Acids in Human Platelets. *Nutr Res.* **1993,** *13,* 1105–1111.

13. Sébédio, J.L.; S.H. Vermunt; J.M. Chardigny; B. Beaufrere; R.P. Mensink; R.A. Armstrong; W.W. Christie; J. Niemela; G. Henon; and R.A. Riemersma. The Effect of Dietary trans α-Linolenic Acid on Plasma Lipids and Platelet Fatty Acid Composition: The TransLinE Study. *Eur. J. Clin. Nutr.* **2000,** *54* (2), 104–113.

14. Beyers, E.C.; and E.A. Emken. Metabolites of cis,trans, and trans,cis Isomers of Linoleic Acid in Mice and Incorporation into Tissue Lipids. *Biochim. Biophys. Acta* **1991,** *1082,* 275–284.

15. Berdeaux, O.; J.L. Sébédio; J.M. Chardigny; J.P. Blond; T. Mairot; J.M. Vatèle; and J.P. Noël. Effects of trans n-6 Fatty Acids on the Fatty Acid Profile of Tissues and Microsomal Metabolism in the Rat. *Grasas Aceites* **1996,** *47,* 86–99.

16. Koletzko, B. Trans Fatty Acids May Impair Biosynthesis of Long-chain Polyunsaturates and Growth in *Man. Acta Paediatr.* **1992,** *81* (4), 302–306.

17. Ratnayake, W.M.N.; and G. Pelletier. Positional and Geometrical Isomers of Linoleic Acid in Partially Hydrogenated Oils. *J. Am. Oil Chem. Soc.* **1992,** *69* (2), 95–105.

18. Boue, C.; N. Combe; C. Billeaud; C. Mignerot; B. Entressangles; G. Thery; H. Geoffrion; J.L. Brun; D. Dallay; and J.J. Leng. Trans Fatty Acids in Adipose Tissue of French Women in Relation to Their Dietary Sources. *Lipids* **2000,** *35* (5), 561–566.

19. Boue, C.; N. Combe; C. Billeaud; and B. Entressangles. Nutritionnal Implications of trans Fatty Acids During Perinatal Period in French Pregant Women. *OCL* **2001,** *8*, 68–72.

20. Grandgirard, A.; A. Piconneaux; J.L. Sébédio; S. O'Keefe; E. Semon; and J.L. Le Querre. Occurence of Geometrical Isomers of Eicosapentaenoic Acids in Liver Lipids of Rats Fed Heated Linseed Oil. *Lipids,* **1989,** *24*, 799–804.

21. Acar, N.; J.M. Chardigny; O. Berdeaux; S. Almanza; and J.L. Sébédio. FA Composition of Plasma, Red Blood Cell, and Liver Phospholipids of Newborn Piglets Fed trans Isomers of α-Linolenic Acid. *Lipids* **2002,** *37* (9), 849–852.

22. Acar, N.; J. Chardigny; M. Darbois; B. Pasquis; and J.L. Sébédio. Modification of the Dopaminergic Neurotransmitters in Striatum, Frontal Cortex and Hippocampus of Rats Fed for 21 Months with trans Isomers of α-Linolenic Acid. *Neurosci. Res.* **2003,** *45*, 375–382.

23. Acar, N.; J.M. Chardigny; O. Berdeaux; S. Almanza; and J.L. Sébédio. Modification of the Monoaminergic Neurotransmitters in Frontal Cortex and Hippocampus by Dietary trans α-Linolenic Acid in Piglets. *Neurosci. Lett.* **2002,** *331*, (3), 198–202.

24. Acar, N.; J.M. Chardigny; B. Bonhomme; S. Almanza; M. Doly; and J.L. Sébédio. Long-term Intake of trans (n-3) Polyunsaturated Fatty Acids Reduces the b-Wave Amplitude of Electroretinograms in Rats. *J Nutr.* **2002,** *132* (10), 3151–3154.

25. Chardigny, J.M.; J.P. Blond; L. Saget; J.L. Sébédio; T. Eynard; D. Poullain; J.M. Vatèle; and J.M. Noël. Influence of a Δ9 trans Ethylenic Bond on the Δ6 Desaturation of Linolenic Acid. In *21st Congress of the Internationnal Society for Fat Research*; P.J. Barnes: The Hague, 1995.

26. Voss, A.; M. Reinhart; and H. Sprecher. Differences in the Interconversion Between 20- and 22-Carbon (n-3) and (n-6) Polyunsaturated Fatty Acids in Rat Liver. *Biochim. Biophys. Acta* **1992,** *1127*, 33–40.

27. Loï, C.; O. Berdeaux; J.M. Chardigny; D. Poullain; J.P. Noël; and J.L. Sébédio. Effects of trans 20:4 and 20:5 Isomers on Rat Platelet Aggregation. In *Proceedings of the 4th International Congress on Essential Fatty Acids and Eicosanoids*; R.A. Riemersma, R. Armstrong, and R. Wilson, Eds.; AOCS Press: Champaign, IL, 1998.

28. Privett, O.S.; E.M. Stearns; and E.C. Wickell. Metabolism of the Geometrical Isomers of Linoleic Acid in the Rat. *J. Nutr.* **1967,** *92*, 303–310.

29. Berdeaux, O.; J.P. Blond; L. Bretillon; J.M. Chardigny; T. Mairot; J.M. Vatèle; D. Poullain; and J.L. Sébédio. In vitro Desaturation or Elongation of Monotrans Isomers of Linoleic Acid by the Rat Liver Microsomes. *Mol. Cell. Biochem.* **1998,** *85*, 17–25.

30. Cunnane, S.C.; and M.J. Anderson. The Majority of Dietary Linoleate in Growing Rats is β-Oxidized or Stored in Visceral Fat. *J. Nutr.* **1997,** *127* (1), 146–152.

31. Bretillon, L.; J.M. Chardigny; J.L. Sébédio; D. Poullain; J.P. Noël; and J.M. Vatèle. Oxidative Metabolism of [1-14C]Mono-trans Isomers of Linoleic and α-Linolenic Acids in the Rat. *Biochim. Biophys. Acta* **1998,** *1390* (2), 207–214.

32. Loreau, O.; A. Maret; D. Poullain; J.M. Chardigny; J.L. Sébédio; B. Beaufrere; and J.P. Noel. Large-scale Preparation of (9Z,12E)-[1-13C]Octadeca-9,12-dienoic Acid, (9Z,12Z,15E)-[1-13C]-Octadeca-9,12,15-trienoic Acid and Their [1-13C] All-cis Isomers. *Chem. Phys. Lipids* **2000**, *106* (1), 65–78.

33. Loi, C.; J.M. Chardigny; O. Berdeaux; J.M. Vatele; D. Poullain; J.P. Noel; and J.L. Sébédio. Effects of Three trans Isomers of Eicosapentaenoic Acid on Rat Platelet Aggregation and Arachidonic Acid Metabolism. *Thromb. Haemostasis* **1998**, *80* (4), 656–661.

34. Chardigny, J.M.; J.L. Sébédio; A. Grandgirard; L. Martine; O. Berdeaux; and J.M. Vatèle. Identification of Novel trans Isomers of 20:5 n-3 in Liver Lipids of Rats Fed a Heated Oil. *Lipids* **1996**, *31*, 165–168.

35. Armstrong, R.A.; J.M. Chardigny; B. Beaufrère; L. Bretillon; S.H.F. Vermunt; R.P. Mensink; A. Macvean; R.A. Elton; J.L. Sébédio; and R.A. Riemersma. No Effect of Dietary trans Isomers of α-linolenic Acid on Platelet Aggregation and Haemostatic Factors in European Healthy Men: The TRANSLinE Study. *Thrombosis Res.* **2000**, *100*, 133–141.

36. Loi, C.; J.M. Chardigny; C. Cordelet; L. Leclere; M. Genty; C. Ginies; J.P. Noel; and J.L. Sébédio. Incorporation and Metabolism of trans 20:5 in Endothelial Cells. Effect on Prostacyclin Synthesis. *Lipids* **2000**, *35* (8), 911–918.

37. Chardigny, J.M.; B. Bonhomme; J.L. Sébédio; P. Juanéda; J.M. Vatèle; M. Doly; and A. Grandgirard. Effects of Dietary trans n-3 Polyunsaturated Fatty Acids on the Rat Electroretinogram. *Nutr. Res.* **1998**, *18* (10), 1711–1721.

38. Zimmer, L.; G. Durand; D. Guilloteau; and S. Chalon. n-3 Polyunsaturated Fatty Acid Deficiency and Dopamine Metabolism in the Rat Frontal Cortex. *Lipids* **1999**, *34* (Suppl.), S251.

39. Rojo, J.A.; and E.G. Perkins. Cyclic Fatty Acid Monomers Formation in Frying Fats. I. Determination and Structural Study. *J. Am. Oil Chem. Soc.* **1987**, *64* (3), 414–421.

40. Sébédio, J.L.; J. Prevost; and A. Grandgirard. Heat Treatment of Vegetable Oils. I. Isolation of the Cyclic Fatty Acid Monomers from Heated Sunflower and Linseed Oils. *J. Am. Oil Chem. Soc.* **1987**, *64* (7), 1026–1032.

41. Christie, W.W.; E.Y. Brechany; J.-L. Sébédio; and J.-L. LeQuere. Silver Ion Chromatography and Gas Chromatography–Mass Spectrometry in the Structural Analysis of Cyclic Monoenoic Acids Formed in Frying Oils. *Chem. Phys. Lipids* **1993**, *66*, 143–153.

42. Combe, N.; M.J. Constantin; and B. Entressangles. Absorption intestinale des espèces chimiques nouvelles (E.C.N.) formées lors du chauffage des huiles. *Rev. Fr. Corps Gras* **1978**, *1*, 27–28.

43. Dobson, G.; W.W. Christie; E.Y. Brechany; J.L. Sébédio; and J.L. LeQuere. Silver Ion Chromatography and Gas Chromatography–Mass Spectrometry in the Structural Analysis of Cyclic Dienoic Acids Formed in Frying Oils. *Chem. Phys. Lipids* **1995**, *75*, 175–182.

44. Le Quere, J.L.; J.L. Sébédio; R. Henry; F. Couderc; N. Demont; and J.C. Prome. Gas Chromatography–Mass Spectrometry and Gas Chromatography–Tandem Mass Spectrometry of Cyclic Fatty Acid Monomers Isolated from Heated Fats. *J. Chromatogr.* **1991**, *562* (1–2), 659–672.

45. Martin, J.C.; C. Caselli; S. Broquet; P. Juaneda; M. Nour; J.L. Sébédio; and A. Bernard. Effect of Cyclic Fatty Acid Monomers on Fat Absorption and Transport Depends on Their Positioning Within the Ingested Triacylglycerols. *J. Lipid Res.* **1997**, *38* (8), 1666–1679.

46. Mossoba, M.M.; M.P. Yurawecz; J.A. Roach; H.S. Lin; R.E. McDonald; B.D. Flickinger;

and E.G. Perkins. Rapid Determination of Double Bond Configuration and Position Along the Hydrocarbon Chain in Cyclic Fatty Acid Monomers. *Lipids* **1994,** *29* (12), 893–896.

47. Joffre, F.; J. Martin; M. Genty; L. Demaison; O. Loreau; J. Noel; and J.L. Sébédio. Kinetic Parameters of Hepatic Oxidation of Cyclic Fatty Acid Monomers Formed from Linoleic and Linolenic Acids. *J. Nutr. Biochem.* **2001,** *12* (10), 554–558.

48. Joffre, F.; A. Roy; L. Bretillon; B. Pasquis; J.P. Sergiel; O. Loreau; J.M. Chardigny; and J.L. Sébédio. In vivo Oxidation of Carboxyl-labelled Cyclic Fatty Acids Formed from Linoleic and Linolenic Acids in the Rat. *Reprod. Nutr. Dev.* **2004,** *44* (2), 123–130.

49. Ribot, E.; A. Grandgirard; J.L. Sébédio; A. Grynberg; and P. Athias. Incorporation of Cyclic Fatty Acid Monomers in Lipids of Rat Heart Cell Cultures. *Lipids* **1992,** *27* (1), 79–81.

50. Artman, N.R. The Chemical and Biological Properties of Heated and Oxidized Fat. *Adv. Appl. Lipid Res.* **1969,** *7*, 245–330.

51. Perkins, E.G. Chemical, Nutritional and Metabolic Effects of Heated Fats. 2) Nutritional Aspects. *Rev. Fr. Corps Gras* **1976,** *23*, 313–322.

52. Crampton, E.W.; R.H. Common; F.A. Farmer; F.M. Berryhill; and L. Wiseblatt. Studies to Determine the Nature of the Damage to the Nutritive Value of Some Vegetable Oils from Heat Treatment. II. Investigations of the Nutritiousness of the Products of Thermal Polymerization of Linseed Oil. *J. Nutr.* **1951,** *44* (1), 177–189.

53. Crampton, E.W.; R.H. Common; F.A. Farmer; A.F. Wells; and D. Crawford. Studies to Determine the Nature of the Damage to the Nutritive Value of Some Vegetable Oils from Heat Treatment. III. The Segregation of Toxic and Non-toxic Material from the Esters of Heat-Polymerized Linseed Oil by Distillation and by Urea Adduct Formation. *J. Nutr.* **1953,** *49* (2), 333–346.

54. Iwaoka, W.T.; and E.G. Perkins. Nutritional Effects of the Cyclic Monomers of Methyl Linolenate in the Rat. *Lipids* **1976,** *11* (4), 349–353.

55. Damy Zarambaud, A.; and A. Grandgirard. Detoxification by the Rat of Compounds Formed During Thermal Polymerization of Linseed Oil. II. Effects of Discontinuous Administration of the Heated Oil on the Urinary Excretion of Glucuronides, Liver Weight and the Tissue Content of Cyclic Monomers. *Reprod. Nutr. Dev.* **1981,** *21* (3), 409–419.

56. Siess, M.H.; M.F. Vernevaut; A. Grandgirard; and J.L. Sébédio. Induction of Hepatic Drug-Metabolizing Enzymes by Cyclic Fatty Acid Monomers in the Rat. *Food Chem. Toxicol.* **1988,** *26* (1), 9–13.

57. Saito, M.; and T. Kaneda. Studies on the Relationship Between the Nutritive Value and the Structure of Polymerized Oils. *Jpn. J. Oil Chem. Soc.* **1976,** *25* (2), 13–20.

58. Sébédio, J.L.; J.L. LeQuere; O. Morin; and A. Grandgirard. Heat Treatment of Vegetable Oils. III. GC–MS Characterization of Cyclic Fatty Acid Monomers in Heated Sunflower and Linseed Oils After Total Hydrogenation. *J. Am. Oil Chem. Soc.* **1989,** *66* (5), 704–709.

59. Sébédio, J.L.; J.L. LeQuere; E. Semon; O. Morin; J. Prevost; and A. Grandgirard. Heat Treatment of Vegetable Oils. II. GC–MS and GC–FTIR Spectra of Some Isolated Cyclic Fatty Acid Monomers. *J. Am. Oil Chem. Soc.* **1987,** *64* (9), 1324–1333.

60. Saito, M.; and T. Kaneda. Studies on the Relationship Between the Nutritive Value and the Structure of Polymerized Oils. 11) Mechanisms of Toxicity of Heat-Polymerized Oils.

*Yukagaku* **1976,** *25,* 842–847.
61. Farmer, F.A.; E.W. Crampton; and M.I. Siddall. The Effect of Heated Linseed Oil on Reproduction and Lactation in the Rat. *Science* **1951,** *113* (2937), 408–410.
62. Potteau, B. Influence of Heated Linseed Oil on Reproduction in the Female Rat and on the Composition of Hepatic Lipids in Young Rats. *Ann. Nutr. Aliment.* **1976,** *30* (1), 67–88.
63. Potteau, B. Transference of Cyclic Monomeric Acids into the Milk of Female Rats Eating Thermopolymerised Linseed Oil. *Ann. Nutr. Aliment.* **1976,** *30* (1), 89–93.
64. Sébédio, J.L.; and A. Grandgirard. Cyclic Fatty Acids: Natural Sources, Formation During Heat Treatment, Synthesis and Biological Properties. *Progr. Lipid Res.* **1989,** *28,* 303–336.
65. Sébédio, J.L.; J.M. Chardigny; P. Juaneda; M.C. Giraud; M. Nour; W.W. Christie; and G.A. Dobson. Nutrional Impact and Selective Incorporation of Cyclic Fatty Acid Monomers in Rats During Reproduction. In *21st World Congress of the International Society for Fat Research*; P.J. Barnes & Associates: The Hague, The Netherlands, 1995.
66. Joffre, F. Effets nutritionnels des monomères cycliques issus de l'acide α-linolénique chez l'animal. Université de Bourgogne: Dijon, France, 2001; p. 153.
67. Lamboni, C.; J.L. Sébédio; and E.G. Perkins. Effect of Cyclic Fatty Acid Monomers from Dietary Heated Fats on Rat Liver Enzyme Activity. *inform* **1995,** *6,* 464.
68. Lamboni, C.; J.L. Sébédio; and E.G. Perkins. Cyclic Fatty Acid Monomers from Dietary Heated Fats Affect Rat Liver Enzyme Activity. *Lipids* **1998,** *33* (7), 675–681.
69. Martin, J.C.; F. Joffre; M.H. Siess; M.F. Vernevaut; P. Collenot; M. Genty; and J.L. Sébédio. Cyclic Fatty Acid Monomers from Heated Oil Modify the Activities of Lipid Synthesizing and Oxidizing Enzymes in Rat Liver. *J. Nutr.* **2000,** *130* (6), 1524–1530.
70. Bretillon, L.; S.E. Alexson; F. Joffre; B. Pasquis; and J.L. Sébédio. Peroxisome Proliferator-Activated Receptor α Is Not the Exclusive Mediator of the Effects of Dietary Cyclic FA in Mice. *Lipids* **2003,** *38* (9), 957–963.
71. Athias, P.; E. Ribot; A. Grynberg; J.-L. Sébédio; and A. Grandgirard. Effects of Cyclic Fatty Acid Monomers on the Function of Cultured Rat Cardiac Myocytes in Normoxia and Hypoxia. *Nutr. Res.* **1992,** *12,* 737–745.

# 11

# The Chemistry and Nutrition of Nonnutritive Fats

**William E. Artz[a], Louise L. Lai, and Steven L. Hansen[b]**

*[a]Department of Food Sciences and Human Nutrition, University of Illinois, Urbana, Illinois, 61801-4726, and [b]Cargill Analytical Services, Minnetonka, MN*

As countries around the world have experienced improved economic conditions, the resulting increase in affluence has resulted in an increase in per capita calorie and fat consumption and a concomitant increase in the problem of obesity. As a result, health concerns have increased, particularly those related to cardiovascular disease. As a response to health concerns and consumer demands in the United States, the food industry has developed a wide variety of low-fat and/or low calorie products. The consumer continues to demand food products that are low-fat (or fat free), but still have the taste and mouth feel of the full-fat products they enjoy. To help meet that demand, fat-based fat substitutes were developed.

Fat substitutes can be divided into three broad categories: carbohydrate-based, protein-based, and lipid-based (1). As a food ingredient, any synthetic food additive must undergo extensive studies to be deemed safe for consumption. Many of the carbohydrate- and protein-based fat substitutes have received GRAS (Generally Recognized As Safe) status from the U.S. Food and Drug Administration (FDA). However, GRAS status for a compound approved for use as a food additive at low concentrations does not guarantee approval for use as a macronutrient substitute. All synthetic macronutrient substitutes face extensive safety testing prior to FDA approval.

Research on fat substitutes has resulted in several formulations with functions closer to that of natural fat. The carbohydrate- and protein-based fat substitutes currently available cannot be used for frying, so this discussion will focus on the fat-based, heat-stable, low- to non-caloric, synthetic fat substitutes. Synthetic substitutes are unique in that they contribute little to the caloric content of the food, yet they retain most of the important functional attributes associated with traditional fats (2). Widespread use of synthetics could substantially reduce the amount of fat calories in the U.S. diet, which would have important positive implications in terms of heart disease and other cardiovascular-related problems. Early developments in this area are discussed elsewhere (3–5).

The most important factor in consumer food selection is taste (90%), followed by nutrition (75%), product safety (72%), and cost (71%) (4). Fat is closely associated with the desirable taste, texture, and palatability found in many foods. People are particularly desirous of fat substitutes that produce a taste comparable to traditional products (6,7). While they believe that products prepared with fat substitutes are more healthful than traditional products, they also feel that "natural" fat substitutes (e.g., Simplesse) are more healthful than "synthetic" fat substitutes (e.g., olestra). In an effort to provide food products with the desired taste, texture, and palatability, several companies are developing synthetic, heat-stable fat substitutes (8). The market for these products is substantial. In the February 2002 issue of the Chemical Market Reporter, the U.S. market for fat replacers was estimated at $414 million (9) and is predicted to continue to steadily increase.

Several important issues concerning synthetic fat and oil substitutes must be answered (11–13) prior to approval. Since fat substitutes are macroingredients, the specification and purity of the sample must be well defined in terms of its chemical stability and metabolic fate. The effect of the unabsorbed (synthetic) fat on gastrointestinal epithelium and lymphatic tissues, microbial populations, bile acids, and fat-soluble vitamins, and drug absorption must also be considered.

## Chemistry of Synthetic Fat Substitutes

The following discussion of the reduced calorie and synthetic fat substitutes only includes a limited number of synthetic fat substitutes and is not meant to be all-inclusive. An extensive review is published elsewhere (8). The reduced-calorie fat-based fat substitutes will be discussed first.

Salatrim, an acronym for short- and long-chain acyl triacylglycerol molecules, is a mixture of at least one long-chain fatty acid (usually stearic acid) and one or two short-chain fatty acids (Nabisco Food Group, Parsippany, NJ). It was renamed Benefat when responsibility for marketing the fat substitute was assumed by the Cultor Food Science Co. Benefat is formed by the interesterification of triacetin, tripropionin, tributyrin, and hydrogenated vegetable oils. It is used in products like confections, baked goods, and dairy products. Salatrim should not be used for deep fat frying (9,16,17), since release of short-chain fatty acids will cause a very undesirable flavor effect. It is completely digestible, but it has fewer calories than normal fats or oils.

Another digestible, but reduced-calorie fat is caprenin. Caprocaprylobehenic is a triacylglycerol that is also known as caprenin. It is formed when glycerol is esterfied with caprylic, capric, and behenic fatty acids (Proctor & Gamble Co.). Caprenin is a reduced-calorie triglyceride that provides ~5 kcal/g of energy due to the partial absorption of the behenic fatty acids. It is used in soft candies and confectionery coatings (1,16,17).

Medium-chain triglycerides (MCT) are esters made from vegetable oils that are hydrolyzed, then fractionated, and finally re-esterified to form triacylglycerols (Stepan Company and Abitec Corp.). They contain 8–12 carbon saturated fatty acids and provide 7 to 8 kcal/g of energy. Although they do contribute to energy intake, they have little tendency to accumulate in the body as fat and have little effect on cholesterol levels. MCT, available based on GRAS self-determination, are stable at a wide range of temperatures. Since the 1950s they have been used clinically to treat patients that have lipid absorption, digestion, or transport disorders. Recently, they have expanded in use to sports nutrition food products (9,16,93). Noebee MLT-B (Stepan Co.) is a low-calorie, medium-chain triglyceride fat substitute. It has a solid fat index similar to partially hydrogenated vegetable oil. It provides slightly fewer calories than normal vegetable fats and oils (84,94).

Vegetable oil mono- and diglyceride emulsifiers have been used as fat substitutes. They can be used to replace shortening in baked goods, icings, and vegetable dairy products. They still contribute to calorie intake; however, they are generally used in smaller amounts. Some examples of emulsifier-based fat substitutes include Dur-Lo® and EC™-25 (17,95).

One of the earliest fat substitutes contained a mixture of completely and partially short-chain fatty acid esterified to pentaerythritol (Heyden Chemical Corp., New York) (14). For example, polypentaerythritol can be esterified with saturated short-chain fatty acids, such as acetic, propionic, and butyric acids. In 1960, similar pentaerythritol esters were investigated for their effect on weight control and blood lipid concentrations (Heyden Chemical Corp.) (15). The esters were neither toxic nor digestible. They could replace some of the fat or oil in shortening and salad dressing, or they could be used as a cooking medium.

Polyvinyl alcohol (PVA) fatty acid esters, such as polyvinyl oleate (PVO), were investigated as potential fat substitutes (Nabisco Brands, Inc., East Hanover, NJ) (18). Polyvinyl alcohol fatty acid esters can be prepared with two different methods: direct esterification of low-molecular weight (MW) PVA with an excess (30%) of fatty acids, acid chlorides, or acid anhydrides aqueous solutions to form esters (polymerization of vinyl esters), or they can be prepared using a transvinylation method. PVO is prepared by acid-catalyzed homogenous esterification of PVA with oleic acid. PVA, for the most part, is soluble only in water. This hinders preparation of fatty acid esters using PVA. Work done by Chetri and Dass (19) claimed to have solved this problem by combining PVA with the catalyst $C_2H_5ONO_2$.DMSO (EN.DMSO), which allowed the use of organic solvents.

Hamm (20) investigated the synthetic fat replacer trialkoxy-glyceryl ether (TGE). Structurally, it contains an ester bond that is the reverse of that contained in a normal triacylglycerol, since fatty alcohols are esterified to a polycarboxylate or acid backbone. Trialkoxy-glyceryl ether production on a large scale is difficult and time-consuming, since purification is required for the removal of xylene, diethers, and the large amounts of moderately polar materials that include mono- and diglyceryl ethers, fatty alcohols, and fatty mesylates. Trialkoxy-glyceryl ether exhibits a viscosity and surface tension in the same range as vegetable oils at room temperature. The interfacial tension of TGE indicates it may be suitable for mayonnaise emulsions.

Dialkyl dihexadecylmalonate (DDM) (Frito-Lay, Inc., Dallas, TX) is a fatty alcohol ester of malonic and alkylmalonic acids, which could be used in high-temperature applications (21–23). Fatty acids most preferred for preparation are myristic, palmitic, stearic, oleic, and linoleic. Suitable acids include malonic acid and monoalkyl and dialkyl malonic acids. Dialkyl dihexadecylmalonate is synthesized by reacting a malonyl dihalide with a fatty alcohol. Test results indicate that a blend of DDM and soybean oil produced potato and tortilla chips that were less oily and as crisp as if fried in vegetable oil (24). This blend resulted in a 33% reduction in calories and a 60% reduction in fat intake. Dialkyl dihexadecylmalonate was also evaluated as a fat substitute in mayonnaise and margarine-type products. The lower MW DDM could be used in mayonnaise and margarine, while the higher MW DDM could be used as a frying oil (22–24, 25).

Nabisco Brands, Inc. developed a partially digestible, carboxylate ester-based fat substitute synthesized with fatty acids or fatty alcohols (26). For example, fatty alcohols can be esterified to citric acid (27). Suitable alcohols include *n*-hexadecyl alcohol, oleyl alcohol, and *n*-octadecyl alcohol. Although citric acid is not recommended for some applications due to its relatively poor thermal stability, the three carboxyl groups and the hydroxyl group make it very useful. Three different types of citric acid fat substitutes (citric-isocitric ether, di-isocitric ether, and di-citrate ether) were synthesized. Functionality of the synthetic fat can be controlled by the choice of fatty alcohol. Possible applications include margarine, mayonnaise, and baked goods.

Esters of fatty alcohols and/or fatty acids with other hydroxycarboxylic acids were also developed for use as fat replacers (Nabisco Brands, Inc.) (28,29). Dioleyl 1-myristoyloxy-1,2-ethanedicarboxylate was prepared and evaluated using animal studies. It was only slightly digestible and has only 4% of the potentially available calories. Possible applications include butter, ice cream, vanilla wafers, coconut oil mimetic, and crackers.

Short- and long-chain diol diesters can be used as fat substitutes (Nabisco Brands, Inc.) (30,31). They are partially digestible (20% after 3 h).

Polyol fatty acid esters are potential fat substitutes for use in calorie-reduced foods (Unilever N.V., Rotterdam, The Netherlands) (32). Preparation involves reacting one or more fatty acid esters with a polyol containing at least four hydroxyl groups in the presence of a basic catalyst. Polyol examples include sorbitol, raffinose, trehalose, and stachyose (16). A fat substitute, composed of indigestible polyol fatty acid polyesters and conventional triacylglycerols, was developed for use in low-fat spreads and margarines (Unilever N.V.) (33).

Glycerol esters of α-branched carboxylic acids can be partial (10%) or complete (100%) fat substitutes in products such as cakes, bread, mayonnaise, dairy products, salad oils, cooking oils, and margarine (Procter & Gamble, Cincinnati, OH) (34). Preparation involves esterifying (stepwise or *trans*) glycerol with an α-branched carboxylic acid. The α-branched carboxylate structure prevents the ester from being hydrolyzed by pancreatic lipase in the intestine.

Esterified epoxide-extended polyols (EEEP) (ARCO Chemical Technology, Inc.) can be used as a replacement for fat or used as cooking oil itself. EEEP are not absorbed, low-caloric, and non-digestible. They are similar to vegetable oils and fats, and are considered non-toxic (35). One subgroup of EEEP includes the fatty acid esterified propoxylated glycerols (EPG) (ARCO Chemical Co., Newtown Square, PA), which were developed as fat substitutes. Glycerol is propoxylated with propylene oxide to form a polyether polyol, which is then esterified with fatty acids (21,22, 24,36–43) to form a modified triglyceride with a polyether extensive between the fatty acids and the glycerol, which is very resistant to lipase activity. Research has been published on model systems that include EPG-08 soyate, oleic acid-esterified propoxylated glycerol (EPG-08 oleate), and linoleic acid-esterified propoxylated glycerol (EPG-08 linoleate) (44,45). The resulting triacylglycerol is similar to natural fats in structure and functionality. Fatty acid EPG is a low- to non-caloric oil that is heat-stable and only very slightly digestible. It can be used in table spreads, ice cream, frozen desserts, salad dressings, bakery products, salad and cooking oils, mayonnaise, and shortenings (35). Suitable polyols include sugars or glycerides, which are reacted with $C_3$–$C_6$ epoxides to form epoxide-extended polyols. The in vivo threshold for nondigestibility occurs when $n$ equals 4. The preferred fatty acids are in the $C_{14}$–$C_{18}$ range. Feeding experiments with rats (38) and mice (41) indicated no toxicity. Preparation of propoxylated glycerides for use as fat substitutes involved transesterifying propoxylated glycerol with esters of $C_{10}$–$C_{24}$ fatty acids in a solvent-free, nonsaponifying system (ARCO Chemical Technology Limited Partnership, Wilmington, DE) (42) to avoid reagents unacceptable in food systems.

Acylated glycerides could replace normal triacylglycerols in fat-containing foods (Procter & Gamble) (46). As the number of acylated groups increases, the acylated glyceride is less digestible by pancreatic enzymes. There are three possible methods for producing acylated glycerides. One is to react glycerol with the methyl ester of a protected α-hydroxy fatty acid. Once the protective group is removed, the glyceride is esterified with the acid chloride of a fatty acid, such as palmitic, stearic, oleic, linoleic, or linolenic acid. Second, glycerol is reacted with an α-methoxy-ethyl ester of a protected α-hydroxy fatty acid. After removing the protecting group, esterification of the α-hydroxy glyceride is completed with a fatty acid or anhydride. Finally, glycerol can be reacted with the methyl esters of unprotected α-hydroxy fatty acids or fatty acids. The α-acylated glyceride may provide partial to total replacement (10–100%) of fat in a food system. Possible applications include salad and cooking oils, plastic shortening, prepared cake and icing mixes, mayonnaise, salad dressing, and margarine.

Dialkyl glycerol ethers have been developed for use as a fat substitute (Swift and Co.) (47). A mixture of didodecyl glycerol ether (20%), cis-9-octadecenyl octadecyl ether (35%), cis-9-octadecenyl tallow alcohol glycerol ether (35%), and dioctadecyl glycerol ether (10%) may be used as a shortening substitute in bakery products or margarines.

Tris (hydroxymethyl) alkane esters are different from most engineered fats in that the esters are partially hydrolyzed in vivo and may contribute from 0.5–6 kcal/g, as compared to 9 kcal/g for conventional fats (Nabisco Brands, Inc.) (48,49). Mixtures of fatty acids obtained from nonhydrogenated or hydrogenated soybean, safflower, sunflower, sesame, peanut, corn, olive, rice bran, canola, babassu nut, coconut, cottonseed, and/or palm oils can be used. Fatty acid derivatives can be used, such as halogenated or dicarboxylate-extended (malonic, succinic, glutaric, or adipic acid) fatty acid residues. Monomeric and dimeric *tris* (hydroxymethyl) alkane esters have been suggested for use in combination, in partial or full replacement of fat in frozen desserts, puddings, pie fillings, margarine substitutes, pastries, mayonnaise, filled dairy products, flavored dips, frying fats and oils, meat substitutes, whipped toppings, compound coatings, frostings, cocoa butter replacements, fatty candies, and chewing gum.

Phenyldimethylpolysiloxane (PDMS) (Dow Corning Corp., Midland, MI) was investigated as a non-caloric frying medium (50) and fat replacer (51). Morehouse and Zabik (50) examined samples of PDMS with four different viscosities for frying fish patties, french fries, and doughnut holes. Phenyldimethylpolysiloxane is thought to be a suitable frying medium due to its thermal and oxidative stability, minimal change in viscosity over a wide temperature range, water repelling ability, and biological inertness. The heat transfer characteristics of PDMS depend on the food composition and temperature. Frying in low-viscosity PDMS resulted in a lighter, less red product with a softer and more consistent texture than products fried in corn oil. The fat content of the PDMS-fried foods was one-fourth or less than that of corn oil-fried foods.

Phenylmethylpolysiloxane (PS) (Dow Corning 550 Fluid, Contour Chemical Co., North Reading, MA) was also studied as a possible fat substitute (51). Phenylmethylpolysiloxane is chemically inert, nonabsorbable, non-caloric, and nontoxic. The viscosity is dependent on the MW of the polymers. Obese female Zucker rats lost weight while on a PS (22% w/w) low-fat (LF) diet, when compared to a LF diet only. The PS-fed rats did not compensate for their caloric reduction by increasing food intake (52). Comparison of the PS diet to cellulose (CE) (22% w/w) showed a reduction in adipocyte size and total carcass fat for PS-diet rats. Fat absorption (percent of actual dietary fat intake) was greater for CE-fed rats than for PS-fed rats.

According to Dow Corning Corp. (53), the safety and lack of toxicity of PDMS and related organosilicones has been established through research. Applications include baked goods, mayonnaise, peanut butter, cereals, icings, and frying oils (54). At very small concentrations, however, silicone-based antifoams are deleterious to cake texture and potato chip and doughnut quality. Other possible applications include dairy products and shortening substitutes (55). Rats fed a diet containing 6.5 wt% PDMS exhibited an undesirable anal leakage. Anti–anal leakage (AAL) agents should be used, consisting of fatty acids with a melting point of $\geq 37°C$. Other AAL agents include particulate silica with a surface area greater than 10 $m^2/g$, preferably 300–400 $m^2/g$ (53,55). A third type of AAL agent is edible, nondegradable, water-insoluble plant fiber.

A membrane lipid, diether phytanyl phosphatidyl glycerol phosphate from *Halobacterium halobium*, was evaluated for its potential as a non-caloric fat substitute (56). The alkyl portion is stable since it is saturated. Ether-linked alkyl groups are stable over a wider pH range than are ester-linked fatty acids and are also less susceptible to hydrolysis. The L-configuration of the alkyl portion of the lipid provides additional stability, since all known phospholipases in nature are specific for the D-isomers of phosphatides. Due to limited amounts of the sample, only acute toxicity tests and emulsion tests were completed. No additional functional characteristics were determined.

There has been extensive interest in carbohydrate fatty acid esters as potential fat substitutes. The most well-known carbohydrate fatty acid ester is the sucrose fatty acid ester called olestra (Olean®), which is composed of hexa-, hepta-, and octa-fatty acid esters that are formed when sucrose and long-chain fatty acids isolated from edible oils are combined (Procter & Gamble Co., Cincinnati, OH). Olestra contains mostly octaesters ($\geq 70\%$). Olestra also contains modest amounts of sucrose heptaesters and small amounts of hexaesters of fatty acids, primarily palmitic (16:0), stearic (18:0), oleic (18:1 *cis*), and linoleic (18.2 *cis*) acids (58–62). On January 24, 1996, olestra use for savory snacks was approved by the FDA after an Olestra Post-Marketing Surveillance Study (OPMSS) was completed by Procter & Gamble (63,64). These experiments proved that olestra is not a toxin, carcinogen, mutagen, disease-causing agent, allergen, or a reproductive toxin (65–67). There have been some studies to show that sucrose polyesters such as olestra could be used medicinally. Work done by Moser and associates have shown that non-absorbable dietary fat substitutes like olestra could potentially help in the removal of lipophilic contaminants such as polychlorobenzene dibenzo-*p*-dioxins and dibenzofurans (PCDD/F), polychlorinated biphenyls (PCB), and hexachlorobenzene (HCB) (68). It has no known effects on drug absorption and no effects on nutrient absorption, except for an effect on lipophilic dietary components such as serum vitamin E, which can easily be offset by supplementation of olestra with *d*-α-tocopherol (69). As a result, the FDA required food products containing olestra to be supplemented with vitamins A, D, E, and K (61,70,67). Pancreatic lipases in the gut are unable to hydrolyze olestra, and thus it is a non-caloric fat substitute (71). Sucrose fatty acid polyesters are synthesized from a variety of fat sources, including partially hydrogenated soybean oil, olive oil, canola oil, corn oil, milk fat, etc. (72–78). Also, due to their lipid-like nature and ability to withstand high temperatures, sucrose fatty acid polyesters are effective frying oil substitutes in fried snack foods (79).

Reports on the synthesis and analysis of carbohydrate-fatty acid esters were published by several groups (57,73,80). Swanson's research group at Washington State University (81,82) published extensively on carbohydrate fatty acid esters synthesized from several carbohydrate sources using several different catalytic conditions. They reported a synthesis with yields of 99.8% utilizing a solvent-free, one-stage process for synthesis of sucrose octaoleate. Synthesis of several carbohydrate/fatty acid-derived fat substitutes, including glucose fatty acid esters, sucrose fatty acid esters, raffinose fatty acid esters, and even larger polysaccharide oligomers such as stachyose and verbascose fatty acid esters, were reported. Even though carbohydrate fatty acid esters do not seem to be susceptible to enzymatic degradation in the gut, there have been enzymatic methods developed for the synthesis of carbohydrate–fatty acid fat substitutes. Enzymatic methods for the synthesis of carbohydrate fatty acid esters were discussed in detail by Riva (83). One of the most promising enzymes tested, particularly for the fatty acid esterification of the alkylated glucosides, was a lipase from the yeast *Candida antartica*, which was immobilized on macroporous resin beads. In addition, *Candida cylindraceae* has a promising lipase that can be used for the synthesis of carbohydrate fatty acid esters (84). Only a limited number of successful enzymatic syntheses have been reported; more research is needed.

Mono- and oligomeric carbohydrate esters were also prepared from sugars or other carbohydrates and saturated fatty acids and/or ester-forming fatty acid derivatives (Hoechst AG, Frankfurt, Germany) (85,86) containing branched or unbranched carboxylic acids of various chain lengths. For example, glucose laurate was examined for its acceptability as a fat substitute in chocolate.

Alkyl glycoside fatty acid esters can replace fat (from 5–95%) in frying oils and white or Italian salad dressings (Curtice-Burns, Inc., Rochester, NY) (87). Alkyl glycosides are formed by the reaction of a sugar with a monohydric alcohol. Mono-, di-, and trisaccharides can be used. The preferred alcohols are alkyl alcohols from 1 to 24 carbons. The hydroxyl groups on the alkyl glycosides are esterified to form lower MW (less than 6 carbon acyl group) acyl ester alkyl glycosides. Then lower MW acyl ester alkyl glycosides are reacted with lower MW alkyl fatty acid esters. Soybean, safflower, corn, peanut, and cottonseed oils are preferred, since they contain $C_{16}$–$C_{18}$ fatty acids that do not volatilize at the temperatures used for interesterification. The preferred alkyl glucosides are the reaction products between glucose, galactose, lactose, maltose, and ethanol and propanol. Blending unsaturates and saturates (> 25% $C_{12}$ or larger) produces a heterogeneous alkyl glycoside fatty acid polyester, which does not induce undesirable anal leakage. Anal leakage may be prevented if the alkyl glycoside fatty acid polyesters have a melting point ≥37°C. Another method to synthesize alkyl glycoside fatty acid esters is through the use of enzymes. Work done by Akoh and Mutua (84) showed that the best synthesis results are obtained when using a combination of lipases from *Candida antarctica* and *Candida cylindraceae* through the process of transesterification. The advantages of this method include the enzymes localized method of attack, relatively few reaction conditions to control, as well as relatively few by-products.

A patent was awarded for the incorporation of the alkyl glycoside fatty acid polyester into food products (Curtice-Burns, Inc.) (88). The fat substitute could replace "visible fats" in shortening, margarine, butter, salad and cooking oils, mayonnaise, salad dressings, and confectionery coatings or "invisible fats" in oilseeds, nuts, dairy, and animal products. Visible fats are fats and oils isolated from animal tissues, oilseeds, or vegetable sources. Invisible fats are the fats and oils contained in the food product that are consumed along with the protein and carbohydrate components of the sources as they naturally occur.

Substitution at 10–100% is possible; however, less than 100% is preferred; the desired range is 33–75%. The method to synthesize alkyl glycoside fatty acid polyester fat substitutes was patented August 27, 1996 by R.S. Meyer, M.L. Campbell, D.B. Winter, and J. M. Root. To synthesize alkyl glycoside fatty acid polyesters, lower acyl ester alkyl glycoside and an alkaline metal catalyst is combined into a reaction mixture and exposed to the reaction conditions. The fat substitute contains at least four fatty acid ester groups that have four twenty-four carbons (89).

Glucosides containing from 1–50 alkoxy groups can be used as fat substitutes at substitution ranges of 10–100% in low-calorie salad oils, plastic shortenings, cake and icing mixes, mayonnaise, salad dressings, and margarines (Procter & Gamble) (90). The alkoxylated alkyl glucoside (i.e., propoxylated or ethoxylated methyl glucoside) is esterified with 4–7 $C_2$–$C_{24}$ fatty acids. For example, ethoxylated glucoside tetraoleate was prepared by reacting oleoyl chloride with Glucam E-20 (ethoxylated methyl glucoside obtained from reacting methyl glucoside with ethylene oxide).

ARCO Chemical Technology Limited Partnership patented a fatty acid esterified polysaccharide (PEP) that is partially esterified with fatty acids. The PEP can be used as a fat substitute in salad oils, cooking oils, margarine, butter blends, mayonnaise, and shortening at a substitution range of 10–100% (91). It is non-absorbable, indigestible, and nontoxic. Suitable oligo/polysaccharide materials include xanthan gum, guar gum, gum arabic, alginates, cellulose hydrolysis products, hydroxypropyl cellulose, starch hydrolysis products (n < 50), karaya gum, and pectin. The degree of esterification is controlled by the length of the acyl ester chain and the total number of hydroxyl groups available for esterification. The preferred level of esterification involves one or more hydroxyl groups per saccharide unit with one or more $C_8$–$C_{24}$ fatty acids. The preferred fatty acid sources are soybean oil, olive oil, cottonseed oil, corn oil, tallow, and lard. Preparation may involve direct esterification or transesterification, with the metal catalyzed (sodium methoxide, potassium hydroxide, titanium isopropoxide, or tetraalkoxide), transesterification preferred due to the charring of saccharides that occurs during direct esterification.

Linear polyglycerol esters were reported as potential cooking oil substitutes (92). The compounds resist hydrolysis by digestive enzymes and are poorly absorbed. For the small amounts that are absorbed. however, hydrolysis should eventually occur. Although it has a slightly greater viscosity, the oil is similar to natural triacylglycerols in color, odor, taste, and other physical characteristics.

Sorbestrin is a low-calorie, heat stable, liquid fat substitute. It is a sorbitol polyester that contains fatty acid esters of sorbitol and sorbitol anhydrides. It contributes 1.5 kcal/g. Its applications include fried foods, salad dressing, mayonnaise, and baked goods (16,93,96).

# Effects of Heating

During deep fat frying, a variety of thermoxidative changes occur in the oil, which results in a variety of nonvolatile and volatile decomposition products. The decomposition rate, as well as the products formed, are primarily dependent upon the fatty acid composition of the triacylglycerols (TAG). If a synthetic fat substitute is designed appropriately for frying, it should have a similar stability and produce the same compounds in similar concentrations as TAG with a similar fatty acid composition. New compounds may be produced as well, however, which may or may not affect product flavor, oil stability, and product safety, depending on the specific compounds produced and the amount formed.

Some of the earliest fundamental studies on the quantitation and identification of nonvolatile decomposition products (NVDP) were reported in 1973 by Perkins et al. (97) on heated vegetable oils. As determined by HPSEC, the early eluting fractions consisted of polymerized and oxidized TAG, while the later eluting fractions consisted primarily of unaltered TAG. Nonvolatile decomposition products formed during the frying of regular fats and oils included TAG with polar and nonpolar cyclic monomer fatty acids, noncyclic monomer fatty acids, fatty acid dimers, and cross-linked TAG polymers of greater MW (98,99). Dimeric TAG polymers are the major non-volatile decomposition products (NVDP) formed in thermally oxidized fats (100). Concurrent with the formation and accumulation of NVDP, there is an increase in the viscosity, red and yellow color intensity, extent of foaming, free fatty acid (FFA) content, and polymer content of the frying oil. The amounts and types of NVDP produced are dependent on the frying time, frying temperature, type of heating (intermittent vs. continuous), amount and type of food fried, and degree of fatty acid unsaturation (99). High-MW compounds (polymeric TAG) may be one of the best indicators of the accumulated thermal abuse of native fats and oils, since accumulation is steady and it occurs without evaporative loss (100,101). The percentage of polymeric TAG is directly related to the percentage of oxidized fatty acids and the percentage of polar compounds produced during heating (102), although the percentage of polymeric TAG is very dependent upon the amount and type of food fried. Frying relatively less food will favor polymer formation as compared to free fatty acid formation. Some consider chromatographic analysis using HPSEC a more rapid analysis that provides more specific information with respect to oil quality than the percent polar analysis as determined with column chromatography (103,104).

Extensive reports on heating studies have been published for only two fat substitutes, olestra and fatty acid esterified propoxylated glycerol. The first report on olestra was published in 1990 (105). Olestra was referred to as sucrose polyester prior to that report. Two samples were heated and compared; the first was a sample of heated olestra, and the second was a sample of a heated mixture of olestra and partially hydrogenated soybean oil. The 100% olestra sample was heated at 190°C for six 12 h days in a 15 lb capacity fryer, while the mixture was heated at 185°C for seven

12 h days in a 15 lb fryer. Raw, french style, cut potatoes were heated in the mixture throughout the seven days of heating (49 batches/d at 380 g/batch). As expected, the relative percentage of polyunsaturated fatty acids decreased after heating, while the relative percentage of saturated fatty acids increased after heating for both samples, due to the degradation of unsaturated fatty acids.

The dimers in the heated olestra were fractionated with preparative thin layer chromatography, while the dimers contained in the heated mixture of olestra and soybean oil were fractionated with preparative size-exclusion chromatography. Fractions containing suspected dimers were transesterified. Methyl ester fatty acid dimers were separated with capillary gas chromatography and tentatively identified based on their retention times relative to methylated dimer fatty acid standards (Emery Chemical Co., Henkel-Emery Group, Cincinnati, OH) (105). In addition, mass spectral analysis after gas-chromatographic separation, was completed on the methyl ester fatty acid dimers. The polymer linkages identified in the fat substitute were the same as those occurring in thermally oxidized vegetable oils containing the same fatty acids.

Nuclear magnetic resonance (NMR) and infrared spectra were obtained on intact olestra monomer and dimer isolates. Plasma desorption mass spectrometry (PDMS) of intact olestra dimers (105), as well as the olestra monomer (106), was completed. During heating there was an increase in the high MW components, as indicated by HPSEC analysis. The PDMS characterization of the high MW olestra fraction isolated with HPSEC indicated that the fraction was an olestra dimer. In addition, PDMS analysis indicated the presence of an olestra-triglyceride component in the heated olestra-triglyceride mixture. The infrared, carbon-13 NMR, and proton NMR spectra of the olestra monomer were virtually identical to those of the olestra dimer, indicating that changes in the sucrose backbone structure had not occurred and that the high MW components were olestra dimers. Gas chromatography/mass spectrometry (GC/MS) analysis indicated the presence of dehydro dimers of methyl linoleate in the methylated dimer fatty acid fraction isolated from dimerized olestra and the HPSEC-isolated olestra-triglyceride fraction, indicating that linkages were the same between olestra dimers and the olestra-triglyceride component (105). The results strongly suggested that the changes were occurring on the polyunsaturated fatty acids, and not the sucrose backbone.

In 1992 additional articles (107,108) were published on the analysis of heated olestra. The first report (107) described an analytical scheme developed for detailed qualitative comparisons of heated fats and oils. The oils (soybean oil and olestra) were transesterified, and the fatty acid methyl esters (FAME) were isolated. Fatty acid methyl esters were separated on the basis of their polarity by adsorption chromatography and solid-phase extraction. Mass spectrometry and infrared spectroscopy were used to identify specific structural components in the compounds separated with capillary GC. The FAME (and other components soluble in the hexane phase) were first fractionated with silica gel column chromatography into four fractions: fractions I and II were unaltered FAME, fraction III was altered FAME, and fraction IV was polar material. Fraction III was further separated using solid phase extraction ($C_{18}$) into fractions IIIA (oxidized FAME) and IIIB (FAME dimers). Gas chromatographic separation of fraction IIIA detected FAME of various chain lengths with aldehydic, hydroxy, epoxy, and keto functional groups. GC separation of fraction IIIB detected three plant sterols and dimer FAME. High-performance liquid chromatography separation of fraction IV produced components consisting of oxidized di- and triglycerides, monoglycerides, and two very polar FAME, one of which was tentatively identified as 9-hydroxyperoxy-10,12-octadecadienoate.

The second report (108) used the fractionation methods presented in the first paper (107) to provide a detailed qualitative comparison of the transesterified fatty acids from heated olestra and heated soybean oil. Olestra was produced from partially hydrogenated soybean oil. Olestra and soybean oil were used separately to fry potatoes. The heated oils were transesterified and fractionated into four fractions based on polarity, with the fractionation scheme presented in the first paper. Analysis indicated that fatty acid components of both oils had undergone similar chemical reactions and changes during frying, and that the altered FAME found were similar to those found in heated fats and oils by other investigators (108).

The oxidative stability index (OSI) is commonly used in the food industry to measure oil quality, particularly as an estimate of the expected life of a frying oil during heating. Akoh (109) compared eight different oils, including some fat substitutes, with an Omnion OSI instrument. Four vegetable oils were compared: crude soybean oil, crude peanut oil, refined bleached and deodorized (RBD) soybean oil, and RBD peanut oil. The four oil substitutes were analyzed, including a soybean oil-derived sucrose polyester, a high-oleate/stearate sucrose polyester, a butterfat-derived sucrose polyester, and a methyl glucoside polyester of soybean oil. As determined by the OSI, the crude oils were the most stable, second were the RBD oils, and the least stable were the oil substitutes, possibly due to the complete absence of tocopherols and other anti-oxidants in the fat substitutes.

In 1997, the first report of a series of studies on the fat-based fat substitute, fatty acid esterified propoxylated glycerol (EPG), was published (111). Fatty acid EPG are modified TAG that contain a polyether extension between the fatty acids and the glycerol backbone that prevents lipase removal of the fatty acids, rendering them nondigestible and non-caloric. Glycerol is first propoxylated, then fatty acids are esterified to the modified glycerol moiety. The samples evaluated included EPG-08 oleate (oleic acid esterified propoxylated glycerol with a mole ratio of propylene oxide to glycerol of 8) and EPG-08 linoleate (linoleic acid esterified propoxylated glycerol with a mole ratio of propylene oxide to glycerol of 8). Samples were heated for ~12 h per day at approximately 190°C in a small deep fat fryer until the TAG polymer content had reached or exceeded 20%. The free fatty acid content, total acid value, p-anisidine value, and food oil sensor values were used as indicators to determine the extent of thermal oxidation of the oil samples after heating. Oil samples were heated without added food samples to simplify interpretation of the results, particularly with respect to the formation and quantitation of novel oxidation products and volatile compounds. The degradation rates of the substrate (the modified triacylglycerol moiety) were determined with capillary supercritical fluid chromatography (SFC). EPG-08 oleate contained ≥ 20% polymer content point after 36 h of heating, while EPG-08 linoleate contained ≥ 20% polymer after only 24 h of heating, indicating, as expected, substantially less stability for the sample containing linoleic acid than for the sample containing oleic acid (111).

In another study, a comparison of EPG-08 oleate and triolein was presented. Oil samples were heated for 12 h per day at approximately 190°C in a small deep fat fryer until the polymer content was ≥ 20%. The oil samples were then fractionated by supercritical fluid fractionation (SFF) to separate a dimer triacylglycerol (TAG) (or fat substitute dimer) fraction from the rest of the sample. After *trans*-esterification, the dimer TAG (or fat substitute dimer) fraction was analyzed, and comparisons were made (GC-MS of dimer fatty acids, as well as NMR of the intact molecule) to determine if the oxypropylene backbone had any role in TAG dimer formation or formation of other thermoxidation products during deep fat frying. The results of the GC-MS analysis, as well as the proton and carbon-13 NMR analysis, indicated that it was the fatty acid composition that affected the dimer formation and the oxypropylene backbone was not involved in the formation of degradation products (112).

The volatile components in the heated oil samples, as well as the fat substitutes, were identified and quantitated. Samples of heated EPG-08 oleate were collected and the volatile components were separated using capillary GC, then the components were identified and quantitated using infrared spectroscopy and mass spectrometry. For comparison, a model compound of triolein was heated under the same conditions. For both oil sample types, the major volatile components found originated from the thermoxidative decomposition of the oleate fatty acid component. The major volatile components found were heptane, octane, heptanal, octanal, *trans*-2-decenal, nonanal, and *trans*-2-undecenal. There were a few compounds found at low concentrations unique to the fat substitute. For example, the volatile components unique to EPG-08 oleate included 1,2-propanediol, 1-hydroxy-2-propanone (hydroxyacetone), 1-acetoxy-2-propanone (acetoxyacetone), 2-heptyl-4-methyl-1,3-dioxolane, 2-ethyl-4-methyl-1,3-dioxolane, 2,2,4-trimethyl-1,3-dioxolane, and 4-methyl-2-octyl-1,3-dioxolane, although all of these have been found in other food products (113).

The volatile components found in heated samples of EPG-08 linoleate (as well as the corresponding model compound, trilinolein) were determined using the same methods used for the analysis of the EPG-08 oleate. The major volatile components found were pentane, hexanal, 2-heptenal, 1-octen-3-ol, 2-pentylfuran, 2-octenal, and 2,4-decadienal (114).

In 1999, studies on esterified propoxylated glycerol were expanded to include propoxylated glycerol soybean oil or soyate (EPG-08 soyate, a soy oil-based fat substitute). The fat substitute stability, as well as the amount and type of oxidation products formed in EPG-08 soyate, were compared to *trans*-esterified soybean oil. The *trans*-esterified soybean oil had slightly greater stability than the fat substitute EPG-08 soyate (115). Another set of oil samples that were compared and examined included *trans*-esterified soybean oil, EPG-08 soyate, and a 50:50 ratio mix of the two. All samples were heated for sufficient time to produce 20% or more polymer. The major volatile components identified included hexanal, heptanal, *trans*-2-heptenal, *trans*-2-octenal, and *trans,trans*-2,4-decadienal. These same volatile components were found in both the heated soy oil-based fat substitute and the heated soybean oil (116).

## Nutritional Implications

Substitution or removal of a significant portion of the dietary lipids from the U.S. diet has substantial implications in terms of health. A substantial reduction in fat consumption would be the single most important positive change in the American diet, according to the U.S. Surgeon General (110). Dietary lipids are associated with an increased risk of cardiovascular disease, arteriosclerosis, obesity, and a variety of related health problems. A significant reduction in the percentage of fat in the diet would have a positive influence on all of these health problems. Another positive effect achieved by the presence of an indigestible oil substitute in the digestive tract might be a decrease in absorption of dietary cholesterol, which would reduce the serum cholesterol concentration (117). In contrast, substantial fat substitute consumption could reduce fat-soluble vitamin (43) absorption, and drug absorption could be depressed. In some cases, laxative effects could be induced. The nutritional implications of fat substitutes have been reviewed (43,118). Some studies indicated that caloric intake may be increased to compensate for a reduction in fat intake (119,120).

Until the last few years, regulatory agencies had limited experience in evaluating macronutrient substitutes; as a result, they were compelled to take a very cautious approach with respect to approval. There are many more limitations to the safety testing of macronutrient substitutes than with food additives used in small concentrations because the same level of safety guarantees cannot be attained. Guidelines are now available that are of particular interest for companies intending to request regulatory approval for macronutrient substitutes (13). The exact chemical structure of the substitute must be known and documented. If it is composed of more than one compound, each of these must be known. The stability should be well documented, especially during food production, storage, and final food preparation. If partial degradation occurs, the by-products should be identified. If any harmful impurities are present, there should be limits on the concentration of these components. Detailed exposure assessment for each population type should be included, which would provide toxicological data demonstrating that the use of the macronutrient substitute, as well as any of the degradative by-products, would not cause harm, at least at the levels of exposure for intended use. Additional tests may be required to provide pharmakinetic, metabolic, and/or nutritional data not normally required for other additives. It is necessary to determine whether the macronutrient substitute is absorbed. If it is poorly absorbed, it must be determined whether large concentrations of the material in the gut affect the morphology, physiology, biochemistry, or normal flora of the gastrointestinal tract. Since some of the normal testing procedures for the evaluation of micronutrients are either inappropriate or provide insufficient data, an innovative or novel approach may be required. The selection of animal models, experimental protocols, and measurement techniques must be carefully planned to provide data that can accurately assess the safety of the substitute. Sufficient data must be provided to demonstrate that the substitute or its metabolites does not interfere with absorption or metabolism of essential nutrients. If interference does occur, it must be determined whether it can be safely offset with nutrient fortification. If the compound is adsorbed at a measurable rate, there must be data indicating the impact, if any, on the nutrient status of the animal. In general, traditional assessments of nutritional status should be sufficient for nutrient status assessment.

Another area of concern is the effect of macronutrient substitutes on selected population segments that are particularly sensitive to the proposed substitute. The effect of the macronutrient substitute on drug absorption or activity should be assessed. Some oil substitutes have a laxative effect, and since this effect can often be greater in children than in adults, it may require additional evaluation. In addition, some consumers may not be aware of the relationship between substitute consumption and laxative effects, as they are with prunes, for example. Special labeling is needed to address the problem. Postmarket surveillance is needed in most cases. One long-term potential problem of particular concern is that of consumption of several macronutrient substitutes concurrently, since it is likely that several of those materials may be approved in the future (13).

## References

1. Akoh, C.C.; K.D. Long; W.P. Flatt; B.S. Rose; and R.J. Martin. Effects of a Structured Lipid, Captex, and a Protein-Based Fat Replacer, Simplesse, on Energy Metabolism, Body Weight, and Serum Lipids in Lean and Obese Zucker Rats. *J. Nutitional Biochemistry* **1998,** *9* (5), 267–275.
2. Miraglio, A.M. Nutritional Subsititutes and Their Energy Values in Fat Substitutes and Replacers. *American Journal of Clinical Nutrition* **1995,** *62* (5 Suppl. S), S1175–S1179.
3. Haumann, B.F. Getting the Fat Out—Researchers Seek Substitutes for Full-Fat Fat. *J. Amer. Oil Chem. Soc.* **1986,** *63,* 278–288.
4. LaBarge, R.G. The Search for a Low-Calorie Oil. *Food Technol.* **1988,** *42* (1), 84–90.
5. Harrigan, K.A.; and W.M. Breene. Fat Substitutes: Sucrose Esters and Simplesse. *Cereal Foods World* **1989,** *34,* 261–267.
6. Anonymous. Consumers Receiving Conflicting Messages. *INFORM* **1992,** *3,* 676–677.
7. Shukla, T.P. Low-Fat Foods and Fat Substitutes. *Foods World* **1992,** *37,* 452–453.
8. Artz, W.E.; and S.L. Hansen. Other Fat Substitutes. In *Carbohydrate Polyester as Fat Substitutes*; C.C. Akoh and B.G. Swanson, Eds.; Marcel Dekker: New York, 1994; pp. 197–236.

9. De Guzman, D. Bunge Captures Top Oilseed Spots with $1.4 Billion Cereol Acquisition (Oils, Fats & Waxes). *Chem. Market Reporter* **2002,** *261* (8), 17–18.
10. Bruhn, C.M.; A. Cotter; K. Diaz-Knauf; J. Sutherlin; E. West; N. Wightman; E. Williamson; and M. Yakkee. Consumer Attitudes and Market Potential for Foods Using Fat Substitutes. *Food Technol.* **1992,** *46* (4), 81–86.
11. Munro, I.C. Issues to Be Considered in the Safety Evaluation of Fat Substitutes. *Food Chem. Toxicol.* **1990,** *28,* 751–753.
12. Borzelleca, J.F. Macronutrient Substitutes: Safety Evaluation. *Reg. Toxicol. Pharmacol.* **1992,** *16,* 253–264.
13. Vanderveen, J.E. Regulatory Status of Macronutrient Substitutes: What FDA Needs to Assure Safety; Paper No. 15-1 presented at the Institute of Food Technologists Meeting, Atlanta, GA, July 25–29, 1994.
14. Barth, R.H.; R. Park; and C. Burrell. Polyhydric Alcohol Esters. U.S. Patent 2,356,745, 1944.
15. Minich, A. Dietetic Compositions. U.S. Patent 2,962,419, 1960.
16. Akoh, C.C. Fat Replacers. *Food Technol.* **1998,** *52* (3), 47–53.
17. Napier, K. *Fat Replacers: The Cutting Edge of Cutting Calories;* 1997; pp. 1–35.
18. D'Amelia, R.P.; and P.T. Jacklin. Polyvinyl Oleate as a Fat Replacement. U.S. Patent 4,915,974, 1990.
19. Chetri P.; and N.N. Dass. Development of a New Method for Synthesis of Poly(vinyl oleate) from Poly(vinyl alcohol). *Polymer* **1996,** *37* (23), 5289–5293.
20. Hamm, D.J. Preparation and Evaluation of Trialkoxytricarballyate, Trialkoxycitrate, Trialkoxyglycerylether, Jojoba Oil, and Sucrose Polyester as Low Calorie Replacements of Edible Fats and Oils. *J. Food Sci.* **1984,** *49,* 419–428.
21. Gillis, A. Fat Substitutes Creat New Issues. *J. Am. Oil Chem. Soc.* **1988,** *65,* 1708–1712.
22. Dziezak, J.D. Fats, Oils, and Fat Substitutes. *Food Technol.* **1989,** *43,* 66–74.
23. Fulcher, J. Synthetic Cooking Oils Containing Dicarboxylic Acid Esters. U.S. Patent 4,582,927, 1986.
24. Anonymous. Fat Substitute Update. *Food Technol.* **1990,** *44,* 92–93.
25. Truswell, A.S. *Dietary Fat Some Aspects of Nutrition and Health and Product Development.* ILSI Europe Concise Monograph Series; ILSI Europe: Brussels, Belgium, 1995; pp. 1–38.
26. Anonymous. Quest for Fat Substitutes Taking Many Routes. *INFORM* **1991,** *2,* 115, 118–119.
27. Huhn, S.D.; P.S. Given, Jr.; and L.P. Klemann. Ether Bridged Polyesters and Food Compositions Containing Ether Bridged Polyesters. U.S. Patent 4,888,195, 1989.
28. Klemann, L.P.; and J.W. Finley. Low Calorie Fat Mimetics Comprising Carboxy/Carboxylate Esters. European Patent 303,523, 1989.
29. Klemann, L.P.; and J.W. Finley. Low Calorie Fat Mimetics Comprising Carboxy/Carboxylate Esters. International Patent 89/01293, 1989.
30. Klemann, L.P.; J.W. Finley; and A. Scimone. Diol Lipid Analogs as Edible Fat Replacements. European Patent 405,874, 1991.
31. Klemann, L.P.; J.W. Finley; and A. Scimone. Long Chain Diol Diesters as Low Calorie Fat Substitutes. European Patent 405,873, 1991.
32. Unilever, N.V. Preparation of Polyol Fatty Acid Polyesters. Netherlands Patent 8,601,904, 1988.

33. Cain, F.W.; F.R. DeJong; and F. Roelof. Edible Fat-Containing Products Containing Indigestible Polyol Fatty Acid Polyesters. European Patent 304,130, 1989.

34. Whyte, D.D. Triglyceride Esters of α-Branched Carboxylic Acids. U.S. Patent 3,579,548, 1971.

35. Artz, W.E.; Soheili K.C.; and Arjona I.M. Esterified Propoxylated Glycerol, a Fat Substitute Model Compound, and Soy Oil After Heating. *J. Agric. Food Chem.* **1999**, *47*, 3816–3821.

36. White, J.F.; and M.R. Pollard. Nondigestible Fat Substitutes of Low-Caloric Value. U.S. Patent 4,861,613, 1989.

37. Arciszewski, H. Fat Functionality, Reduction in Baked Foods. *INFORM* **1991**, *2*, 392–399.

38. White, J.F.; and M.R. Pollard. Non-digestible Fat Substitutes of Low-Calorie Value. European Patent 325,010, 1989.

39. Duxbury, D.D.; and N.M. Meinhold. Dietary Fats and Oils. *Food Processing* **1991**, *52*, 58–59.

40. White, J.F.; and M.R. Pollard. Esterified Epoxide-Extended Polyols as Nondigestible Fat Substitutes of Low-Caloric Value. European Patent 254,547, 1988.

41. White, J.F.; and M.R. Pollard. Low-Calorific and Non-digestive Substitute of Fat/Oil. Chinese Patent 1,034,572, 1989.

42. Cooper, C.F. Preparation of Propoxylated Glycerides as Dietary Fat Substitutes. European Patent 353,928, 1990.

43. Hassel, C.A. Nutritional Implicators of Fat Substitutes. *Cereal Foods World* **1993**, *38*, 142–144.

44. Artz, W.E.; J.A. Reiling; C. Bernaski-Abrassart. Volatiles in a Fat-Based Fat Substitute Model Compound, Esterifired Propoxylated Glycerol Soyate. *J. Food Lipids* **2001**, *8* (3), 191–204.

45. Artz, W.E.; S.L. Hansen; and M.R. Myers. Heated Fat-Based Oil Substitutes, Oleic and Linoleic Acid-Esterified Propoxylated Glycerol. *J. Am. Oil Chem. Soc.* **1997**, *74*, 367–374.

46. Volpenhein, R.A. Acylated Glycerides Useful in Low Calorie Fat-Containing Food Compositions. U.S. Patent 4,582,715, 1986.

47. Trost, V.W. Low Calorie Fat Substitutes. Canadian Patent 1,106,681, 1981.

48. Klemann, L.P.; J.W. Finley; and A. Scimone. Tris-hydroxymethyl Alkane Esters as Low Calorie Fat Mimetics. U.S. Patent 4,927,658, 1990.

49. Klemann, L.P.; J.W. Finley; and A. Simone. Tris-hydroxymethyl Lower Alkane Esters as Fat Mimetics. U.S. Patent 4,927,659, 1990.

50. Morehouse, S.E.; and M.E. Zabik. Evaluation of Polydimethylsiloxane Fluids as Non-caloric Frying Media. *J. Food Sci.* **1989**, *54*, 1062–1065.

51. Bracco, E.F.; N. Baba; and S.A. Hashim. Polysiloxane: Potential Noncaloric Fat Substitute; Effects on Body Composition of Obese Zucker Rats. *Am. J. Clin. Nutr.* **1987**, *46*, 784–789.

52. Labell, F. Co-crystallization Process Aids Dispersion and Solubility. *Food Processing* **1991**, *52*, 60–63.

53. Ryan, J.W. Food Composition Containing a Siloxane Polymer and a Particulate Silica. U.S. Patent 4,925,692, 1990.

54. Frye, C.L. Fat and Oil Replacements as Human Food Ingredients. European Patent

205,273, 1986.

55. Ryan, J.W. Prevention of Anal Leakage of Polyorganosiloxane Fluids Used as Fat Substitutes in Foods. European Patent 368,534, 1989.

56. Post, F.J.; and N.F. Collins. A Preliminary Investigation of the Membrane Lipid of *Halobacterium halobium* as a Food Additive. *J. Food Biochem.* **1982,** *6,* 25–29.

57. Akoh, C.C.; and B.G. Swanson. Fat Substitutes in Foods: Growing Demand and Potential Markets. In *Carbohydrate Polyesters as Fat Substitutes*; Marcel Dekker: New York, 1994; Chapter 12.

58. Bergholz, C.M. Safety Evaluation of Olestra, a Nonabsorbed, Fatlike Fat Replacement. *Crit. Rev. Food Sci. Nutr.* **1992,** *32,* 141–146.

59. Hunt, R.; N.L. Zorich; and A.B.R. Thomson. Overview of Olestra: A New Fat Substitute. *Can. J. Gastroenterol.* **1998,** *12* (3), 193–197.

60. McManus, G.G.; G.W. Buchanan; H.C. Jarrell; R.M. Epand; R.F. Epand; and J.J. Cheetham. Membrane Perturbing Properties of Sucrose Polyesters. *Chem. Phys. Lipids* **2001,** *109* (2), 185–202.

61. Peters, J.C.; K.D. Lawson; S.J. Middleton; and K.C. Triebwasser. Assessment of the Nutritional Effects of Olestra, a Nonabsorbed Fat Replacement: Summary. *J. Nutr.* **1997,** *127* (Suppl. 8), S1539–S1546.

62. Deis, R.C. *Food Product Design* **1996,** (Sept.).

63. Patterson, R.E.; A.R. Kristal; J.C. Peters; M.L. Neuhouser; C.L. Rock; L.J. Cheskin; D. Neumark-Sztainer; and M.D. Thornquist. Changes in Diet, Weight, and Serum Lipid Levels Associated with Olestra Consumption. *Arch. Intern. Med.* **2000,** *160* (17), 2600–2604.

64. Akoh, C.C. New Developments in Low Calorie Fats and Oils Substitutes. *J. Food Lipids* **1996,** *3* (4), 223–232.

65. Cooper, D.A.; J. Curran-Celentano; T.A. Ciulla; B.R. Hammond; R.B. Danis; L.M. Pratt; K.A. Riccardi; and T.G. Filloon. Olestra Consumption Is Not Associated with Macular Pigment Optical Density in a Cross-Sectional Volunteer Sample in Indianapolis. *J. Nutr.* **2000,** *130* (3), 642–647.

66. Balasekaran, R.; J.L. Porter; C.A. Santa Ana; and J.S. Fordtran. Positive Results on Tests for Steatorrhea in Persons Consuming Olestra Potato Chips. *Ann. Intern. Med.* **2000,** *132* (4), 279–282.

67. Burks, A.W.; L. Christie; K.A. Althage; J.M. Kesler; and G.S. Allgood. Randomized, Double-Blind, Placebo-Controlled, Food Allergy Challenge to Olestra Snacks. *Regul. Toxicol. Pharmacol.* **2001,** *34* (2), 178–181.

68. Moser, G.A.; and M.S. McLachlan. A Non-absorbable Dietary Fat Substitute Enhances Elimination of Persistent Lipophilic Contaminants in Humans. *Chemosphere* **1999,** *39* (9), 1513–1521.

69. Nuck, B.A.; T.G. Schlagheck; and T.W. Federle. Inability of the Human Colonic Microflora to Metabolize the Nonabsorbable Fat Substitutes, Olestra. *J. Indust. Micro.* **1994,** *13,* 328–334.

70. Thornquist, M.D.; A.R. Kristal; R.E. Patterson; M.L. Neuhouser; C.L. Rock; D. Neumark-Sztainer; and L.J. Cheskin. Olestra Consumption Does Not Predict Serum Concentrations of Carotenoids and Fat-Soluble Vitamins in Free-Living Humans: Early Results from the Sentinel Site of the Olestra Post-Marketing Surveillance Study, *J. Nutr.* **2000,** *130* (7), 1711–1718.

71. Hunt, R.; N.L. Zorich; and A.B.R. Thomson. Overview of Olestra: A New Fat Substitute. *Can. J. Gastroenterol.* **1998,** *12* (3), 193–197.

72. Volpenhein, R. Synthesis of Higher Polyol Fatty Acid Polyesters Using Carbonate Catalysts. U.S. Patent 4,517,360, 1985.

73. Rios, J.J.; M.C. Pérez-Camino; G. Márquez-Ruiz; and M.C. Dobarganes. Isolation and Characterization of Sucrose Polyesters. *J. Am. Oil Chem. Soc.* **1994,** *71*, 385–390.

74. Márquez-Ruiz, G.; M.C. Pérez-Camino; J.J. Rios; and M.C. Dobarganes. *J. Am. Oil Chem Soc.* **1994,** *71*, 1017.

75. Chase, G.W., Jr.; C.C. Akoh; and R.R. Eitenmiller. Evaporative Light Scattering Mass Detection for High Performance Liquid Chromatographic Analysis of Sucrose Polyester Blends in Cooking Oils. *J. Am. Oil Chem. Soc.* **1994,** *71* (11), 1273–1276.

76. Drake, M.A.; L. Ma; B.G. Swanson; and G.V.B. Canovas. Rheological Characteristics of Milkfat and Milkfat-Blend Sucrose Polyesters. *Food Res. Int.* **1994,** *27*, 477–481.

77. Drake, M.A.; T.T. Boutte; F.L. Younce; D.A. Cleary; and B.G. Swanson. Melting Characteristics and Hardness of Milkfat Blend Sucrose Polyesters. *J. Food Sci.* **1994,** *59*, 652–654.

78. Drake, M.A.; T.T. Boutte; L.O. Luedecke; and B.G. Swanson. Milkfat Sucrose Polyesters as Fat Substitutes in Cheddar-Type Cheeses. *J. Food Sci.* **1994,** *59*, 326–327.

79. Thomson, A.B.R.; R.H. Hunt; and N.L. Zorich. Review Article: Olestra and Its Gastrointestinal Safety. *Aliment. Pharmacol. Ther.* **1998,** *12* (12), 1185–1200.

80. Drake, M.A.; C.W. Nagel; and B.G. Swanson. Sucrose Polyester Content in Foods by a Colorimetric Method. *J. Food Sci.* **1994,** *59*, 655–656.

81. Akoh, C.C.; and B.G. Swanson. Optimized Synthesis of Sucrose Polyesters: Comparison of Physical Properties of Sucrose Polyesters, Raffinose Polyesters and Salad Oils. *J. Food Sci.* **1990,** *55*, 236–243.

82. Akoh, C.C. Synthesis of Carbohydrate Fatty Acid Polyesters. In *Carbohydrate Polyesters as Fat Substitutes*; C.C. Akoh and B.G. Swanson, Eds.; Marcel Dekker: New York, 1994; pp. 9–35.

83. Riva, S. Enzymatic Synthesis of Carbohydrate Esters. In *Carbohydrate Polyesters as Fat Substitutes*; C.C. Akoh and B.G. Swanson, Eds.; Marcel Dekker: New York, 1994; pp. 37–64.

84. Osborn, H.T.; and C.C. Akoh. Structured Lipids—Novel Fats with Medical Nutraceutical, and Food Applications. *Compr. Rev. Food Sci. Food Safety* **2002,** *1*, 93–103.

85. Mieth, G.; A. Weiss; H. Behrens; J. Pohl; and J. Brueckner. Verfahren zur Herstellung von Kohlenhydrat-fettsaeureestern. GDR Patent 3,156,283, 1982.

86. Deger, H.M.; W. Fritsche-Lang; A. Reng; M. Schlingman; and C.J. Lawson. Verfahren zur Herstellung von Gemischen aus mono- und oligomeren Kohlenhydratestern, die so erhaeltlichen Kohlenhydratestergemische und ihre Verwerdung. GDR Patent 3,639,878, 1988.

87. Meyer, R.S.; J.M. Root; M.L. Campbell; and D.B. Winter. Low Caloric Alkyl Glycoside Fatty Acid Polyester Fat Substitutes. U.S. Patent 4,840,815, 1989.

88. Winter, D.B.; R.S. Meyer; J.M. Root; and M.L. Campbell. Process for Producing Low Calorie Foods from Alkyl Glycoside Fatty Acid Polyesters. U.S. Patent 4,942,054, 1990.

89. Meyer, R.S.; M.L. Campbell; D.B. Winter; and J.M. Root. Alkyl Glycoside Fatty Acid Polyester Fat Substitute Food Compositions and Process to Produce the Same. U.S. Patent 5,550,220, 1996.

90. Ennis, J.L.; P.W. Kopf; S.E. Rudolf; and M.F. van Buren. Esterified Alkoxylated Alkyl Glycosides Useful in Low Calorie Fat-Containing Food Compositions. European Patent 415,636, 1991.
91. White, J.F. Partially Esterified Polysaccharides (PEP) Fat Substitutes. U.S. Patent 4,959,466, 1990.
92. Dobson, K.S.; K.D. Williams; and C.J. Boriack. The Preparation of Polyglycerol Esters Suitable as Low-Caloric Fat Substitutes. *J. Am. Oil Chem. Soc.* **1993,** *70* (11), 1089–1092.
93. Weber, N.; and Mukherjee, K.D. Solvent-Free Lipase-Catalyzed Preparation of Diacylglycerols. *J. Agric. Food Chem.* **2004,** *52* (17), 5347–5353.
94. Zind, T. *Prepared Foods* **2003,** (Oct.).
95. American Dietetic Association. Position of the American Dietetic Association: Fat Replacers. *J. Am. Dietetic Assoc.* **2005,** *105* (2), 266–275.
96. Livesey, G.; and J. Pokorny. In *Study on Obesity and Functional Foods in Europe, Cost Action 918*; A. Palou, M.L. Bonnet, and F. Serra, Eds.; European Commission: Brussels, Belgium; pp. 317–330.
97. Perkins, E.G.; R. Taubold; and A. Hsieh. Gel Permeation Chromatography of Heated Fats. *J. Am. Oil Chem. Soc.* **1973,** *50*, 223.
98. Brooks, D.D. Some Perspectives on Deep-Fat Frying. *INFORM* **1991,** *2*, 1091–1095.
99. White, P.J. Methods for Measuring Changes in Deep-Fat Frying Oils. *Food Technol.* **1991,** *45*, 75–80.
100. Paquette, G.; D.B. Kupranycz; and F.R. van de Voort. The Mechanisms of Lipid Autoxidation. 2. Non-volatile Secondary Oxidation Products. *Can. Inst. Food Sci. Technol. J.* **1985,** *18*, 197–206.
101. Paradis, A.J.; and W.W. Nawar. Evaluation of New Methods for the Assessment of Used Frying Oils. *J. Food Sci.* **1981,** *46*, 449–451.
102. Billek, G.; G. Guhr; and J. Waibel. Quality Assessment of Used Frying Fats: A Comparison of Four Methods. *J. Am. Oil Chem. Soc.* **1978,** *55*, 728–733.
103. White, P.J.; and Y.C. Wang. A High Performance Size-Exclusion Chromatographic Method for Evaluating Heated Oils. *J. Am. Oil Chem. Soc.* **1986,** *63*, 914–920.
104. Husain, S; G.S.R. Sastry; and N. Prasada Raju. Molecular Weight Averages as Criteria for Quality Assessment of Heated Fats and Oils. *J. Am. Oil Chem. Soc.* **1991,** *68*, 822–826.
105. Gardner, D.R.; and R.A. Sanders. Isolation and Characterization of Polymers in Heated Olestra and an Olestra/Triglyceride Blend. *J. Am. Oil Chem. Soc.* **1990,** *67*, 788–796.
106. Sanders, R.A; D.R. Gardner; M.P. Lacey; and T. Keough. Desorption Mass Spectrometry of Olestra. *J. Am. Oil Chem. Soc.* **1992,** *69*, 760–771.
107. Gardner, D.R.; R.A. Sanders; D.E. Henry; D.H. Tallmadge; and H.W. Wharton. Characterization of Used Frying Oils. Part 1: Isolation and Identification of Compound Classes. *J. Am. Oil Chem. Soc.* **1992,** *69*, 499–508.
108. Henry, D.E; D.H. Tallmadge; R.A. Sanders; and D.R. Gardner. Characterization of Used Frying Oils. Part 2: Comparison of Olestra and Triglyceride. *J. Am. Oil Chem. Soc.* **1992,** *69*, 509–519.
109. Akoh, C.C. Oxidative Stability of Fat Substitutes and Vegetable Oils by the Oxidative Stability Index Method. *J. Am. Oil Chem. Soc.* **1994,** *71*, 211–216.
110. U.S. Surgeon General. The Surgeon General's Report on Nutrition and Health; U.S. Department of Health and Human Services: Washington, DC, 1988; DHHS (PHS)

Publication No. 88-50210.

111. Artz, W.E.; S.L. Hansen; and M.R. Myers. Heated, Fat-Based, Oil Substitutes, Oleic and Linoleic Acid Esterified Propoxylated Glycerol. *J. Am. Oil Chem. Soc.* **1997,** *74,* 367–374.

112. Hansen, S.L.; W.J. Krueger; L.B. Dunn; Jr. and W.E. Artz. Nuclear Magnetic Resonance and Gas Chromatography/Mass Spectroscopy Analysis of the Nonvolatile Components Produced During Heating of Oleic Acid Esterified Propoxylated Glycerol, a Fat Substitute Model Compound, and Trioleylglycerol. *J. Agr. Food Chem.* **1997,** *45,* 4730–4739.

113. Mahungu, S.M.; S.L. Hansen; and W.E. Artz. Volatile Compounds in Heated Oleic Acid-Esterified Propoxylated Glycerol. *J. Amer. Oil Chem. Soc.* **1998,** *75,* 683–690.

114. Mahungu, S.M.; S.L. Hansen; and W.E. Artz. Identification and Quantitation of Volatile Compounds in Two Heated Model Compounds, Trilinolein and Linoleic Acid Esterified Propoxylated Glycerol. *J. Agr. Food Chem.* **1999,** *47,* 690–694.

115. Artz, W.E.; K.C. Soheili; and I.M. Arjona. Esterified Propoxylated Glycerol Soyate, a Fat Substitute Model Compound, and Soy Oil After Heating. *J. Agr. Food Chem.* **1999,** *47,* 3816–3621.

116. Artz, W.E.; J.A. Reiling; and C. Bernaski-Abrassart. Volatiles in a Fat-Based Fat Substitute Model Compound, Esterified Propoxylated Glycerol Soyate. *J. Food Lipids* **2001,** *8* (3), 191–204.

117. Mattson, F.H.; C.J. Glueck; and R.J. Jandacek. The Lowering of Plasma Cholesterol by Sucrose Polyester in Subjects Consuming Diets with 800, 300, or Less Than 50 mg of Cholesterol per Day. *Am. J. Clin. Nutr.* **1979,** *32,* 1636–1644.

118. Mela, D.J. Nutritional Implications of Fat Substitutes. *J. Am. Diet. Assoc.* **1992,** *92,* 472–476.

119. Caputo, F.A.; and R.D. Mattes. Human Dietary Responses to Cover Manipulation of Energy, Fat and Carbohydrate in a Midday Meal, *Am. J. Clin. Nutr.* **1992,** *56,* 36–43.

120. Rolls, B.J.; P.A. Pirraglia; M.B. Jones; and J.C. Peters. Effects of Olestra, a Noncaloric Fat Substitute, on Daily Energy and Fat Intakes in Lean Men. *Am. J. Clin. Nutr.* **1992,** *56* (1), 84–92.

# PART 4

# Application

# 12

# Dynamics of Frying

**Frank T. Orthoefer[a] and Gary R. List[b]**
[a]Germantown, TN, [b]USDA ARS NCAUR, Peoria, IL

Frying, one of the most important methods of food preparation, is widely used by both the food industry and consumers. Its popularity continues to grow, even with the media now stressing reduced dietary oil consumption and reduction in *trans* fatty acids in our diets. Fats and oils have unique properties that add to the flavor and mouthfeel most desirable in overall food palatability. Deep fat frying is a popular restaurant preparation technique because it is fast and convenient. Properly fried foods have a dry, nongreasy appearance and taste. When well prepared, the food surface is crisp outside and moist or tender inside. Flavor, color, and odor are all important attributes.

Frying is deceptively simple, yet it is one of the least understood methods of food preparation. It continues to be more of an art than a science. Types of frying include pan/griddle, deep fat, and industrial/continuous frying. In all cases, the oil provides an effective medium for energy transfer from the heat source to the food.

Proper frying practice as well as the most appropriate frying oil or shortening is generally determined by experience. Production of desirable characteristics in fried food depends on the heat capacity of the frying medium, thermal conductivity of the food, proper temperature differentials between the oil and food, dehydration on the surface of the food, and oil absorption and interaction with food components to develop the desired texture and flavor. In this chapter, heat transfer and mass transfer are discussed, along with mass changes during frying.

## The Frying Process

In deep fat frying, the food is completely surrounded by the frying fat. Fritsch (1) outlined the principal events and the mechanism of the frying process (Fig. 12.1). Heat is transferred to food and aids in developing color and flavor. The following events occur: i) moisture either on the surface or in the interior of the food forms steam; ii) the temperature of the frying oil decreases as a result of food addition, while additional heat is supplied from the heat source; iii) high temperatures promote reactions between the food components; and iv) dehydration of the surface and fat absorption, primarily on the surface, imparts a desirable taste and texture to the food. The intimate contact between the food and the oil makes frying a more efficient

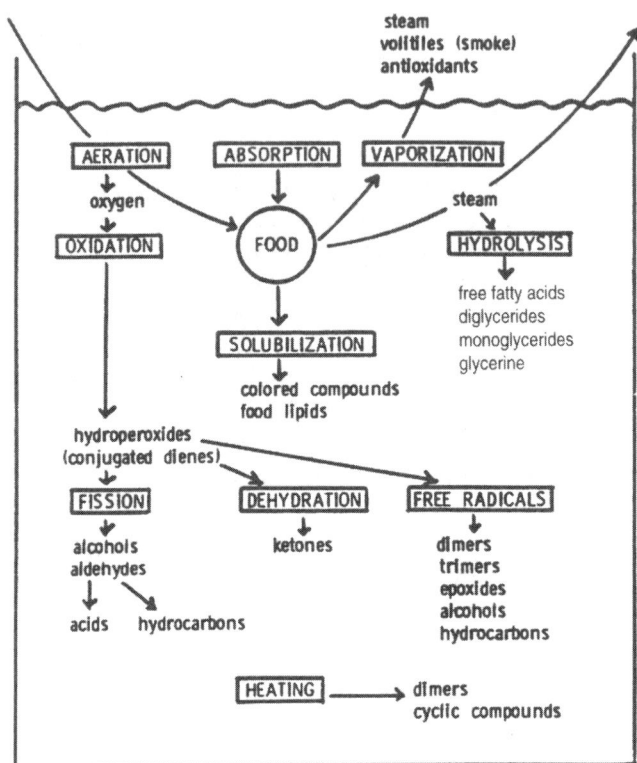

**Fig. 12.1.** Changes occurring during deep fat frying (1).

process than dry oven or wet steam methods. Absorption of oil, surface dehydration of food (crust formation), development of surface color, and creation of flavors account for the desirable taste of fried food. The occurrence of these measurable (sensible) events during frying is portrayed in Fig. 12.2.

Steam is emitted during the entire cooking process until the food is removed from the fryer. Evolution of steam indicates that the pressure inside the food is greater than that in the kettle. Steam limits the penetration of oil through the surface to the interior of food. Temperatures used for frying are generally in the 162–196°C range. Major considerations for frying temperatures are excessive time to fry at the lower temperature and degradation of oil at higher frying temperatures.

The uniformity of frying depends somewhat on uniformity of food pieces being fried. Obviously, small pieces of food fry more quickly, whereas larger pieces require longer times for heat penetration. Larger pieces generally require lower fry temperatures to prevent overcooking or burning of the food surface. Obtaining the optimal end point depends on many quality attributes, including heat capacity of

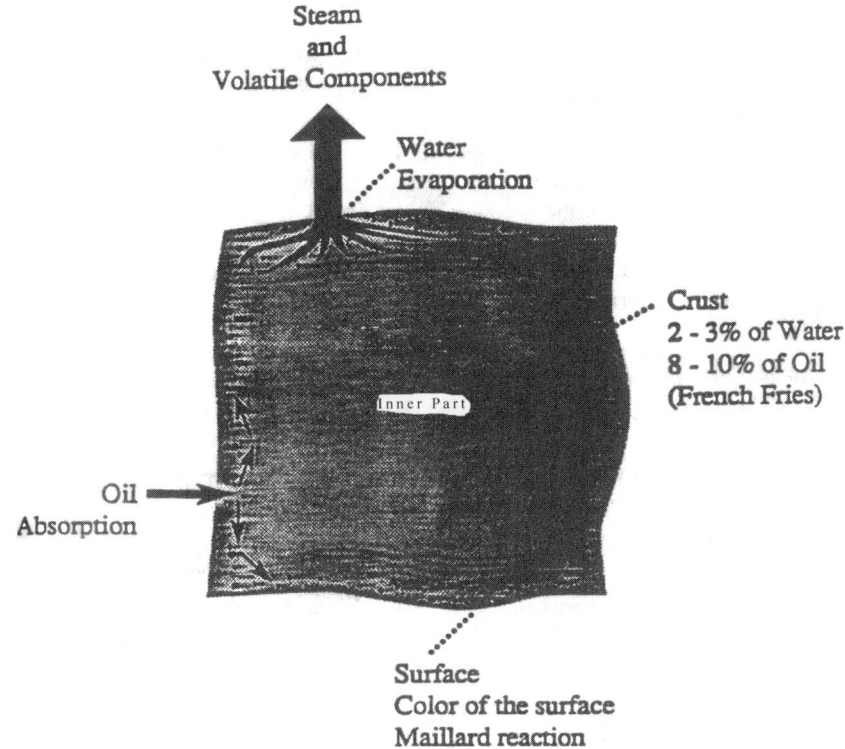

Steam
and
Volatile Components

Water
Evaporation

Crust
2 - 3% of Water
8 - 10% of Oil
(French Fries)

Inner Part

Oil
Absorption

Surface
Color of the surface
Maillard reaction

**Fig. 12.2.** Structure of cross section of french-fried potato (formation of crust of crisp) (2).

frying medium, thermal conductivity of food, and proper temperature differentials between oil and food.

# Heat Transfer

## Principles of Heat Transfer

Deep fat frying involves heat transfer. Heat transfer is the movement of energy from one point to another or from the heat source to the oil and finally to the food being fried. The temperature difference determines the rate of heat transfer. The thermal properties of the oil and food consist of specific heat, thermal conductivity, and heat of fusion. Consideration of each factor is required for the optimal design of frying equipment (3,4).

The amount of heat involved in frying is measured either in calories or Btu (British thermal units). One calorie is the quantity of heat required to raise the temperature of 1.0 g of water 1°C, and Btu is the amount of heat necessary to raise 1 lb water 1°F. At

20°C, 1 Btu is equal to 251.634 calories.

The heat capacity, $C_p$ ($p$ indicates a constant pressure), is used to define the amount of heat necessary to cause a given mass to change 1°. Specific heat is a dimensionless ratio equal to the heat capacity since the heat capacity for water is 1.0 cal/g °C or 1.0 Btu/lb °F. The specific heat of a substance is the ratio of the heat capacity of a substance to that of water.

There are three basic mechanisms of heat transfer: conduction, convection, and radiation. Heat transfer by conduction is described by Fourier's first law, which states that positive heat flow occurs in the direction of decreasing temperature. According to Fourier's first law of heat transfer,

$$q/A = -k\,(dT/dx)$$

where $q$ is the rate of heat flow, $A$ is the area through which heat is transferred, and the $q/A$ ratio is the rate of heat transfer per unit area (heat flux).

Thermal conductivity is a physical property of a material. The thermal conductivity of a substance is defined as the quantity of heat flow per unit time across a unit area perpendicular to the direction of heat flow. It measures how well heat is conducted through a material. The higher the thermal conductivity of a material, the greater the heat flux; therefore positive heat flow occurs in the direction of decreasing temperature.

Heat transfer by convection occurs when molecules move from one point to another and exchange energy. Bulk molecular motion is involved. Heat transfer by convection is the rate of heat exchange at the interface between a fluid and a solid. The rate of heat transfer is proportional to the temperature difference. The rate of heat transfer by convection is greater than conduction. Convection is governed by Newton's law of cooling. It is directly proportional to the heat transfer area and the temperature difference between the hot and cold fluids. The quantitative relationship is expressed as

$$g = hA\Delta T$$

where $g$ is rate of the heat transfer, $A$ is the heat transfer area, $\Delta T$ is the temperature difference, and $h$ is a proportionality constant (also called the heat transfer coefficient). The magnitude of $h$ depends on the properties of the fluid, the nature of the surface, and the velocity of flow past the heat transfer surface. Because oil has a high $h$ value, it is an efficient medium for convective heat transfer.

In actual practice during deep frying, heat transfer occurs by combined convection and conduction. Heat transfer from frying oil involves convective heat transfer, and conductive heat transfer occurs through the food.

**TABLE 12.1**
**Specific Heat of Simple Saturated Triglycerides (5)**

| Material | Temperature (°C) | Specific Heat |
|---|---|---|
| Trilaurin | 66.0 | 0.510 |
| | 73.7 | 0.515 |
| | 81.9 | 0.519 |
| | 89.5 | 0.524 |
| | 97.1 | 0.530 |
| Trimyristin | 58.4 | 0.514 |
| | 65.3 | 0.518 |
| | 85.3 | 0.530 |
| | 91.9 | 0.534 |
| Tripalmitin | 65.7 | 0.519 |
| | 72.8 | 0.525 |
| | 86.8 | 0.533 |
| | 96.0 | 0.53 |
| Tristearin | 79.0 | 0.530 |
| | 88.8 | 0.536 |
| | 98.5 | 0.542 |

**TABLE 12.2**
**Changes of Specific Heat of Two Vegetable Oils with Temperature (5)**

| Fully hydrogenated cottonseed oil (Iodine value: 6.5) | | Soybean oil (Iodine value: 128.3) | |
|---|---|---|---|
| Temperature (°C) | Specific Heat | Temperature (°C) | Specific Heat |
| 79.6 | 0.520 | 80.4 | 0.493 |
| 119.8 | 0.544 | 130.9 | 0.526 |
| 160.4 | 0.570 | 172.3 | 0.558 |
| 201.4 | 0.584 | 209.6 | 0.590 |
| 219.4 | 0.595 | 240.2 | 0.617 |
| 270.3 | 0.643 | 271.3 | 0.666 |

## Thermal Properties of Oils

The specific heat of fatty acids or glycerides increases with an increase in fatty acid chain length and decreases as the fat becomes more unsaturated. There is a progressive increase in specific heat as the temperature increases. For liquid oils, the specific heat from 27–57°C is calculated as:

**TABLE 12.3**
**Heat Content of Fat and Fatty Acids (5)**

| | Heat Content (btu/lb) | | | |
|---|---|---|---|---|
| Temperature (°F) | CO[a] (61) | PHCO[b] (61) | HHCO[c] (61) | TFA[d] (69) |
| 20 | 83.0 | 44.0 | 37.8 | – |
| 30 | 92.4 | 53.5 | 41.0 | – |
| 40 | 99.3 | 63.5 | 44.9 | 0.0 |
| 50 | 106.7 | 72.4 | 48.9 | 6.0 |
| 60 | 113.8 | 82.0 | 53.2 | 12.3 |
| 70 | 117.8 | 92.0 | 57.7 | 19.0 |
| 80 | 123.6 | 102.0 | 62.8 | 27.2 |
| 90 | 128.5 | 111.2 | 67.7 | 36.8 |
| 100 | 133.4 | 119.2 | 72.5 | 49.2 |
| 110 | 138.3 | 127.2 | 77.7 | 67.0 |
| 120 | 143.2 | 133.2 | 83.8 | 72.8 |
| 130 | 148.1 | 338.0 | 93.7 | 78.6 |
| 140 | 153.0 | 144.3 | 116.8 | 84.4 |
| 160 | 162.8 | 154.4 | 175.8 | 96.0 |
| 180 | — | — | 186.5 | — |
| 200 | — | — | 197.4 | — |

[a]CO = Cottonseed oil.
[b]PHCO = Partially hydrogenated cottonseed oil.
[c]HHCO = Highly hydrogenated cottonseed oil.
[d]TFA = Tallow fatty acids.

$$Cp = 0.4914 + 0.004t$$

and, for liquid hydrogenated oils from 47–87°C:

$$Cp = 0.4715 + 0.00117t$$

where $Cp$ is specific heat and $t$ is temperature. As shown, the specific heat of pure triglycerides and natural fat increases as the temperature increases (Tables 12.1 and 12.2).

The heat of fusion of various fats ranges from 45–52 cal/g. Heat of fusion increases with increasing chain length of the fatty acid and higher degrees of saturation. The heat content of various fats is given in Table 12.3.

Thermal conductivity is difficult to measure. Fat is considered to be a poor conductor of heat, or, in other words, it is an insulator. For example, thermal

**TABLE 12.4**
**Mean Heat Capacity of Foodstuffs Between 32–212°F (6)**

| Foodstuff | Water Content (%) | Heat Content (Btu/lb °F) |
|---|---|---|
| **Fish** | | |
| Fresh | 80 | 0.86 |
| Fried | 69 | 0.72 |
| **Fat** | | |
| Butter | 15 | 0.50 |
| Margarine | 12 | 0.40 |
| Vegetable Oil | — | 0.40 |
| **Fresh Vegetable** | | |
| Carrots | 88 | 0.92 |
| Cucumber | 97 | 0.98 |
| Mushroom | 90 | 0.94 |
| Potatoes | 75 | 0.81 |
| White Cabbage | 91 | 0.98 |
| **Fruit** | | |
| Fresh | 75–92 | 0.80–0.90 |
| Dried | 30 | 0.50 |
| **Meat** | | |
| Beaf, fat | 51 | 0.69 |
| Beaf, lean | 72 | 0.82 |
| Pork, fat | 39 | 0.62 |
| Pork, lean | 57 | 0.73 |
| **Miscellaneous** | | |
| White Bread | 45 | 0.67 |
| Salt | 15 | 0.28 |
| Macaroni | 13 | 0.44 |
| Rice | 12.5 | 0.43 |
| Egg White | 87 | 0.92 |
| Egg Yolk | 48 | 0.67 |
| Water | — | 1.00 |

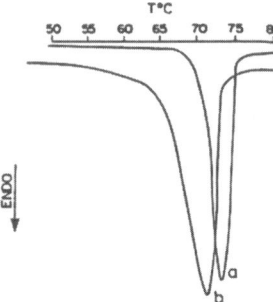

**Fig. 12.3.** Characteristic thermograms by differential scanning calorimetry. a) Tristearin without SPAN 60; b) Tristearin with SPAN 60. SPAN 60 is sorbitan monostearate (8).

conductivity of olive oil in cal/sec cm °C is 0.00040 at 19°C and 0.000385 at 71°C or 0.097 and 0.093 Btu/h ft °F, respectively. The thermal conductivity of oils on average is 0.079 Btu/h ft °F.

## Thermal Properties of Food

The amount of heat required to change the temperature of a unit mass one degree is described by the equation

$$Cp = Q/MT$$

where $Cp$ = specific heat in cal/g – °C or Btu/lb °F, $Q$ = amount of heat in cal or Btu, $M$ = mass in g or lb, and $T$ = temperature change in °C or °F.

Calorimetric techniques are used to determine heat capacity. Differential scanning calorimetry (DSC) is a direct, quick, and dynamic method (Fig. 12.3). Measurement of specific heats at frying temperatures is difficult.

Examples of the mean heat capacity of various foods are given in Table 12.4. Several expressions are used to predict the heat capacity of food products (6,7). Charm (6) proposed an expression that is dependent on the composition of food, consisting of

$$Cp = 2.094X_F = 1.256X_S = 4.187X_H$$

where 2.094 = specific heat of the fat, 1.256 = specific heat of the solids, and 4.187 = specific heat of the water. Generally, these equations are less accurate for low-moisture products than they are for high-moisture ones.

Since foods consist of multiple phases, the "apparent" thermal conductivity is determined. It is dependent on i) the thermal conductivity of each phase, ii) the volumetric proportions of each phase, iii) the structural arrangement or geometrical distribution of constituent phases, and iv) the temperature. Bulk density is also a factor affecting apparent thermal conductivity.

**TABLE 12.5**
Thermal Conductivity of Various Foods at Different Temperatures (9)

| Food | Temperature (°C) | Thermal Conductivity W/(m °K) |
|---|---|---|
| Beef | 5 | 0.5106 |
| | 10 | 0.527 |
| Carrot | — | 0.6058 |
| Egg White | — | 0.3380 |
| Egg Yolk | 2.8 | 0.5435 |
| Lamb | 5.5 | 0.4777 |
| Oatmeal, dry | — | 0.3380 |
| Olive Oil | 5.6 | 0.1887 |
| | 100 | 0.1627 |
| Onions | 8.6 | 0.5746 |
| Peanut Oil | 3.9 | 0.1676 |
| Cowpea | 16.7 | 0.3115 |
| Pork | 6 | 0.4881 |
| Poultry | 59.3 | 0.5400 |
| Potato | — | 0.5540 |
| | — | 0.4119 |
| Tomato | — | 0.5279 |

Most expressions to predict thermal conductivity considered the product as a two-phase system: water and solid components. Others rely on components of the food. The thermal conductivity of several foods is shown in Table 12.5.

Foods are subject to variability in both composition and structure; this results in varying thermal conductivities. Thermal conductivity may be calculated from the composition of the pure components and the volume fraction of each component. The thermal conductivities of water, ice, protein, fat, and carbohydrates are in values of $W/mK$ as follows:

$$kw = 0.57109 + 0.0017625T - 6.7306 \times 10^{-6}T^2 \text{ (water)}$$
$$kic = 2.2196 - 0.0062489T = 1.0154 \times 10^{-4}T^2 \text{ (ice)}$$
$$kp = 0.1788 + 0.00119958T - 2.7178 \times 10^{-6}T^2 \text{ (protein)}$$
$$kf = 0.1807 - 0.0027604T - 1.7749 \times 10^{-7}T^2 \text{ (fat)}$$
$$kc = 0.2014 + 0.0013874T - 4.3312 \times 10^{-6}T^2 \text{ (carbohydrates)}$$

At any given temperature, oil has a lower thermal conductivity compared to water (4).

**TABLE 12.6**
**Major Mass Changes During Frying of Frozen Prefried Food (11)**

| Prefried Food | Chemical Composition | Before Frying | After Frying |
|---|---|---|---|
| Potatoes | Moisture | 6.9 | 3.2 |
| | Lipids | 4.6 | 8.8 |
| | Dry matter, lipid free | 28.5 | 29.1 |
| | Total | 100.0 | 61.1 |
| Battered hake | Moisture | 58.2 | 41.8 |
| | Lipids | 6.6 | 13.8 |
| | Dry matter, lipid free | 35.2 | 35.6 |
| | Total | 100.0 | 91.2 |

**TABLE 12.7**
**Oil Absorption by Potato Products After Deep Fat Frying (2)**

| Product | Size (cm) | Oil Absorption[a] (g oil/100 g food) |
|---|---|---|
| Potato Chips | — | 40 |
| French Fries | $10 \times 10 \times 70$ | 9 |
| | $6 \times 6 \times 70$ | 13 |

## Heat Transfer During Deep Fat Frying

Heat transfer from the frying oil (fluid) involves convective heat transfer from the fluid and conductive heat transfer through the food (solid). Both frying oil and water from the food are major factors in heat transfer. Oil provides an effective medium for heat transfer toward the food, whereas water is an effective medium for heat transfer within the food. Water is a better conductor of heat than the fat, protein, and carbohydrate portions of the food. In addition, water is evaporated to steam on contact with the hot frying oil. Steam carries off thermal energy from the oil surrounding the frying food. As a result, although the temperature of the oil may be as high as 196°C, the temperature of food being fried is only about 100°C. This effectively protects the food from being charred by the hot frying oil (10).

From a practical standpoint, heat balance in a fryer is necessary to permit the preparation of nongreasy fried food. The heat balance refers to the relationship between the heat input and heat requirement. The ability of the fryer to heat shortening is generally measured in Btu/h and is inherent in the fryer design. The requirement for heat is controlled by the fry cook. Satisfactory performance of a fryer is dependent on keeping the demand for the heat below that which the heating system can supply.

The heat requirement is a function of how rapidly moisture is removed from food. The addition of cold food to the fryer reduces the oil temperature. Evaporation of moisture by boiling further cools the oil. The heating system must be able to replace the heat at a rate faster than what is lost. Oil temperature is to be maintained during frying with a continuous evolution of steam. The heat requirement is determined

**TABLE 12.8**
Percent Fat and Moisture Changes in Fresh and Prefried Potatoes (13)

| | Regular Cut (3/8″) Potatoes | | |
| --- | --- | --- | --- |
| | Fresh | Frozen Prefried | Frozen Prefried Finished Fried |
| % Moisture | 75–85 | 60–70 | 50–60 |
| % Fat | 0–1 | 5–6 | 10–12 |

by the relationship between the amount of moisture removed and availability of moisture. Most moisture is removed from the surface of the food, as previously shown for french fries (Fig. 12.2). The size of the piece is important for determining total moisture removed. For example, a fry basket of large pieces of food may have the same total moisture as a basket of small food pieces. Small pieces have a larger surface area per unit weight than the larger pieces. The heat penetration into the smaller pieces is more rapid than for larger pieces. Moisture loss is more rapid, and therefore a higher rate of heat recovery is required for small pieces.

## Frying Capacity

In the operation of a fryer, the capacity is dependent on the type of food being fried. Excess water must be drained from food to prevent possible overproduction of steam, resulting in oil overflowing from the sides of the fryer. Simply increasing the temperature of the fryer does not compensate for fryer capacity limitations. Higher oil temperatures tend to overcook the surface of food, leaving the interior undercooked. A low temperature setting provides insufficient heat capacity to recover heat loss, and food becomes soaked with oil, leaving the product with a greasy appearance.

# Mass Transfer

Mass transfer in food operations includes dehydration, distillation, liquid extraction, and leaching. All are involved in deep fat frying. Dehydration occurs with the continuous evolution of steam. Distillation of components from the food occurs, giving the characteristic odors of food being fried, and extraction and leaching of components from food to frying oil occurs.

Frying fat is absorbed into fried food. The degree of absorption of frying fat for potato products is shown in Table 12.6. Potato chips have a high surface area and absorbed the highest quantity of fat (Table 12.7). The majority of commercially prepared fried food is preblanched or prefried. Prefrying decreases the frying time necessary to obtain a thoroughly cooked final product with good color. A comparison reflecting oil absorption and moisture content (evaporation/dehydration) of fresh potatoes and prefried potatoes is given in Table 12.8.

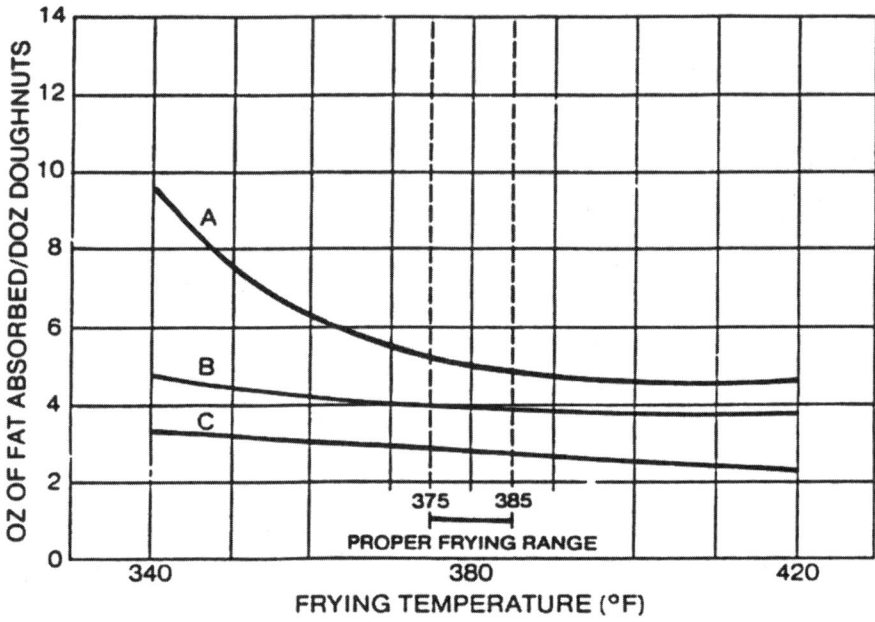

**Fig. 12.4.** Absorption in relation to fryng temperature (13).

## Water Loss

Water on the surface of frying food evaporates on contact with hot oil. This allows water in food to migrate outward, which is a continuous process. The term "pumping" is used to describe the flow of water from interior to exterior (12). This effectively protects against the charring of food from hot frying oil. Steam should constantly be emitted during the entire cooking process until the food is removed from the fryer. The evolution of steam indicates that the pressure inside the food is greater than that in the kettle. Therefore, it also limits penetration of frying oil through the surface to the interior of the food.

## Oil Uptake

The conventional way to describe oil uptake is as a percentage of total weight of the product (wet basis). To alleviate the dependency of oil uptake on water replacement, Pinthus et al. (14) introduced an alternative oil uptake criterion, expressing the weight ratio between oil uptake and moisture loss. This criterion is believed to be instrumental in assessing the effectiveness of reducing oil uptake during deep fat frying.

Many factors were reported to affect oil uptake, including oil quality, frying temperature and duration, product shape, product composition (e.g., moisture, solids, fat, and protein), porosity, prefrying treatments (e.g., drying and blanching),

**TABLE 12.9**
**Effects of Frying Conditions on Percent Fat of Finished Fried Potatoes (13)**

| Type of Frying | Size of Cut | Frying Conditions | % Fat on Finished Weight |
|---|---|---|---|
| Raw to Done | 1/4" shoestring | 4 min at 300°F | 7–9 |
| | 1/4" shoestring | 4 min at 350°F | 15–17 |
| | 1/4" shoestring | 4 min at 400°F | 15–17 |
| | 3/8" regular cut | 5 min at 300°F | 12–14 |
| | 3/8" regular cut | 5 min at 350°F | 11–13 |
| | 3/8" regular cut | 5 min at 400°F | 11–13 |
| Blanched 2 min at 350°F | 1/4" shoestring | Not browned | 8–10 |
| | 1/4" shoestring | 2 min at 300°F | 19–20 |
| | 1/4" shoestring | 2 min at 350°F | 16–18 |
| | 1/4" shoestring | 2 min at 400°F | 16–18 |
| | 3/8" regular cut | Not browned | 6–7 |
| | 3/8" regular cut | 3 min at 300°F | 12–14 |
| | 3/8" regular cut | 3 min at 350°F | 11–13 |
| | 3/8" regular cut | 3 min at 400°F | 11–13 |
| Frozen, preblanched | 1/4" shoestring | Not browned | 7–9 |
| | 1/4" shoestring | 3 min at 300°F | 18–20 |
| | 1/4" shoestring | 3 min at 350°F | 16–18 |
| | 1/4" shoestring | 3 min at 400°F | 16–18 |
| Frozen, preblanched | 3/8" regular cut | Not browned | 5–6 |
| | 3/8" regular cut | 4 min at 300°F | 12–14 |
| | 3/8" regular cut | 4 min at 350°F | 10–12 |
| | 3/8" regular cut | 4 min at 400°F | 10–12 |
| | 3/8" regular cut | Thawed before frying, not browned | 5–6 |
| | 3/8" regular cut | 4 min at 350°F | 6–18 |
| | 3/8" regular cut | 4 min at 350°F Cooled to room temp., redipped 2 min at 350°F | 18–20 |
| Frozen, preblanched | 3/8" crinkle cut | Not browned | 6–7 |
| | 3/8" crinkle cut | 4 min at 350°F | 11–13 |
| | Steak house cuts | Not browned | 3–4 |
| | Steak house cuts | 6 min at 350°F | 7–10 |

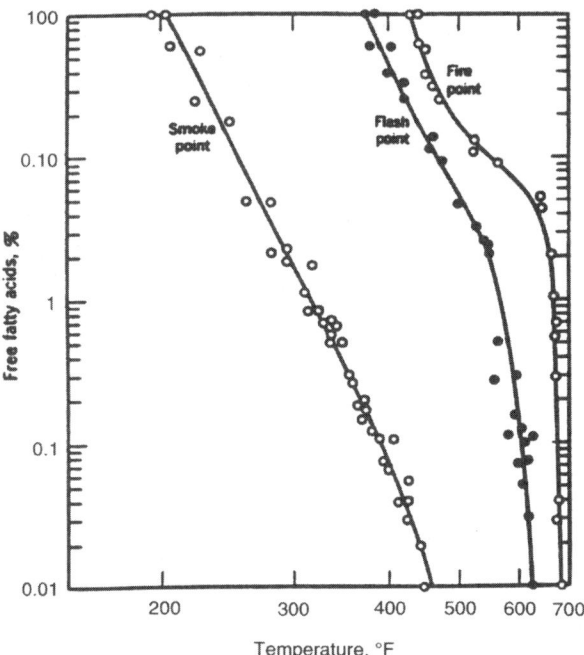

**Fig. 12.5.** Smoke, fire, and flash points or miscellaneous crude and refined fats and oils, as a function of the content of free fatty acids (5).

coating, and others (15). Overall, oil absorption is controlled to a large extent by the surface area exposed during frying and is also affected by the previous history of the food. Excess oil absorption may result from low frying temperatures or overloading the fryer beyond design capacity (Fig. 12.4). At low temperatures, there is a tendency to cook food longer to obtain the desired browning of the food. Oil absorption, therefore, increases. Table 12.9 shows how different frying conditions affect the percentage of fat in french-fried potatoes. In contrast, cooking doughnuts for 3 min at 175, 185, and 200°C gave no significant differences in fat absorption, but the color of the doughnuts was quite different (16). The smoke point of the fat also affects oil absorption. A significant negative correlation between oil absorption by doughnuts and smoke point of frying oil was reported (16). A linear relationship between oil uptake and water removal was also reported (17). Oil absorption is not dependent on the type of oil utilized during frying, but is related to the state and temperature of the oil. A high initial moisture content results in elevated oil uptake (18).

## Physical Properties of Oil Related to Frying

Oils are water insoluble, hydrophobic substances consisting predominantly of glycerol esters of fatty acids or triglycerides. Characteristics permitting their use for deep fat

**TABLE 12.10**
Vapor Pressure of Triglycerides (Temperature Corresponding to Specified Pressures of Mercury (5)

| Compound | Temperature (°C) | |
|---|---|---|
| | 0.05 mm | 0.001 mm |
| Tributyrin | 91 | 45 |
| Tricaproin | 135 | 85 |
| Tricaprylin | 179 | 128 |
| Tricaprin | 213 | 159 |
| Trilaurin | 244 | 188 |
| Trimyristin | 275 | 216 |
| Tripalmitin | 298 | 239 |
| Tristearin | 313 | 253 |
| Soybean Oil | 308 | 254 |
| Olive Oil | 308 | 253 |
| 2-Oleyl-1,3-distearin | 315 | 254 |
| 1-Myristyl-2-palmityl-3-stearin | 297 | 237 |
| 1-Palmityl-2-lauryl-3-stearin | 290 | 232 |
| 1-Myristyl-2-lauryl-3-stearin | 282 | 223 |
| 1-Palmityl-2-captyl-3-stearin | 280 | 223 |
| 1-Capryl-2-lauryl-3-myristin | 249 | 189 |

**TABLE 12.11**
Vapor Pressure of Monoglycerides (5)

| Monoglyceride | Presure (mm) | Temperature (°C) |
|---|---|---|
| Monocaprin | 1 | 175 |
| Monolaurin | 1 | 186 |
| Monomyristin | 1 | 199 |
| Monopalmitin | 1 | 211 |
| Monostearin | 0.2 | 190 |
| Monoolein | 0.2 | 186 |

frying are high vapor pressures and resistance to decomposition. Decomposition of oil relates primarily to hydrolysis and oxidative/thermal decomposition.

## Vapor Pressure

Vapor pressure, relating to boiling point and heat of vaporization, is the most important characteristic in the application of fats to fried food preparation. Triglycerides have extremely low vapor pressures, so evaporation does not occur. Examples of vapor pressures are given in Table 12.10 for both synthetic triglycerides and natural fats. Partial esters and fatty acids have considerably higher vapor pressures. Monoglycerides

**TABLE 12.12**
**Boiling Point of Saturated Fatty Acids (5)**

| Pressure (mm) | Boiling Point (°C) | | | | | | |
|---|---|---|---|---|---|---|---|
| | Caproic | Caprylic | Capric | Lauric | Myristic | Palmitic | Stearic |
| 1 | 61.7 | 87.5 | 110.3 | 130.2 | 149.2 | 167.4 | 183.6 |
| 2 | 71.9 | 97.9 | 121.1 | 141.8 | 161.1 | 179.0 | 195.9 |
| 4 | 82.38 | 109.1 | 132.7 | 154.1 | 173.9 | 192.2 | 209.2 |
| 8 | 94.6 | 121.3 | 145.5 | 167.4 | 187.6 | 206.1 | 224.1 |
| 16 | 107.3 | 1346 | 159.4 | 181.8 | 202.4 | 221.5 | 240.0 |
| 32 | 120.8 | 149.2 | 174.6 | 197.4 | 218.3 | 238.4 | 257.1 |
| 64 | 136.0 | 165.3 | 191.3 | 214.6 | 236.3 | 257.1 | 276.8 |
| 128 | 152.5 | 183.3 | 209.8 | 234.3 | 257.3 | 278.7 | 299.7 |
| 256 | 171.5 | 203.0 | 230.6 | 256.6 | 281.5 | 303.6 | 324.8 |
| 512 | 192.5 | 225.6 | 254.9 | 282.5 | 309.0 | 332.6[a] | 355.2[a] |
| 760 | 205.8 | 239.7 | 270.0 | 298.9 | 326[a] | 351.5[a] | 376.1[a] |

[a]Values obtained by extrapolation.

are readily distilled. Examples of their vapor pressures are shown in Table 12.11.

Fatty acids are more volatile than the corresponding mono-, di-, or triglycerides. They are a source of smoke arising from used frying oil. The vapor pressures or boiling points of selected saturated fatty acids are shown in Table 12.12. The longer the chain length of the fatty acid, the higher the vapor pressure.

## Smoke, Fire, and Flash Points

Smoke, fire, and flash points are indirect measures of the thermal stability of a fatty material when heated in contact with air. Smoke point is the temperature at which smoke is first detected in a laboratory apparatus. Flash point is the temperature at which volatile products are evolved at a sufficient rate to be ignited. Fire point is the temperature at which volatile products support continued combustion.

Fatty acids are more volatile than glycerides; therefore, smoke, flash, and fire points of glycerides depend principally on the content of free fatty acids (FFA) (Fig. 12.5). The unsaturation of an oil has little or no effect on the smoke, flash, or fire points.

## Hydrolysis

Triglycerides, under proper conditions of water miscibility, hydrolyze to FFA and glycerol:

$$C_3H_5 (OOCR)_3 + H_2O \rightarrow C_3H_5 (OH)_3 + 3HOOCR$$

**TABLE 12.13**
**Physical and Chemical Changes of Corn Oil During Deep Fat Frying and Continuous Heating (22)**

| | Oil Used for Frying (h) | | | | | | | Oil Continuously Heated (h) |
|---|---|---|---|---|---|---|---|---|
| | 0 | 3 | 6 | 12 | 30 | 60 | 90 | 90 |
| FFA[a] | 0.12 | 0.13 | 0.13 | 0.17 | 0.30 | 0.88 | 1.37 | 0.32 |
| P.V.[b] (meq/kg) | 1.34 | 1.53 | 1.63 | 2.75 | 1.92 | 2.41 | 2.94 | 2.20 |
| I.V.[c] (Wijs) | 128.00 | 128.00 | 127.00 | 126.00 | 126.00 | 123.00 | 124.00 | 122.00 |
| R.I.[d] (40°C) | 1.4625 | 1.4675 | 1.4680 | 1.4681 | 1.4681 | 1.4681 | 1.4681 | 1.4681 |
| Color[e] | 2.86 | 3.26 | 3.92 | 4.58 | 5.26 | 8.04 | 8.56 | 12.47 |
| Viscosity[f] | 39.7 | 40.0 | 40.3 | 43.2 | 42.3 | 44.9 | 43.9 | 50.4 |
| Foaming (mL) | — | — | — | — | — | — | — | 200.0 |

[a]FFA = Free fatty acids (% oleic acid).
[b]P.V. = Peroxide value.
[c]I.V. = Iodine value.
[d]R.I. = Refractive Index.
[e]Photometric.
[f]In centisokes, 37.7°C.

An equilibrium is reached if reactants and products are not removed from the reaction. High temperature, pressure, water, and contaminants accelerate fat splitting. Soaps and alkaline cleaning agents promote hydrolysis of oil.

## Flavor Development

Lipids in food undergo a variety of reactions as a result of heat treatment. These are important to consumers because of their significance in food flavor, nutrition, and safety. The chemistry of lipid oxidation at elevated temperatures is complex. Both thermolytic and oxidative mechanisms occur in a fryer with combined heat and air exposure. These are discussed in detail in other chapters.

A complex series of compounds results from heated oil even when saturated triglycerides are heated for 1 h at 180°C. Alkanes, alkenes, fatty acids, ketones, oxypropyl esters, propene and propanediol esters, and diacylglycerols are produced. The reaction occurs homogeneously within the oil phase. Specificity as to fatty acid type or position is unknown. A preferential release in favor of unsaturated and shorter chain fatty acids was noted (19,20).

In unsaturated fatty acids, formation of dimeric and cyclic compounds dominates the thermolytic reaction. The occurrence of many compounds is explained on the basis of formation or combination of free radicals from homolytic cleavage of carbon–carbon linkages near the double bond. Alicyclic monoene and diene dimers and saturated dimers are produced. Polymerization of unsaturated fatty acids can also

**TABLE 12.14**
Physical and Chemical Changes of Triglycerides During Simulated Deep Fat Frying (22)

|  | Trilinolein (C:18:2=) | | Triolein (C18:1=) | | Tristearin (C18:0=) | |
|---|---|---|---|---|---|---|
|  | Before Treatment | After Frying | Before Treatment | After Frying | Before Treatment | After Frying |
| Color[a] | 3.55 | 76.00 | 5.8 | 62.5 | 1.26 | 12.04 |
| FFA[b] | 0.44 | 2.60 | nil | 3.9 | nil | 4.00 |
| I.V.[c] | 176.00 | 155.40 | 85.0 | 78.1 | 0.0 | 0.5 |
| P.V.[d] | 25.80 | 4.70 | 0.9 | 3.4 | 0.0 | 3.2 |
| Visc.[e] | 36.20 | 200.60 | 56.2 | 101.8 | 16.0 | 21.1 |
| R.I.[f] | 1.4728 | 1.4793 | 1.4632 | 1.4655 | 1.4402 | 1.4420 |
| NAF esters | — | 26.30 | — | 10.8 | — | 4.2 |

[a]Photometric.
[b]In percent.
[c]Wijs.
[d]meq/kg.
[e]Viscosity, centistokes, 30°C. Note that viscosity and refractive index of tristearin were measured at 80°C.
[f]40°C.

occur via Diels–Alder reactions.

Even saturated fatty acids when heated at a high temperature (150°C) undergo lipid oxidation. Oxidative products consist of carboxylic acid, 2-alkanones, $n$-alkanols, lactones, $n$-alkanes, and 1-alkenes. The principal mechanism apparently involves formation of monohydroperoxides and oxygen attack occurring at methylene groups of the fatty acid. Stepwise oxidation gives rise to a series of lower acids and decomposition products.

Unsaturated fatty acids are much more susceptible to oxidation. The principal chemical reaction involves the formation and decomposition of hydroperoxide intermediates. Isolation of intermediates is difficult. In a study by Lomans and Nawar (21) at 180°C (356°F), peroxide values (P.V.) were followed with time as follows: 5 min, 237; 10 min, 251; 20 min, 119; 30 min, 80; and 60 min, 44. Hydroperoxide decreased even more rapidly at higher temperatures, resulting in low P.V. after only a short heating time. Complex decomposition patterns occur with peroxide decomposition. The number of compounds that form are in the hundreds.

In terms of flavor development, the thermal oxidation of frying oil is the result of the formation of a complex pattern of decomposition products (22). The products reflect the decomposition of the component fatty acids. After the initial 30 min of deep frying, primary volatile oxidation products are detected. Quantities vary widely, depending on the type of oil, food, and heat treatment. Formation generally reaches a plateau as a result of simultaneous evaporation, decomposition, and formation. The amounts of some volatiles produced from corn oil are shown in Table 12.13. The

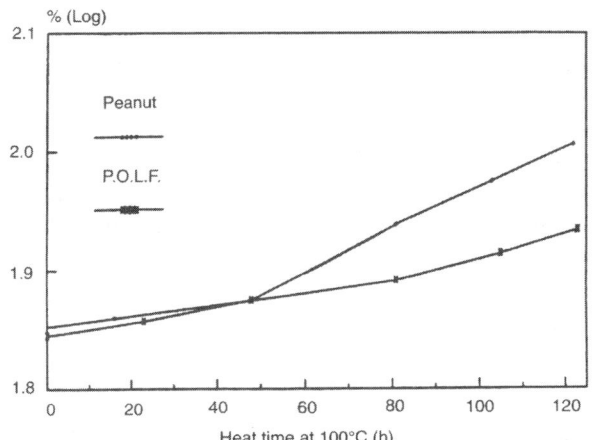

**Fig. 12.6.** Increase of viscosity during frying (28).

**TABLE 12.15**
**Viscosity Increase in Fresh and Used Oil (Doughnuts): Comparison of Fat Heated in Air and in Nitrogen for 48 and 96 h at 375°C (24).**

|  | Viscosity at 212°F (centistokes) | Titratable acidity (% as oleic) | Iodine Number | Fat Absorption % |
|---|---|---|---|---|
| Unheated control | 8.25 ± 0.10 | 0.04 ± 0.03 | 56.4 ± 0.2 | 18.0 ± 0.2 |
| Heated in air (48h)[a] | 10.45 ± 0.20 | 0.99 ± 0.05 | 50.0 ± 0.2 | 22.0 ± 1.0 |
| Heated in nitrogen (48h)[a] | 8.25 ± 0.10 | 0.10 ± 0.03 | 56.3 ± 0.2 | 18.6 ± 0.2[b] |
| Heated in air (96h)[a] | 14.95 ± 0.20 | 2.09 ± 0.06 | 44.7 ± 0.2 | 25.1 ± 1.0 |
| Heated in nitrogen (96h)[a] | 8.25 ± 0.10 | 0.10 ± 0.03 | 56.0 ± 0.2 | 18.6 ± 0.2[b] |

[a]2-lb fat/5-L flask, specific surface area = 0.34 cm²/g.
[b]Difference in fat absorption between this sample and the unheated control was significant in the 99% probability range.

amount of volatiles formed cannot be used as an indication of the extent of use of a commercial frying oil.

Other changes that occur in a frying oil during use include increase in viscosity, polarity, FFA content, dark color, decrease in iodine value, change in refractive index, decrease in surface tension, and increase in tendency of oil to foam (Table 12.13).

# Deterioration of Frying Oil

Both frying oil and lipids in food undergo many reactions as a result of heat treatment (Table 12.14), including decomposition, polymerization, and oxidation. The complex physicochemical changes affect both the frying medium and the food being fried. Some changes are desirable. An example is normal flavor development. Others are undesirable because of their effects on the performance of the frying oil, nutritional quality, off-flavor formation, and food safety.

Oil deteriorates slowly when heated. When food is fried, the evaporating steam acts to strip some decomposition products, delaying deterioration of fat. Fat held at frying temperatures, but without frying, breaks down rapidly. Lowering the oil temperature when it is not being used for frying delays the deterioration of the fat.

The turnover rate is probably the most important factor in retaining the fat in a satisfactory condition. General guidelines call for daily replenishment of the amount of oil absorbed by food as a percentage of the total oil capacity of the fryer. The turnover rate is a function of the absorption of fat by the food being fried. Some oil is also lost during filtration. Various factors are influential in oil absorption. The addition of fresh fat to used fat in good condition maintains the ability of the fat to contribute good color and flavor to fried food.

## FFA Development

Free fatty acids form during frying due to the presence of moisture in the foods exposed to oil. Normally, 0.5–0.8% FFA is encountered. If the frying rate is insufficient to drive water from food at a sufficient rate, the FFA content increases rapidly. High-FFA frying oil smokes at a progressively lower temperature. Anything that restricts the loss of water from the frying kettle has the same effect. Fry kettles should not be covered during frying without exhaust provisions. Specificity as to fatty acid type or position in the triglyceride is unknown. A preferential release in favor of unsaturated and shorter chain fatty acids was noted (19,20).

## Polymerization

During frying, some polymerization of the fat may occur. In some cases, this leads to foam formation. Polymerization, thermal or oxidative, can affect the greasiness of fried food. Polymerization is shown as an increase in oil viscosity (Fig. 12.6, Table 12.15). As viscosity increases from polymerization, oil absorption increases. With doughnuts, this can be desirable. Oil is often conditioned or "broken in" for up to 48 h to give better frying quality. A 30–40% increase in fat absorption in other fried food, however, is undesirable.

An estimate of the degree of polymerization is made by measuring the percentage of nonurea-adduct-forming fatty acids (NUAF). Urea forms a complex or adducts with normal straight-chain fatty acids, but does not form a complex with fatty acids altered by branching, cyclizing, or polymerizing. Rock and Roth (24) found a direct relationship between the amount of NUAF and the increase in viscosity of a fat used for frying. Nonurea-adduct-forming fatty acid compounds are suggested to be a cause of loss of nutritional value in heated fat (25). Values from 0.5–5.36% were reported in commercial samples of used frying oil (26,27).

## Foam Development

Foaming generally occurs when a fat has polymerized. Antifoams (for example, silicones) break foam as it forms. In addition, silicones may act as antioxidants for fat at frying temperatures (29).

Oil-soluble components, as well as fat from the food being fried, affects fryer performance. The leaching of lecithin from egg yolk batters or doughnut mixes causes foaming. Oil-containing fatty acid soaps also cause foaming. Chicken, meat, and fish are high in fat exchange with the fry medium.

# Summary

Frying food, although seemingly simple, follows defined principles of heat and mass transfer. During frying, there is a convective heat transfer from frying oil and a conductive heat transfer through food. As a result, mass transfer occurs. Water evaporates out of the food, and frying oil penetrates the food.

Mass transfer and heat transfer are the two basic forces that drive most frying reactions, leading to many physicochemical changes in both oil and food. Some mass changes, such as crust formation, surface browning, and flavor development, are desirable. Others are undesirable, for example, increases in FFA, foam formation, oil uptake by the food, and a decrease in the smoke point, all lead to the deterioration of the frying oil and the off flavor of the fried food.

Both the frying oil and the food being fried play major roles in obtaining desired characteristics of fried food. This is because they differ in such thermal properties as specific heat, heat of fusion, thermal conductivity, vapor pressure, and smoke point. Other quality factors include proper fryer capacity, uniform size of food pieces, and proper water content.

Procedures for oil handling in a frying operation have been reviewed (30). Initial quality specifications, oil loading at the processing plant, oil receiving and unloading at the frying plant, and storage are covered in detail. Fryer operations including start-up, shutdown, steady state operation, temporary shutdown, used oil storage, and sanitation, all affect the quality of frying oil and the foods fried in them.

Industrial fryers fall into three types: batch fryers, continuous fryers, and vacuum fryers. Batch fryers resemble restaurant fryers, except they are designed to handle several hundred pounds of products per batch and are used for special products such as kettle-fried potato chips. Large scale production of fried snacks is most often carried out in continuous fryers. Vacuum fryers are used primarily for fruits or vegetables, where color retention and a minimum amount of browning is desired in the product. Although expensive, vacuum fryers have distinctive features and advantages over batch and continuous fryers. Fryer capacity data have been published, as well as oil turnover times in continuous fryers (31).

The dynamics of frying in restaurant operations have been reviewed extensively and should be consulted for further information (32).

The technology of frying coated foods has been reviewed (33), as well as the interaction of fried foods with packaging materials (34).

## References

1. Fritsch, C.W. *J. Am. Oil Chem. Soc.* **1981**, *58*, 272.
2. Guillaumin, R. In *Frying of Food: Principle Changes, New Approaches*; E.G. Varela, A.E. Bender, and I.D. Morton, Eds.; VCH Publishers: New York, 1988.
3. Heldman, D.R.; and R.P. Singh. *Food Process Engineering*, 2nd ed.; AVI Publishing: Westport, CT, 1981.
4. Toledo, R.T. *Fundamentals of Food Process Engineering*, 2nd ed.; Van Nostrand Reinhold: New York, 1991.
5. Swern, D. *Bailey's Industrial Oil and Fat Products*; Wiley Interscience: New York, 1982.
6. Charm, S.E. *The Fundamentals of Food Engineering*, 3rd ed.; AVI Publishing: Westport, CT, 1978.
7. Dickerson, R.W. In *The Freezing Presentation of Foods*, 4th ed.; D.K. Tressler, W.B. Van Arsdel, and M.J. Copley, Eds.; AVI Publishing: Westport, CT, 1969; Vol. 2.
8. Schlichter, J.; N. Garti; and S. Sarig. *J. Am. Oil Chem. Soc.* **1986**, *63*, 788.
9. Polley, S.L.; O.P. Snyder; and P. Kotnour. *Food Technol.* **1980**, *34*, 76.
10. Blumenthal, M.M. *Food Technol.* **1991**, *45*, 68.
11. Perez-Camino, M.C.; G. Marquez-Ruiz; M.V. Ruiz-Mendez; and M.C. Debarganes. *J. Food Sci.* **1991**, *56*, 1644.
12. Lydersen, A.L. *Mass Transfer in Engineering Practice*; John Wiley & Sons: New York, 1983.
13. Lawson, H.W. *Standards for Fats and Oils*; AVI Publishing: Westport, CT, 1985.
14. Pinthus, E.J.; P. Weinberg; and I.S. Saguy. *J. Food Sci.* **1992**, *58*, 204.
15. Selman, J.D.; and M. Hopkins. *Technical Memorandum 475*; Camden Food and Drink Research Association: Chippingampden, Gloucestershire, U.K., 1989.
16. Lowe, B.; P.M. Nelson; and J.H. Buchanan. *The Physical and Chemical Characteristics of Lards and Other Fats in Relation to Their Culinary Value. III. For Frying Purposes*; Agricultural Experiment Station, Iowa State College of Agriculture and Mechanic Arts: Ames, IA, 1940; Research Bulletin 279.
17. Gamble, M.H.; P. Rice; and J.D. Selman. *Int. J. Food Sci. Technol.* **1987**, *22*, 233.
18. Gamble, M.H.; and P. Rice. *Lebensm. Wiss. Technol.* **1988**, *21*, 62.
19. Buziassy, C.; and W.W. Nawar. *J. Food Sci.* **1968**, *33*, 305.
20. Noble, A.; C.C. Buziassy; and W.W. Nawar. *Lipids* **1967**, *2*, 435.
21. Lomans, S.S.; and W.W. Nawar. *J. Food Sci.* **1982**, *47*, 744.
22. Chang, S.S.; R.J. Peterson; and C.-T. Ho. *J. Am. Oil Chem. Soc.* **1978**, *55*, 718.
23. Nawar, W.W. In *Flavor Chemistry of Fats and Oils*; American Oil Chemists' Society: Champaign, IL, 1985.
24. Perkins, E.G. *Food Technol.* **1960**, *14*, 508. Rock, S.P., and H. Roth. *J. Am. Oil Chem. Soc.* **1964**, *41*, 228.
25. Firestone, D.; W. Horvitz; L. Friedman; and G.M. Shue. *J. Am. Oil Chem. Soc.* **1961**, *38*, 253.
26. Sahasrabudhe, M.R.; and V.R. Bhalerao. *J. Am. Oil Chem. Soc.* **1963**, *40*, 711.
27. Thompson, J.A.; M.M. Paulose; B.R. Reddy; R.G. Krishnamurthy; and S.S. Chang. *Food Technol.* **1967**, *21*, 405.

28. Bracco, U.; A. Diefenbacher; and L. Kolarovic. *J. Am. Oil Chem. Soc.* **1981,** *58,* 6.
29. Freeman, K. *J. Food Sci.* **1973,** *50,* 101–103.
30. Gupta, M.K. In *Frying Technology and Practices*; M.K. Gupta, K. Warner, and P.J. White, Eds.; AOCS Press: Champaign, IL, 2004; pp. 50–75.
31. Gupta, M.K.; R. Grant; and R.F. Stier. In *Frying Technology and Practices*; M.K. Gupta, K. Warner, and P.J. White, Eds.; AOCS Press: Champaign, IL, 2004; pp. 91–109.
32. Gupta, M.K. In *Frying Technology and Practices*; M.K. Gupta, K. Warner, and P.J. White, Eds.; AOCS Press: Champaign, IL, 2004; pp. 110–124.
33. Sasiela, R.J. In *Frying Technology and Practices*; M.K. Gupta, K. Warner, and P.J. White, Eds.; AOCS Press: Champaign, IL, 2004; pp. 125–155.
34. Marsh, K.S. In *Frying Technology and Practices*; M.K. Gupta, K. Warner, and P.J. White, Eds.; AOCS Press: Champaign, IL, 2004, pp. 156–177.

# ·13·

# Foodservice Frying

**Don Banks**

*Edible Oil Technology 8155 San Leandro, Dallas, Texas 75218*

Before countertop and floor-model fryers, which provide the foundation for foodservice frying, were introduced in the 1930s, frying in restaurants, hotels, and institutions was a kitchen-stove process. The fryers, or cookers as they were commonly called, consisted of kettles with concentric inner liners that had small, evenly spaced drain holes. The liners were the forerunners of frying baskets and were the subject of some of the earliest U.S. patents on frying equipment. These patents addressed the benefits that the liners provided, including ease of loading and unloading the cooker, and straining the food so that the kettle did not have to be removed from the stove between batches. The claims also included uses of the liner to cook/fry doughnuts, peanuts, chicken, and fish.

Using a kettle on a stove to fry food provided a challenge. Without a good method of controlling temperature or rate of heating, it was very difficult to consistently achieve the rich golden color, crisp texture, and characteristic flavor that make fried foods so appealing. Before the introduction of commercial fryers, the quality and consistency of fried food depended on the experience and skill of the cook. When commercial fryers, such as the first Wells Automatic Electric Fryer (1932) became available, the benefits to cooks and customers were soon realized. Batch fryers with thermostatic controls took the guesswork out of proper frying temperatures and greatly simplified frying. The cook could adjust the thermostat and begin frying when the set point was reached. By using a set temperature, the frying time for uniform batches was constant, allowing the cook to simply load the fryer and remove the fried product at the right time. Optimization of time and temperature is key to preparing consistent, high quality fried food from batch to batch.

The introduction of batch fryers gave frying a new identity in the kitchen. Although the fryers only occupied small spaces in the kitchen, they produced large volumes of high-quality food quickly and easily. As a result, rapid growth in foodservice frying occurred in the 1940s. By that time gas and electric batch fryers were available from a number of equipment manufacturers, adding support to the growth. Foodservice frying has continued to expand and grow over the years, providing new and interesting foods that are well received by customers. Year after year, fried foods, led by french-fried potatoes, account for several of the top ten best-selling menu items.

Batch fryers are well suited for the task of preparing a wide variety of foods.

Other than dividing the food into uniform pieces, applying breading or batter as appropriate, and frying, little preliminary preparation is necessary. After the time and temperature requirements are set, frying becomes automatic.

In foodservice, frying is typically conducted on demand and, as a result, will peak and ebb throughout the day. Depending on the menu and local eating patterns, fryers will usually be operated at full capacity for a few hours a day, intermittently for a few hours, and idle the remainder of the time. Operating a fryer on an intermittent basis is part of foodservice frying, but it is one of the primary reasons that frying oil has to be downgraded and replaced periodically. If foodservice fryers were operated batch after batch without interruption and if the oil was filtered regularly, frying oil would rarely need to be downgraded or discarded.

In practice, foodservice fryers typically remain idle as much or even more of the time than they are used. During idle and low-production periods, the oil is subjected to thermal and oxidative stress. Without the benefits from active frying, which include steam stripping, make-up oil addition, and generation of fresh product flavor, fryer oil idled at frying temperature will rapidly degrade.

## Frying Oil Turnover

Frying oil turnover is a good indicator of how much stress oil is subjected to in daily operation. Turnover is one of several important factors that influence the quality of the frying oil, but it is such a critical factor that it will be addressed first and in some detail below.

Oil turnover is defined as the time required for oil usage to equal the total volume of oil in a fryer and can be calculated for foodservice fryers by the equation:

$$\text{Oil Turnover} = \frac{\text{Total pounds of oil in the fryer}}{\text{Average pounds of oil used per hour}}$$

A key element in the calculation is that average oil usage is based on the total time the fryer is "on" during the workday. If the fryer is turned off during idle periods, the lay time would not be counted. If the fryer remains on, however, even if set to a lower temperature, the idle time would be counted.

Oil absorption during frying will vary over a wide range, depending on temperature, the product being fried and operational practices. The range can vary from a low of about 2% for frying peanuts to 36% or higher for potato chips, but 8–25% is a more common range for foodservice items. As oil is absorbed by the product and carried out of the fryer, make-up oil must be added periodically to maintain the proper level.

The frying capacity of a batch fryer is usually rated based on the maximum pounds of french fries that can be produced per hour using partially fried potatoes. Since "par-fries," as they are commonly called, will absorb about 8% oil during finish frying, the capacity rating can be used to calculate the oil turnover for continuous operation. For example, a fryer that holds 32 lb of oil and can produce 50 lb of french fries/h requires the addition of 4 lb of make-up oil each h. After 8 h, 32 lb, or one fryer volume of oil, will need to be added. In this example, the fryer has an 8 h oil turnover.

The term "oil turnover" is widely used but can be misleading. An 8 h turnover, or 100% turnover for an 8 h shift, does not mean that all of the original oil in the fryer has been replaced at the end of 8 h. Although a quantity of new oil equivalent to the capacity of fryer has been added, the fryer will contain a mixture of the original oil and make-up oil. Every time make-up oil is added, it commingles with the oil already in the fryer. As more product is fried, the oil absorbed by the product reflects the blend of the original and make-up oils. At the completion of one oil turnover, one-half of the original oil remains in the fryer, and, after two oil turnovers, one-fourth of the original oil remains. The progression continues, but essentially all of the original oil will have been replaced following four oil turnovers.

Based on fryer specifications, most batch fryers will have turnover rates in the 5–12 h range during continuous production. A 5 h turnover is very rapid and will be sufficient to maintain good oil quality under most frying conditions. Extending the turnover to 12 h places much more stress on the oil, but good quality can still be maintained in many applications. Unfortunately, few batch fryers are operated continuously or even almost continuously.

In practice, fryers may be heated to temperature 30 min–1 h (or more) before needed for frying. After start-up, fryer usage will be variable, with continuous usage required a few times a day, separated by slow or idle periods. As a result, the actual oil turnover for most foodservice fryers will be much slower than the rate calculated for continuous frying. Turnover rates of 20 h and above are not uncommon.

Fryer oil quality cannot be maintained in batch fryers with a turnover rate approaching or exceeding 20 h. Even with careful control of all other aspects of frying, including temperature, moisture, crumbs, seasonings, and filtration, the oil will deteriorate after a few days of usage and need to be replaced. Fresh oil addition will be too slow to maintain fryer oil quality.

Oil turnover should be a primary deciding factor when contemplating any change in the operation of a fryer. Changes that slow the rate of turnover increase oil stress and reduce fry-life, while changes that accelerate turnover reduce stress and improve fry-life.

## Fryer Selection

A wide range of fryers and special features are available to meet foodservice frying needs. The choice is, perhaps, greater than for any other type of kitchen equipment and requires careful review for selection of the best equipment for each operation. Some general considerations to support fryer selection are reviewed here.

One of the first steps in fryer selection is reviewing the types and quantities of food to be prepared. Leading products such as french fries, which are usually fried in dedicated fryers to optimize quality, should be totaled separately. Foods that can be fried in the same fryer without compromising quality can be grouped together to calculate volume requirements. Products that may adversely affect other products should also be totaled separately. The objective is to assess the peak frying demand for each of the foods and food combinations to determine volume and fryer requirements.

The decision to fry different products in the same fryer should be carefully considered for each operation and made on an individual basis. Foods that exchange fat, transfer flavor, change oil color or impart other detectable characteristics should be tested and assigned to separate fryers as necessary.

Fryers also must be matched to the type of product to be fried. Some products, such as french fries, can be fried in deep, narrow baskets that hold a number of servings. Other products such as chicken and seafood fry best if allowed ample room for separation during frying. Doughnuts are usually fried in wide, shallow fryers designed specifically for this product.

Fryers with cool zones (reservoirs of oil below heating surfaces that remain well below frying temperature) should be considered for applications where accumulation of breading, batter, or crumbs is a problem. Almost all gas fryers and a number of electric fryers have cool zones. The zones effectively limit scorching and carbonization of particulates. Cool zones are particularly important when crumbs accumulate rapidly or contain ingredients, such as sugar or lecithin, which quickly degrade oil. If a fryer with a cool zone is needed, a means of controlling particulates at their source is also needed. Cool zones do not provide perfect protection and do not serve as an alternative for effective crumb control. Cool zones also add extra oil volume not used for frying, so benefits from limiting the scorching of crumbs should be balanced against the effects of a slower oil turnover.

The size of a fryer is usually specified according to the number of pounds of frying fat it holds. In foodservice, fryers usually range in size from about 15–45 lb for counter models and from 30–200 lb for floor models. Thirty to forty lb fryers are the most common. A good generalization for sizing fryers is to use as many small fryers as are needed and as few large fryers as necessary to prepare food in a timely manner. A more specific guideline is to determine the smallest fryers that can be used based on the product(s) to be fried, serving size, and frequency of orders throughout the day. Multiple fryers, including mixed sizes if needed, can then be installed to meet peak volume frying requirements. Smaller fryers allow more flexibility in matching capacity to frying requirements during the day. Fryers can be placed in service as needed or held in reserve, thereby protecting the oil from unnecessary heat stress, reducing heating costs, and improving the rate of oil turnover. Using smaller fryers also allows greater flexibility in dedicating fryers for specific items in order to protect against flavor, seasoning, or fat exchange.

Divided fryers are also available and can be used to fry different products concurrently. A divided fryer prevents oil transfer across the fryer but may not prevent heat transfer unless separated by efficient insulation. If both sides of the fryer are operated at the same time and in a similar manner, thermal separation may not be critical; but should be considered, however, if the fryer compartments are operated at different times or under different conditions. Determine the need for thermal separation when considering a divided fryer and make selections accordingly.

Choices for fryer heating are gas and electricity. If economy of operation is critical, local energy cost, fryer heat input, and efficiency of heat transfer should be used to calculate comparative operating costs. Fryer manufacturer can provide information on energy input and heating efficiency for each of their fryers. Reliability of energy supply and the programs that utility companies use for reducing or interrupting service during high-demand periods should also be considered.

Electric fryers, with immersion resistance heaters located in the frying kettle, are efficient, with energy conversion ratings that approach 100%. Electric fryers offer a number of conveniences. Installation is uncomplicated, and the smaller fryers are portable. They are easily moved for cleaning or servicing. Heating elements in most models are pivot-mounted and self-cleaning. Oil pans in many models can be removed to facilitate cleaning and oil filtration.

When using fryers with electric heaters, it is essential that the elements be completely immersed during heating or frying. A fire hazard can develop any time during operation that the heating element is not completely covered by oil, because the exposed part of the element could overheat. The risk can result from a low fat level, an improperly positioned heating element, or, in the case of solid frying fat, cavities developing around the element during melting.

Gas-heated fryers offer a number of different types of burners and fryer configurations from which to choose, with most models incorporating cool zones. Installation requires compliance with regulations regarding gas piping, combustion air, and venting of combustion gases. Atmospheric gas burners are the most common but infrared burners and catalytic heaters are also available. Gas fryers with heating tubes and atmospheric burners are about 40% efficient in transferring heat to oil. Incorporating a "turbulator" or baffles in the tubes can increase the heating efficiency to about 50%. Infrared and catalytic burners further improve heating efficiency to the order of 70%.

In general, electrically heated fryers are easier to install and have an advantage when convenience is a major consideration, particularly when smaller fryers are used. Gas fryers are usually preferred for high performance and heavy-duty usage.

Fryers with automatic melting cycles are available on a number of gas and electric fryers. Melt cycles pulse heat for short periods of time followed by delay periods so that the heating surfaces are not excessively heated. Pulsed heating can limit oil scorching and moderate heat stress on the fryer, but does not eliminate risks. The fryer still must be carefully packed when loaded with shortening. If shortening bridging over heating surfaces occurs, oil scorching and/or equipment damage can still result. Fryers with melt cycles should be considered when fats that are solid at ambient temperatures are used.

Other fryer features, including programmable computer systems for frying various products, automatic basket lifts, heat monitors to prevent temperature overshoot, timers, and electronic ignitions are also available.

## Oil Filters

Filter systems should be considered at the same time that fryers are selected. Filter operation should be made as convenient as possible to support frequent use. Built-in filtration systems that will serve either one fryer or an entire bank of fryers are available. Portable filters are also available from most of the fryer manufacturers as well as from a number of independent foodservice equipment companies.

Oil filtration is an important component of fryer oil management and proper usage can help extend oil fry-life. Crumbs and other particulates should be removed from the oil in a timely manner to minimize their contribution to dark color, high fatty acid, and scorched or burned flavors, among other problems. In foodservice operations, oil filtration can support extended fry-life but should not necessarily be expected to allow for continuous oil usage without ever changing oil. The slow rate of oil turnover, accumulation of contaminants, and buildup of degradation components necessitate renewal of the oil on a periodic basis in most operations.

Filtration equipment and filter aids should be selected based on the extension of oil fry-life they provide. Tests should be conducted to verify performance improvement and guard against any adverse results. A basic filter system that is convenient to use should be tested first. More elaborate systems and filter aids can then be tested and selected as warranted. The extension of fry-life that can be achieved depends on the specifics of each operation, but if a filter system can provide a 20–30% improvement, that is sufficient to justify cost for most operations.

## Selecting Frying Fat

A large number of fats and oil products are used for frying, from heavy-duty shortenings to light-duty oils. Selecting a specific frying oil for a particular application requires a balanced approach with due consideration for ease of handling, fry-life, specific product requirements, eating qualities, product appearance, and cost. General comments are presented below to aid in the selection of frying fat.

### Fry-Life

Heavy-duty frying shortenings with Oil Stability Index (OSI) values in the 85–125 h range (or AOM stability in the 200–300 h range) and melting points generally in the 103–110°F range provide long fry-life. Fatty acid components of these shortenings are very stable and do not contribute to development of objectionable flavor/aroma during normal fry-life. Heavy-duty shortenings are designed for extended use but still require optimum fryer management and frequent filtration to achieve maximum performance.

High-stability liquid oils, with frying performance similar to that of the heavy-duty shortenings, are available. Historically, partial hydrogenation and fractionation have been used to produce the high-stability oils. More recently, new varieties of oil seeds with high-oleic (i.e. 75–90%) content have been developed that provide good performance in heavy heat-stressed applications.

Regular frying shortenings generally have OSI values in the 20–40+ h range. Included in this group are the all-purpose shortenings, those made for both frying and baking applications. These shortenings have good oxidative and flavor stability but provide shorter service than do the heavy-duty shortenings.

Oils with OSI values around 20 h are liquid counterparts to regular frying shortenings. Historically, these oils have been produced by partial hydrogenation and winterization, but with recent advances in seed breeding, new varieties of oilseed yielding mid-oleic (i.e. 50–75%) oils with good oxidative stability are now available.

Fluid, opaque shortenings, formulated by incorporating a small percentage of higher-melting fat in a liquid oil base, are being used in some foodservice operations. The key to the fry-life of an opaque shortening is the oxidative stability of its liquid component, which can have an OSI value ranging from as low as 9 h to over 20 h. The lower stability fluid shortenings have limited fry-life and can develop objectionable flavors relatively quickly when subjected to heavy heat stress in batch frying applications. Salad oils are also being used in foodservice frying, primarily as minimum-cost options. Salad oils can develop objectionable flavors rapidly. These oils may be functional for single or short-term use, but are not well suited for use in regular foodservice frying.

Another aspect that should be considered when selecting frying fat is the time and effort required for cleaning fryers. The lower stability oils, particularly salad oils, develop more polymers than high-stability oils do. Polymerized oil builds up on the fryer and frying equipment, making cleaning much more difficult and expensive.

A number of synthetic antioxidants are approved for use in edible oils and can be added to increase oxidative stability prior to use. These antioxidants primarily provide protection for oil during shipping and storage, but during frying, steam distillation can reduce concentrations to an ineffectual level of a few parts per million after a few hours of continuous usage.

Food-grade methyl silicone (dimethylpolysiloxane) is often added to frying fats and can extend fry-life. Silicone was originally added to frying oil to control foaming, but it was soon noted that the additive extended fry-life. Silicone was originally added to frying oil to control foaming, but it was soon noted that the additive could also support a modest improvement in fry-life. The result is that the addition of methyl silicone to frying fat represses foaming and increases fry-life.

## Special Requirements

When selecting frying fats, specific performance requirements that may be needed for a particular type of food should be taken into account. Special performance requirements are application-specific and may involve product appearance, seasoning adhesion, mouthfeel/eating quality, and fat retention, among others. For example, when doughnuts are prepared and served fresh, good-quality pourable frying oil can be used. Pourable oil is easy to load into the fryer, heating does not require special attention, and frying performance is good. But if the doughnuts are packaged for later consumption, a complication develops. Some of the liquid oil can slowly separate from the doughnuts and pool in the package. Packaged doughnuts should be fried in shortening that solidifies at room temperature.

### Convenience and Ease of Handling

Shortenings are commonly supplied in 50 lb cubes that can be difficult to divide and pack into fryers. The use of shortening requires careful attention to detail during fryer loading and melting to prevent possible damage to both fryer and shortening as noted above. Some shortenings are available in 5 lb bricks that make handling and fryer loading somewhat easier.

### Cost

High-performance frying fats are premium priced but can be cost-effective when properly maintained. High-stability pourable fats are very convenient to use but usually cost more than solid fats with a similar fry-life. Low-stability oils have the lowest purchase price, but they have a short fry-life and can develop objectionable flavors after only limited use. In general, the potential for producing some food with strongly objectionable flavor toward the end of oil fry-life increases as the cost of oil decreases.

### Specifications

Comprehensive specifications should be developed for each frying fat to be used. The specifications should include composition, chemical, physical, and functional requirements as needed to ensure identity and performance. Product brand names provide a convenient reference but should not be used in place of purchasing specifications. Specifications are needed to address any performance, nutritional, or labeling issues that arise. The use of specifications also provide purchasing flexibility and supports competitive pricing.

## Oil Receipt and Storage

It is essential to keep good records of frying oil receipts, including supplier, oil type, lot number, code date, and date of use. Frying fat is a food item as received and becomes part of the fried product served to the consumer. Good records support good order of business, rotation of stock, and provide a track to resolve any performance issues that arise.

Frying fat should be stored under cool conditions, normally below 85°F. Air in the storage area must be dry and free of aromas or undesirable vapors. Plastic containers and polyethylene-lined boxes commonly used to package solid frying fat do not serve as gas barriers. Aromas from seasonings, cleaning agents, fuels, chemicals and combustion gases that have entered oil storage areas have all been known to contaminate frying fat and render it unusable.

A moderate inventory of frying fat should be maintained to ensure supply during heavy usage or possible delivery delays, but the reserve should not be excessive. Under good storage conditions, frying fat does not deteriorate rapidly, but the potential for a problem increases the longer fat is stored. Maintaining a moderate inventory, as needed to ensure supply, supports fresh stock and limits risk after receipt.

## Fryer Oil Loading and Melting

The proper fill level for oil at frying temperature is marked and labeled in most batch fryers. The fryer is designed and performance specifications are established based on frying with the proper fill level. Overfilling the fryer extends turnover time and can contribute to poor oil quality. Under-filling can cause uneven frying and, in extreme cases when a heating element is exposed, can be a fire hazard.

Loading fryers with fluid oil is very convenient, and heating is uncomplicated. Liquid oil is easily poured into the fryer and it heats uniformly, aided by convection currents that move freely over heating surfaces. During fryer filling, allow for a 10% increase in volume as the oil heats from room temperature to frying temperature.

Loading and melting solid shortening in a fryer needs to be done cautiously. When solid fat is used, it is essential to pack the shortening tightly around the heating elements in electric fryers and tightly against all heating tubes or the heating surfaces in gas fryers. During melting, any cavity that forms around a heating element or any shortening bridging that develops over a heating surface can cause localized overheating and can damage both fryer and oil. Shortening cavity formation is a major concern with electric fryers that have heating elements located high in the fryer. If a cavity forms around the heating element and leaves it exposed, the element can overheat and either burnout or ignite the oil.

With gas tube fryers, the major concern during shortening melting is forming a bridge over a heating tube. The exposed tube can overheat, expand disproportionately with the rest of the fryer, and stress the joint between tube and fryer sidewall. Repeated stress can cause cracking in the joint and lead to oil burning inside a fire tube. The first indication of a crack in a tube is usually a rapid increase in free fatty acids and a scorched flavor/aroma in the oil. A stream of bubbles may also be present.

When a fryer with a melt cycle is used, the frying fat temperature will usually be just below 200°F when fully melted. When making adjustments in the fill level, allow for a 6% expansion, which occurs when the fat is heated from 200 – 350°F to initiate frying.

# Frying Temperature

The normal temperature range for foodservice frying is 325–375°F, and the extended range is 300–400°F. When frying in the normal temperature range, most foods cook rapidly; develop a golden color, crisp texture, and good flavor; and absorb 8–25% oil. When the frying temperature is lowered, the trend is toward extended fry time, lighter color, less flavor development, and increased oil absorption.

Frying at higher temperatures tends to reverse these trends and to bring about thinner crusts and less oil absorption. Frying at higher temperatures can also cause the crusts to cook faster than the interiors of some products. If a lower temperature is not selected, crusts of these products will need to be over-fried for the interior to be properly cooked.

For frying potato chips, the use of two different temperature profiles provides an example of how temperature can affect product attributes. Although not in the mainstream of foodservice frying, regular and kettle-fried potato chips provide a readily available reference. Potato slices fried at 350°F, with a normal temperature drop and recovery, will have a delicate texture and light crunch, characteristic of regular potato chips. If the same potato slices are placed in a batch fryer at 318°F and controlled in such a manner that the temperature cycles down to 287°F and slowly recovers to 318°F, the crusts will be thicker and internal structures will be coarser. The chips will have a hard texture and firm crunch in comparison to regular chips.

In batch frying, the oil temperature drops from the set point as frying is initiated, primarily in response to batch size, temperature of product, and moisture content. The temperature drop is normally in the area of 30–40°F, but can be greater, particularly for frozen products. There are no firm rules, but a temperature drop of 50°F or greater should prompt a review. The best guideline is that the temperature should recover to the set point at least by the end of the frying cycle so that the fryer will be ready to fry the next batch.

When selecting frying temperatures, the effects on oil oxidation and fry-life should be considered. The general guideline for chemical reactions is that increasing the temperature by 18°F (or 10°C) doubles the rate of reaction. During the early stages of free radical oxidation, fats closely follow the reaction-rate guideline; therefore, the difference in the rate of fat oxidation at any two temperatures can be calculated with the following formula,

$$R = 2^{(t_1 - t_2)/18}$$

Where,
R = increased oxidation rate
$t_1$ = the higher temperature (in Fahrenheit)
$t_2$ = the lower temperature (in Fahrenheit).

For example, the difference in the rate of oil oxidation resulting from increasing the frying temperature from $325-350°F$ is $2^{1.39}$, which equals 2.6. Therefore, at 350°F, oil oxidizes more than twice as quickly as it does at 325°F. It is apparent that the frying temperature, even within the normal frying range, should be carefully selected to limit oxidation and maximize fry-life.

The midpoint of the normal frying range, 350°F, is a good starting point to establish the frying temperature for new products. If a lower temperature can be used to produce the desired product, oil oxidation is minimized and oil fry-life is extended. If a higher temperature is needed to achieve specific product benefits, the benefits should be sufficient to offset the reduction in oil fry-life.

Foodservice is a highly innovative segment of the food industry. Each year new fried foods are developed and added as menu items, particularly by progressive restaurants and fast-food chains. In support of new product development, any frying temperature that achieves the best flavor, texture, or eating qualities for a product should be used. If the frying requirements are outside normal limits, a fryer designed to meet these specialized requirements should be considered and developed as appropriate.

In discussing frying temperatures, accuracy also must be addressed. Temperature controllers and regulating systems in most fryers are very reliable, but accuracy should be verified regularly. A number of digital thermometers with calibration devices are available for this purpose. Fryers should be checked on a regular schedule to ensure accurate temperature control.

# Batch Frying Practices

## Fryer Start-Up

Heating fat to the appropriate temperature earlier than needed for frying stresses the oil, causing unnecessary oxidation and thermal degradation. To minimize stress, fat should not be heated earlier than needed for frying. Ideally frying should begin as soon as the fat reaches the set temperature.

## Variable Frying Demand

The demand for fried food varies throughout the day, depending on menu and customer eating patterns. Frying demand should be monitored, and a schedule should be developed for fryer heating. Anticipate demand, bring fryers on-line as needed, and take fryers off-line when the demand decreases.

## Maximum Batch Size

Maximum batch size for each product should be set based on food serving size and the rate of temperature recovery in the fryer. Batch size should be adjusted so that the oil temperature recovers to the set point at least by the end of the frying cycle.

There is a tendency in foodservice to raise the temperature set point in an attempt to fry larger batches faster. The practice, however, is not beneficial. The maximum heat input of the fryer, which determines the rate of recovery, is set by fryer design.

If batch size is adjusted so that the temperature just recovers at the completion of the frying cycle, the fryer is operating at its maximum heating capability. Raising the temperature and increasing batch size causes a larger temperature drop, longer recovery time, and oilier product.

## Fryer Idling

During slack periods, fryers that are not needed for frying should be turned off. Idling a fryer by lowering the temperature set point to 200–250°F is not recommended, especially if the fryer is not going to be used for several hours. Maintaining oil temperature any higher than necessary causes unnecessary oxidation and polymerization.

## Utility Fryer

It is not always possible to use separate fryers to prevent the transfer of flavors or other traits. This limitation can occur during high demand, when food must be prepared quickly, or during low demand, when fewer fryers are being used. When such products must be fried in a common fryer, one fryer should be designated as a utility fryer and used accordingly.

## Crumb Control

Effective crumb control is essential in providing good fried-product quality, maintaining oil quality, and achieving maximum oil fry-life. In most cases, careful product handling provides good control, but in some instances, reformulation of the breading or batter may be needed. Procedures required for separating scrap, draining excess batter, removing excess breading, and controlling other particulates before they enter the fryer should be part of routine quality control.

Controlling crumbs before they enter the fryer helps minimize the oil degradation to which they contribute. Filtering crumbs after they have accumulated in the fryer can limit further damage to the oil but does not recover oil quality that has already been lost. All breading, food pieces, and crumbs that separate from food and float during frying should be skimmed from the oil at the end of each batch.

## Oil Filtration

Frying fat should be filtered as often as necessary to prevent crumbs from degrading in the fryer. As crumbs degrade, they contribute to dark oil color, high fatty acid content, scorched or burned flavor, and short fry-life. Generally, oil should be filtered once a day when crumb accumulation is minimal, once a shift in most operations with moderate crumb accumulation, and two or more times a shift as needed when crumb accumulation is heavy.

## Fryer Oil Level

The oil level should be maintained at the normal frying level during production, with regular addition of make-up oil as needed. Toward the end of production, the oil level should be allowed to decrease to the minimum acceptable for frying. This practice allows for fresh oil addition at the beginning of the next production period to enhance fryer oil quality.

## Fryer Shutdown

Fryers should be turned off promptly after the last batch of product is fried. Any delay in turning off the fryer causes unnecessary oil oxidation. After shutdown, the oil should be filtered according to established procedures, and the fryer should be cleaned. Filtered oil may be stored in a separate container or returned to the fryer. If it is returned to the fryer, a cover should be placed over the fryer to prevent the possibility of contamination by foreign material. The cover should be used any time the fryer is not in use.

## Oil Quality Monitoring

Monitoring fryer oil quality and determining when the oil should be changed is critical to maintaining fried product quality. Foodservice operations usually do not have the facilities for conducting standard laboratory tests. As an alternative to standard testing, quality indicators are usually used. Indicators may be based on any aspect or attribute of the oil that can be related to quality, including hours of use, number of times oil has been used, color, smoke evolution, and foam height.

# ·14·

# Industrial Frying

**Don Banks**

*Edible Oil Technology, 8155 San Leandro, Dallas, Texas 75218*

Deep fat frying as an independent business began in the late 1890s, primarily to produce potato chips as a snack food. Initially the chips were fried in open kettles heated on kitchen stoves. As the businesses grew, production moved from home kitchens to garages and later to small factories. When the first factories were built, the same open-kettle concept was used to build freestanding fryers. Fryers were simple in design, consisting of square, rectangular, or round kettles fitted with open gas burners. Brick fireboxes and air blowers were later added to increase combustion efficiency and direct heat to the bottom of the kettles.

In 1929 Freeman McBeth of the J.D. Ferry Co. developed a fryer with a channel design that produced chips in a continuous manner. Chips were moved through the fryer by an elliptically actuated rake system that submerged the chips and moved them forward a few inches at a time. The introduction of the Ferry continuous potato chip cooker marked a major advance in frying technology and inaugurated frying on an industrial scale.

The term "industrial frying" is used to describe high-volume, deep fat frying conducted as a primary business. Industrial fryers have production capacities ranging from about 300 lb/h to thousands of pounds per hour. Production is normally continuous and highly automated. Large batch-frying operations are included in the industrial frying classification.

## Fryer Design

After hot oil circulation and remote heating were added to the McBeth fryer, the basic design features for both direct-fired and remotely heated fryers commonly used today were in place. Subsequently, a wide variety of industrial fryers have been designed to produce essentially every type of fried food from savory snacks, nutmeats, doughnuts, and vegetables to breaded products and tempuras. In addition to traditional snack food and par-fry (french fry) fryers, there are also pressure fryers, vacuum fryers, multizone fryers, shaped product fryers, and a number of others that are now commercially available in a wide variety of sizes and configurations.

Although numerous new features have been built into fryers since the introduction of the McBeth fryer, until recently frying by immersion in hot oil has been the common principle for designing continuous fryers. In the past few years a new type of fry has been introduced that can be used in a traditional manner for immersion frying and can also be used for frying with curtains of oil, as will be discussed in more detail below.

The new methodology involves frying by conveying product through multiple sheets or curtains of hot oil. The oil curtains in the fryer, patented by Heat and Control, are formed by laminar flow from "overflow weirs" (see below) located above the fryer conveyor belt. The hot oil flows over the product, passes through a porous conveyor belt, and flows down to a pan from where it is recirculated. To support uniform color development on the bottom of some products, a perforated baffle plate under the product conveyor can be employed as a "puddler" panel to dynamically retaining a 1/8−3/16 in. depth of oil from the flowing curtains.

This method can be used to fry a variety of products, including delicate items coated with breading or tempura. The fryer design provides for product items to be loaded onto a conveyor belt, moved through the fryer without contacting other product pieces, and discharged with relatively little loss of coating or surface disruption. For products such as nutmeats, firm collets, and others that are not subject to contact damage, the fryer belt can be loaded as heavily as desired up to the capacity of the fryer. Frying can then be conducted with minimum oil using the multiple oil curtains system, or alternatively, oil can be retained in the fryer oil pan and adjusted to any appropriate level for partial or full immersion frying. Frying times are generally extended about 25% using the oil curtain methodology as compared to full-immersion frying.

The overflow weirs in the Heat and Control fryer extend across the full width of the fryer and have a cross-sectional shape similar to a rounded capital letter "M." Oil is continually supplied to the middle of the "M" and as it overflows, two curtains of oil uniformly flow over the product being fried. The number of overflow weirs to be built into in a fryer will depend on the products that are to be fried. Similarly, the spacing between the weirs can be designed as needed.

For breaded products an overflow weir spacing of 12 in. is commonly used. The spacing provides for a brief lay period as product is conveyed from weir to weir. The lay period allows for steam evolution in a somewhat gentler manner than would occur with continuos immersion frying and can reduce loss of breading. For frying peanuts and other nutmeats, closer spacing between overflow weirs can be used.

When the fryer is setup for oil curtain frying, the total volume of oil in the system is about half that of a comparable immersion fryer. The reduced volume supports a rapid rate of oil turnover. With the inclusion of a full-flow oil filtration unit and a good fryer management program, oil quality may be sustained on a continuing basis.

Residence time in the fryer can be set as needed, generally from 20 seconds to about eight minutes. Fry time can be adjusted for full frying or par-frying to meet product requirements. Par-frying is commonly used for breaded meats and other products that may be processed further before packaging and then stored under refrigeration. The products are then ready for distribution to restaurants and fast-food operations for finish cooking and serving.

# Frying Overview

Before fryer selection and operating procedures are addressed, the process of industrial frying is reviewed below. For focus, savory snack frying is used as a general reference, and potato chips, the leading product in the category, serves as a more specific reference. Oil uptake by the fried product is also briefly addressed.

For this discussion, an industrial fryer is viewed as a series of zones where product handling and processing conditions are controlled to regulate frying. On a somewhat arbitrary basis, six zones or fryer areas can be designated as follows: (i) entry, (ii) case hardening, (iii) shape firming, (iv) cooking, (v) finish frying, and (vi) takeout.

## Fryer Entry Area

During the first few seconds the product is in the fryer, it is rapidly heated and the starch begins to gelatinize, first on the surface and later through to the interior. Within the next few seconds, the product is covered with small steam bubbles when moisture on the surface of the product reaches its boiling point. Initially, the steam bubbles that form and break away from the surface are quite uniform in size and distribution. The boiling action helps keep the product pieces separated and, along with the flow of the oil, prevents clump formation. Mechanical pressure, or even pressure due to excessive oil flow, that would force the product into close contact must be avoided in the entry area. The entry of the fryer can be viewed as a free frying zone where product pieces move freely with the aid of a smooth oil flow and the boiling action of steam evolution. At the end of the free-frying zone, a "breaker bar" or other device may be used to separate any clumps that have formed.

## Case-Hardening Zone

As the product leaves the entry area, a paddle wheel (or other device) is used to facilitate uniform product distribution across the fryer, regulate flow, and briefly submerge the product. As frying continues, the bubble pattern on the product surfaces begins to change. Some areas begin to emit larger and more frequent steam bubbles, while other areas exhibit slower bubbling.

The larger bubble sites mark surface points where steam channels have erupted. The steam emitted from the subsurface cells generates pressure within the product. As pressure builds, intercellular separation occurs, forming channels that break through to the surface and allow the steam to escape. (When intercellular adhesion is weak, bilateral separation/delaminating occurs and blisters form.)

The smaller bubble points, which show gradual reduction in steam emission, reflect the dehydration of cells on and close to the surface. In potato chips, these cells dehydrate, flatten in a brick-like arrangement, and begin forming the outer layer of a crust. As more surface cells dehydrate and flatten, the chips begin to case harden. When case hardening is complete, product pieces have less tendency to stick together, but they are still pliable and malleable.

## Shape-Firming Zone

Additional paddle wheels are used to further regulate product flow, supporting product "bedding" across the fryer, and provide for surface frying, which protects against product deformation and clumping. In this zone of the fryer, the large bubble sites on the product surface continue to emit steam, while the smaller bubble points further diminish. Crust building occurs as an additional few layers of cells dehydrate, flatten, and add to the crust structure. Crust building is influenced by frying temperature and is a few (flattened) cells thick when higher temperatures are used; it can be twice as thick with lower temperatures.

As frying continues, cells below the crust layer dehydrate and become fixed. Frying temperature and temperature profile significantly influence how they become fixed. Under nominal frying conditions, cell walls are distorted but retain a generally cellular structure, interlaced with some voids. When a temperature profile, typical of kettle frying, is used, the interior loses almost all of its original cellular structure and is changed into a porous, cavernous structure with irregular wall thicknesses. The walls can be several (flattened) cells thick in one area and taper to single-cell thickness in others.

## Cooking/Moisture Reduction Zone

In this zone, the surface of the product is no longer sticky, and the shape is relatively firm. The product can now be held below the surface of the oil, usually with a submerger belt, to facilitate cooking and moisture reduction.

As frying continues, steam evolution slowly subsides. The temperature of the product, which until now has been maintained at the boiling point by steam evolution, begins to increase, initiating the final phase of frying.

## Finish Frying Area

Finish frying is a very dynamic and complex part of production, with regard to fryer control and changes that occur within the product. Finish frying begins as the product approaches the end of the submerger belt, and continues until the product is removed from the fryer by the takeout conveyor.

If product were removed from the fryer just as finish frying begins, it would have little flavor, light color, poor texture, and low oil content in comparison to standard product. It is the dehydration of the product and the final temperature rise that support the reaction chemistry generating much of the flavor of the finished product. Movement of the product through the finish frying area of the fryer must be carefully controlled to regulate flavor, support texture, and influence oil uptake.

### Product Takeout

Active frying is completed when the product is removed from the oil, but the product remains hot for a period of time and continues to cook. The final product moisture, texture, flavor, color, and oil content are all influenced by how the product is handled on the takeout conveyor.

## Oil Uptake

Oil uptake by fried foods involves several mechanisms including surface wetting, capillary action, and vacuum absorption. Upon entering the fryer, the surface of a product is coated with oil by wetting action, and, depending on the structure of the surface, some oil may be absorbed by capillary action. As the product is heated and begins to emit steam, the escaping gas (water vapor) sweeps oil away from the surface and possibly, albeit temporarily, reverses some oil uptake.

When steam evolution subsides, oil can again be absorbed through capillary action and surface wetting. As the product is removed from the fryer, oil wetting the surface is carried along. The product, depending on its structure, shape, orientation, and loading the takeout conveyor can also carry out additional oil.

As the product is first removed from the fryer, the oil on the surface is readily visible, giving the product a wet appearance. As cooling occurs, the residual water vapor in the product contracts, creating a vacuum that pulls the surface oil inward. Upon further cooling, the product develops a noticeably dryer appearance.

A significant portion of oil uptake stems from the surface oil carried out of the fryer with the product and the vacuum associated with chip cooling. Managing the surface oil allows either an increase or decrease of the oil content in the finished product. To increase uptake, additional oil can be sprayed on the product to wet the surface a second time. As the product cools further, the additional oil is vacuum-absorbed. If reduced oil content is desired, the product can be maintained at temperature as it leaves the fryer, and the surface oil can be largely removed by blowing with a stream of hot high-velocity gas. Some of the early commercial processes utilized a stream of heated air to remove surface oil, but they caused rapid oil oxidation and reduced product shelf life. Units developed more recently utilize a dry-steam process in conjunction with an oxygen-free chamber. Centrifugal force is also used in some operations to reduce oil content.

## Fryer Selection

The fryer is the heart of an industrial frying operation, and its selection plays a major role in controlling the product. There are many fryers available to choose from, to fry almost any product, from salty snacks and par-fries to vegetables and breaded entrees. Fryers in any one category may appear similar, but there can be definite differences in configuration and operating capabilities.

When requirements for the fried product are general in nature and the attributes have nominal ranges, then most fryers in a given category can produce an acceptable product. As the list of required attributes for a product becomes more detailed and as the acceptable ranges narrow, fryers with more specific configurations and operating capabilities are required. Special requirements may involve any aspect of the fryer, including temperature profile, multiple heating zones, product shaping, extended free frying, and monolayering, as well as various combinations of clump breaker bars, spinners, paddle wheels, rakes, submergers, takeout conveyors, and so forth.

For optimal fryer selection it is essential to know exactly what product is to be produced and to be able to measure its attributes. Until the target product is defined in terms of measurements and numbers, it is not sufficiently defined to direct optimal fryer selection. To define the product, critical quality attributes must be listed along with target values, control limits, and the test methods to be used for analysis. The frying requirements needed to produce the target product should then be outlined, identifying how each of the attributes is to be developed and controlled during frying. Fryer selection then becomes a process of choosing the best match between fryer and product.

Fryer manufacturers offer a choice of fryers, ranging from general purpose to highly sophisticated product-specific fryers. For most applications, a fryer that can produce the desired product is commercially available. If a suitable fryer is not available, consideration should be given to designing an appropriate fryer and having it custom built.

## Sizing the Fryer

After fryer configuration and operating requirements have been established, it is essential to size the fryer properly. The best quality products are produced when fryers operate at full capacity and on a continuous basis. Operating a fryer outside of its design capacity can cause both unacceptable product quality and equipment failures, with risks increasing in proportion to the deviation from design capacity.

Operating a fryer above its rated capacity changes the frying temperature profile, which can adversely affect flavor, resulting in a relatively bland product with a baked flavor characteristic. Operating a fryer above the rated capacity also forces excessive heat input, which can lead to early equipment failure.

Operating a fryer below its rated capacity reduces oil turnover and increases oxidative stress. Extended operation below the rated capacity causes fryer oil quality problems, product flavor problems, and reduced finished product shelf life.

Some allowance in sizing a fryer is needed to provide flexibility in production and future market growth. The flexibility should be planned around the number of hours a day and the number of days a week the fryer will be operated, but all production should be based on operating the fryer at its designed capacity.

## *Fryer Oil Turnover*

Oil turnover is usually expressed as the number of hours required for the make-up oil added during frying to equal the total quantity of oil in the frying system. The make-up oil is added to replace the oil taken out of the fryer as part of the fried product. Oil turnover can be calculated as follows:

$$\text{Fryer Oil Turnover} = \frac{\text{total pounds oil in frying system}}{(\% \text{ oil in product})(\text{fryer output: pounds/hour})}$$

Industrial fryers commonly have oil turnover rates ranging from 5 – 10 h, with some fryers ranging a few hours higher. Oil turnover is a good general indicator of the oxidative stress that oil is subjected to during production and is an important consideration in fryer selection. A specific rate of turnover is not very meaningful, however, until other factors are considered.

The product fried, type of oil, fryer design, heating system, and operating conditions all must be considered before the turnover rate can be used to assess oil stress. Some frying systems operate without difficulty with an oil turnover rate of 12 h, while other operations, involving different products and fryers, encounter problems when the turnover rate exceeds 7 h.

The oil turnover rate should not be misconstrued as the time required to replace all the oil in the fryer with fresh oil. Commingling of the fresh make-up oil with the fryer oil precludes an exchange on a one-for-one basis. After one turnover, one-half of the original oil remains in the fryer and is further halved with each subsequent turnover.

Usually, after four turnovers, the oil analytical values begin to equilibrate. The equilibration reflects blending of the fryer oil and make-up oil at a constant rate, resulting in steady state frying. The turnover rate and time required for oil equilibration are essential considerations when conducting either oil tests or finished product evaluations.

# Selection of Frying Oil

A number of factors, including product fried, design of fryer, rate of oil turnover, flavor and eating characteristics of the product, shelf life requirements, and historical usage influence the selection of frying oil. The different aspects of oil selection can each be individually assessed and then used collectively for oil selection.

Evaluating frying oil performance requires extended test frying and analytical evaluation. When fresh oil is first placed in the fryer, it has its full compliment of antioxidants (synthetic and/or natural), and is free of degradation products. Evaluating the oil or testing the finished product before the oil reaches frying equilibrium, as indicated by steady state analytical values, gives misleading results.

In most industrial frying operations, the fryer oil analytical values begin to approach equilibrium after four oil turnovers, as noted previously. Maintaining an analytical record of the oil as it approaches equilibrium can provide useful preliminary information, but before collecting product for flavor and shelf life testing it is essential to wait until the oil has fully equilibrated. The analytical testing usually includes color, flavor/aroma, free fatty acid content, and oxidative stability. Additional testing, including polymers, carbonyls, para-anisidine, oxidized fatty acids, conjugated dienes, and others may also be used.

When there are no unusual conditions inherent in a frying operation, a number of oils may be candidates for usage. Under these conditions, the oxidative stability of the oil, shelf life of the product, and rate of fryer oil turnover are the initial considerations for oil selection.

After oils meeting the basic requirements for frying have been identified, oil flavor and eating qualities must be assessed. Frying oil contributes flavor components and flavor precursors to the food being fried; however, most commercial frying oils are processed to be as bland as possible. Due to this bland flavor, fresh oil can actually dilute and diminish the flavor of fried foods during the early stages of frying.

As frying continues, product flavors and aromas are concentrated in the oil, and it soon smells and tastes like product being fried. At the same time, the frying oil is subjected to hydrolytic, oxidative, and thermal changes, which can contribute to the flavor profile of the product. With a good fryer and favorable operating conditions, the frying oil carries and supports the full, rich flavor and aroma of the product. The oil type, design of the frying system, and operating conditions can significantly affect, for better or worse, the flavor contribution from the oil.

In addition to flavor, oil significantly affects other important eating characteristics, including lubricity, mouth feel, flavor release, and a rich-eating quality. Lubricity and mouth feel reflect the physical nature of the frying medium, ranging from juicy for liquid oil, through creamy for soft fats, to dry and then waxy for fats with higher solid contents and elevated melting points.

Melting characteristics of the frying fat also influence the rate of flavor release. When the fat is liquid, the full flavor of the product is readily available at the beginning of mastication. When the fat in the food is solid, the flavor intensity and speed of release is diminished, depending on composition and melting characteristics of the particular fat used.

Product shelf life requirements vary, depending on the nature of the product, packaging, ambient conditions, and code date policies. Some generalizations can be made for preliminary screening, but matching frying fat to product shelf life requirements still must be done on a case-by-case basis. Flavor and oxidative changes are important in the selection, but physical considerations, such as oil migration, fat crystal changes, and related appearance attributes, also must be evaluated.

# Bulk Oil Storage

Bulk oil storage is an integral part of an industrial frying operation. Bulk handling provides a cost savings of several cents per pound and is not complicated. The systems must be designed and installed properly, but support is readily available.

One of the most important considerations in using a bulk oil system is to size the storage tank for rapid oil turnover and frequent fresh oil shipments. When oil shipments are received on a one- or two-week basis, rapid turnover almost guarantees good storage oil quality. When reasonable storage conditions are maintained, oil quality problems rarely occur.

The Peroxide Value (PV) can be conveniently used to monitor storage-tank oil quality, provided the oil was thoroughly evaluated on receipt and that the initial PV was below 1.0 meq/kg, preferably below 0.5 meq/kg. With proper attention to temperature control, and avoiding aeration or contamination, the PV should normally be below 1.0 at the end of the first week and below 2.0 meq/kg at the end of the second week. Commingling new shipments of oil with existing inventory should be monitored and managed but is not a major limitation when the PV of storage oil is less than 2.0 meq/kg.

When oil is routinely stored longer than two weeks, the potential for quality problems increase. Three weeks represents the dividing line, when system design and operating procedures become much more critical. Nitrogen protection should be included for bulk systems when storage time is expected to approach or exceed three weeks.

With adequate tank insulation and heating, it is rarely necessary to mix or circulate bulk oil. Circulating oil during storage rapidly increases oxidation and should be avoided. Nitrogen protection can limit the risk, but if any oxygen enters the system, rapid oxidation results. If mixers are required for a particular installation, they should be operated by interval timers and set for minimum mixing and maximum lay times that can be used.

# Preparation for Fryer Start-Up

## Fryer Equipment Check

With the exception of the oil circulation pump, essentially all equipment associated with a fryer can be checked and tested prior to filling the fryer with oil. It is important to test conveyors, paddle wheels, spinners, submergers, takeout conveyors, and related equipment before oil loading to avoid unnecessary aeration and oxidation. All equipment should be turned off after testing and remain off until needed for production.

The fryer oil circulation and heating system should be checked only as part of the production start-up procedure. Loading, circulating, and heating oil earlier than necessary to start production is highly detrimental and should be avoided.

## Fryer Oil Loading

To avoid unnecessary risk of contamination or abuse, fryer oil should not be loaded earlier than needed to meet the schedule for initial oil heating. When the fryer is filled, oil loading should be controlled manually to a designated level that allows for both thermal expansion and displacement by the product at the start of production. The automatic oil level control should not be used until production starts to avoid overfilling the fryer. When correctly loaded, the oil level will be just above the minimum acceptable frying level as product is first produced. Minimum oil loading supports fresh oil addition at the beginning of production and avoids the protracted oil turnover that would result from overfilling.

## Oil Heating

Oil heating instructions for the fryer should always be carefully followed. Fryer manufacturers usually include instructions for preheating the oil as a precaution to remove any moisture in the system. The procedure usually requires heating the oil to 95–100°C and maintaining the temperature until all moisture boils away.

Preheating should be scheduled and coordinated as part of the start-up procedure. After preheating, the oil should be heated directly to frying temperature. There should be no delays in starting production, since any interruptions cause unnecessary oil aeration and degradation. The time required to complete oil heating should be determined for each fryer and specified as part of the start-up procedure. The oil should reach frying temperature just in time to start production on schedule.

## Production Start-Up

Heating and circulating oil in preparation to startup production is a period of heavy stress. To minimize oil oxidation and degradation, it is essential to start production as soon as the oil reaches frying temperature. The automatic oil level control system should be turned on as part of the start-up procedure. During fryer filling, the oil level should be managed as noted above so that fresh make-up oil will begin entering the fryer as soon as production begins.

## Fryer Control

Fryer control has long been an art practiced by the fryer operator. When fryers are of moderate size, a skilled operator can monitor the frying system and make adjustments based on experience and analysis of the finished product to maintain control of the fryer. The practice is far from perfect but does provide satisfactory control for many operations.

One of the main difficulties with operating a fryer by art is that control is focused on fryer output and lags behind any changes at the input of the fryer. A skilled operator can make some adjustments based on changes at the input end of the fryer, but control remains primarily a reactive type of operation based on variation of the finished product. Since the size of fryers has increased over the years, operating by art is more difficult, and the risk to product quality has correspondingly increased. When chip fryers have capacities above 3,000 lb/h (par-fry french fry operations have capacities above 30,000 lb/h), the risk of producing even a few minutes of product that does not meet quality standards is not acceptable. Better methods of controlling fryers were needed and have been developed.

Leading fryer manufacturers have developed proprietary software to run fryers by computer control and some food companies have developed their own systems. The programs are designed to maintain fryers in a steady state condition, balancing variations in the raw product and feed rate with control of operating parameters. Focusing on input to the fryer and making adjustments concurrent with changing demands can produce more consistent products. Under optimal computer control conditions, the exit of the fryer becomes a quality-monitoring station rather than a fryer control point, yielding a more consistent and higher quality product.

In the absence of computer control, some companies have developed a set of systematic operating procedures for improved control of larger fryers. Energy requirements and operating parameters are determined for the full range of variation in raw materials and translated into fryer settings. During production, operating parameters can be adjusted to the predetermined settings as changes in the raw materials occur. Systematic fryer operation can work quite well and provides an economical method to improve product quality.

Arranging for production with large, uniform lots of raw materials should not be overlooked as a method for improving fryer control. With a consistent feedstock, the fryer can be set up for optimal quality and then remain essentially unchanged throughout the entire production period. Depending on the product, many methods can be used to control feedstock, including the use of exacting raw material specifications, sequencing closely matched lots, and blending small lots into large, uniform lots. The key is to actively manage the feedstock rather than accepting random variation. Managing the feedstock requires little expense and pays large returns in both production efficiency and finished product quality.

## *Fryer Operation*

Production should be adjusted to the designed capacity of the fryer at start-up and maintained at capacity during operation. Operating a fryer below capacity, above capacity, or on an intermittent basis all lead to problems. When fryers are operated intermittently or below capacity, excessive oxidation occurs and problems involving frying oil, product flavor, and shelf life result. Operating a fryer above capacity increases heat-stress, leading to early equipment failure and causes inconsistent flavor due to product overloading. The highest quality and most consistent product is produced when fryers are operated at full capacity on a continuous basis, stopping production only for scheduled maintenance and cleaning at the end of each week.

When continuous production for a full week is not warranted, the next best option is to operate continuously as long as necessary to manufacture all product needed for the week, and then stop production.

For business and staffing reasons, new operations commonly schedule production on the basis of an eight-hour day. Start-up and shutdown of a fryer on a daily basis induces much more stress on the oil than would occur under continuous production and causes more variation in product quality. When a daily schedule is used, comprehensive oil and fryer management programs must be carefully developed and fully utilized to produce quality product. Operating a fryer a few hours a day or a few hours every few days is not recommend and should be avoided if possible.

## Interruptions in Production

Any interruption in production, when the oil is circulated at frying temperature and no product is being fried, creates a high-risk period for the oil and finished product. The risk increases sharply as downtime extends. In industrial operations, increased consumer complaints can be correlated with product made following interruptions in production exceeding 20 min. Every interruption in production, regardless of whether the fryer is directly involved or not, places the frying oil and product quality at risk. Each interruption should be managed as a potential fryer problem. When it is apparent that an interruption will occur and last longer than a few minutes, a partial fryer shutdown should be initiated.

Specific procedures must be tailored for each operation, but one of the first steps to be taken is to stop the addition of make-up oil. Continued oil circulation is necessary, but oil heating should be discontinued. When the last product is removed from the fryer, all paddle wheels and similar equipment not required for oil circulation should also be stopped. Check to ensure that no product remains in the fryer. Next, determine how long the downtime will last and make a decision about proceeding with a full fryer shutdown.

The preferred management option is to avoid downtime. An example of an interruption that can be avoided is the downtime that results when all workers begin and finish a shift at the same time. Changing workers in relay or overlapping shifts can easily avoid the downtime. When downtime cannot be avoided, it should be limited to as short a time as possible. For example, if a product changeover regularly occurs that requires an equipment change, a procedure with specific time requirements, not exceeding 20 min, should be employed.

## Fryer Shutdown

Fryer shutdown is another period of high stress. The oil is hot, circulating, and no product is being produced. To minimize oil degradation, many steps can be taken before, during, and after production.

One of the first steps to take is to discontinue the addition of make-up oil at an appropriate time before the end of production so that the oil is close to the minimum level for frying when production is completed. This procedure limits the volume of oil that is subjected to oxidation during shutdown, storage, and the next start-up.

As the end of production approaches, maintain oil circulation but discontinue heating. The latent heat in the system will maintain the oil temperature for a period of time and can support normal frying for several minutes. With some experimentation, the proper time to discontinue heating so that the oil is at the minimum frying temperature as the last product leaves the fryer can be determined. This procedure prevents unnecessary heating at the end of production, avoids a temperature spike, and promotes rapid cooling of the frying system following shutdown.

When the last product leaves the fryer, maintain oil circulation, but turn off all paddle wheels, submergers, and other equipment not needed for circulation to avoid unnecessary aeration of the oil. The air blower in the heating system should remain on to facilitate fryer cooling. Ensure that no residual product remains in the fryer. Briefly operate the product conveying equipment if necessary to clear any residual product.

The next step is one of the most important. Oil circulation must be stopped as soon as the allowed temperature limit is reached. The highest shutdown temperature that the fryer manufacturer allows should be used, as any additional circulation causes unnecessary oil abuse. The responsibility for stopping oil circulation at the proper time should be specifically assigned, and a documented procedure should be used. The quality of the oil and the quality of the product produced at the start of the next production are largely determined by the length of time the oil is circulated at the end of the previous production.

Circulating the oil after the end of production is not done to benefit the oil but as a requirement of the fryer manufacturer to prevent possible warping or other damage to the frying system. From the perspective of oil quality, a brief period of circulation to ensure that all of the oil-handling system is at or below the frying temperature is the only circulation that benefits the oil. Do not make any changes in the post-frying oil cooling procedure without the full consent of the fryer manufacturer.

## Oil Cooling

Oil coolers (water-to-oil heat exchangers) are now being used to rapidly cool and facilitate oil transfer to storage for some of the larger fryers. The coolers provide additional protection against oxidation, but their use also ensures that fryer operators actively manage the oil until it is safely in storage. The quality benefits from using the coolers are moderate in comparison to a well-managed fryer shutdown program, however, due to time and management benefits associated with their use, oil coolers should be considered when the cost can be justified by reducing risks and improving product quality.

## Oil Transfer

If production is to be restarted the day following shutdown, oil can be left in the fryer. If production is not to be resumed within 24 h, the oil should be transferred to the fryer oil storage tank as soon as it reaches the transfer temperature. The transfer temperature depends on the equipment, piping, and procedures used for each operation. Filtration of the oil during transfer provides a number of benefits and is recommended.

## Oil Filtration

Industrial fryers normally have crumb removal belts or screens, located ahead of the oil circulation pumps that are used to separate large particulates and protect heat exchanger tube bundles from becoming clogged. The basic crumb control systems do not remove small particulates from the oil. Historically, with the notable exception of par-fry french fry operations, continuous fryer oil filtration has not been widely used. More recently, the use of fryer oil filtration systems has become more common. Increasingly, filters are being purchased for use with new fryers and for retrofitting existing operations. Driven by competition, companies have installed filters to improve product appearance and to gain general quality improvements. In addition to improving product appearance and oil color, filtration can support lower free fatty acid content, lighter fried flavor, and shelf life improvement. In general, the use of filters in industrial frying can make a good operation better, but should not be expected to control major problems inherent in a frying operation.

When considering the installation of a filter system, the source of particulates should first be investigated and controlled to the extent possible by making processing and procedural changes, then filtration can be assessed to achieve further improvement. Most industrial fryers will tolerate a relatively heavy crumb load before problems, such as high fatty acid, limit production. When filter systems are evaluated, the potential downside risks, including increased oil volume (that extended oil turnover), aeration, and possible addition of contaminants, must be evaluated to guide proper filter selection.

## Fryer Oil Storage

The PV of the fryer oil storage tank provides a good indication of how carefully fryer shutdown and oil transfer were managed at the end of the previous production. The PV should be less than 5.0 meq/kg after the tank has cooled to ambient temperature. A value below 5.0 meq/kg, preferably about 3.0 meq/kg, reflects good shutdown practices and management. If the PV is greater than 5.0 meq/kg but less than 10.0 meq/kg, the oil was subjected to abuse, the cause of which should be identified and corrected. If the PV is above 10 meq, the oil was heavily abused, and a thorough investigation is needed to correct the problem and prevent recurrence. The oil should be placed under a quality hold pending comprehensive analysis and quality assessment to direct disposition.

# 15

# Practical Foodservice Frying: Troubleshooting

**Michael D. Erickson**

*Cargill, 600 N. Gilbert Street, Fullerton, CA 92837*

Of the billions of pounds of frying fats produced annually, more than half are used in restaurants. Since frying fats are central to the success of most restaurants, there exists a need to provide dependable, high-quality products that also reflect an appreciation of growing consumer awareness. Restaurants that are most successful recognize the genuine need for relying on their suppliers, not only for consistently high-quality frying fats but also for sound technical service. For frying applications, a large part of technical service amounts to troubleshooting and solving field problems.

Although the literature offers volumes of information and data related to laboratory frying or heating experiments (1–7), only limited information drawn from actual field experience is available. The objective of this chapter is to offer a practical, systematic approach for identifying and resolving perceived frying problems commonly encountered in foodservice frying. Most frying problems are operational and can be categorized as follows: premature foaming, premature smoking, premature darkening, off flavors and odors, atypical frying performance, and other situations. Although these categories suggest individual causes for frying problems, it is rare that the source of a problem affects only one of these manifestations.

## Premature Foaming

Often a distinction between excessive bubbling and true foaming must be made. When bubbling action associated with normal frying becomes exaggerated and mistaken for foaming, it is usually due to a combination of thawed product fried in a fryer filled with shortening well above the recommended fill level. Thawed product puts less of a demand on the fryer than normal, allowing faster temperature recovery. This fairly rapid temperature recovery serves to accelerate the elimination of surface moisture on the product creating large, "boiling" bubbles. An overfilled vat adds to the situation by containing more shortening to bubble up. When excessive bubbling occurs with frozen product in fryers filled correctly, it is usually due to either overloading the baskets, frying above the recommended temperature, or a combination of both.

True foaming is best described as resembling beer foam. When the shortening integrity begins to diminish, the bubbles associated with normal frying activity are accompanied by small, densely packed pockets of amber-yellow foam. These small pockets of foam then persist when the product is removed from the fryer. If there is more foam than bubbles when the product is being fried, clearly the shortening has passed the point of discard. Further substantiation of this is usually seen in the quality of the finished product.

## Salt

In most premature foaming complaints involving a product requiring salting, the fryer nearest the salting station is perceived to have the problem. When normal "rushes" (surges in patronage through out the day) are at their peak, liberal applications of salt afford the opportunity to inadvertently introduce salt into the fryer.

One possible explanation that would account for the detrimental effect of introducing salt into a fryer is that salt ionizes in the presence of moisture liberated from the product being fried. This ionization provides the opportunity to form soaps when it contacts acidic material resulting from coincidental hydrolysis. Furthermore, salt is a potential source of trace metals known to accelerate deterioration.

## Polymerized Oil

Polymerized oil is the brown or amber gum-like material that usually accumulates on temperature-sensing probes, heating elements of electric fryers, around the perimeter of the fryer at the fill line, and even on frying baskets. This material is highly broken-down oil resulting from a combination of inadequate cleaning and prolonged exposure to frying temperature. Exhaust fans over fryers allow volatile material liberated from the surface of the shortening to condense on filter screens and on the inside lining of the hood. If left unaddressed, condensation could accumulate to a point where it begins to drip back into the fryer. This would be analogous to adding a concentrated form of spent (discarded) oil to usable shortening.

## Boil-Out Compound Residue

Thorough rinsing after boil-out is as important as eliminating the polymerized oil it was designed to remove. To achieve complete removal of any residue, rinse with copious amounts of water. Ensure that there are no visible signs of light-colored film or chalk-like material remaining anywhere on the fryer. In cases where overtreatment has made it difficult to remove the last traces of surface residue, a dilute vinegar solution (10:1; water/vinegar) followed by generous amounts of water may be helpful. Leaving boil-out residue in the vat is comparable to adding soap to shortening.

## Exposure to Copper and Brass

Copper, and its most common alloys, brass and bronze, are well-known oxidation catalysts. Table 15.1 (8) shows that copper is an effective oxidation promoter, in

**TABLE 15.1**
**Catalytic Effect of Certain Metals (ppm)**[a]

| | |
|---|---|
| Copper | 0.05 |
| Manganese | 0.60 |
| Iron | 0.60 |
| Chromium | 1.20 |
| Nickel | 2.20 |
| Vanadium | 3.00 |
| Zinc | 19.60 |
| Aluminum | 50.00 |

Source: Sonntag, N.O.V. (8).

[a]To reduce the keeping time of lard by 50% at 98°C.

addition to the effect of other metals.

Operators should inspect thermocouples and frying baskets daily because they may be plated copper or brass. If that is the case, normal wear and tear can create opportunities for exposure if the plating chips.

### Topping Off with Used Shortening

Maintaining proper frying levels through addition of fresh shortening plays an integral role in obtaining maximum fry-life. By topping off with used shortening, the amount of breakdown material contributing to foaming is, at best, maintained and in some cases probably increased. Although it seems harmless, and at times very convenient, it should always be avoided.

### Overheating

Overheating, or frying at higher than recommended temperatures, serves to accelerate the processes responsible for shortening breakdown, giving the perception of reduced fry-life. The temperature of the shortening should be measured routinely to verify the accuracy of the thermostat.

## Premature Smoking

The smoke point of an unused oil or fat is directly related to the amount of free fatty acid present. Table 15.2 shows this relationship (9). Although there are other breakdown materials that volatilize when subjected to frying temperatures, free fatty acids, in their many forms, contribute substantially to the material liberated from the surface of the shortening appearing as smoke.

**TABLE 15.2**
Effect of Free Fatty Acid Content on Smoke, Flash, and Fire Points

| Free fatty | Smoke point | | Flash point | | Fire point | |
|---|---|---|---|---|---|---|
| acid (%) | (°C) | (°F) | (°C) | (°F) | (°C) | (°F) |
| 0.04 | 218 | 425 | 327 | 620 | 366 | 690 |
| 0.06 | 210 | 410 | | | | |
| 0.08 | 205 | 400 | | | | |
| 0.10 | 200 | 390 | 313 | 595 | 363 | 685 |
| 0.20 | 190 | 375 | | | | |
| 0.40 | 177 | 350 | | | | |
| 0.60 | 171 | 340 | | | | |
| 0.80 | 165 | 330 | | | | |
| 1.00 | 160 | 320 | 307 | 585 | 360 | 680 |

Source: Weiss, T.J. (9).

### Inadequate Filtration and Skimming

This is perhaps the single biggest cause of premature smoking. While product remnants remain in the shortening at frying temperatures, they continue to cook until they char and liberate smoke.

# Boil-Out Compound Residue

Because of its ingredients, boil-out compound promotes breakdown material that, relative to usable shortening, is volatile and vaporizes at frying temperatures.

# Questionable Quality of Product Being Fried

This is most applicable to battered or breaded products. Irregularities in the coating can release excessive amounts of what is often referred to as "dust" into the shortening. Hash-brown potatoes and tortilla products are notorious for their shedding.

### Topping Off with Used Shortening

Used shortening already contains a certain amount of breakdown material. Adding it to usable shortening gives the appearance of premature smoking.

### Overheating

Overheating is another major contributor to premature smoking. Just like any other chemical reaction, adding heat increases reaction speeds. Therefore, the higher the temperature, the faster the shortening breaks down. Often, the cause is faulty temperature-sensing probes or a thermostat in need of recalibration.

# Premature Darkening

## Inadequate Filtration and Skimming

Burned product remnants, if allowed to accumulate in the fryer, stain the oil, giving the perception of premature darkening.

## Overheating

Again, overheating, or frying at higher than normal temperatures, accelerates breakdown. Associated with shortening breakdown is, of course, shortening darkening.

## Improper Fryer Loading: Solid Shortening

This essentially amounts to overheating. Aside from safety considerations, improper loading of the fryer with fresh shortening can result in premature breakdown. Most gas fryers have a relatively deep well, within which temperature sensing probes are located. If the well is not completely packed with shortening, or, in the case of pourable shortening, not immediately filled to the proper level, the probe requires more heat, until the air around the probe reaches the desired temperature. By this time, the temperature at the surface next to the fire tubes in gas fryers, or heating elements in electric fryers, is high enough to make the shortening smoke, and if left unattended may even burst into flames.

# Off Flavors and Odors

## Topping Off with Used Shortening

This not only contributes to imparting undesirable oxidized flavors and odors, but also to the development of those flavors and odors of the product that was fried in it.

## Improper Filtration Sequence

Another opportunity to develop off flavors and odors is connected with filtration practices, especially in those operations that filter mechanically. Fryer proximity should not be the main criterion for determining the proper sequence. It is not recommended for an operation that fries a fish product to filter the fish vat first, as one could expect to obtain some uniquely flavored french fries. Fish vats or vats used to fry highly spiced products should always be filtered last. Operations that combine products in a fryer (cannot use dedicated fryers) can protect the consistency of the product mix by filtering the highest-flavored product fryer last.

**TABLE 15.3**
**Relative Rates of Reactivity of Common Fats and Oils Due to Inherent Stability and Iodine Value (IV)**

|  | Inherent stability[a] | Calculated IV[b] |
|---|---|---|
| Safflower | 7.6 | 149 |
| Soybean | 7.0 | 132 |
| Sunflower | 6.8 | 136 |
| Corn | 6.2 | 128 |
| Low linolenic soybean[c] | 6.2 | 115 |
| Canola | 5.5 | 120 |
| Cottonseed | 5.4 | 110 |
| Rapeseed (high erucic acid) | 4.1 | 99 |
| Peanut | 3.7 | 100 |
| Lard | 1.7 | 62 |
| Olive | 1.5 | 82 |
| Palm | 1.3 | 50 |
| Tallow | 0.86 | 44 |
| Palm Kernel | 0.27 | 13 |
| Coconut | 0.24 | 8 |

*Source*: Erickson, D.R. and G. List (10).
[a]Decimal fraction of fatty acids multiplied by the relative rate of reactivity with oxygen for each fatty acid (fatty acid/reactivity rate): oleic (C18:1)/1; linoleic (C18:2)/10; linolenic (C18:3)/25.
[b]Calculated IV for C18:1, C18:2, and C18:3.
[c]Calculated from average range of C18:1, C18:2, and C18:3. Analyses conducted by Kraft Food Ingredients, Memphis, TN, 1993.

## Cross-Contamination with Different Shortenings

Different shortenings have flavor profiles unique to that shortening. This is implied in Table 15.3 (10) in that if different oils have different inherent stabilities, it then follows that their breakdown products, the substances responsible for the oil's unique flavor, can contribute to flavor problems. Beef tallow and blends of beef tallow and vegetable oil (animal/vegetable = A/V) are selected primarily for their unique flavor contribution (profile) and inherent stability. Mixing with 100% vegetable shortening dilutes the capacity of A/V shortening to impart its typical flavor.

Furthermore, when shortenings begin to break down, there are flavors generated that are unique to the particular shortening breaking down. The breakdown flavors of some vegetable shortenings could account for off flavors and odors in A/V shortening, and the reverse when mixed. This is also true when commingling different vegetable-based shortenings. Fryers should always be dedicated to using one type of shortening if opportunities to generate off odors and flavors are to be minimized.

### Exposure to Copper or Brass

This has been identified as being responsible for the development of acrid, biting, bitter, and even sour flavors and odors.

### Questionable Quality of Product Being Fried

A typical cyclic complaint, usually occurring in late autumn or early winter, has been described by customers as reminiscent of the odor emitted from a potato found well past its suitability for use. The same thing can occur with french fries that have been cut from potatoes that were processed toward the end of their "cellar" storage. One reason why it may have gone undetected before frying is because they are frozen. Because odor tends to linger in the fryer, usually the shortening is identified as the sole cause.

### Atypical Frying Performance

This complaint usually amounts to the finished product being perceived as uncharacteristically light- or dark-colored, or too greasy. All three perceptions are usually temperature related. There is a common misconception that the amount of absorption is strictly a function of the shortening's composition. Absorption, relative to usable shortening, is a function of temperature. Greasiness is an indication of improper frying temperature, shortening that has surpassed its usability, and in some cases, product of questionable quality.

## Other Situations

### Green Shortening

Occasionally a customer may complain that recently purchased shortening has turned green. The usual circumstance is that the fryer was just cleaned and loaded with a cube (block) of unused shortening. Invariably, whoever loaded the fryer with fresh shortening neglected to remove the blue poly (plastic) bag that lines the inside of the cardboard container. Blue and yellow, of course, make green.

### Reduced Fry-Life but "Not Doing Anything Different"

From the perspective of most restaurant managers experiencing sudden upsets in their shortening programs, nothing has changed operationally. These situations tend to test the limits of a supplier's diplomacy. Also typical is the lack of immediate evidence suggesting that operational changes have occurred that could affect frying performance. Some slight, seemingly insignificant alteration has indeed occurred, however.

Many complaints are accounted for by broken or severely worn baskets. When a basket breaks during use, an employee obtains a replacement from the supply room. He or she returns with one that was previously removed from operation because it too was broken or worn!

Another good example of "not doing anything different" involves a mechanical filter. A store manager was convinced that poor-quality shortening was responsible for reduced fry-life based on problems with consecutive deliveries of different shortening lots. This assumption is not unreasonable since the problem did not go away with different lots of shortening. After exhaustive investigation, it was determined that the mechanical filtering machine was damaged during previous use. Apparently, the person responsible for filtering the fryers accidentally broke off the return nozzle. The manager, upon discovery, proceeded immediately to the local hardware store where he purchased an appropriate length of *copper* tubing to replace the missing stainless steel nozzle. Obviously, copper tubing was an inappropriate choice of metal for repairing the filtering machine.

The point of the previous examples is that sometimes, when the obvious fails to uncover the source of the problem, it is necessary to pursue a critical, but objective evaluation of a store's operation.

## Normal (Expected) Seasonal Lulls in Business

Nearly every restaurant experiences at least one period per year when business slows considerably. This reduction in patronage results in a corresponding reduction in the volume of product going through a fryer. This in turn further results in shortening subjected to thermal abuse without the prolonging effects of frying itself (removal of breakdown material by steam from the product being fried) or the required frequent replenishment with unused shortening due to absorption. It is far more deleterious to idle a shortening at frying temperatures for prolonged periods than it is to fry in it frequently.

## New Fryer, Different Performance

Concern about different frying performance from new fryers often arises from an operation that has recently purchased a fryer with split vats. Most industrial gas fryers hold 22.7 kg (50 lb) in each of two wells. A split vat is further divided to hold 11.3 kg (25 lb) each. By not adjusting the normal product volume accordingly, a larger than normal demand is put on the fryer, which in turn can affect temperature recovery. This in turn would account for a perceived difference in finished product integrity.

For the same reasons mentioned in the discussion of seasonal effects on fry-life, the split vat fryer could be considered to reduce fry-life complaints. If 22.7 kg of shortening is affected by idling at frying temperatures, it then follows that one-half that amount would be more susceptible to premature breakdown under the same or similar circumstances.

## Determining Shortening Discard Point

Perhaps the question asked most often is "When should shortening be discarded?" Because frying is so complex, there is no quantitative scientific test to determine when a shortening either is approaching the end of suitability for continued use or is indeed ready to be discarded. The more successful restaurants invest considerable resources in training staff to recognize the system's criteria of finished product acceptability. Some managers, viewing this as jeopardizing consistency, address this potentially difficult decision by discarding shortening on an established schedule customized for their individual operation.

Although several shortening test kits, color monitors (by comparing to a standard), and electronic monitoring devices are commercially available, none provides the flexibility necessary to serve all operations under all circumstances. However, there is one consideration for deciding if a routine testing is beneficial or not for an operation. Routine testing requires sampling on an established frequency which, in turn, causes operators to pay more attention to the shortening. For example, while at the fryer performing the test, an operator also has the opportunity to observe such things as excessive food particles floating on the surface and the need for skimming, or the level is exceptionally low. Both examples address know causes of premature breakdown. Experience has shown that operational guidelines set forth by corporate operations management are just that: guidelines. Each store is different. It has unique circumstances connected with its daily operations that cannot be addressed with general operational policy and procedure. This includes subtle, but acceptable, variations in unused shortening or oil, variations in products being fried, different generations of fryers, different types of fryers (e.g., gas versus electric, size), occasional circumstantial departure from approved operational procedure, and unintentional employee errors that occur regardless of a manager's supervisory prowess. These factors suggest that a precise test to determine when a shortening should be discarded under any circumstance is most likely impractical. Test kits may have utility, however, if they result in managers, supervisors, or individual operators paying more attention to shortening management and maintenance in general.

## Summary

Achieving maximum fry-life with minimal frying problems can be accomplished through the following practices:

1. Adequate filtration and frequent skimming.
2. Frequent thermostat calibration.
3. Maintain proper fryer levels with unused shortening.
4. Idle fryers at reduced temperatures when feasible.
5. Cover fryers during slow periods.
6. Use dedicated fryers.

7. Boiling-out frequently.
8. Frequently inspect and remove exhaust hood condensates.
9. Daily inspect fryers and other equipment.

Adherence to this relatively simple routine not only minimizes frying-related problems, it also helps to ensure consistent, high-quality fried foods.

## References
1. Chang, S.S.; R.J. Peterson; and C.-T. Ho. *J. Am. Oil Chem. Soc.* **1973,** *55*, 718.
2. Frankel, E.N. *Progress in Lipid Research*; Pergamon Press: New York, 1982; Vol. 22, pp. 1–33.
3. Frankel, E.N.; L.M. Smith; C.L. Hamblin; R.K. Creveling; and A.J. Clifford. *J. Am. Oil Chem. Soc.* **1984,** *61*, 87.
4. Fritsch, C.W. *J. Am. Oil Chem. Soc.* **1981,** *58*, 272.
5. Paulose, M.M.; and S.C. Chang. *J. Am. Oil Chem. Soc.* **1978,** *55*, 375.
6. Perkins, E.G. *Food Technol.* **1960,** *14* (10), 508.
7. Perkins, E.G. *J. Am. Oil Chem. Soc.* **1965,** *42*, 782.
8. Sonntag, N.O.V. In *Bailey's Industrial Fat Products*, 4th ed.; D. Swern, Ed.; John Wiley & Sons: New York, 1982; Vol. 1, p. 152.
9. Weiss, T.J. *Food Oils and Their Uses*; AVI Publishing: Westport, CT, 1983; p. 16.
10. Erickson, D.R.; and G. List. In *Bailey's Industrial Fat Products*, 4th ed.; T. Applewhite, Ed.; John Wiley & Sons: New York, 1982; Vol. 3, p. 267.

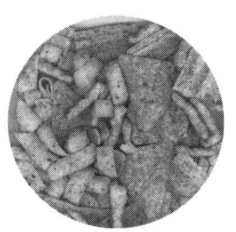

# PART 5

# Evaluation

# 16

# General Considerations for Designing Laboratory Scale Fry-Tests

**Don Banks**

*Edible Oil Technology, 8155 San Leandro, Dallas, Texas 75218*

Laboratory fry-tests are commonly used for a number of purposes including the evaluation of new oils, assessing the performance of alternate oils for existing applications, screening candidate oils for a new application, and/or to investigate frying oil problems.

In contrast to analytical tests, there is no standard procedure for conducting a laboratory fry-test, and test protocols are developed according to need. Two basic types of test design are commonly used. One utilizes a general frying protocol and the other uses a protocol developed for a specific application. The general protocol is used to develop an overview profile of oil performance throughout its fry-life; however, the data derived from testing are not considered definitive and have to be interpreted within the context of the test procedure used. Simply put, a general frying protocol only provides general information. To assess oil performance for a specific application (i.e. application-specific testing), a test protocol closely matching the conditions that oil will be subjected to during commercial usage has to be designed and carefully followed.

Most work is conducted with bench-top batch fryers. Fryers are usually rated according to the maximum number of pounds of french fries they can produce per hour using partially fried potatoes (par-fries). Electrically heated fryers, which typically hold about two gallons of oil and are rated at 30 lb/h, are commonly used but other appropriately sized fryer can also be used. Some testing is also done with specially designed small, continuous fryers, but limited availability restricts usage.

The discussion that follows addresses considerations for conducting lab fry-tests and provides a basis for developing test procedures to meet various needs. An overview of a general fry-test is included at the end of the chapter along with a description of an application-specific fry-test.

## Frying versus Oil Heating

One of the most important considerations in designing a test is to insure that an appropriate amount of frying is included. A test protocol that stipulates heating oil and holding it at temperature with only occasional frying constitutes more of a heat stress test rather than a fry-test.

For testing to simulate frying in industrial and well-managed fast-food operations, lab frying should be conducted with only the minimal delay between batches that may be needed for heat recovery and fryer loading. For disruptions in frying lasting more than a few minutes, such as during a break between morning and afternoon frying sessions, fryer oil heating should be discontinued.

In some commercial operations, oil is held at frying temperature during idle periods. A fry-test can be designed to incorporate idle periods to simulate practices used in a specific operation, but oil sampling and testing should be included to document the oxidative degradation that occur. Test results can then be used to review frying practices and make recommendations for changes to support improved frying practices and oil quality.

## Oil Selection and Supply

After deciding which oil (or oils) will be tested, the next consideration is to obtain an ample supply from a single lot of production. Estimate the amount that will be needed and obtain an adequate quantity to avoid the possibility of running out of oil before the test is completed. If the supply should be exhausted during testing, and if additional oil from the same lot of production cannot be readily obtained, the test should be terminated to avoid introducing variables that could influence test results. Always analyze an unused sample of the oil when it is received to verify that the quality is in full compliance with the product specification.

## Application-Specific Test Design

When preparing to test oil for a specific application, first conduct a preliminary investigation to observe the procedures used for all aspects of the operation. The principal focus for the investigation is on the overall frying operation including fryer oil loading, oil heating, frying procedures, operating parameters, daily production schedule, rate of production, make-up oil addition, oil turnover, oil quality testing, and finished product handling and/or packaging and distribution. However, associated elements such as oil purchasing records, storage and handling procedures, and used oil handling and filtration, should also be audited to gain added insight to the overall operation. This preliminary investigation provides the basis for developing an application-specific test design.

# Selection of Product for Frying

Potato products are commonly used for general testing, but care should be taken in selecting the product for frying to avoid unintended oil blending. For example, par-fried french fries contain about 7% oil from absorption during initial frying. If used for fry-testing, some of the par-fry oil will leach into the frying oil and can affect product flavor and oil quality. Sliced and cubed potato products that have not been par-fried are commercially available and can be used for testing. Good quality fresh potatoes, as well as a number of other fresh or prepared products, can also be used for testing.

For application-specific testing, the product to be fried should be the product that is normally produced. A number of products, including par-fries, chicken, fish, nutmeats, whole corn masa, and others, contain oil that will leach and blend with the frying oil, but as part of the normal operation, the blending cannot be avoided. However, it should be noted that after extended periods of frying a product with high oil content, the oil inherent in the product can account for almost half of the oil in the fryer. In such a case, the timing for taking frying oil and finished product samples for evaluation will have to be set according to what best reflects the objectives of the test.

# Oil Conditioning

Edible oil processing is designed to remove essentially all flavor and aroma from oils and yield completely bland products. As a result, when fresh oil is first used for frying, it will act as a diluent, somewhat diminishing the flavor of the fried product. As frying continues, flavor accumulates in the oil and it soon begins to enhance the flavor of the finished product. During the initial period of frying, heat reaction flavors, derived from components in the product being fried, account for essentially all of the flavor that accumulates in the oil.

As frying continues, the oil itself can begin to contribute flavor notes to the fried product, depending on the heat and oxidative stress it's exposed to. The first of these notes are commonly derived from minor components in the oil. Some oils generate characteristic flavors that can either enhance or detract from the overall flavor of the finished product.

With continued frying, flavor notes derived from oil oxidation can accumulate in the oil and contribute to the flavor of the fried product. For some oil and production combinations, lightly oxidized oil can enhance product flavor. With heavier stress and extended use, oxidative degradation of oil leads to objectionable flavors/aromas and to eventually rancidity.

With good frying practices and rapid oil turnover that support equilibrium conditions, essentially the entire finished product flavor is derived from the product being fried. These conditions are commonly achieved with the newer, high production, continuous fryers that have rapid oil turnover. With older continuous fryers, flavor contribution from minor components in the oil and from the accumulation of secondary oxidation products, derived from the degradation of fatty acids, tends to be greater.

In batch frying operations, oil turnover is rarely sufficient to support steady-state frying conditions. As a result, the quality of the frying oil does not equilibrate; rather it continually changes and degrades during usage. With good frying practices and careful oil management, extended fry-life can be achieved, but, as noted, equilibrium conditions can rarely be realized. During commercial production, careful oil and product monitoring is required to support timely replacement of the oil or objectionable flavors and eventually rancidity will occur.

For general fry-testing, when the objective is to profile the complete fry-life of oil, oil conditioning is important but not a central focus of the testing. Samples of oil and product will be taken throughout the test and the corresponding amount of oil conditioning should be taken in to account to aid with the interpretation of results.

For application-specific testing, oil conditioning is essential to support representative results. Normally, oil conditioning is continued until analytical monitoring indicates that oil quality closely matches the values associated with commercial production. The evaluation of product samples collected as the oil becomes fully conditioned will provide good insight as to the quality that will be obtained during commercial production.

## Make-up Oil

Make-up oil is oil that is used to replenish the fryer during frying. The addition of this oil is needed to maintain the frying level during extended testing. However, it is important to be aware that this addition affects both oil quality and product flavor. To mediate the impact of adding fresh make-up oil, it should be added at the start of testing for the day and sampling for analytical testing should be scheduled toward the end of the day's production.

Ideally, the same amount of make-up oil would be added at the start of production each time it is needed, but that is not always the case. In practice, oil usage can vary from day to day and contribute to anomalous variations in test data. The situation can be particularly troublesome if it occurs when conducting parallel tests with different oils. When different quantities of make-up oil have to be added to each fryer, there is no good way to compare and contrast oil test results.

An alternative procedure the author has used to conduct tests of limited duration is to fill fryers to the upper limit of the acceptable oil level at the start and not to add make-up oil during the test. Taking into account design considerations, tests lasting several days can be completed with the oil level remaining in the acceptable range.

## Oil and Product Sampling

Oil and product samples for analytical and sensory testing should be collected just before shutdown at the end of frying periods. The reason for the timing is to avoid the effects of oxidative stress associated with start-up, as well as the residual effects from the previous fryer shutdown.

When testing is being conducted to profile the full fry-life of an oil, samples are usually collected on a daily basis. When testing is being conducted to produce optimal product for sensory evaluation, samples are commonly collected periodically as needed to monitor oil quality during the oil conditioning period. Monitoring is increased as the oil becomes fully conditioned and then principal samples of oil and product are collected for extensive analytical and sensory testing.

# Oil Analysis

A number of tests can be used to track the state of the oil analytically when testing the full fry-life of oils. Basic testing usually includes flavor, color, free fatty acid, polar compounds, and polymers. Additional tests, including p-Anisidine, Oil Stability Index (OSI), and fatty acid composition, to name a few, can be included to provide further insight as to changes in oil quality that occur during the course of frying. For application-specific fry-tests, analytical testing is usually focused on monitoring oil attributes to achieve a close match with conditions that represent commercial production.

# Product Testing

Testing the finished product is an especially important component of a fry-test. While this chapter addresses fry-test technology, the overriding consideration is that the purpose of frying is to produce products for consumer consumption. Therefore, achieving favorable sensory test results is key to commercial success.

For fry-and-serve products, sensory testing has to be coordinated so that product is presented to panelists in a manner consistent with commercial practices. Delaying sensory evaluation, especially if product storage and reheating are involved, risks the potential for introducing uncontrolled variables and should be avoided.

For savory snacks, product handling should closely match commercial practices including packaging, storage, distribution, and store shelf display, and the timing for sensory testing should be consistent with consumer product purchasing and consumption patterns.

# Overview: General Fry-Test

As noted above, there are no standard procedures for conducting laboratory scale fry-tests, and testing protocols are commonly developed to meet specific needs. The following is provided as an overview for a general fry-test procedure:

•   Obtain an ample supply of oil and verify compliance with product

specifications.
- Select product to be fried and obtain adequate supply.
- Setup fryer, verify cleanliness and absence of any residual cleaning agents.
- First day: Fill fryer to "full" mark and start heating. On subsequent days, add make-up oil to restore fill level, record amount and start heating.
- Heat oil to frying temperature (i.e. 350°F) and verify temperature.
- As soon as oil reaches frying temperature, begin frying first batch.
- For morning session, sequentially fry batches for 3.5 hours, with minimal delay between batches as needed for heat recovery and product handling.
- Discontinue frying and turn fryer off for one hour at midday.
- For afternoon session, heat oil to frying temperature and sequentially fry batches for 3.5 hours.
- During last half-hour of frying collect sample of product for evaluation.
- At conclusion of testing, turn fryer off, collect oil sample for testing.
- Repeat daily procedure until the end of oil fry-life is reached as determined by analytical and organoleptic assessment of oil and product.
- Testing can usually be completed within 7 – 14 days.

## Application-Specific Fry-Test

Testing for a specific application is similar to the procedure above for the general fry-test, except that oil and product samples are taken only as needed for monitoring quality attributes. Daily frying sessions are continued until the oil and product attributes reach a close match with attributes for commercial production. Sufficient product is then produced as needed for testing and evaluation.

Oil conditioning for application-specific testing commonly takes about three to five days and then product is produced for testing. The quantity of product needed for testing can usually be produced within one to two days.

# 17

# Designing Field Frying Tests

**Michael D. Erickson**

*Cargill, 600 N. Gilbert Street, Fullerton, CA 92837*

The design of a meaningful field test begins with establishing a practical objective. The incentive to conduct a field test is usually driven by economics. Typical reasons include claims of longer fry-life, reduced initial cost of the shortening and, most recently, trans fatty acid reduction or elimination. Regardless of the incentive, increasing the value of that part of the operation is the real objective. Shortening with claims of longer fry-life may command a higher initial cost. If, however, the extended fry-life offsets the initial outlay because of reduced discards, then the shortening presents a better value than the one currently in use. Conversely, if the initial cost of the new shortening is less than the shortening currently in use, then the issue is either comparable fry-life or reduced cost even with increased discards. In both examples, however, it is assumed that finished product quality is unaffected and is the ultimate criteria for acceptability. Moreover, as the difficulty of used oil discard increases, longer fry-life will likely become more important, in the future.

As previously mentioned, a clear, well-presented objective for conducting a field test and sincere recognition of the integral role of the restaurant staff from the beginning will help lay the foundation for a successful test. This, combined with close, consistent monitoring throughout the test, maximizes the returns on the significant investment of time, effort and resources necessary to conduct a meaningful field test.

It is important to establish a frame of reference for the entire field test first. That is, what does "normal" or "routine" mean for this store? This is also the appropriate time to establish criteria for acceptability. It is recommended that this be done by monitoring two frying cycles before evaluating the new shortening. This baseline evaluation is then followed by two frying cycles using the test shortening. Finally, another two frying cycles with the standard shortening are monitored. These last two cycles are used to substantiate that the initial baseline evaluation reflects two "normal" frying cycles. This approach is summarized as follows: phase 1 is baseline evaluation, phase 2 is new product evaluation, and phase 3 is follow-up baseline evaluation.

Appearance, flavor and odor (how the product looks, tastes and smells) are the usual criteria for the acceptability of fried foods. Chemical tests are useful for potential correlation opportunities unique to a store but should not be relied on for sole acceptability criteria. There is no better method of assessing quality of shortening than by a sensory evaluation of the fried products by well-trained tasters, who, incidentally, are usually restaurant managers and their staff.

It is essential to maintain the same operating conditions during all phases of the test. These conditions include product consumption, shortening consumption, and equipment reliability. Calibrate all fryers used in the test before starting each phase. Successful execution of the field test ultimately depends on how well the objective is communicated to those actually performing the field test.

The most critical variable in a field test is the unavoidable departure from routine store operations. It can also be the hardest variable to manage. As a result, it is reasonable to anticipate concern (or even slight reluctance) from store managers when asked to participate in activities that disrupt standard operating procedures. Furthermore, the required scrutiny of the store's overall operation during a field test often invites additional anxiety for managers and supervisors, which could influence test results.

Another critical aspect of these types of evaluations is the establishment of "normal" restaurant operation. Every successful store manager wants the normal operation of his or her store to reflect procedures set forth by their systems operations management. In actual practice, though, every restaurant has unique inherent circumstances (usually a function of the store's market) that diminish a manager's ability to operate exclusively "by the book." Consequently, blanket claims (e.g., that a restaurant system will experience the same new shortening performance throughout that system) should be met with skepticism. Data supporting a decision to expand a test become available when new shortening performance meets or exceeds expectations. This is the primary incentive for instilling the importance of maintaining normal operation of the store during the test period. Any intentional changes will jeopardize the credibility of the results.

Stressing potential benefits of a test is also important. A dedicated store manager welcomes the opportunity to contribute meaningfully to the system. Sincere reassurance that the objective is to evaluate the performance of a new frying shortening and not the overall performance of the store's operation helps foster cooperation. Conveying the message that credible results ultimately rest with a manager's familiarity with the routine operation of his or her particular store is another motivational tool.

## Conducting the Field Test

### Phase 1: Baseline Evaluation

Fry-life is determined by frequency of frying. Keeping weekly receipts for products and shortening is a convenient way of monitoring average product consumption and mix, in addition to calculating product-to-shortening usage ratios. This is important in understanding the effect of shortening replenishment resulting from absorption by products being fried. (Monitoring weekly shipments of shortening need not be limited to two weeks prior to the actual test. It may be beneficial to go back as far as practical and attempt to further substantiate the consistency of usage before the scheduled test period.) All things being equal, frequent frying translates to longer fry-life. There are two reasons for this apparent dichotomy. First, moisture in the product being fried converts to steam in the fryer, which serves to effectively "strip out" (deodorize, in a sense) breakdown material that would normally accumulate and contribute to accelerating the breakdown process. Second, absorption by the product being fried requires replenishment with unused shortening. This has been referred to as the "dilution solution" for prolonging fry-life.

As previously mentioned, analytical testing can be useful in assessing overall shortening performance. The more complicated the field test, however, the more potential there is for disrupting the store's routine. To minimize disruption due to analytical sampling, a "user-friendly" sampling format is recommended. This includes providing predated, clearly labeled sample containers and a means of safely removing samples from the fryer.

Sampling frequency usually depends on the type and number of analyses deemed necessary. Daily sampling is typical. Because of daily replenishment, samples can be gathered both before and after filtration; this, however, should be kept to a minimum of *after* filtration but *before* topping off.

Many analyses are available for characterizing performance differences between two shortenings during a given fry cycle. The most common ones include free fatty acid (FFA), color, iodine value (IV), total polars, polymer profiles, and soap.

Polymer profiles are done by high-performance liquid chromatography (HPLC), which requires expensive equipment. Of the previously mentioned analyses, the most convenient are FFA and color. It is important to mention again that the criteria for acceptability should be determined by sensory evaluation. Analytical testing may or may not correlate with the sensory evaluation. One important non-sensory aspect of shortening performance can be correlated with analytical testing. Premature smoking is typically associated with high apparent FFA.

It is important to be aware of the limitations of analytical testing. Most of the traditional analyses that evaluate used frying fats are intended for analyzing unused fats and oils. Therefore literal interpretation of the analytical results can be misleading.

Frying is arguably the most complicated and complex application of edible fats and oils because of the nearly infinite number of potential side reactions resulting from oxidation. When, for instance, an FFA analysis is performed on unused refined, bleached, and deodorized oil, it is reasonable to assume that any contribution to detectable acidity is the result of unreacted (caustic refining) or unremoved (deodorization) FFA during processing. Furthermore, results are typically calculated using the formula weight for oleic acid. In oil subjected to thermal abuse in the presence of oxygen, it is probable that many compounds other than free fatty acids are contributing to the results. Therefore results obtained for used frying oils are better described by the terms "apparent free fatty acid" or "relative acidity."

Soap analysis would be analogous to the explanation of FFA analysis previously described. Therefore "apparent soap" or "relative alkalinity" may be more appropriate than soap when results are obtained from used frying oils.

Most important, though, is the realization that any results obtained are relative to that particular store's operation and should be interpreted as such. It should not be assumed that the results could be interpolated.

Another important aspect of pre-monitoring a store is the opportunity to "rehearse" for phase 2, which is the evaluation of the new shortening. It is better to discover weaknesses in the program during the preliminary evaluation than to do so during the new shortening evaluation.

## Phase 2: New Product Evaluation

When phase 1 is satisfactorily completed, evaluation of the new shortening begins. Phase 2 should not be any different from phase 1, except for the new shortening. Perhaps the most important part of phase 2 is ensuring that the same lot of shortening is used throughout the evaluation. If, for some unforeseen reason, the lot of shortening is depleted before completing the two fry-cycle evaluations, the test must be terminated. As a rule, 1.5 times as much shortening consumed in phase 1 should be made available for phase 2. If the reason for initiating the test was improved fry-life claims, then at least an equal amount of shortening should be used. Furthermore, an unexpected increase in business during the test will require additional shortening inventories.

## Phase 3: Follow-Up Baseline Evaluation

At this point, the field test is essentially complete, assuming the protocol was followed in phases 1 and 2. The main reason for phase 3 is demonstrating that the test period reflects consistency of operation. Every operation experiences slow periods, which translate into opportunities to skew the results. Shortening consumption typically increases during slow business periods because the frequency of frying decreases. It is important to establish that this did not contribute to poor performance of a shortening that showed enough promise to warrant an evaluation in the first place. All things being equal, frequent frying translates to longer the fry-life. There are two reasons for this apparent dichotomy. First, moisture in the product being fried converts to steam in the fryer, which serves to effectively "strip out" (deodorize, in a sense) breakdown material that would normally accumulate and contribute to accelerating the breakdown process. Second, absorption by the product being fried requires replenishment with unused shortening. This has been referred to as the "dilution solution" for prolonging fry-life.

## Interpretation of Results

Generally the decision to proceed to an expanded field test is not a difficult one, provided the criteria for determining when the shortening should be discarded remain consistent. If the test shortening performs better, phase 2 will be longer than phases 1 and 3.

Establishing the true shortening use at a store is another important aspect of interpreting results. One common misconception regarding shortening usage is that doubling fry-life reduces shortening cost by one-half. Fry-life is also determined by frequency of discard.

Figure 17.1 shows results of reducing shortening discards by one-half based on the following formula:

$$\text{Consumption} = \text{Discards + Absorption}$$
$$\text{(Annual Purchases)} \quad \text{(Shortening Sold with Food)}$$

CONSUMPTION    =    DISCARDS + ABSORPTION
(Annual Purchases)        (Shortening Sold with Food)

| | |
|---|---|
| CONSUMPTION: | 20 million (mM) lbs/year |
| Number of Stores: | 900 |
| Av. fryer capacity/store: | 150 lbs. (3′ 50 # fryers) |
| Average fry-life: | 7 days |

DISCARDS per year:    52 (365/7)

Therefore,

ABSORPTION (#/yr) = CONSUMPTION – DISCARDS
= 20 mM – (52 discards)(150 #/discard)(900 stores)
= 13 mM #/yr

DISCARDS (#/yr) = 20 mM – 13 mM
= 7 mM #/yr

Reducing the discards by one-half:

(7 mM #/yr)/2 = 3.5 mM #/yr
CONSUMPTION = ABSORPTION + DISCARDS
= 13 mM #/yr + 3.5 mM #/yr
= 16.5 mM #/yr

% Reduction in Consumption:

% reduction in consumption = [(20 mM #/yr –16.5 mM #/yr)/20 mM] #/yr′ 100
= 17.5%

**Fig. 17.1.** Hypothetical usage calculation.

This calculation demonstrates that a 50% increase in fry-life translates to a 17.5% decrease in total shortening consumption. This is an example of how the true potential savings should be calculated when determining if a demonstrated increase in fry-life has value compared to the cost of the shortening currently being used.

# 18

# Evaluation of Used Frying Oil

**Frank T. Orthoefer[a] and Gary R. List[b]**

[a]Germantown, TN, [b]USDA ARS NCAUR, Peoria, IL

The popularity of fried food has been discussed extensively. The consumer prefers the flavor, appearance, and texture of food prepared this way. Frying is a chemically complex process that is not well understood (1,2). Frying is still considered by many to be an art rather than a science. Several changes occur in food during frying, such as starch gelatinization, protein denaturation, water vaporization, and textural changes. During frying, oil is absorbed by the food, and water is volatilized. The quality of the oil (the frying medium) and of the food fried in that oil are intimately bound.

Frying oil quality affects oil absorption and the types of by-products and residues absorbed by food. The type of food being fried also affects fry-life. Changes in oil and its assessment throughout the frying cycle are detailed in this chapter.

## Changes in Oil During Frying

Frying oil changes with use, going from fresh through its optimal state to a degraded condition. An overview of the types of reactions that occur during frying is shown in Chapter 12, Fig. 12.1 (3). Alterations of the oil were described by several investigators (4–6). Methods used to quantitate changes in oil were summarized by White (7). The main quality parameters were discussed by Jacobson (8). The types of degradation reactions are thermal, oxidative, and hydrolytic (3). These reactions produce a variety of physical and chemical changes in the oil, including increases in viscosity, volatile materials, polarity, free fatty acid (FFA) content, color development, and tendency of the oil to foam. Decreases in iodine value, refractive index, and surface tension also occur.

Some changes that occur in oil during frying are visible (7). The color of the oil darkens, and the oil appears thicker (has a higher viscosity) and has an increased tendency to smoke. From a sensory perspective, the odor and flavor of frying oil, as well as food fried in the oil, also change.

Perhaps the most noticeable change in the oil during use is darkening. Food, when fried, can introduce various components to the oil, such as carbohydrates, phosphates, sulfur compounds, and trace metals. Many of these compounds contribute to color formation by reacting with fat or its breakdown products (8).

Foods fried with particularly dark-colored frying oil appear darker. The color of

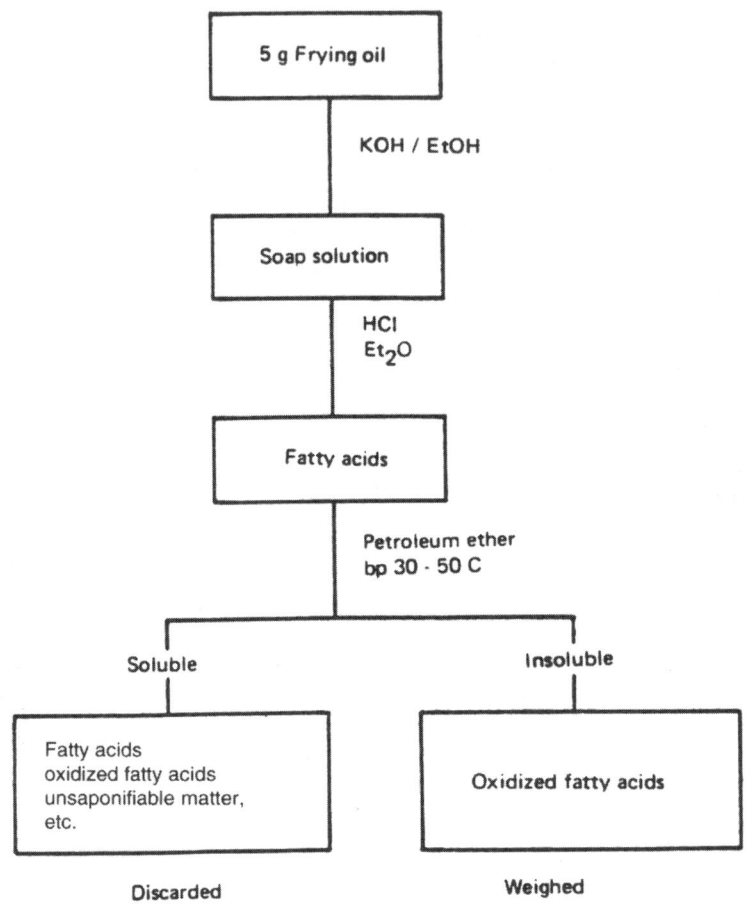

**Fig. 18.1.** Outline for oxidized fatty acid analysis (23).

these fried foods is gray, dull, or unevenly colored. The rate of color formation can also be influenced by the type of food being fried. For example, potatoes contribute little color to oil during frying, while scallops cause rapid oil darkening. Breading also may accelerate darkening of the oil. Breading with a high reducing sugar content (glucose in the form of honey, dextrose, or corn syrup) forms color very rapidly (8).

The source of some of the color development in frying oil, as well as flavor development, is generally believed to be nonenzymatic or Maillard browning reactions (5). During the Maillard reaction, amino groups, normally from a free amino acid or a side chain on a protein, combine with a carbonyl carbon to form an amino compound (8). Furthermore, lipid-derived aldehydes react with Maillard reaction intermediates to form long-chain alkyl-substituted pyrazines, long-chain alkyl-substituted heterocyclic sulfur-containing compounds, and reaction products from

**TABLE 18.1**
Major Volatile Compounds Identified from a "Cotton Ball" Model System Containing Cysteine, Proline, and Lactose Fried in Corn Oil

| I$_E$(DB-1)$^a$ | Structure | Area (%) |
|---|---|---|
| 355 | Pentanol | 0.53 |
| 392 | Hexanal | 2.60 |
| 551 | 2-trans-Heptenal | 4.54 |
|  | 1-Octen-3-ol | 0.61 |
| 620 | 2-Pentylfuran | 1.21 |
| 625 | 2,4-Heptadienal | 0.51 |
| 654 | 2-trans-Octenal | 1.21 |
| 759 | 2-trans-Nonenal | 1.01 |
| 863 | 2-trans-Decenal | 2.13 |
| 894 | 2-cis-4-trans-Decadienal | 3.76 |
| 897 | 2-trans-4-trans-Decadienal | 43.03 |
| 967 | 2-trans-Undecenal | 3.41 |
| 970 | 4,5-Epoxy-2-trans-decenal | 1.56 |
| 972 | 4,5-Epoxy-2-cis-decenal | 0.71 |
|  | Palmitic acid | 6.91 |
| Sum |  | 73.73% |

$^a$Linear retention indices on DB-1 column calculated in relation to ethyl ester of carboxylic acids.

Schiff-base formation with amino acids or peptides in the Maillard reaction (9).

Lipids also contribute both desirable and undesirable flavors. The major decomposition products from frying oil in a model system of cysteine, proline, and lactose are shown in Table 18.1 (9). The major decomposition compound in this system is 2,4-decadienal, a linoleate decomposition product that occurs via the 9-hydroperoxide. The flavor and odor of this compound is described as "deep-fried" (10). Other degradation products that contribute to flavor and odor include hexanal, 2-pentylfuran, and 4,5,epoxy-2 cis/trans decenal.

The food being fried also has an effect on the frying oil. The food may produce its own volatiles, interact with the frying oil, and contribute to oil darkening. Studies were performed to determine changes in the nutritive value of the oil, possible toxic effects of the heated oil, and the sensory quality of the oil and of the food fried in it.

A model systems approach using purified or synthesized triglycerides was used to identify some alterations that occur in frying oil (Chapter 10, Table 10.15) (11). Many analytical differences observed when comparing oil before and after frying are noteworthy. However, performing these analyses to assess the quality of the oil during frying operations is not very practical.

# Chemistry of Frying Oil Deterioration

Chemical changes in the oil underlying the visible changes consist of three different types of reactions: oxidation, polymerization, and hydrolysis (7).

## Oxidation

Frying oil may undergo both oxidative and thermolytic degradation. Oil in the fry kettle is exposed to air at least on the surface of the oil. Oxygen in the air reacts with the heated fat to form hydroperoxides, conjugated dienoic acids, epoxides, hydroxides, and ketones. These compounds may undergo fission into smaller fragments or remain in the triglyceride and result in cross-linking that leads to the formation of dimeric or high polymeric triglycerides (5).

Oxidation products are classified as volatile, monomeric, or polymeric compounds (7). Some volatile decomposition products (VDP) are removed from the oil by steam that evolves during frying of food. Other reaction products may remain in oil as nonvolatile decomposition products (NVDP). Oxidation proceeds rapidly at frying temperatures. The higher the temperature, the more rapid the oxidation.

The rate of oxidation of the frying fat may be affected by factors other than temperature, such as turnover rate; surface exposure to air; presence of prooxidant metals, such as iron or copper; presence of high-temperature antioxidants; presence of silicone antifoams; and quality of frying fat (12).

Very high temperatures (200–700°C) are required to produce significant nonoxidative deterioration of saturated fatty acids (5). Alkanes, fatty acids, ketones, oxopropyl esters, acrolein, and $CO_2$ were found after heating a simple triacylglycerol 1 h at 180°C.

With unsaturated fatty acids, formation of dimeric and cyclic compounds is the dominant reaction (13). Their formation results from homolytic cleavage of carbon–carbon linkages near the double bond (5). Dimeric compounds are believed to result from the combination of allylic radicals from hydrogen abstraction at the methyl groups alpha to the double bond. The radicals then undergo intermolecular or intramolecular addition to the carbon–carbon double bonds. Dimerization and polymerization of unsaturated fatty acids can also occur via Diels-Alder reactions forming cyclic dimers or trimers. Aromatic and cyclic monomers also form as a result of cyclization and aromatization of individual unsaturated fatty acids.

Saturated fatty acids and esters are relatively stable in the presence of oxygen. At temperatures higher than 150°C, however, even saturated lipids undergo oxidation (5). The major oxidative products are carboxylic acids, 2-alkanones, n-alkanals, lactones, n-alkanes, and 1-alkenes. In model studies, compounds produced in the absence of oxygen were the same as those produced in the presence of air. The amounts formed, however, were greater in the presence of air.

The mechanism of thermal oxidation of saturated fatty acids involves the formation of monohydroperoxides (14). The oxygen attack occurs at all methylene groups. Hydroperoxides break down rapidly at elevated temperatures. The principal pathways of breakdown occur via the formation and decomposition of hydroperoxide intermediates. At higher temperatures, such as at frying temperatures (180°C), decomposition occurs very rapidly (Table 18.2) (15). The primary decomposition products are also unstable and undergo further oxidative decomposition. Literally hundreds of products were identified (11).

## Polymerization

Excessive oxidation of the frying fat results in polymerization. Fats can form new carbon-carbon bonds in the absence of oxygen. If these bonds are formed within one fatty acid, cyclic fatty acids are produced (7). Dimeric acids and triglycerides result from bonds between two fatty acids either within the same triglyceride or between two separate triglyceride molecules. Polymerization is one cause of foaming. It also

**TABLE 18.2**
Peroxide Values (P.V.) (meq/kg) for Ethyl Linolenate After Different Time-Temperature Treatments

|       |      | 6 h   | 24.5 h | 45 h   | 69 h   |        |
|-------|------|-------|--------|--------|--------|--------|
| 70°C  | P.V. | 1777  | 1058   | 505    | 283    |        |
|       | Time | 5 min | 10 min | 20 min | 30 min | 60 min |
| 180°C | P.V. | 237   | 251    | 119    | 80     | 44     |
|       | Time | 3 min | 5 min  | 10 min | 20 min | 30 min |
| 250°C | P.V. | 44    | 77     | 198    | 67     | 0      |

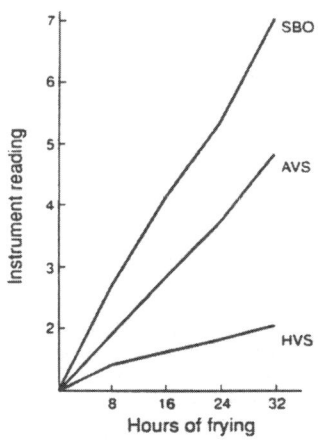

**Fig. 18.2.** Instrument readings representing an increase in dielectric constant and therefore a decrease in the quality of an oil during the frying of potatoes in soybean oil (SBO), animal-vegetable shortening blend (AVS), and hydrogenated vegetable shortening (HVS) (6,26,27).

causes gumming of the oil.

Foaming may occur when small bubbles arise along the sides of the fry kettle. Oil that foams excessively should be discarded because it creates safety and fire hazards. Gumming of the kettle results from polymerized fat. Gumming appears on the sides of the kettle, utensils, and baskets used for frying. In addition to being a problem for cleanup, many of the polymeric materials that form are very bitter (8).

Paradis and Nawar (16) reported that high-molecular-weight compounds are reliable indicators of fat abuse, since these compounds steadily increase and are not volatile. The high-molecular-weight compounds are responsible for physical changes in the frying fat, such as increases in viscosity and foaming. They are also responsible for chemical changes, such as FFA, carbonyl value, hydroxyl content, and saponification value (5). NVDP also result in a decrease in unsaturation and an increase in the formation of high-molecular-weight products (17). VDP from oxidized oils are quantitatively small. The major portion of decomposition products are nonvolatile and include dimeric and polymeric compounds, cyclic monomers, and high molecular weight polar compounds.

In the presence of oxygen, the primary decomposition products are alkyl hydroperoxides and dialkyl hydroperoxides, which further decompose to form oxy and peroxy radicals (5). The radical combination leads to the formation of oxydimers and polymers that possess hydroperoxide, hydroxide, epoxide, and carbonyl groups.

It is difficult to analyze the polymers involved in complex mixtures. The precise structures, as well as effects of various oxidative reactions leading to their formation, are not clear.

With natural fats and frying oils, decomposition is a complex array of reaction products (11). The products formed in each oil reflect the decomposition of its constituent fatty acids. Oxidation products of unsaturated fatty acids are the dominant compounds. Both volatile and nonvolatile decomposition products form.

After only 30 min of heating at fry temperature (180°C), primary volatile oxidation products can be detected (5). The amount of these oxidation products varies depending on the type of oil, food, and heat treatment. Volatile oxidation products may plateau, probably the result of their evaporation or decomposition. The amounts of volatiles produced cannot be used as a measure of the immediate quality of commercial frying oil (5). Elaborate analytical systems are required for the collection, identification, and quantitation of VDP (11). More practical measures of heat abuse in frying oil are available.

## Hydrolysis

The reaction of water with the frying oil results in the development of FFA and partial glycerol esters (7). Development of FFA arises partly from hydrolysis and partly as an end-product of oxidation (3).

The presence of FFA in frying oil was shown to catalyze further hydrolysis of triglycerides (18). Generally, the FFA level does not correlate with fried food quality (3).

The amount of FFA formed is directly proportional to the amount of steam released by food into the fat. Frying large quantities of high-moisture food increases the rate of FFA development. There is a preferential release of unsaturated and shorter chain fatty acids, likely due to their greater water solubility (13,19).

The presence of excessive crumbs in the frying oil also accelerates the rate of FFA development. Frequent skimming of the oil minimizes FFA development. High turnover, however, keeps frying oil in better condition as measured by color, oxidation, polymerization, or FFA development. Quantitatively, the smoke point of used frying fat decreases in proportion to the FFA present (Chapter 12, Fig. 12.5) (20).

# Methods for Assessing the Heat Abuse of Frying Oils

Nonspecific methods have been used to determine the heat abuse of frying oils. Many methods were used because of their convenience or because an unskilled operator could perform them easily. Traditional methods include color, FFA, foam, smoke point, and viscosity. More sophisticated methods have since evolved, which are useful in estimating heat abuse in frying oils. These methods are polar compounds, oxidized fatty acids, conjugated dienes, and fatty acid analysis. Several quick tests also emerged. They include dielectric constant, colorimetric tests, RAU test, test strip, and alkaline contaminant materials. Other methods also emerged based on the determination of specific compounds formed during frying. These are gas–liquid chromatography of dimers and size-exclusion chromatography of dimers and other polymers.

## Traditional Methods

### Color

The color of frying oil darkens with use and eventually affects the color of the fried product (12). Color comparison kits are available that permit the quick assessment of oil color. Color charts, graded from acceptable to unacceptable fried food appearance, were developed by several foodservice companies.

More accurate color determinations can be made with a Lovibond tintometer. Color of oil alone is not adequate to determine the acceptability of frying oils. Different frying oils and foods being fried in these oils will darken the oil at different rates.

### Free Fatty Acids (FFA)

Development of FFA in oil tends to parallel other degradative reactions during frying (3). Free fatty acids develop from both hydrolysis and oxidation. Specific end points for a frying oil depend on the type of oil in the fryer and on the food being fried. However, the FFA content of the oil does not correlate well with frying oil quality.

### Foam Test

During frying, excessive foaming in the oil that does not dissipate or disperse is an indication that the oil should be discarded. Foaming is a safety hazard.

*Smoke Point*
The smoke point of an oil decreases with use because of an increase in low–molecular weight compounds, mainly FFA. The smoke point is not easily determined in a food-service environment.

*Viscosity*
Increase in viscosity is due to the formation of high molecular weight compounds in oil. Food fried in oil with a high viscosity tends to absorb more frying oil. The food then appears soggy, greasy, and less appetizing. Viscosity may be determined by various types of viscometers or simply by timed flow through an orifice. Since viscosity is dependent on temperature, measurements must be made at a standardized temperature.

## Standardized Methods

*Polar Components*
Determination of total polar components is an approved method of IUPAC (21) and the Association of Official Analytical Chemists (AOAC) (22). Total polar material is determined by dissolving 2.5 g oil in a petroleum ether/diethyl ether (87:13) mixture. The sample is eluted on a silica gel column, where polar compounds are absorbed (23). After evaporation of the elution solvents, the nonpolar material is weighed, and total polar material is determined by the difference. In some countries, a level of 27% total polar material is the upper limit for frying oil (16).

*Oxidized Fatty Acids*
An oxidized fatty acid method is used in Germany to evaluate frying oil quality (23). The procedure is outlined in Fig. 18.4. Oxidized fatty acids are petroleum ether–insoluble fatty acids. The frying oil is considered to be significantly deteriorated at 0.7–1.0% oxidized fatty acids. The relative accuracy of this method is reported to be poor (23).

*Conjugated Dienes*
Formation of reactive intermediates occurs during the deterioration of the frying oil. Some methods used to determine these intermediates include carbonyl values, conjugated dienes, or anisidine values. Official methods for these determinations are available (21,22).

Conjugated dienes occur when polyunsaturated fatty acids are oxidized. One double bond shifts to form the conjugated diene. This shift is measurable by the absorbance at 232 nm. During frying, absorbance of oil at 232 nm increases, then plateaus. This test is only useful for polyunsaturated oil (25).

In general, the measurement of intermediates is not a satisfactory method for determining the progressive deterioration of an oil.

*Fatty Acid Composition*
Fatty acid composition changes during frying. In particular, polyunsaturated fatty acids decrease, and total saturated fatty acids increase. As expected, the most significant decreases (>25%) occur in the most highly unsaturated fatty acids (26).

## Quick Tests

*Dielectric Constant*
As an oil degrades during frying, there is an increase in the polar materials that affect the dielectric constant of the oil (Fig. 18.2) (7,27,28). Generally, the dielectric constant depends on the oil type, water content, amount of oil leached from food being fried, and the polar materials from oil deterioration. A food oil sensor, an instrument marketed by Ebro Int., formerly Northern Instruments Corporation (Lino Lakes, MN), may be used to measure the dielectric constant of an oil. The instrument readings are reported to correlate well with chemical changes in FFA, color, total polar material, peroxide value, and dienes (Table 18.3) (28).

*Colorimetric Tests*
A spot test was developed by Robern and Gray (29) utilizing a silica gel glass slide impregnated with alkaline bromocresol green. The presence of FFA or oxidized lipids caused the indicator to change color from blue to green to yellow.

*RAU Test*
This method uses a bromothymol blue indicator. It is similar to the colorimetric test and is marketed under the Oxifrit-Test trade name (30,31). The test involves mixing a fat sample that has been dissolved in an organic solvent with the color indicator. The resulting color change is compared to four color grades progressing from turquoise blue to blue-green, green, and olive green to brown. The color grade

**TABLE 18.3**
**Correlation Between Instrument Response and Other Analyses of Combined Data of Three Shortenings with Different Stabilities Deteriorated Both by Frying Potatoes and by Heating Without Frying**

| Analysis | Number of Samples | Correlation Coefficient |
|---|---|---|
| Total polar materials | 24 | 0.991 |
| Decrease in iodine value | 18 | 0.947 |
| Color | 24 | 0.785 |
| Peroxide value | 24 | 0.773 |
| Diene content | 24 | 0.745 |
| Free fatty acids | 24 | 0.569 |

correlates with analytical bench methods for oxidized fatty acids and polar and nonpolar components.

*Test Strip*
A test strip first marketed in 1986 was developed for determining when oil should be discarded (32). The strip was formulated to determine the FFA content in a kitchen environment. The correlation between titratable FFA of an oil and the color of the test strip was 0.89. Correlations between test strip FFA and the taste of various foods were reported to be a practical means of determining frying oil quality.

*Alkaline Contaminant Materials (ACM)*
Alkaline contaminant materials are generated from oil-processing, fried food, and oil degradation products from use in frying (1). Alkaline contaminant materials raise the pH of the oil due to the formation of alkaline materials, particularly soap. Bromophenol blue is used as the indicator. Color changes from yellow to green to violet occur with an increase in ACM or soap concentrations. ACM was then correlated with the number of days the frying oil was used. The increase in soap corresponding to the days of frying and frying oil quality does not necessarily apply to all frying oils and products fried in them.

## Complex Methods

Various techniques have been used to measure the formation of specific products in abused frying oils. Generally, these techniques are complex and are useful only in laboratory situations.

*Gas-Liquid Chromatography*
This may be used to separate and quantify oxidative and thermal dimers in heated oils (33). Dimers are correlated with heat abuse. Cyclic monomers were also determined with this method (34).

*Size-Exclusion Chromatography*
Low molecular weight material, as well as unchanged and polymeric glycerides, may be separated by molecular size using gel permeation chromatography (23,35–37). The major problem with size exclusion chromatography is the length of time required for separation. A procedure using a high-performance size-exclusion chromatograph (HPSEC) was also described (38). High-performance size-exclusion chromatographic analysis can be completed in 20 min.

Dobarganes et al. (39) initially separated frying fat with the IUPAC (21) method and then applied it to HPSEC. Separation of nonaltered triglycerides gave a more accurate measurement of dimers, oxidized triglycerides, diglycerides, and fatty acids.

## Methods Evaluation

Several studies compared procedures for evaluating frying oil abuse. Results of tests for polar materials generally show the best reproducibility. Polar content also relates to the dielectric constant, a much simpler, rapid test convenient for on-site analyses. The method chosen for evaluating frying oil abuse depends on accuracy desired, time allowed to run the test, and cost. Testing requirements for quality control purposes differ markedly from those for a development chemist or oil researcher.

## Summary

Many complex reactions occur in frying fat during use. These may be classified into oxidation, polymerization, and hydrolysis. Some deterioration reactions are viewed as creating the desirable flavor, color, and texture of fried food. Others are undesirable from the perspectives of flavor, nutrition, and toxicology. Because of the complexity of deterioration, a universal analytical determination for end point detection is not available. Unacceptability of fried food is an obvious end point to the oil's fry life, particularly in foodservice preparation. The amount of total polar materials in the frying oil have shown the best correlation between food quality and frying oil deterioration. The dielectric constant of frying oil also correlates to oil degradation and acceptability of fried food.

## *trans* Fats, Acrylamide, Stabilization, Safety Issues

The decade following the publication of the first edition of this book has seen several issues brought before the American public with respect to used frying oils, i.e. *trans* fats and acrylamide formation in fried foods. (39,40) As of Jan. 1, 2006, *trans* fatty acids must be listed on food labels as a separate line. The snack food industry responded before this deadline by switching to naturally stable oils such as corn, cottonseed, and mid-oleic sunflower oils for the deep fat frying of snacks and potato chips. Although thermal heating of hydrogenated frying fats does not produce appreciable increases in *trans* fatty acids, the desire to eliminate them from the diet has stimulated much research into alternatives to hydrogenated fats, including the use of structurally modified fats in food products and as frying media. These include high-oleic soybean oil, low and ultra-low linolenic soybean oil, high-oleic corn oil, mid-oleic sunflower oil, and high-oleic canola oil. Studies concluded in the laboratory have shown that they perform equal to or better than hydrogenated products (41–50).

A second subject receiving much attention during the past 10 years is acrylamide. It is a widely used monomer suspected of being genotoxic, carcinogenic, and producing neuropathy. It is metabolized to the epoxide, glyciamide, also considered to be neurotoxic. An excellent review of the analysis and mechanism of acrylamide formation in fried foods appeared in 2002 and should be consulted for further information (40).

In order to minimize acrylamide formation, it is recommended that fry operators lower the frying temperature, control the temperature closely, and never fry at temperatures above 175°C. Also, deep fat frying oils should not contain silicones (dimethylpolysiloxane, DPMS). In frying potatoes and products therefrom, older potatoes should not be used because they contain higher contents of reducing sugars which are precursors to acrylamide formation (40).

Another area receiving much attention over the past decade is that of stabilizing frying oils with natural antioxidants and the effects of frying conditions on the minor constituents including tocopherols, sterols and phenolic acids (49,50). Virgin olive oil, sesame seed, and rice bran oils contain a number of natural antioxidants that, when blended with high-oleic sunflower oil, unhydrogenated rapeseed, or soybean oils, perform in frying tests equal to palm olein, a very stable frying fat commonly used in Europe (51–52).

In many European countries, frying fats and oils are regulated, either by federal or local laws, having been in place from 1973–1993, not so much from the point of view of health issues, but from a desire to improve the quality of fried foods. Frying fats should be discarded when polar compounds exceed 24–27%. Acid values of 1–4% are considered unacceptable and smoke points below 170°C are unacceptable. Frying fats are not regulated in the United States, but free fatty acids above 2% are considered unusable (53). Although frying fats are subject to control under the provisions of the Federal Food, Drug, and Cosmetic Act, the FDA believes it has not been determined that frying fats are injurious to health (53). Acrolein is one of the many volatile breakdown products from frying fats. Exposures to 170–430 ppb results in respiratory irritation. The Occupational Safety and Health Administration (OSHA) has established limits of 0.1 ppm (100 ppb) in workroom air during an 8-h day, 40-h work week.

## References

1. Blumenthal, M.M.; J.R. Stockler; and P.J. Summers. *J. Am. Oil Chem. Soc.* **1985,** *62,* 1373.
2. Clark, W.; and G. Serbia. *Food Technol.* **1991,** *45,* 84.
3. Fritsch, C.W. *J. Am. Oil Chem. Soc.* **1981,** *58,* 272.
4. Nawar, W.W. In *Flavor Chemistry of Fats and Oils*; D.B. Min and T.H. Smouse, Eds.; American Oil Chemists' Society: Champaign, IL, 1985; pp. 39–60.
5. Blumenthal, M.M. *Food Technol.* **1991,** *45,* 68.
6. Perkins, E.G. *Food Technol.* **1960,** *14,* 508.
7. White, P.J. *Food Technol.* **1991,** *45,* 75.
8. Jacobson, G.A. *Food Technol.* **1991,** *45,* 72.
9. Zhang, Y.; W.J. Ritter; C.C. Barker; P.A. Traci; and C. Ho. In *Lipids in Food Flavors*; American Chemical Society Series 558; C. Ho and T. Hartman, Eds.; American Chemical Society: Washington, DC, 1994; pp. 49–60.
10. Patton, S.; I.J. Barnes; and L.E. Evans. *J. Am. Oil Chem Soc.* **1959,** *36,* 280.
11. Chang, S.S.; P.J. Peterson; and C. Ho. Chemical Reactions Involved in the Deep-Fat

Frying of Foods, *J. Food Sci.* **1978**, *55*, 718.

12. Lawson, H.W. *Standards for Fats and Oils*; AVI Publishing: Westport, CT, 1985; pp. 55–57.
13. Noble, A.C.; C. Buziassy; and W.W. Nawar. *Lipids* **1967**, *2*, 435.
14. Selke, E.; W.K. Rohwedder; and H. Dutton. *J. Am. Oil Chem. Soc.* **1975**, *52*, 232.
15. Lomanno, S.S.; and W.W. Nawar. *J. Food Sci.* **1982**, *47*, 744.
16. Paradis, A.J.; and W.W. Nawar. *J. Food Sci.* **1981**, *46*, 449.
17. Perkins, E.G. *Food Technol.* **1967**, *21*, 125.
18. Krishnamurthy, R.G. In *Bailey's Industrial Oil and Fat Products*, 4th ed.; D. Swern, Ed.; John Wiley & Sons: New York, 1982; Vol. 2, pp. 315–341.
19. Buziassy, C.; and W.W. Nawar. *J. Food Sci.* **1968**, *33*, 305.
20. Formo, M.W. Smoke, Fire, and Flash Points. In *Bailey's Industrial Oil and Fat Products*, 4th ed.; D. Swern, Ed.; John Wiley & Sons: New York, 1979; Vol. 1, pp. 210–212.
21. IUPAC Applied Chemistry Commission on Oils, Fats, and Derivatives. Method 2.505: Evidence of Purity and Determination from Ultraviolet Spectrophotometry; and Method 2.507: Determination of Polar Components in Frying Fats. In *Standard Methods for Analysis of Oils, Fats and Derivatives*, 7th ed.; C. Paquot and A. Hautfenne, Eds.; Blackwell Scientific: Oxford, U.K., 1987.
22. Association of Official Analytical Chemists (AOAC). Method 18.074: Polar Components of Frying Fats. In *Official Methods of Analysis*, 13th ed.; AOAC, Washington, DC, 1984. Dobarganes, M.C.; M.C. Perez-Camino; and G. Marquez-Ruiz. *Fat Sci. Technol.* **1988**, *90*, 308.
23. Billek, G.; G. Guhr; and J. Waibel. *J. Am. Oil Chem. Soc.* **1978**, *55*, 728–733.
24. American Oil Chemists' Society (AOCS). Method Ti 1a-64: Spectrophotometric Determination of Conjugated Dienoic Acid; and Method Cd 8-53: Peroxide Value. In *Official Methods and Recommended Practices of the American Oil Chemists' Society*, 3rd ed.; American Oil Chemists' Society: Champaign, IL, 1983.
25. Peled, M.; T. Gutfinger; and A. Letan. *J. Sci. Food Agric.* **1975**, *26*, 1655.
26. Miller, L.A.; and P.J. White. *J. Am. Oil Chem Soc.* **1988**, *65*, 1324.
27. Graziano, V.J. *Food Technol.* **1979**, *33*, 50.
28. Fritsch, C.W.; D.C. Egberg; and J.S. Magnuson. *J. Am. Oil Chem. Soc.* **1979**, *56*, 746.
29. Robern, H.; and L. Gray. *Can. Inst. Food Sci. Technol. J.* **1981**, *14*, 50.
30. Berger, K.G. *The Practice of Frying*; PORIM, Ministry of Primary Industries: Selangor, Malaysia, 1984.
31. Croon, L.B.; A. Rogstad; T. Leth; and T. Kiutamo. *Fette Seifen Anstrichm.* **1986**, *88*, 87.
32. Mlinar, J. *J. Am. Oil Chem. Soc.* **1986**, *63*, 838–839.
33. Firestone, D. *J. Am. Oil Chem Soc.* **1970**, *40*, 247.
34. Rojo, J.A.; and E.G. Perkins. *J. Am. Oil Chem. Soc.* **1987**, *64*, 414.
35. Aitzetmuller, K. *Fette Seifen Anstrichm.* **1972**, *74*, 598.
36. Guhr, V.G.; and J. Waibel. *Fette Seifen Anstrichm.* **1979**, *81*, 511.
37. Perkins, E.G.; R. Taubold; and A. Hsieh. *Food Technol.* **1973**, *50*, 223.
38. White, P.J.; and Y.-C. Wang. *J. Am. Oil Chem. Soc.* **1986**, *63*, 914.
39. Anon., *Fed. Regist.* **2003**, *68* (July 11), 41433–41506.
40. Gertz, C.; and S. Klostermann. *Eur. J. Lipid Sci. Technol.* **2002**, *104*, 762–771.
41. Warner, K.; and M.K. Gupta. *J. Am. Oil Chem. Soc.* **2003**, *80*, 275–280.
42. Su, C.; M.K. Gupta; and P.J. White. *J. Am. Oil Chem. Soc.* **2003**, *80*, 171–176.

43. Suhelli, K.C.; W.E. Artz; and P. Tippaywat. *J. Am. Oil Chem. Soc.* **2002,** *79,* 1197–1200.
44. Warner, K.; and S. Knowlton. *J. Am. Oil Chem. Soc.* **1997,** *74,* 1317–1322.
45. Gupta, M.K. *inform* **1998,** *9,* 1150–1154.
46. Warner, K.; and M.K. Gupta. *J. Food Sci.* **2005,** *70,* 5395–5400.
47. Gunstone, F.D. *J. Oleo Sci.* **2001,** *50,* 269–279.
48. Matthaus, B. *Eur. J. Lipid Sci. Technol.* **2006,** *108,* 200–211.
49. Kiatsrichart, S.; M. Brewer; K.R. Cadawallader; and W.E. Artz. *J. Am. Oil Chem. Soc.* **2003,** *80,* 479–483.
50. Erickson, M.; and N. Frey. *Food Technol.* **1994,** *48,* 63–68.
51. Kochar, S.P. *Eur. J. Lipid Sci. Technol.* **2000,** *102,* 552–559.
52. Silkeberg, A.; and S.P. Kochhar. U.S. Patent 6,033,706, 2000.
53. Stier, R.F. In *Frying Technology and Practices*; M.K. Gupta, K. Warner, and P.J. White, Eds.; AOCS Press: Champaign, IL, 2004; pp. 178–199.

# 19

# Evaluation of Passive and Active Filter Media

**Robert A. Yates**

*Dallas Group of America, Inc., Fabricon Blvd., Jeffersonville, IN 47130*

Among the eight recommendations for optimal frying made by the 3rd International Symposium on Deep Fat Frying (March 2000, Hagen-Halden, Germany), was the statement: "One of the basic tools to ensure food and oil quality is the use of filtration. Filter materials should be used to maintain oil quality as needed" (1). Wide varieties of materials are promoted as filter media for used frying oils and the determination of the best filter medium for an application is not simple. It is important to understand the differences between passive and active filtration. Passive filtration is the removal of particulates with an inert filter medium. There are two types of passive filters: absolute (or surface) and depth (2).

Absolute filters are usually characterized by a single thin layer of regular openings of known porosity (Fig. 19.1). Generally, the smaller the openings, the finer the particle size retained. Absolute filtration with a steel or plastic screen, cartridge, or paper filter can remove particles down to about 5 µm. The main concern with absolute filters is the ratio of filtering surface area to the number of particles to be removed. When particles begin to block the openings, the liquid flow rate through the filter slows. When most or all of the openings are blocked, the liquid flow stops, and the filter is said to be "blinded."

In depth filtration, a thick layer of porous material containing irregular, or "tortuous" channels, will trap particles 1 µm in diameter, or even smaller, throughout the matrices (Fig. 19.2). A depth filter can be a pad made from a thick nonwoven fabric, or a cake of filter-aid materials such as diatomaceous earth or perlite, which are inert powders. Diatomaceous earth (often referred to as DE or diatomite) consists of the shells of hydrous silica secreted by diatoms, which are small prehistoric aquatic plants related to algae. Deposits of DE have been found on all continents. Perlite is the generic name for naturally occurring siliceous volcanic rock consisting of fused sodium potassium aluminum silicate. Depth filters do not blind as easily as absolute filters because they have a greater porosity and thus a greater filtering surface area.

Deep fried food absorbs a significant amount of oil and carries it out of the fryer. Fresh oil must be added to the fryer to maintain the proper oil level. In systems like doughnut frying where there is a very high oil turnover rate, the oil quality (and thus the food quality) seldom deteriorates to the point where the oil must be discarded. Oil from this type of frying system is filtered through a passive filter medium to remove particles that might otherwise deposit on the food.

In frying operations where the oil turnover rate is not sufficiently high, the concentrations of free fatty acids (FFA) (total acidic material), polymers, and other triacylglycerol (fat or oil) degradation products increase throughout the frying cycle. Batters, breading, and the food itself can leach undesirable impurities into the oil. Improper cleaning with cleansers and boil-out compounds can leave behind trace metals that can contaminate the oil. All of these are oil-soluble contaminants that cannot be removed by passive filtration. When oil-soluble impurities are allowed to build up in a fryer, the oil quality deteriorates over time until the oil is unfit for use and discarded.

Active treatment removes a portion of the oil-soluble impurities that cannot be removed by passive filtration. There are two types of active treatment: phase separation/emulsion breaking and adsorption. In phase separation/emulsion breaking, the oil is put in contact with an aqueous solution. Impurities are extracted into the water layer, the water layer is drained away, and the oil layer is returned to the fryer (3). This process is not practical for most foodservice applications.

In adsorptive treatment of frying oil, the adsorbent (or a blend of two or more adsorbents) attracts the polar products of triacylglycerol degradation and/or prooxidant metals and holds them for removal by filtration. Adsorbents are used as loose powders or as powders embedded in a fiber matrix. By design, they also serve as depth filters to remove very fine particulates. By removing oil-soluble contaminants (which can act as catalysts for further degradation reactions) as well as particles, active filtration is more effective in slowing the rate of oil decomposition than passive filtration. Controlled frying studies indicate that after several days of frying and filtration with an adsorbent-type filter aid, the concentration of total polar compounds in the oil is lower than the level that can be explained by adsorption alone. Adsorptive filtration has also been shown to have a positive effect on the nutritional value of used edible oil (4).

Adsorption is a surface phenomenon resulting from the characteristics of the outermost layer of atoms or molecules on the solid. There are two types of adsorption: physical adsorption (*physisorption*), which involves only relatively weak intermolecular forces, and chemical adsorption (*chemisorption*), which essentially involves the formation of a chemical bond between the sorbate molecule and the surface of the adsorbent. Characteristics that distinguish physical and chemical adsorption are shown in the Table 19.1.

Table 19.2 is a list of U.S. patents that illustrate the variety of materials, both natural and synthetic, that are promoted for active filtration of used edible oil. Table 19.3 shows the typical analysis of some of these materials. The pH values shown in Table 19.3 are those of a 5% suspension in deionized water. Surface area was

**TABLE 19.1**
**Distinguishing Adsorption Characteristics**

| Physical adsorption | Chemical adsorption |
|---|---|
| Low heat of adsorption | High heat of adsorption |
| (< 2 or 3 times latent heat of evaporation) | (> 2 or 3 times latent heat of evaporation) |
| Nonspecific | Highly specific |
| Monolayer or multilayer | Monolayer only |
| No dissociation of adsorbed species | May involve dissociation |
| Only significant at relatively low temperatures | Possible over a wide temperature range |
| Rapid, non-activated, reversible | Activated; may be slow and irreversible |
| No electron transfer, although polarization | Electron transfer leading to bond formation |
| of sorbate may occur | between sorbate and surface |

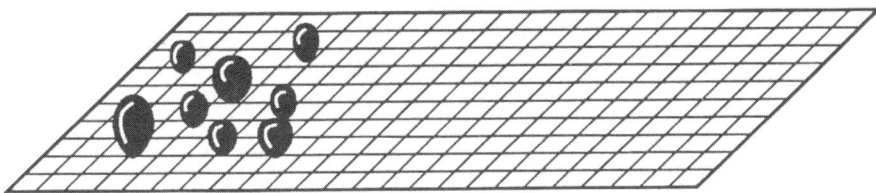

**Fig. 19.1.** Absolute filter.

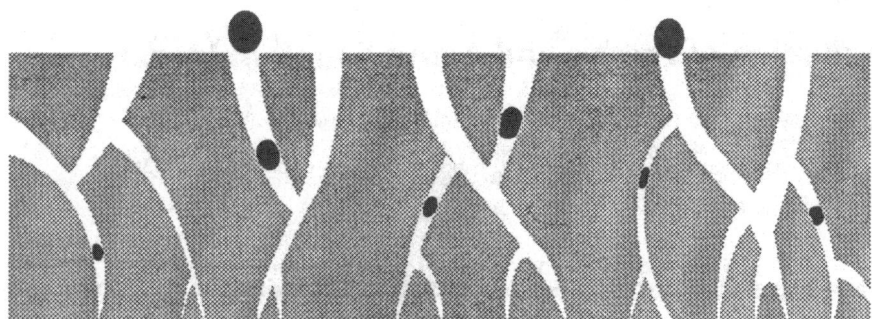

**Fig. 19.2.** Cross section of depth filter.

**TABLE 19.2**
**U.S. Patents for Oil Treatments and Materials**

| U.S. patent # | Year | Material(s) |
| --- | --- | --- |
| 3,231,390 | 1966 | Magnesium oxide |
| 3,947,602 | 1976 | "Food-compatible acid" with diatomaceous earth, activated carbon, silica, or activated alumina. |
| 3,954,819 | 1976 | Molecular seive |
| 3,968,741 | 1976 | Activated carbon |
| 3,976,671 | 1976 | Molecular seive |
| 4,112,129 | 1978 | Mixture of diatomite, synthetic calcium silicate, and synthetic magnesium silicate. |
| 4,330,564 | 1982 | "Food-compatible acid" and rhyolite carrier. |
| 4,349,451 | 1982 | "Food-compatible acid" and rhyolite carrier. |
| 4,681,768 | 1987 | Synthetic magnesium silicate. |
| 4,734,226 | 1988 | Acid-treated amorphous silica. |
| 4,735,815 | 1988 | Mixture of gel-derived alumina and acid-activated bentonite or synthetic magnesium silicate. |
| 4,764,384 | 1988 | Mixture of synthetic amorphous silica, synthetic magnesium silicate, and diatomaceous earth. |
| 4,880,652 | 1989 | Synthetic amorphous silica. |
| 4,988,440 | 1991 | Mixture of carbon and synthetic calcium or magnesium silicate. |
| 5,151,211 | 1992 | Neutral bleaching clay and chelating polycarboxylic acid. |
| 5,252,762 | 1993 | "Base treated inorganic porous adsorbents." |
| 5,348,755 | 1994 | Activated carbon |
| 5,391,385 | 1995 | Mixture of alumina and amorphous silica. |
| 5,560,950 | 1996 | Mixture of cyclodextrin and an adsorbent. |
| 5,597,600 | 1997 | Mixture of synthetic magnesium silicate and an alkaline component. |
| 6,187,355 | 2001 | Mixture of synthetic magnesium and calcium silicates with citric acid and perlite. |
| 6,248,911 | 2001 | Metal substituted silica xerogel. |
| 6,638,551 | 2003 | Mixture of silica, acidic alumina, clay, and (optionally) citric acid |

determined by nitrogen adsorption (5) with a Model 2200 high-speed surface area analyzer (Micromeritics Instrument Corp., Norcross, GA).

## Chemical Additives

From the list in Table 19.2, it is clear that some filter aids promoted for both passive and active oil filtration include chemical additives such as citric acid. While of significant use in the preparation of edible oils, the benefit of their addition to used frying oils is questionable. Citric acid is a well-known sequestering agent used to control prooxidant metals in the refining of edible oils. Citric acid decomposes at 175°C (347°F), minimizing its effectiveness in a foodservice operation.

Antioxidants such as tocopherol are added to edible oils with the intention of improving shelf life. Antifoams are used to suppress foaming during frying. These materials are added to the fryer as the oil lost to absorption by the food is replaced ("addback" or "make-up" oil). There is no documented evidence that further addition of these materials is beneficial to oil life.

## Methods of Evaluation of Filtration Media

Several methods are used to evaluate filter media for foodservice applications. The methods vary greatly in time, cost, and levels of technical expertise required to perform them.

## Controlled Frying Study

The most desirable way to evaluate filter media is via a controlled frying study. In

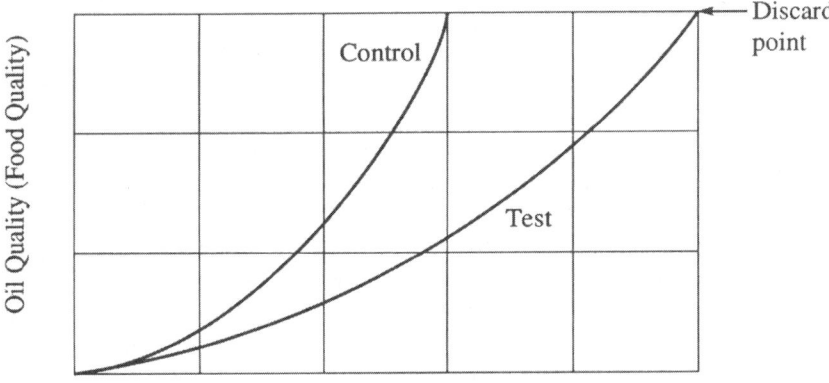

**Fig. 19.3.** Typical controlled frying study; y axis can be %FFA, color, dielectric constant, total polar compounds, or organoleptic response.

**TABLE 19.3**
Common Physical and Chemical Properties of Adsorbent Materials

| Material | Trade name | pH | %LOD[a] | %LOI[b] | Surface Area m²/g |
|---|---|---|---|---|---|
| Activated carbon | Darco T-88[c] | 8.00 | 4.8 | 90.6 | 824 |
| Acidic pH alumina | Brockmann I[d] | 4.30 | <1.0 | 2.8 | 167 |
| Neutral pH alumina | ABA 6000[e] | 6.55 | 11.5 | 18.1 | 312 |
| Basic pH alumina | A-2[f] | 9.55 | 0.1 | 7.7 | 787 |
| Bleaching earth #1 | Filtrol 105[g] | 3.80 | 15.2 | 5.6 | 311 |
| Bleaching earth #2 | Tonsil Supreme[h] | 3.55 | 12.3 | 4.9 | 400 |
| Calcium silicate | Silasorb[i] 10.45 | 6.4 | 12.5 | 168 | |
| Diatomaceous earth | FW-18[j] | 9.60 | 0.1 | 0.2 | <10 |
| Magnesium silicate | Magnesol XL[k] | 8.50 | 10.8 | 10.9 | 619 |
| Silica | TriSyl[l] | 2.75 | 65.6 | 3.9 | 955 |

[a] LOD = Loss on drying at 105°C for 2h.
[b] LOI = Additional loss on ignition at 900°C for 1.5h (% of dry basis weight).
[c] American Norit Co., Jacksonville, FL.
[d] Aldrich Chemical Co., Milwaukee, WI.
[e] Selecto, Inc., Kennesaw, GA.
[f] LaRoche Chemicals, Baton Rouge, LA.
[g] Harshaw Filtrol, Cleveland, OH.
[h] LA Saloman, Port Washington, NY.
[i] Celite Corp., Lompoc, CA.
[j] Eagle Picher, Reno, NV.
[k] Dallas Group of America, Inc., Liberty Corner, NJ.
[l] WR Grace & Co., Baltimore, MD.

such a study, duplicate frying/filtering cycles are performed, with the only variable being the filter medium. It is best to perform the cycles side by side, but for industrial frying, where very large fryers are used, the trials may be run sequentially.

Thoroughly clean the fryers, accurately calibrate their temperature controls including thermometers, and fill each with frying oil from the same production lot. Cook the same weight of food (preferably from the same production lot) in each oil, each day, for several days. The weight of food cooked per pound of oil in the fryer should be equal to a typical day's frying volume.

Filter the oils at regular intervals: one as a control, and one with the filter aid under test. For example, at the end of each day filter the control oil with diatomaceous earth (an inert filter powder), and filter the test oil with an active filter aid. Sample the oils at the same time(s) each day and analyze for the quality parameters important to the operation. Typical parameters used to evaluate oils in a frying study are total polar compounds (IUPAC method 2.507) (6), free fatty acids (AOCS method Ca 5a-40) (7), photometric color (AOCS method Cc 13c-50) (7), and dielectric constant (8). For each fryer, record the weight of oil used to replace oil lost to samples, absorbed by food, and absorbed by the filtration media. If possible, the food fried in each of the oils should be evaluated by a trained taste panel.

Any criteria of oil or food quality can be used as the end point of the test. Graph the oil and taste evaluation data versus time or pounds of food cooked (Fig. 19.3). Evaluate the two systems on the basis of oil quality, food quality, the amount of food cooked per pound of oil used, and the overall cost of each system per pound of food prepared.

A controlled frying study gives the most complete and accurate evaluation of filter medium, but it is also the most costly. To evaluate every different possible filter media by this method would not be practical. Screening must be done so that only the one or two most promising materials are evaluated in the controlled study. There are several ways to evaluate filter media in the laboratory (9–11).

## Laboratory Batch Treatments

The simplest and least expensive way to evaluate adsorbents in the laboratory is in small-batch treatments. Split a sample of degraded oil from a frying operation into several portions large enough to analyze important chemical and physical oil quality parameters. Treat and/or filter each oil portion with the different filter media

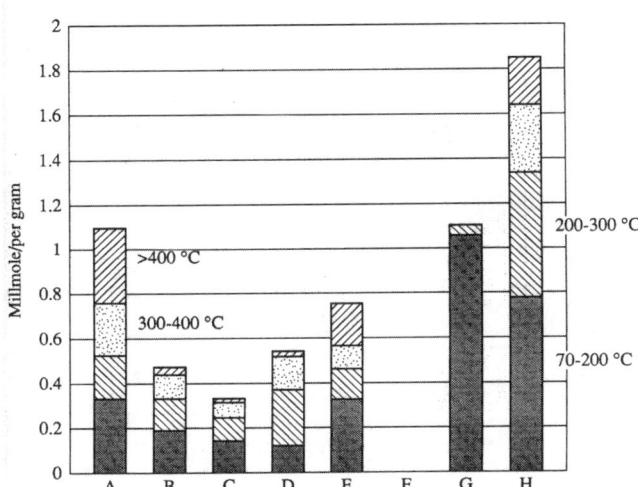

**Fig. 19.4.** Surface concentrations and their intensity distributions for basic adsorption sites of materials tested: A, activated carbon; B, alumina (basic pH); C, alumina (pH neutral); D, alumina (acidic pH); E, bleaching earth; F, diatomaceous earth; G, silica; H, synthetic magnesium silicate. [See Zhu, Z.Y., R.A. Yates, and J.D. Caldwell, J. Am. Oil Chem. Soc. 71:189 (1994).]

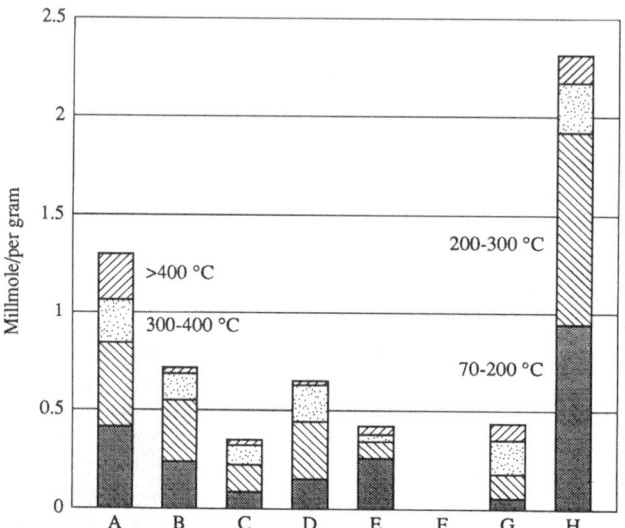

**Fig. 19.5.** Surface concentrations and their intensity distributions for acidic adsorption sites of materials tested. Letter allocations along the x-axis and source are as in Fig. 19.4.

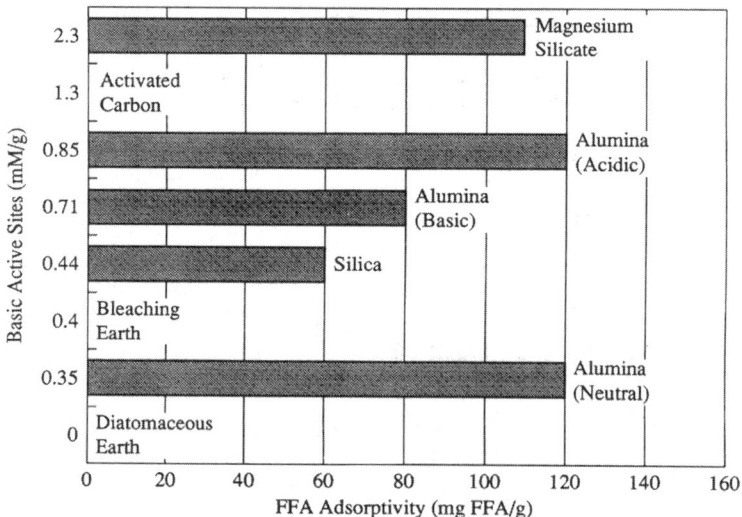

**Fig. 19.6.** Bar graph of conentrations of basic adsorption sites versus adsorptivity of free fatty acids. Source as in Fig. 19.4.

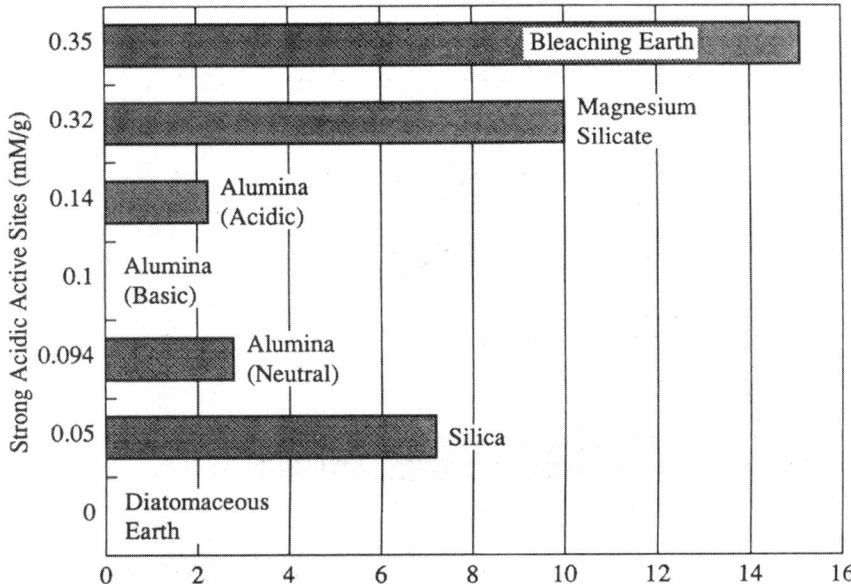

**Fig. 19.7.** Bar graph of concentrations of strong acidic sites versus adsorptivity of color. Source as in Figure 19.4.

under conditions similar to those to be used in the full-scale filtration, for example, temperature, contact time, and percent filter aid.

For evaluation of inert (passive) filter media, analyze the oils for particulate remaining after filtration (AOCS method Ca 3a-46) (7). For active filter aids, analyze the oils for important oil quality parameters such as % FFA (7) and photometric color (7), among others.

Available analytical methods may not be sensitive enough to measure the change made in an oil quality parameter by a single small percentage of adsorbent treatment. Therefore, it may be helpful to either perform a treatment with a larger percentage of the adsorbent, to perform several sequential treatments, or both.

Table 19.4 shows results of the evaluation of two adsorbents by large, sequential adsorbent treatments. Portions of used oil were treated with 5 wt% adsorbent at 150°C for 30 min. Enough oil was removed from each filtrate to analyze for % FFA, color, and dielectric constant. Although this first treatment showed a measurable difference between the two materials, the remaining filtrate was weighed and again treated with 5 wt% adsorbent. This procedure was repeated five times. After the five treatments, the data clearly showed that adsorbent 1 significantly reduced FFA concentration, photometric color, and dielectric constant. Adsorbent 2 reduced the color somewhat, but did not significantly affect FFA concentration or dielectric constant.

In any filter medium evaluation, one should never rely on the analysis of a single oil

**TABLE 19.4**
Example of Multiple Treatment Study Data Table[a]

| Sample | Treatment | %FFA | Photo color | Dielectric constant |
|---|---|---|---|---|
| Initial oil | — | 8.56 | 82.0 | 3.47 |
| Adsorbent #1 | 1 | 7.83 | 60.2 | 3.42 |
| Adsorbent #1 | 2 | 7.12 | 53.7 | 3.26 |
| Adsorbent #1 | 3 | 6.40 | 48.6 | 3.18 |
| Adsorbent #1 | 4 | 5.56 | 44.1 | 3.10 |
| Adsorbent #1 | 5 | 4.71 | 38.3 | 3.04 |
| Initial oil | — | 8.56 | 82.0 | 3.47 |
| Adsorbent #2 | 1 | 8.53 | 76.0 | 3.57 |
| Adsorbent #2 | 2 | 8.43 | 68.9 | 3.46 |
| Adsorbent #2 | 3 | 8.45 | 63.9 | 3.43 |
| Adsorbent #2 | 4 | 8.40 | 59.1 | 3.41 |
| Adsorbent #2 | 5 | 8.37 | 55.5 | 3.38 |

[a] Five sequential 5 wt% treatments; treatment contact time, 5 min; treatment temperature, 150°C.

quality parameter. The composition of the filter aid can affect the analytical evaluation of frying oil quality in either laboratory or field evaluations. Highly alkaline filter aids can greatly lower the titratable FFA value of used oil by a combination of adsorption of FFA and chemical reaction to form metal soaps. Addition of an oil-soluble acid to the filter aid may temporarily mask some metals from measurement of the titratable soap value of the oil (AOCS method Cc 17-79) (7). Note: AOCS Method Cc 17-79 is not a direct determination of metal soaps, but is a measure of total relative alkalinity. The method was developed for the determination of very low residual concentrations of alkaline metals in refined oils, and its accuracy in the determination of values over 50 ppm is questionable.

# Extraction of Adsorbed Polar Compounds from Filter Cakes

The filter cake that results from the treatment of used frying oil with an adsorbent is composed of the filter aid, the absorbed oil (nonplar triacylglycerol), and the adsorbed polar compounds. To accurately quantify the amount of each polar compound actually removed from the oil by the adsorbent, it is necessary to extract all of the polar compounds from the filter cake. Extraction from the filter cake is analogous to the

**TABLE 19.5**
**Total Weight in Grams of Compounds Extracted from Filter Cakes by Polar Solvents.**

| Adsorbent | Pol.[b] | Tg[c] | Dg[d] | LMW[e] | TPC[f] | mg polars/ga |
|---|---|---|---|---|---|---|
| Activated carbon | 0.00 | 0.50 | 0.83 | 1.90 | 2.74 | 67 |
| Alumina (acidic) | 0.06 | 0.07 | 0.10 | 3.09 | 3.25 | 79 |
| Alumina (neutral) | 0.05 | 0.10 | 0.15 | 3.34 | 3.54 | 86 |
| Alumina (basic pH) | 0.15 | 0.29 | 0.44 | 3.03 | 3.63 | 89 |
| Bleaching earth | 1.43 | 0.72 | 0.87 | 2.27 | 4.47 | 109 |
| Diatomaceous earth | 0.04 | 0.02 | 0.01 | 0.05 | 0.10 | 2 |
| Magnesium silicate | 1.37 | 0.98 | 1.44 | 5.33 | 8.14 | 199 |
| Silica | 1.22 | 1.72 | 2.04 | 1.88 | 5.14 | 125 |

Source: Yates, R.A., and J.D. Caldwell, J. Am. Oil Chem. Soc. 70:507 (1993).
[a] mg Total polar material adsorbed per g of active filter aid.
[b] Pol. = Polar compound content.
[c] Tg. = Triacylglycerol content.
[d] Dg. = Diacylglycerol content.
[e] LMW = Low–molecular weight content (free fatty acids, monoacylglycerols, etc.)
[f] TPC = Total polar compound content.

IUPAC column chromatography method of determining total polar compounds in used frying oil (8). Absorbed oil is removed by extraction with a nonpolar solvent like petroleum ether, leaving only the filter aid and adsorbed polar oil degradation products. Adsorbed impurities are removed by subsequent extraction(s) with polar solvents and quantitatively determined by high-performance size exclusion chromatography (12, 13). By measuring the initial weight of adsorbent, the total weight of filter cake, and the weight of each extraction residue, the amount of each component adsorbed per unit weight of adsorbent can be calculated.

Table 19.5 shows the results of this type of analysis for several common filter materials (14). In that study, 41 g (as received) of several filter aids were used to treat 1.3 gallon portions of the same used frying oil in a restaurant-type recirculating filter. The weight of each species of polar compound adsorbed was determined for each filter medium. It is clear from this data that diatomaceous earth adsorbs very few, if any, impurities from used frying oil and thus is only a passive filter aid. This analysis also shows that some filter media adsorb only one species of impurity from the oil. The best example of this is the aluminas that mainly adsorbed FFA.

# Temperature-Programmed Desorption (TPD)

The characteristics of adsorbents responsible for adsorption of impurities from used oils include surface area; pore size distribution; and number, relative polarity, and intensity distribution of the active adsorption sites (15). Temperature-programmed desorption is a relatively new technology used to determine the number and strength of the active acidic and basic adsorption sites of a material.

In TPD, the adsorbent is placed in a column that is fitted into a gas chromatograph (16) or special TPD instrumentation (17). A probe chemical (usually $n$-butyl amine for the determination of acidic sites or trifluoroacetic acid for basic sites) is introduced into the column until the adsorption sites are saturated. Once saturated, the column is slowly heated to desorb any adsorbed probe chemical. The total quantity of desorbed material and the temperature at which it desorbs are measurements of the number and strength of the adsorptive sites, respectively. Figures 19.4 and 19.5 show the results of this type of analysis for several common adsorbents (18). Figures 19.6 and 19.7 show how these data may relate to the adsorption of impurities from used frying oils.

# Field Evaluation

After screening products in the laboratory and/or test kitchen, it is time to "use it and see how it works." (Some evaluators go directly to this method after reading published literature and product bulletins.) Field evaluation involves monitoring the frying/filtration process in a manner similar to the controlled frying study: Monitor the current filtration system for several cycles as the control, and then put the new system in place and follow it for several fry cycles. It is best to record data for several cycles and to average all data to allow for day-to-day and week-to-week variations in the frying system. These variations can be minimized (but not eliminated) by starting all studies on the same day of the week. Record data for pounds of food fried and pounds of oil used to generate data tables and graphs as in the controlled frying study.

Although marketers like to use phrases like "double the useful life of your frying oil," it is difficult to predict the exact performance of an active filter aid in a particular frying system because no two systems are identical. Variables such as type of oil used, type and amount of food fried each day, and frying temperature all affect the rate of triacylglycerol degradation and thus the frying oil quality. Some known factors that affect the performance of active filter media are explained as follows:

## *Amount of Time Adsorbent Contacts Oil*

Adsorption is not instantaneous, and the optimal contact time will vary depending on a number of factors. Sufficient time must be allowed for the adsorbent to adsorb the maximum amount of impurities. Adsorbent contact time for industrial fryers may be several hours. In restaurant filtrations, the contact time is short, usually about 5 min, to keep labor costs to a minimum. Also, since restaurant-size filters are not heated, the contact time is kept short so that the oil does not cool and become too viscous to pump through the filter.

## Temperature of Oil When Treated

Although some adsorbents may perform better at low oil temperatures, the oil must be kept hot enough to ensure acceptable flow rates through the filter. Bheemreddy et al. state that the optimal temperature for active filtration of a used frying oil is greater than 140°C (19).

## Amount of Adsorbent Used per Treatment

Adsorption is a phenomenon wherein the more adsorbent used in a single treatment, the less effective each unit weight of the adsorbent may be. Therefore, doubling the weight of adsorbent used per treatment will not necessarily remove twice as many impurities.

# Factors to Consider When Choosing a Filter Aid (20)

## Cost of Material

Consider the manufacturer's suggested treatment level and determine the cost per day of filtration. Active filter aids are generally more expensive than passive filter aids; however, the additional expense is often offset by savings resulting from improved oil and food quality, extended oil life, and reduced labor for fryer cleaning.

## Adsorptive Capacity of Material

How much polar compound is adsorbed per unit weight of adsorbent? One adsorbent may be less expensive per pound, but it could be more costly to use because it is less effective. Impregnated filter papers and thin filter pads are easier to use than loose powders supplied in bulk packages or packets, but they may contain too little active material to adsorb a significant amount of impurities.

## Adsorptive Selectivity of Material

Some adsorbent materials are selective and primarily adsorb only one class of polar compounds.

## Amount of Oil Absorbed per Filtration

The weight of oil discarded in the filter medium needs to be replaced with fresh oil. This is not necessarily negative because it can act as a partial discard system to keep the oil in the fryer fresher.

## Filtration Rate

In most foodservice operations with two or more fryers, it is desirable to filter all oils through the same filter cake without stopping to change the filter medium. Some filter media may have a good flow rate for a single fryer volume of oil but have significantly poorer flow rates on subsequent filtrations as the layer of particles builds.

### Investment in New Equipment
Can the new filter medium be used in the current equipment, or will additional equipment purchases be necessary?

*Safety*
Materials to be used for filtration of edible oils, including the binders in filter pads, must be acceptable for contact with food to the governing body for the country in which they are to be used. Manufacturers must supply a current Material Safety Data Sheet (MSDS) to be made available to all employees. Filter aids should be stored according to the manufacturer's instructions on the MSDS and kept separated from cleaning chemicals to avoid contamination. Unless otherwise instructed, filter aids should be stored in the original container with all labels intact.

## Some Final Notes on Optimal Filtration
For operations where several volumes of oil are filtered through the same filter cake, it is best to filter the fryer with oil in the best condition first, and then proceed sequentially, ensuring that the fryer with oil in the worst condition is filtered last. Filtration of the best oil first slightly dilutes the poorer oil with the clean oil trapped in the filter cake. Clean oil filtered through a filter cake that contains dirty oil will become contaminated with impurities from the oil trapped in the pores. Desorption of polar compounds from the adsorbent into the clean oil can also occur. This system may not be possible in operations where undesirable flavors may be carried from one fryer to the next through the filter cake. In such a case, the fryer with the highest potential for contributing "off" flavors should be placed last in the filtering rotation.

Although it is possible, it is not economical to use an active filter aid to rejuvenate severely degraded oil. Active filter aids are most effective when used as a maintenance treatment to remove small amounts of degradation products with each filtration. As part of a good oil maintenance program, they can slow the rate of oil degradation. This results in a more economical operation and better quality fried foods.

## Conclusion
Recent trends in deep frying include a movement toward lower *trans* fatty acid oils that degrade more rapidly in the frying process than their predecessors and may have much shorter useful lives. These newer frying oils are also more expensive, but the cost of the oil versus the cost of the filter medium is not the only consideration. In addition to savings from reduced oil usage due to useful oil life extension, the added benefits of a good filtration system are improved and more consistent food quality, reduced energy usage through improved heat transfer (high-quality oils have greater heat capacity, so less energy is needed for frying), and labor reduction resulting from

reduced fryer cleanings and fewer oil changes.

## References

1. DGF Meeting Offers Frying Oil Recommendations, *inform* **2000,** *11*, 630–631.
2. M.J. Matteson. *Filtration Principles and Practices*; Marcel Dekker: New York, 1986.
3. Lopez, M. U.S. Patent 4,968,518, 1990.
4. Perkins, E.G.; and C. Lamboni. Magnesium Silicate Treatment of Dietary Heated Fats: Effect on Rat Liver Enzyme Activity. *Lipids* **1998,** *33*, 683–687.
5. Brunauer, S.; P.H. Emmett; and E. Teller. *J. Am. Chem. Soc.* **1938,** *60*, 309.
6. Paquot, C.; and A. Hautefenne, Eds. *IUPAC Standard Methods for the Analysis of Oils, Fats, and Derivatives*; Blackwell Scientific Publications: Oxford, U.K., 1987.
7. American Oil Chemists' Society (AOCS). *Official Methods and Recommended Practices of the American Oil Chemists' Society*; AOCS Press: Champaign, IL, 2001.
8. Graziano, V.J. *Food Technol.* **1979,** *33*, 50.
9. Boki, K.; S. Shinoda; and S. Ohno. *J. Food Sci.* **1989,** *54*, 1601.
10. McNeill, J.; Y. Yakuda; and B. Kamel. *J. Am. Oil Chem Soc.* **1986,** *63*, 1564.
11. Mancini, J.; L.M. Smith; R.K. Creveling; and H.F. Al-Sheik. *J. Am. Oil Chem. Soc.* **1986,** *63*, 1452.
12. Perkins, E.G.; R. Taublod; and A. Hseih. *J. Am. Oil Chem. Soc.* **1973,** *50*, 223.
13. White, P.; and Y. Yang. *J. Am. Oil Chem. Soc.* **1986,** *63*, 914.
14. Yates, R.A.; and J.D. Caldwell. *J. Am. Oil Chem. Soc.* **1993,** *70*, 507.
15. Ruthven, D.M. *Principles of Adsorption and Adsorption Processes*; John Wiley & Sons: New York, 1984.
16. Choudhary, V.R. *J. Chromatogr.* **1983,** *268*, 207.
17. Micromeritics Instrument Corp. *The Micro Report*; Micromeritics Instrument Corp.: Norcross, GA, 1991; Vol. 2, No. 3.
18. Zhu, Z.Y.; R.A. Yates; and J.D. Caldwell. *J. Am. Oil Chem. Soc.* **1994,** *71*, 189.
19. Bheemreddy, R.M.; M.S. Chinnan; K.S. Pannu; and A.E. Reynolds. *J. Food* **2002,** *25*, 23–40.
20. Lawson, H.W. *Standards for Fats and Oils*; AVI Publishing: Westport, CT, 1985.

# 20

# Sensory Evaluation of Frying Fat and Deep-Fried Products

**Sharon L. Melton**
*Retired, University of Tennessee, Department of Food Science and Technology, Knoxville, TN*

Sensory evaluation is defined as "a scientific discipline used to invoke, measure, analyze, and interpret reactions to characteristics of foods and materials as they are perceived by the senses of sight, smell, taste, touch, and hearing" (1). According to Mounts and Warner (2), sensory evaluation is a necessary part of both product development and quality control. Penfield and Campbell (3) suggested that sensory evaluation could also be used in research and shelf life studies and that many sensory tests are used in food-science research.

Sensory tests are classified into two basic types: affective and analytical (4). In affective testing (which includes consumer testing), preference, acceptance, or opinions on food products are determined. Usually 50–100 untrained, randomly selected people are asked which one, two, or more products they prefer. Affective tests are important to market-development studies but are not as relevant to flavor research investigations (4). A detailed discussion of consumer sensory evaluation was published by the American Society for Testing Materials (ASTM) (5). Analytical tests evaluate differences between products and rate the quality or intensity of odor or flavor characteristics (4). According to Penfield and Campbell (3), analytical tests discriminate between or among samples (difference tests) or describe or score the quality of a product (descriptive tests). The discriminator tests also include sensitivity tests.

Several tests are used in affective or consumer testing. The most commonly used evaluation technique for measuring acceptability or preference is the hedonic scale (3). Hedonic scales generally have 5–9 points, with descriptive terms denoting likability of product or product characteristic, as shown in Fig. 20.1. Other scales have been used to minimize misinterpretation of the terms in Fig. 20.1. The facial hedonic scale illustrated in Fig. 20.2 (6) is particularly useful for this effort.

A 9-point hedonic scale was used by Fuller et al. (7) with a panel of 20–26 individuals to estimate the acceptability of potato chips fried in different oils, with and without the antioxidant propyl gallate (PG), and stored under different light intensities for up to 20 days. Fig. 20.3 shows that potato chips fried in cottonseed

Taste each of the three coded samples in the order they are presented and mark how much you like or dislike it on the scale below.

Code

Like extremely _____ _____ _____
Like very much _____ _____ _____
Like moderately _____ _____ _____
Like slightly _____ _____ _____
Neither like or dislike _____ _____ _____
Dislike slightly _____ _____ _____
Dislike moderately _____ _____ _____
Dislike very much _____ _____ _____
Dislike extremely _____ _____ _____

Comments:

**Fig. 20.1.** Hedonic score card for consumer acceptability of products.

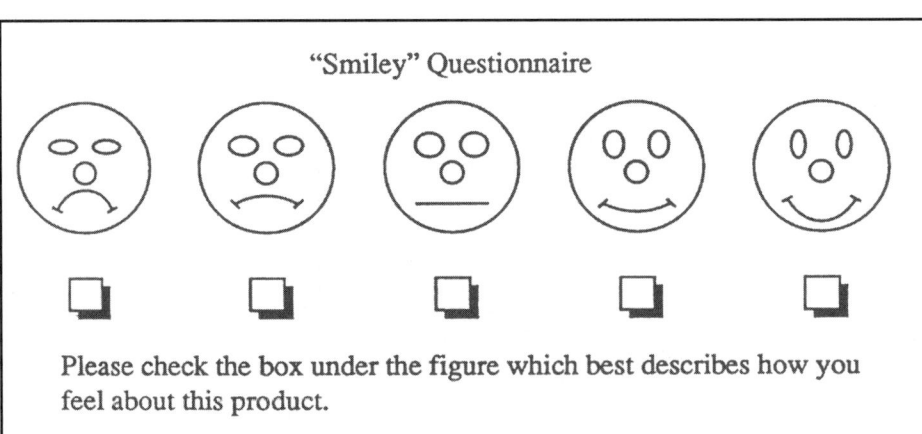

"Smiley" Questionnaire

Please check the box under the figure which best describes how you feel about this product.

**Fig. 20.2.** Facial hedonic scale used by the Continental Can Company (6).

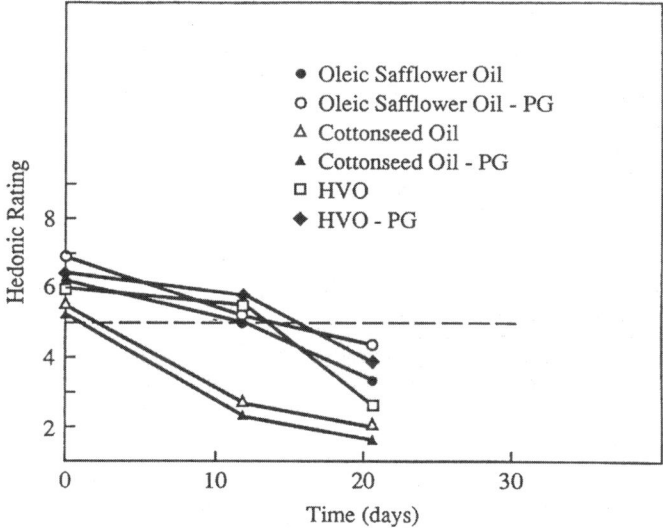

**Fig. 20.3.** Average hedonic ratings of potato chips fried in different oils with and without antioxidant and stored under light with 13 foot-candles intensity (HVO is hydrogenated vegetable oil, and PG is the antioxidant corresponding to 0.01% propyl gallate and 0.005% citric acid in the oil) (7).

oil, with and without PG, received lower scores, initially and throughout the 20-day storage period, than chips fried in other oils.

A consumer panel ($n$ = 119) used an 8-point hedonic scale (1 = dislike extremely to 8 = like extremely) to evaluate flavor and overall likability of fresh potato chips fried in canola and/or cottonseed oils (8). The mean sensory scores for flavor and overall likability did not differ ($P > 0.05$) among the oil types used for chip production and were between "like slightly" and "like moderately" (5.2–5.5).

Consumer-type panels ($n$ = 77), however, using likability hedonic scales showed different degrees of likability for the flavor of codfish fried in fresh vs. used (discarded) commercial frying fat (9). Part of the panel ($n$ = 29) liked the flavor of codfish fried in fresh fat more than that of codfish fried in used fat; part of the panel ($n$ = 24) reversed that trend by liking the flavor of codfish fried in used fat. The rest of the panel ($n$ = 24) liked the flavor of codfish fried in fresh or used fat equally well.

Behavior exhibited by the consumer panel evaluating fried codfish also extended to consumer evaluation of food fried in different oils (9). Actual flavor differences due to types of frying oil certainly exist; however, the degree of likability for individual panelists is dependent on backgrounds and life experiences.

Several specific methods are included in analytical sensory difference tests. The triangle, duo-trio, paired comparison, and ranking or rating difference from a control are examples of such tests. The premise of these tests is to detect differences among samples based on one attribute; if a flavor difference is being evaluated, the test is invalid if panelists differentiate by color, texture, or temperature instead. The triangle

Sample provided          Odd sample

_____          _____

_____          _____

_____          _____

Difference observed: _____

**Fig. 20.4.** Typical questionnaire for the triangle test (3).

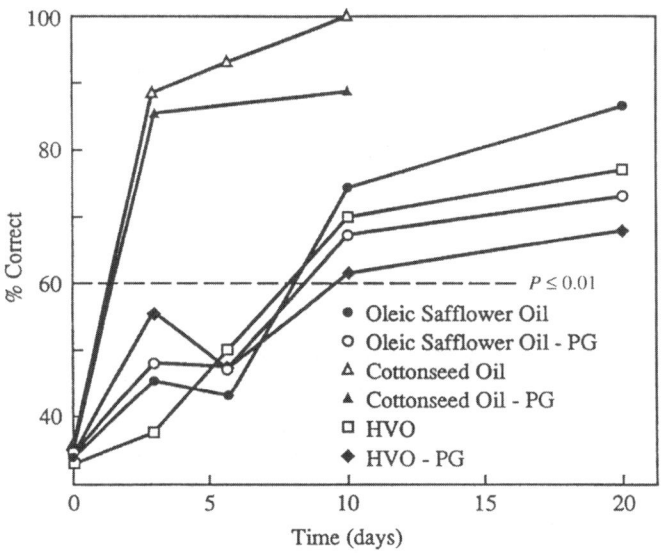

**Fig. 20.5.** Correct identification in the triangle test of identical samples of potato chips fried in different oils with and without antioxidant added and stored in air up to 20 days in 13 foot-candles intensity light (HVO is hydrogenated vegetable oil, and PG is the antioxidant corresponding to 0.01% propyl gallate and 0.005% citric acid in the oil) (7).

test is perhaps the easiest analytical test for inexperienced panelists. In this test, the panelist is asked to determine which two of the three samples presented are identical and which one is different. A typical questionnaire for a triangle test is presented in Fig. 20.4. Data from triangle tests can be analyzed statistically with significant results. Tables for determining the significance have been published elsewhere (3,8).

The triangle test was used by Fuller et al. (7) to determine the time required to cause significant odor changes in potato chips fried in different oils and held under light intensities of 13 and 100 foot-candles. The control for the triangle test was nitrogen-packed potato chips stored in the dark. During sensory evaluation, the control and treated chips were presented as the odd sample an equal number of times. The results of the triangle tests are given in Fig. 20.5 for potato chips fried in different oils, with and without antioxidant, and stored under 13 or 100 foot-candles of light for up to 20 days. Figure 20.5 shows that after only 3 days of storage at 13 foot-candles of light, the panel differentiated between chips fried in cottonseed oil and stored in light and those stored in the dark under nitrogen. By 12 days, the panel differentiated between chips fried in hydrogenated vegetable oil or high oleic sunflower oil and stored in light and chips fried in the same oils but stored in the dark.

The triangle test was also used by Maga (10) to determine if blindfolded, experienced panelists could differentiate between the flavors of light- and dark-colored potato chips during storage. The blindfolded panel could not differentiate flavor or texture between light- and dark-colored potato chips stored up to 2 wk.

The other analytical sensory tests (duo-trio, paired comparison, and ranking or rating difference from a control) are also applicable to fried foods. Details on the use of these tests are in Warner (4) and in ASTM manuals (11). Except for ranking, these tests, including the triangle, can be used for product matching, product improvement, process improvement, and quality control (2). The triangle and duo-trio tests are also used as measures of panelists' discriminatory ability in selection for a trained panel.

Analytical descriptive tests require panelists with more experience and training than do simpler difference tests (4). Panelists must not only be able to detect differences among samples, but they must have a vocabulary to describe their perceptions. The American Oil Chemists' Society (AOCS) has adopted, as official methods, two variations of quality and intensity rating scales for flavor of oil, as shown in Fig. 20.6 and 20.7. The usability of such scales as those listed in Fig. 20.6 and 20.7 depends on panelists being very well trained in identifying various flavors and scoring their intensity. According to Mounts and Warner (2), analytical descriptive tests can be applied to matching products, product improvement, process improvement, cost reduction, and/or selection of new sources of supply, quality control, and correlation of sensory with chemical/physical measures.

Mounts et al. (12) used a trained 15-member sensory panel to evaluate fresh and stored potatoes fried in low-linolenic acid, regular, and partially hydrogenated soybean oil. The panel evaluated the fried potato flavor on overall intensity and for intensity of individual flavors on a 10-point intensity scale, where 10 = bland and 1 = strong. The panel was experienced in testing vegetable oil according to AOCS Recommended Practice CG 2-83 (13). Fishy flavors were significantly lower in potatoes fried in soybean oil with lower linolenic acid content than those fried in regular soybean oil (6.2% linolenic acid) (12). Potatoes fried in low-linolenic acid soybean oil also lacked the hydrogenated flavors of potatoes fried in hydrogenated oil.

Robertson and Morrison (14) used analytical descriptive tests to evaluate the flavor of potato chips fried in different oils and stored under accelerated conditions at 31°C for up to 10 wk. A 10-member trained panel used the following defined flavor descriptors: fresh = good chip flavor, no off flavors; heated = good quality but flavor

| Flavor grade | Description of flavor[a] |
|---|---|
| 10 (excellent) | Completely bland |
| 9 (good) | Trace of flavor but not recognizable |
| 8 | Nutty, sweet, buttery, corny |
| 7 (fair) | Beany, hydrogenated, popcorn, bacony |
| 6 | Oxidized, musty, weedy, burnt, grassy |
| 5 (poor) | Raw, reverted, rubbery, watermelon, bitter |
| 4 | Rancid, painty |
| 3 (very poor) | Fishy, buggy |
| 2 | Intensive objectionable flavors |

[a]Flavor intensity at presented concentration rated slight.

**Fig. 20.6.** AOCS flavor-quality scale (4).

| Flavor grade | Flavor intensity |
|---|---|
| 10 | Bland |
| 9 | Trace |
| 8 | Faint |
| 7 | Slight |
| 6 | Mild |
| 5 | Moderate |
| 4 | Definite |
| 3 | Strong |
| 2 | Very strong |
| 1 | Extreme |

| | Description Intensity | | |
|---|---|---|---|
| Descriptions | Weak | Moderate | Strong |
| _____ | _____ | _____ | _____ |
| _____ | _____ | _____ | _____ |
| _____ | _____ | _____ | _____ |
| _____ | _____ | _____ | _____ |

**Fig. 20.7.** AOCS flavor-intensity scale (4).

indicating some storage; stale = fair, not fresh, flat, or tasteless for aging; off flavor = poor undesirable flavor, such as, grassy; and rancid = bad, old oil flavor. The panel was trained to detect the stages of flavor deterioration using chips that were fresh or heated in a 60°C oven for 3, 6, or 9 days.

The descriptive quality ratings were transformed into numerical scores where 1 = fresh to 5 = rancid, and the data were analyzed statistically. After the panelists were trained to detect stages of flavor deterioration, they were also trained to evaluate intensities of each odor–flavor descriptor as 1 = weak, 2 = moderate, and 3 = strong. These intensity ratings were expressed as flavor intensity values (FIV), where FIV is calculated from the following equation:

$$FIV = [N_w + 2(N_m) + 3(N_s)]/N$$

where $N_w$ = number of weak responses, $N_m$ = number of medium responses, $N_s$ = number of strong responses, and $N$ = number of tasters (14). An average FIV was determined for each descriptor used by the panel to describe the flavor of the chips: fresh, heated, stale, off flavor, and rancid.

The trained panel was used to evaluate the flavor of potato chips fried in cottonseed, sunflower, or palm oil and then stored for up to 10 wk. A coded reference chip also was evaluated at the same time as the stored chips. The reference chip was fresh and was fried in cottonseed oil. The analysis of variance for descriptive quality ratings of potato chips fried in different oils and stored up to 10 wk is shown in Table 20.1. The sensory scores of the potato chips are shown in Fig. 20.8, and FIV for fresh (0-wk storage time) and 10-wk-stored chips fried in different oils are shown in Table 20.2. The trained panel scored the flavor of fresh chips fried in cottonseed oil between very good, fresh (or a value of 1.0) and good, heated (or a value of 2.0). The average sensory score for the flavor of fresh chips fried in palm and sunflower oils was between 2.0 and 3.0. Of all flavor descriptors, fresh received the maximum FIV in 0-wk chips fried in cottonseed or sunflower oil, while the descriptor receiving the maximum FIV for 0-wk chips fried in palm oil was off flavor (Table 20.2). After 10 wk of storage, the stale flavor description received the maximum FIV for chips fried in cottonseed oil, but the rancid flavor had the highest FIV in chips fried in sunflower oil. Chips fried in palm oil and stored 10 wk had the greatest FIV for the rancid and heated descriptors (Table 20.2).

Two other descriptive methods for testing the quality of fried food are the flavor profile method and the quantitative descriptive analysis. In the flavor profile method, perceived factors are called "character notes." A list of these notes is made by each trained panel member during preliminary evaluation of the food being investigated. The lists are then compared by the entire panel, and agreement is reached on which notes will be used for further analysis. Then each panelist evaluates the sample again for the intensity of each character note and the amplitude of overall aroma and taste. The panel then reaches a consensus of opinion on the intensity of each character note

**TABLE 20.1**
Analysis of Variance of Descriptive Quality Ratings for Stored Potato Chips Fried in Different Oils[a]

| Source of variation | Df[b] | Sum of squares | Mean squares | F-value |
|---|---|---|---|---|
| Time | 5 | 147.364 | 29.462 | 21.14[c] |
| Samples | 2 | 22.149 | 11.079 | 7.95[c] |
| Time × samples | 10 | 26.440 | 2.644 | 1.90 |
| Reps (time × samples) | 36 | 50.198 | 1.394 | |
| Panelists (rep × samples) | 486 | 461.500 | 0.949 | |

Source: Robertson, J.A., and W.H. Morrison, III, J. Food Sci. 43:420 (1978).
[a]Reference samples were omitted from the analysis of variance.
[b]Degrees of freedom.
[c]Significant at the P < 0.05 level.

**TABLE 20.2**
Flavor Intensity Values (FIV) of Stored Potato Chips[a]

| Storage time (wk) | Flavor description | Reference cottonseed | Cottonseed | Sunflower | Palm |
|---|---|---|---|---|---|
| 0 | Fresh | 1.27 | 1.80 | 0.83 | 0.53 |
| | Heated | 0.40 | — | 0.20 | 0.13 |
| | Stale | 1.10 | 0.13 | 0.07 | 0.30 |
| | Off-flavor | — | — | 0.37 | 0.87 |
| | Rancid | 0.17 | 0.10 | 0.30 | 0.20 |
| 10 | Fresh | 2.37 | — | — | — |
| | Heated | 0.17 | 0.20 | 0.03 | 0.07 |
| | Stale | — | 1.20 | 0.73 | 0.90 |
| | Off-flavor | 0.07 | 0.43 | 0.70 | 0.67 |
| | Rancid | — | 0.53 | 0.87 | 0.87 |

Source as in Table 20.1.
[a]Chips fried in oils except reference cottonseed oil were stored at 31°C; those fried in reference cottonseed oil were stored at -20°C.

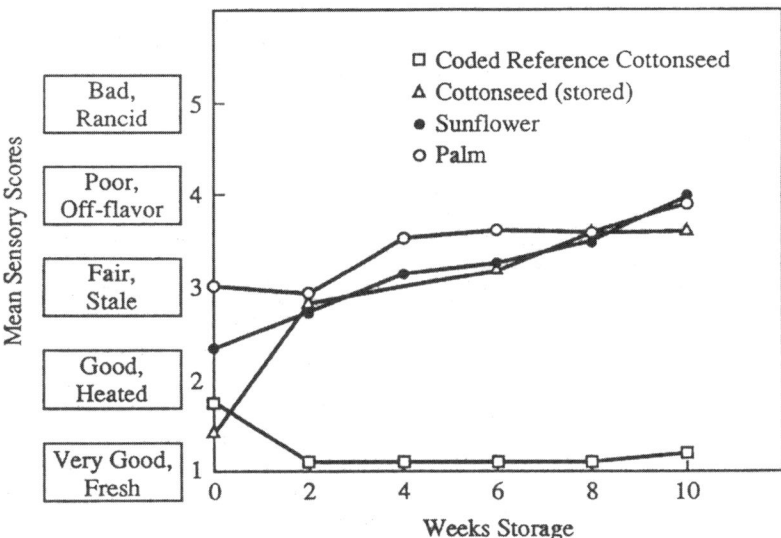

**Fig. 20.8.** Effect of storage on mean (n = 30) sensory scores of potato chips fried in different frying oils (14).

and the amplitude of the overall aroma, taste, and aftertaste. In the original flavor profile method, the results were not subjected to statistical analysis. Although flavor profiling was successful for comparing products, for quality control, and for product development, it was not used extensively in sensory evaluation of fried food or oil. Berry et al. (16) gave a classic application of flavor profiling.

In quantitative descriptive analysis (QDA), the sensory characteristics of a product are identified and quantified by a panel under the guidance of a panel leader (17). After the identification of characteristics to be evaluated by the panel as a group, each panelist individually rates the intensity of each characteristic on a 150-cm horizontal linear scale anchored at the ends by the terms "weak" and "intense." The results of QDA are analyzed statistically using analysis of variance, factor, or regression analysis (2).

Quantitative descriptive analysis was used by researchers analyzing the sensory characteristics of fried food. Stevenson et al. (18) used an eight-member trained QDA panel to assess exterior crispness, interior dryness, intensity of oil flavor, oil off flavor, oily mouthcoat, potato flavor, color, and overall quality of french-fried potatoes cooked in canola oil and soybean fat. Results reported by Stevenson et al. (18) are illustrated in Fig. 20.9, which shows the scores they reported in tabular form. In Fig. 20.9, each line radiating from the center of the diagram represents a sensory characteristic; the distance from the center outward on each line represents the intensity of that characteristic. The point at which the product line intersects each characteristic line is the relative intensity of that characteristic in the product.

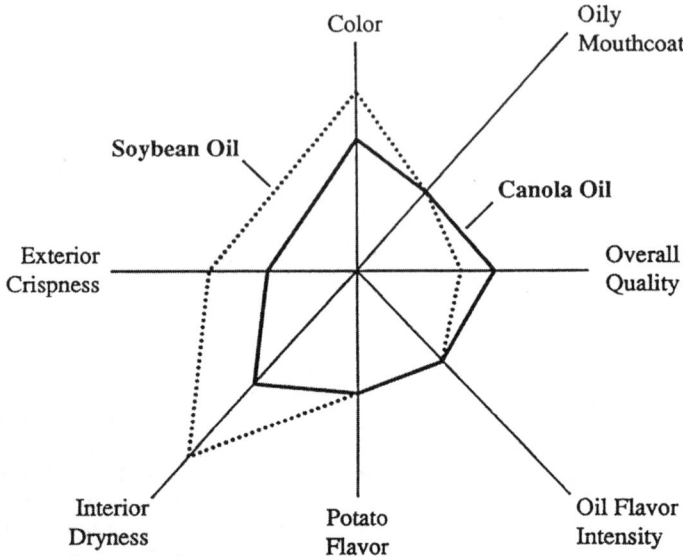

**Fig. 20.9.** Average scores from sensory evaluation (QDA) of french-fried potatoes cooked in canola oil or soybean oil as prepared from data presented by Stevenson et al. (18).

When there is a significant difference in intensity of a characteristic between the two products, there is a difference between the intersections of the product lines with the characteristic line. For example, french-fried potatoes cooked in soybean oil were darker in color, had greater exterior crispness and interior dryness, but had less overall quality than french-fried potatoes cooked in canola oil. French-fried potatoes cooked in either oil had equal intensities of potato and oil flavors and the same level of oily mouth feel (18).

Several sources give extensive details on panel selection and training (19), presentation of samples (20), and sensory evaluation facilities (5,21–23). Warner (4) reported that the qualifications for analytical panelists include an interest in testing oil-containing foods, normal sensory acuity, availability to participate in regularly conducted panels, ability to discriminate known differences, and consistency of response for the same samples. A minimum of 10–12 panelists is recommended for analytical panels if the results will be analyzed statistically (24). Only five highly trained panelists, however, are needed for a flavor profile panel (4); their results are not usually analyzed statistically.

The triangle test is often used to select potential candidates for analytical panels based on their ability to detect flavor differences. To be selected for the panel, a candidate must score at least 60% correct of a total of 24 responses (15). According to Warner (4), attribute scoring or descriptive tests are unsuitable for screening panelists because definite correct responses cannot be established.

In training, panelists should be familiarized with panel room operation, test procedure and products, and scoring or rating methods used. Panelists should then be trained to identify sensory characteristics and to develop a common vocabulary of descriptions for those characteristics. Reference standards for typical odors/flavors are necessary in training a panel, according to Warner (4), who listed reference standards for flavors in oils and oil-containing food. The reference standards also should be characterized by expert judges prior to being used in training. Other chemical standards of odors in food were published by Harper et al. (25).

For presentation to panelists, all samples should be of uniform portion and temperature and should be coded properly. Freshly fried food should be served to panelists at 50°C in covered glass containers (4). The sample size of the fried food should be one-inch cubes, and no more than 2–4 samples should be presented to panelists at any one session (4). If the fried food is a snack food, it may not be possible to have a one-inch cube and it may also be inappropriate to serve it at 50°C. Specific sample recommendations are also given by the ASTM (16). Three-digit random numbers are the most commonly used sample codes for sensory analysis and can be selected from a random number table in a statistics book or sensory evaluation reference book (17).

To avoid bias in sensory analysis experiments with a small number of samples (3 or fewer), all possible orders of sample presentation should be used an equal number of times (3). When there is a large number of samples to be analyzed in an experiment, the order of presentation should be randomized to minimize psychological effects (23).

The environment also is an important factor in the sensory evaluation of food and should be without distractions so that the judges can concentrate on the testing at hand and do so independently (3). The judges should be separated, with each judge in an individual compartment if possible. Detailed information on the requirements for a sensory evaluation laboratory were published by the ASTM (26).

## References

1. Institute of Food Technologists. *Food Technol.* **1981**, *35*, 50.
2. Mounts, T.L.; and K. Warner. In *Handbook of Soy Oil Processing and Utilization*; D.R. Erickson, E.H. Pryde, O.L. Brekke, T.L. Mounts, and R.A. Falb, Eds.; American Oil Chemists' Society: Champaign, IL, 1980; pp. 245–266.
3. Penfield, M.P.; and A.M. Campbell. In *Experimental Food Science*, 3rd ed.; Academic Press: San Diego, CA, 1990; pp. 51–77.
4. Warner, K. In *Flavor Chemistry of Fats and Oils*; D.B. Min and T.H. Smouse, Eds.; American Oil Chemists' Society: Champaign, IL, 1985; pp. 207–221.
5. American Society for Testing Materials (ASTM). STP 682. In *Manual on Consumer Sensory Evaluation*; E.E. Schaefer, Ed.; ASTM: Philadelphia, PA, 1979.
6. Ellis, B.H. *Food Technol.* **1968**, *22*, 583.
7. Fuller, G.; D.G. Guadagni; M.L. Weaver; G. Notter; and R.J. Horvat. *J. Food Sci.* **1971**, *36*, 43.

8.  Melton, S.L.; M.K. Trigiano; M.P. Penfield; and R. Yang. Potato Chips Fried in Canola and/or Cottonseed Oil Maintain High Quality. *J. Food Sci.* **1993,** *58*, 1079–1083.
9.  Melton, S.L.; S. Jafar; D. Sykes; and M.K. Trigiano. *J. Am. Oil Chem. Soc.* **1994,** *71*, 1301.
10. Maga, J.A. *J. Food Sci.* **1973,** *38*, 1251.
11. American Society for Testing and Materials (ASTM). STP 434. In *Manual on Consumer Sensory Testing Methods*; ASTM: Philadelphia, PA, 1968.
12. Mounts, T.L.; K. Warner; G.R. List; W.E. Neff; and R.F. Wilson. *J. Am. Oil Chem. Soc.* **1994,** *71*, 495.
13. American Oil Chemists' Society (AOCS). In *Official Methods and Recommended Practices of the American Oil Chemists' Society*, 4th ed.; D. Firestone, Ed.; American Oil Chemists' Society: Champaign, IL, 1990.
14. Robertson, J.A.; and W.H. Morrison, III. *J. Food Sci.* **1978,** *43*, 420.
15. Caul, J.F. *Adv. Food Res.* **1957,** *7*, 1.
16. Berry, B.W.; J.A. Maga; C.R. Calkins; L.H. Wells; Z.L. Carpenter; and H.R. Cross. *J. Food Sci.* **1980,** *45*, 1113.
17. Stone, H.; J. Sidel; A. Woolsey; and R.C. Singleton. *Food Technol.* **1974,** *28*, 24.
18. Stevenson, S.G.; L. Jeffery; M. Vaisey-Genser; B. Fyfe; F.W. Hougen; and N.A.M. Eskin. *Can. Inst. Food Sci. Technol. J.* **1984,** *17*, 187.
19. American Society for Testing Materials (ASTM). STP 758. In *Guidelines for the Selection and Training of Sensory Panel Members*; ASTM: Philadelphia, PA, 1981.
20. American Society for Testing Materials (ASTM). End Use Products. In *Annual Book of ASTM Standards*; ASTM: Philadelphia, PA, 1989; Vol. 15.07.
21. Larmond, E. *Laboratory Methods for Sensory Evaluation of Food*, Publication 1637; Canadian Department of Agriculture: Ottawa, Ontario, Canada, 1977.
22. Amerine, M.A.; R.M. Pangborn; and E.B. Rossler. In *Principles of Sensory Evaluation of Food*; Academic Press: New York, 1965.
23. Larmond, E., *Food Technol.* **1973,** *27*, 28.
24. Sidel, J.L.; and H. Stone. *Food Technol.* **1976,** *30*, 32.
25. Harper, R.; D.G. Land; N.M. Griffiths; and E.C. Bate-Smith. *Br. J. Psychol.* **1968,** *49*, 231.
26. American Society for Testing Materials (ASTM). In *Physical Requirement Guidelines for Sensory Evaluation Laboratories*, Publication 913; J. Eggert and K. Zook, Eds.; ASTM: Philadelphia, 1986.

# PART 6

# Regulation

# 21

# Regulation of Frying Fat and Oil

**David Firestone**

*Retired, Food and Drug Administration, College Park, MD 20740*

Recently there has been increased recognition that improper use of frying fats degrades the fat and reduces the quality of fried food. There are no worldwide regulations and guidelines for control of frying fat, however. Although no regulations have been established in the United States, a number of European countries promulgated specific laws and regulations. In addition, the Swedish Food Regulations include an ordinance with general advice on how to handle frying fat and oil.

Deep fat frying is a complex method of food preparation in which many reactions occur, resulting in oxidative and hydrolytic degradation and polymerization of the oil. Many studies in the past 50 years showed that thermally oxidized oil affects the growth of test animals. Conversely, additional studies indicated that optimal frying conditions cause no significant change in the fatty acid composition of the frying oil and no toxicity in test animals. Safety and other aspects of deep fat frying technology were discussed at an Institute of Food Technologists symposium in 1990 (1).

## Frying Oil Quality

Recognizing that the quality of fried food is affected by the quality of the frying fat, the German Society for Fat Science (Deutsche Gesellschaft fur Fettwissenschaft, DGF) organized two symposia on frying fats and oils in the 1970s (2,3). Following the 1979 symposium, the DGF proposed that total polar compounds be determined to complement the traditional organoleptic (sensory) evaluation of frying oil quality. This method [AOCS Official Method Cd 20-91 (4)], involving chromatography on a silica gel column, became a standard reference method in many European countries concerned with possible health risks from improper or excessive use of fats and oils for frying. Determination of dimeric and polymeric triglycerides (DPTG) by gel permeation chromatography [AOCS Official Method Cd 22-91 (4)] is also widely used to control frying fat quality and is a routine regulatory test in The Netherlands.

Many laboratory tests were proposed for quality assessment of frying oils (Table 21.1). Quick tests are also available, permitting inspectors and operators to easily screen oils at the fryer. These include the Oxifrit Test (redox indicator) and the Fritest (carbonyl compounds) distributed by E. Merck (Darmstadt, Germany), and the Veri-Fry quick tests available from Libra Laboratories (Metuchen, NJ).

## Regulations and Guidelines in the United States

The U.S. Food and Drug Administration (FDA) has not established specific regulations to control the quality of frying oils, since it has not been determined that frying oils used in deep frying are injurious to health. Frying oils are subject to control, however, under the general provisions of the Federal Food, Drug, and Cosmetic Act, which states that a food is considered to be adulterated if it "contains any poisonous or deleterious substance which may render it injurious to health" [sec. 402(a)(1)], or if it "consists in whole or in part of any filthy, putrid or decomposed substance, or if it is otherwise unfit for food" [sec. 402(a)(3)]. In addition, the FDA's retail food protection code, revised in 1993 (5), contains a set of standards to ensure hygienic practices and adequate operation and maintenance of equipment in food establishments.

The Meat and Poultry Inspection Manual of the U.S. Department of Agriculture, Food Safety and Inspection Service (USDA/FSIS) (6) contains some general guidelines for frying meat and poultry products. These guidelines allow antioxidants and antifoaming agents in frying fat. Excessive foaming, darkened color, and objectionable odor or flavor are evidence of unsuitability, requiring rejection of the fat. Fat or oil should be discarded "when it foams over the vessel's side during cooking or when its color becomes almost black as viewed through a colorless glass container." Serviceable life of fat can be extended by holding the frying temperature below 400°F (204°C), replacing one-third or more each day, filtering as needed, and cleaning the system at least weekly. Addition of an antifoam agent (methylpolysiloxane) to fresh fat is recommended. "Acceptable poultry frying operations should be carried out at about 375°F (190°C) or higher for 10–13 minutes when parts are not precooked. Large amounts of sediment and free fatty acid content in excess of 2% are usual indications that frying fats are unwholesome and require reconditioning or replacement." The manual also contains guidelines for cleanup of frying equipment.

Inquiries were made during 1989–1994 to 35 U.S. cities and all 50 U.S. state health departments and food control agencies to determine the laws and regulations

**TABLE 21.1**
**Laboratory Quality Control Tests**

| | | |
|---|---|---|
| Acid Value | Smoke Point | Epoxides |
| Carbonyl Value | Viscosity | Iodine Value |
| Cyclic Fatty Acids | Volatiles | DPTG[a] |
| Dilectic Constant | Anisidine Value | Refract. Index |
| Total Polar Compounds | Color | TBA Test[b] |
| Fatty Acid Composition | Free Fatty Acids | Diene Value |
| | | Peroxide Value |

[a]DPTG = Dimeric and polymeric triglycerides.
[b]TBA = Thiobarbituric acid.

available for controlling the use of frying fats and oils in restaurants and food-processing establishments. A total of 24 cities and 36 states responded. The replies indicated that there are no specific regulations in U.S. cities and states other than those assuring that the fats and oils used in food-service establishments are obtained from approved sources and are not adulterated. Many health departments responded that there are no specific regulations for frying fat, except as provided in Title 21 of the *Code of Federal Regulations* and the FDA *Food Service Sanitation Manual* of 1976.

The San Francisco Health Department pointed out that, although cooking fats and oils are not addressed in the California Uniform Retail Food Facilities Law, inspectors check for color, sediments, and excessive smoke and odor of oils used in cooking. Corrections are made through the replacement of cooking oils. Inspections by the Philadelphia Department of Public Health include organoleptic comparison of used cooking oil with fresh oil. The Chicago Department of Health regulates fats and oils under the general provisions of Chapter 4-344 of the Municipal Code of Chicago, which addresses sanitation practices in food establishments. Food products are required to have approved labels and be free of rancidity. During routine inspections, frying oils are checked for color, sediments, and foreign objects, as well as excessive smoke. If necessary, oil samples are collected for determination of rancidity by the Kreis test. The Restaurant Inspection Program of Wisconsin's Department of Health and Social Services is primarily concerned that frying fats and oils come from approved sources and are "maintained in a reasonably clean condition." Frying fat and oil are not considered a public health hazard from bacterial contamination because of the high cooking temperatures used in deep fat frying operations. Rather, concerns are related to exhausting frying fumes and addressing consumer complaints of "off" odors and flavors.

In response to an inquiry in 1981 by the Connecticut Department of Health Services, the FDA's State Programs Branch advised the state agency that cooking oils and fats could be reused after filtering as long as the filtering material was clean and the oil looked clear and was properly stored. In 1990, the FDA's State Programs Branch issued the following updated advisory: i) there is no standard frequency for filtering fat used in a deep frying operation; ii) any presence of "off" odors or visible evidence of foreign material, filth, or other adulterants would warrant discarding the fat; and iii) fat must be adequately protected from contamination during use, storage, or filtering. Several states reported an interest in passing legislation or amending regulations requiring food establishments to inform customers of the type of cooking oil used in food preparation and the percentage of saturated fat present in the cooking oil.

## Regulations and Guidelines in Other Countries

During 1990–1994 inquiries were also made about regulations for frying fats and oils and fried food in other countries. Fifty-four countries were contacted (Table

**TABLE 21.2**
**Survey of Frying Fat and Fried Food Regulations: List of Countries Contacted**

| Responded to Survey | Did not Respond to Survey |
| --- | --- |
| Australia | Algeria |
| Austria | China |
| Belgium | China (Taiwan) |
| Brazil | Colombia |
| Canada | Egypt |
| Costa Rica | India |
| Denmark | Indonesia |
| England | Kuwait |
| Finland | Mexico |
| France | Morocco |
| Germany | Paraguay |
| Hong Kong | Peru |
| Hungary | Philippines |
| Iceland | Poland |
| Ireland | Russia |
| Israel | Tunisia |
| Italy | United Arab Emirates |
| Japan | Uruguay |
| Korea | Venezuela |
| Luxembourg | |
| Malaysia | |
| Netherlands | |
| New Zealand | |
| Norway | |
| Pakistan | |
| Portugal | |
| Saudi Arabia | |
| Singapore | |
| South Africa | |
| Spain | |
| Sweden | |
| Switzerland | |
| Thailand | |
| Turkey | |

21.2). Thirty-three countries responded, including 19 of 21 European countries. Austria, Belgium, France, Hungary, Italy, and Spain have specific laws, regulations, or standards for frying oil. Other countries surveyed have no specific laws or regulations for frying fat, although several countries (Germany, The Netherlands, Switzerland, Finland, Norway, and Sweden) enforce measures for practical control in restaurants and fast-food establishments. Additional information on control of frying fats and oils in these and several other countries is outlined.

## Australia

The National Food Authority, established in August 1992, is responsible for setting food standards enforced by the states and territories under their own food laws. Fats and oils for frying are not presently regulated in detail by the 1987 Australian Food Standards Code, but it does prescribe standards for various foods, including frying oil, and states that edible fats and oils used in frying may contain sorbitans and polysorbates as well as not more than 10 mg/kg dimethylpolysiloxane.

Australian Defense Force Specification 5-5-2 (November 1984) requires that deep fat frying be in accordance with good manufacturing practice and comply with state and territory food regulations. Solid fat for deep fat frying should comply with the following: i) moisture, not more than 3 g/kg; ii) free fatty acids, not more than 1 g/kg; iii) slip melting point, not less than 38°C and not more than 49°C; iv) peroxide value, not more than 2 meq/kg; v) gallates, not more than 0.1 g/kg; and vi) clean flavor and free from objectionable odor. Liquid fat for deep frying should comply with similar requirements for moisture, free fatty acids, peroxide value, and gallate content. In addition, saturated fatty acid content should not be more than 500 g/kg total fatty acids. Fats for deep frying should not contain mineral oil or more than 50 g/kg erucic acid.

The Victoria Health Department noted that municipal councils are responsible for monitoring of food premises. Frying oils are subject to collection and analysis for iodine value, saponification value, unsaponifiable matter, acid value, and peroxide value, as well as qualitative tests for adulterants. Some local councils began using Oxifrit Test kits to determine the degree of deterioration of frying fats used in kitchens and bakeries.

## Austria

The Austrian Codex Alimentarius (Austrian Foodstuffs Book), 3rd edn., Chapter B 30, states that frying fat should not exhibit unpleasant odor and taste, unacceptable appearance (dark color, foaming), or high level of carbonaceous residue. Also, frying fats should not have an acid value greater than 2.5, smoke point under 170°C, total polar compounds above 27%, or oxidized fatty acids insoluble in petroleum ether above 1%. Frying fats should not be heated above 180°C.

## Belgium

A royal decree issued in 1974 defined quality standards for edible fats and oils. An additional royal decree issued in 1978 authorized additives in edible oils, including up to 3 mg/kg dimethylpolysiloxane in frying oils. Oils intended for frying must be labeled "Oil for Frying," and dimethylpolysiloxane, if present, must be listed on the label. A law (royal decree) issued in 1988 (7) forbade preparation of fried food in frying fat heated above 180°C or with a free fatty acid content above 2.5%, DPTG above 10%, total polar compounds above 25%, viscosity greater than 37 mPa-sec at 50°C (food fats) or 27 mPa-sec at 50°C (food oils), or smoke point under 170°C. Frying oils and fats may not contain more than 2% linolenic acid. The law specifically forbade preparation of fried food in equipment not provided with temperature control.

## Czech Republic

Guidelines of the National Health Institute effective January 1, 1995, include recommendations that total polar compounds in frying oils should be less than 25% and that DPTG be less than 10%.

## France

A constitutional law in 1905 allowed French authorities to regulate food preparation and to specify conditions for analysis (8). A 1973 regulation specified that deep frying fat should not contain more than 2% linolenic acid. Synthetic antioxidants (BHA, BHT, and gallates) are permitted, as are natural tocopherol concentrates in oil and fats intended for industrial use (minimum 5-kg containers). Silicone additives are prohibited. Decree No. 86-857 of July 18, 1986 (9) specifies that fats and oils with more than 25% total polar compounds are unfit for human consumption.

## Germany

There are no specific laws or regulations in Germany for control of frying fats. Recommendations resulting from the two DGF symposia on frying fats (2,3), however, are generally applied for control of edible fats and oils and frying fats. These recommendations were established following reports of gastrointestinal distress after fried food was eaten. According to A. Seher of the Federal Institute for Fat Research in Münster, an epidemiological study was unable to link abused fat with these episodes, but it revealed that many restaurants were abusing fat, particularly those frying meaty foods.

According to the 1973 DGF recommendations, used frying fats are considered to have deteriorated if i) taste or flavor is unacceptable, ii) smoke point is below 170°C and the content of oxidized fatty acids insoluble in petroleum ether is 0.75% or higher, or iii) the content of oxidized fatty acids insoluble in petroleum ether is higher than 1.0%. After development of the method for determining total polar compounds in 1979, the DGF recommended allowing no more than 27% total polar compounds in food fats. The basic recommendations of the DGF are still valid. Recommendations adopted by the Arbeitskreis Lebensmittelchemischer Sachverstandiger der Lander und des Bundesgesundheitsamtes (ALS) (Working Sector of the Food Chemical Authorities of the Local and State Health Offices) in 1991 are as follows: sensory characteristics (appearance, odor, and taste) of frying fats are of primary importance; petroleum ether–insoluble fatty acids, maximum of 0.7%; total polar compounds, maximum of 24%; smoke point, mininum of 170°C; smoke point difference (from unheated fat), maximum of 50°C; acid value, maximum of 2.0.

## Hungary

There are no mandatory regulations in Hungary for frying fat quality. Standard no. MSz-08-1907-87, valid January 6, 1988, recommends determination of total polar compounds for estimation of fat quality as follows: below 25%, acceptable quality; between 25 and 30%, frying fat should be changed; more than 30%, frying fat is unusable.

The National Institute of Food Hygiene and Nutrition recommended that iron and copper fryers should not be used, frying temperatures should be kept between 160 and 180°C, smoke point below 180°C indicates that the fat has deteriorated and should be discarded, and the surface-to-volume ratio of fryers should be minimized. Sunflower oil may be used for up to 8–10 h of frying, if treated carefully. Corn oil may be used for 10–13 h, and lard, for 18–20 h, if treated carefully. After frying, the oil should be filtered and stored at a low temperature. Deteriorated frying oils should not be used for human consumption or for animal feed.

## Iceland

Guidelines issued by the Environmental and Food Agency of Iceland are as follows:

1. Use fats and oils intended for deep frying. Many types of salad oil do not maintain their quality at the temperatures used for deep frying.
2. Do not mix used fats or oils with new ones, as that would accelerate deterioration.
3. Clean all frying equipment regularly and filter the fat. All dirt and residue of detergents and cleaning products adversely affect the quality of fats and oils. Avoid contact of copper or copper compounds with fat. Do not apply salt to foodstuffs above the frying pan, since metal compounds in salt could result in the deterioration of the fat.
4. The appropriate frying temperature is 165–190°C. Higher temperatures result in dark color, oxidation, hydrolysis, and polymerization. If the temperature is too low, the frying time is too long, affecting the quality of foodstuffs. To minimize the drop in temperature, it is important not to overload the frying pan.
5. When the fat is heated, the temperature should not be set higher than the

temperature to be used for frying.

6.  Durability of fat can be prolonged by keeping the temperature between 90 and 120°C when the fat is not in use.
7.  As the heat transfer in solid fat is low, it should be melted at low temperatures to avoid overheating certain parts of the fat. Slight burning or overheating of fat can accelerate deterioration and spoil all of the fat in the pan.
8.  Remember that spoiled frying fat can have adverse health effects.

## Israel

Although Israel has no specific regulations for cooking and frying oils, guidelines published by the Swedish National Food Administration (10) are recommended for application by Food Control Administration inspectors. Israel's Food Control Administration submitted a request to the Standards Institution of Israel in 1992 that a requirement for total polar compounds be added to the vegetable oil standard, but the request was rejected.

## Italy

The Ministry of Health issued a regulation January 1, 1991, for fats and oils used for frying "to prevent possible risks to consumers from improper or excessive use of fats for frying." The regulation specifies the following: i) use only those oils and fats for frying that are resistant to heat; ii) avoid frying temperatures above 180°C; iii) total polar compounds should not be more than 25 g/100 g; iv) prepare the food to be fried properly, avoiding the presence of water and addition of salt and spices as much as possible which accelerate changes in frying fat; v) allow excess oil to drain from the food after frying to avoid absorption of excessive oil by food; vi) change the oil frequently, check the quality of the oil or fat during frying, and do not use oil too long (indicated by darkened color, viscosity, and tendency to smoke); vii) filter the oil if it will be used and clean the filter and fryer because charred crust, viscous oily residue, or old oil accelerates alteration of oil; viii) avoid "reconditioning" oil (addition of fresh oil), as fresh oil rapidly alters when in contact with used oil; and ix) protect frying oils and fats from light.

## Japan

There are no formal regulations in Japan regarding the quality of frying oil. Concerning food establishments, however, there are the following guidelines for determining when to discard frying oil: i) if smoke point is less than 170°C, ii) if acid value is more than 2.5; and iii) if carbonyl value is more than 50.

## Luxembourg

There are no specific regulations for frying fat in Luxembourg. General regulations in force for all foods, however, also apply to frying fats. For practical control in food establishments preparing fried food, the food inspector uses E. Merck's Fritest. If the Fritest is positive, then frying fat is checked for free fatty acids, total polar compounds, taste, color, odor, and appearance.

## The Netherlands

Food laws in The Netherlands are enforced by 16 food inspection services, each covering an inspection area of about one million people. Inspectors sample the frying oil or fat in restaurants, snack bars, fish shops, and so forth. Samples are brought to the laboratory where they are checked for odor, taste, acid value, and DPTG content. Frying fat or oil is "unfit for human consumption" if the acid value is higher than 4.5 and/or DPTG content is higher than 16%.

## Portugal

There are no specific regulations for frying fats and oils in Portugal. The Ministry of Agriculture's Food Quality Institute, however, examines frying and cooking oil for color and odor and by E. Merck's Fritest and Oxifrit Test and Libra Laboratories' Veri-Fry quick tests. If positive, the oil is analyzed for the content of total polar compounds.

## Scandinavian Countries

Scandinavian countries have no specific laws or regulations applicable to frying fats. General regulations applicable to edible fats and oils apply to frying fat. Norway's laws require foods to be free of pollutants and toxic substances and specify that only tocopherols (an antioxidant) and citric acid may be added to fats and oils. For practical control in restaurants and fast-food establishments, Norwegian inspectors may use organoleptic evaluation or the Fritest. In Sweden, the Oxifrit Test is used as a quick test, and the method for total polar compounds is used as a reference method.

In Finland, the fat is considered spoiled when color, odor, and taste are less than 1 upon evaluation (on a scale of 1 to 5), or when acid value is greater than 2.5 and smoke point is less than 170°C (11). Vegetable oil is spoiled when color, odor, and taste are less than 1 upon evaluation, or when the Fritest is greater than 2, acid value is greater than 2, and smoke point is below 180°C. It is also considered spoiled when the iodine value decrease (compared to that of unused oil) is greater than 16, acid value is more than 2, and smoke point is below 180°C. These guidelines have been used since 1976.

In 1991, the National Food Administration of Finland issued a circular letter to be observed by all public health boards outlining procedures for sampling and analysis of frying fat. Test criteria are as follows: sensory evaluation (smell, taste, color), see previous discussion; total polar compounds, maximum of 25%; acid value, vegetable oil, 2.0; acid value, solid fat, 2.5; smoke point, vegetable oil, minimum of 180°C; smoke point, solid fat, minimum of 170°C; Fritest, vegetable oil, maximum of 2

```
No. and Date _____

Amount of fat in kette_____ Added daily _____

Fat totally renewed, date _____

How often will the fat be renewed totally? _____

_____

Criteria for total renewal (except time) _____

_____

How long daily will fat be kept hot?_____

At the time of sampling, how long kept hot? _____

Fat temperature, reported_____; measured _____

Filtering and storage of fat_____

_____

What will be fried just now?_____

What else will be fried? _____

Date of the kettle _____

_____

Cleaning of the kettle, last (when)_____

_____

How was kettle cleaned? _____

_____

_____


Sensory evaluation of the fried food:

Taste _____ Smell _____ Color _____

Observations on the spot:

Overal Cleanliness _____

Smoke _____ Ventilation _____
```

**Fig. 21.1.** Finland National Food Administration fool laboratory inspection form that is filled out by the health inspector.

(scale 1–3); Oxifrit Test, below 3 (scale 1–4); food oil sensor, below 4 (scale 0–6).

In addition, information is provided on the choice of kettle material (stainless steel is recommended) and the proper use and cleaning of frying equipment. Lastly, an inspection form is presented (Fig. 21.1) to be completed by the health inspector and provided to the food laboratory.

The Swedish National Food Administration (NFA) prepared a document in 1989 presenting advice and guidelines on handling frying fat. A summary of NFA guidelines

**TABLE 21.3**
**Guidelines for Deep Fat Frying from the Swedish National Food Administration**

1. All fat in the fryer must be changed before it smokes or foams. Use tests such as food oil sensor or Oxifrit Test to indicate when it is time to change.

2. Strain the fat and clean the fryer once daily. Rinse carefully after cleaning. Solid material in the fat and detergent residue accelerate breakdown of the fat. Store the strained fat at room temperature or at lower temperatures in a covered stainless steel vessel. If iron pots are used, they should be rinsed only with hot water. Detergents remove the protective film of polymerized fat that builds up during use.

3. Frying temperature should be 160–180°C (320–356°F). At lower temperatures, the product absorbs more fat. At higher temperatures, the fat deteriorates more quickly.

4. Use fat specially intended for frying.

5. Avoid salting or seasoning fried food over the fryer. Salt and seasoning can accelerate breakdown of the fat.

6. Lower the temperature when not frying and protect the fat from light.

7. The fryer should have no iron, copper, or brass parts that come in contact with the heated fat.

8. Keep a constant level of fat in the fryer. Fry a little at a time to keep the temperature as even as possible. Prefry when large amounts will be prepared.

9. Use a separate fryer, if possible, for potatoes. The fat deteriorates more rapidly when meat or fish are fried than when only potatoes are fried.

10. Caution: Do not overheat. If the fat temperature rises above 300°C (572°F), the fat may burn.

is shown in Table 21.3. An English edition was issued in June 1990 (10). Its purpose is to encourage the employees of food establishments to prepare high-quality fried food. The NFA recommends use of the Oxifrit Test as a quick test for kitchen staff or local food control inspectors. According to Food Control Ordinance SLV FS 1990:10, food producers must have some form of quality control program approved by control officials. The NFA recommended use of the Oxifrit Test as part of compulsory quality control programs. In Sweden, antifoam agents, such as silicones, are not permitted in frying oil because they mask natural foaming in deteriorated oil.

## Spain

Royal decrees of 1981 and 1983 regulate transportation, processing, and commerce of edible fats and oils but are not applicable to frying. A later decree protecting consumers (12), however, specifies that frying oils and fats must not contain foreign compounds, must contain less than 25% total polar compounds, should satisfy sensory evaluations, must not alter the quality of fried foods, and must not be sold for subsequent use in preparing food products after use in preparing fried food.

## Switzerland

Food control is carried out by the individual member states (cantons) of the Swiss confederation. Basic food legislation is contained in the federal law of 1905 (13). A 1936 ordinance deals with labeling and advertising of food and food additives, as well as investigation and inspection of food establishments. Laboratories of the large cantons are responsible for food control in each territory.

The Swiss Food Ordinance controls frying oils and fats in restaurants and catering facilities and gives guidelines for food preparation and sale. The Swiss Public Health Office issued a list of permitted additives and maximum levels in foods, including coloring agents, antioxidants, and emulsifiers allowed in food fat and oil (these additives do not improve frying performance and are not used in frying oils). Silicone additives are forbidden.

Food inspectors check frying oil for odor, taste, color, and smoking and observe the state of hygiene in the food establishments. Suspect frying oil quality is checked on the spot by the Fritest. If positive, the oil is checked in the laboratory for the level of total polar compounds. The DGF recommendation that frying oil should not contain more than 27% total polar compounds is generally followed by food control officials.

Frying oil is considered to have deteriorated if odor and taste are objectionable. It is also deteriorated if the odor and taste are not clearly objectionable but smoke point is less than 170°C and total polar compounds are greater than 21%, or if odor and taste are not clearly objectionable but total polar compounds exceed 27%. These criteria are based on the assumption that a careful, experienced cook monitors not only the quality of frying oil, but also the hygiene of the kitchen. If the frying oil does not adhere to the recommendations, owners of establishments deemed sanitary are warned to take care of the frying oil in use. Owners of unsanitary establishments are asked to improve conditions as well as quality of frying oil.

# Summary

Formal laws and regulations for control of frying fat quality have been adopted by only a few countries. Several other countries, however, employ practical guidelines and test procedures to control the quality of frying fats and fried food. In addition, there is increasing awareness that good frying practice and proper control of frying fat improve the quality and acceptability of fried food. Guidelines for handling frying fat and on-site tests to check fat quality are useful tools for overall food sanitation and quality control programs worldwide.

## References

1. The Chemistry and Technology of Deep-Fat Frying. *Food Technol.* **1991,** *45,* 67.
2. Meeting Summary: DGF Symposium on Frying and Cooking Fats, *Fette Seifen Anstrichm.* **1973,** *75,* 49.

3.  Special Issue: DGF Symposium on Frying and Cooking Fats. *Fette Seifen Anstrichm.* **1979,** 493.
4.  American Oil Chemists' Society (AOCS). *Official Methods and Recommended Practices of the American Oil Chemists' Society,* 4th ed.; AOCS Press: Champaign, IL, 1993.
5.  U.S. Food and Drug Administration. *1993 Recommendations of the U.S. Public Health Service*; U.S. Food and Drug Administration: Washington, DC; Food Code PB 94-113941 AS.
6.  Food Safety and Inspection Service. *Meat and Poultry Inspection Manual*; Food Safety and Inspection Service, U.S. Department of Agriculture: Washington, DC, December 1990; Sec. 18.40, p. 125.
7.  Royal Decree on Use of Edible Fats and Oils for Preparation of Fried Foods. *Belgian State J.* **1988,** *20,* 1544–1545.
8.  *Rev. Fr. Corps Gras* **1991,** *38,* 224.
9.  *J. Off. Republ. Fr.* **1986,** *118* (170), 9126.
10. National Food Administration. *General Advice on Handling Frying Fats*; National Food Administration: Uppsala, Sweden, 1990; SLV FS 1990:2.
11. Marcuse, R. *Fette Seifen Anstrichm.* **1979,** *81,* 551–554.
12. Quality Standard for Heated Fats and Oils. *Off. State Bull.* (Spain) **1989,** *26,* 2665–2667.
13. World Health Organization (WHO). *Public Health in Europe 28: Food Safety Services,* 2nd ed.; WHO Regional Office for Europe, WHO: Copenhagen, 1988; pp. 177–180.

# 22

# Environmental Concerns

**Michael J. Boyer**
*President AWT, Atlanta, GA*

This chapter discusses various solid and liquid wastes resulting from the use and consumption of frying fats and assumes that most readers need updates on changing environmental regulations in this area. Other chapters in this book deal with various aspects of frying fat use and chemistry.

Although the trend in North America is toward lower fat consumption, frying still represents a superior product for many foods, and frying fat consumption is measured in millions of pounds per year. Much of that is in the institutional and retail foodservice markets. While a significant portion also goes into retail consumer products, little direct data is available concerning the environmental impact of household use of fat and oil products. Much of this chapter addresses the environmental effects of frying fats used in food processing and food service. Waste from actual fat and oil processing, as well as household use, is briefly addressed.

The author conducted full-scale research and developed solutions to environmental challenges associated with these waste materials from the food processing and foodservice industries. The quantities of fat used in food preparation and left unconsumed must exit the system as waste in one form or another, and the cumulative totals are truly astonishing.

Traditionally, little attention has been paid to the handling of these materials. Initial efforts have been directed to assess and regulate these waste materials in a meaningful way. More recently, the environmental regulatory community increased its focus on this issue, particularly at the municipal sewer-system industrial pretreatment level. In addition, the generators themselves (e.g., restaurants, and foodservice vendors) have become more proactive to ensure that their waste is properly handled. In December of 2005, an entire conference was held directed solely at the issues of fats, oils, and grease management in municipal sanitary sewer systems, grease trap management, and related concerns (1).

## Sources and Characteristics

### Sources

Waste fat and grease products from institutional and commercial sources are categorized into two groups: i) spent fat/grease from food-preparation fryers, usually stored on-site in separate containers (5 gallon cans, tanks, and the like) and collected by companies who further process the material, usually for use in animal feeds; and ii) waste fat and grease that collect in a sewer-system grease trap. Grease traps are cleaned on a periodic basis. This material was historically disposed of as solid waste in municipal sanitary landfills.

Waste fat and grease from households falls into the second category, although no grease traps are involved, and the waste fat flows directly to the sanitary sewer along with the remainder of kitchen sanitation activities. Little direct data is available with respect to wastes and wastewater (kitchen wastes) from households, other than the aggregate impact on publicly owned sanitary sewage plants. Industrial wastewater regulators for large municipal sanitary sewer systems report sewer-line blockage problems from residential areas where high-fat diets are typically consumed. Additionally, these reports are often associated with inadequately designed sewer-line grades or low spots that allow fat to accumulate.

Waste from the actual production of frying fats results from the hydrogenation of refined oil as well as votator and packaging areas. Wastewater is associated with floor and equipment sanitation activities from these areas, along with handling and storage losses for raw materials and finished products (spillage). Oil and grease concentrations in these wastewaters reflect the efficiency of cleanup activities, product mix, and production runs. Wastewater characteristics can vary greatly from plant to plant, and even on a daily basis for a given plant. These waste fats and oils are normally captured in the early primary separation stage of the overall wastewater treatment process and recovered for further processing into animal feed fat. There are no significant solid wastes or air emissions resulting from this phase of the processing.

## Characteristics

There is little published data on those waste products from grease-trap-cleaning activities. Tables 22.1 and 22.2 present a summary of characteristics for commercial and institutional waste. Information on characteristics of grease-trap contents was generated through several in-house studies conducted by the author. Other data are available through private processors and industry groups, such as the National Rendering Association, associations such as the Water Environment Federation (WEF), and through operators of larger municipal sewer systems who studied local management needs in this area. It is important to recognize that waste from commercial and institutional activities is invariably a mixture of vegetable frying fats and cooking oils, as well as animal-based grease and tallow. In this chapter these are collectively referred to as "frying fats." The author's experience suggests that the waste from upscale restaurants tends to contain less solid/hydrogenated frying fat and more liquid oils, such as peanut oil. Waste grease from fast-food establishments is almost entirely solid frying fats. From a practical standpoint, the use of recovered material mixtures, and characteristics of waste from a treatment perspective, make the specific blends or mixtures academic.

**TABLE 22.1**
**Wastewater Characteristics — Grease Processing**

| Parameter[a] | Untreated | Gravity Separator Effluent | Filtered Effluent |
|---|---|---|---|
| BOD,[b] mg/L | 14,000–100,000 | 1,000–7,450 | 150–500 |
| | (15,178–108,420) | (1,084–8,077) | (162–540) |
| Oil and grease, mg/L[c] | 25,000–150,000 | 500–3,240 | 75–250 |
| (as Freon extractable) | (27,105–162,630) | (542–3,512) | (81–271) |
| Total suspended solids, mg/L | 30,000–231,000 | 750–4,200 | 100–300 |
| | (35,526–250,000) | (813–4,553) | (108–325) |
| pH | 3.5–5.0 | 3.9–5.0 | 3.9–5.0 |

[a]Range: low–high as concentration, mg/L(lb/d, based on 130,000 gal/d).
[b]BOD = biochemical oxygen demand.
[c]Raw influent contains from 2.5–15% oil and grease, based on our data and City of Dallas information.

**TABLE 22.2**
**Wastewater By-product Characteristics — Grease Processing**

| Item | Amount |
|---|---|
| Recovered oil, dry weight lb/d | 27,100 |
| Sludge solids, dry weight lb/d with gravity separator | 4,600 |
| With separation and filtration | 5,300 |
| gpd, thickened to 2% solids with gravity separation | 27,600 |
| With gravity separation & filtration | 31,800 |
| As 35% filter cake | 9 yd³/d |

Accumulated material, often referred to as "yellow grease," collected from fryer waste is processed and used primarily for feed fat. Minor amounts are used for the production of feedstock fatty acids. The collected fryer material contains minor amounts of moisture and foreign food particles, but generally has total fatty acid values of 90% or greater. These materials are readily processed using indirect steam heat for drying, as well as by filtration to remove the solids.

As indicated in Tables 22.1 and 22.2, the material pumped from a grease trap can vary greatly, depending on the type of establishment, frequency of sanitation activities, and other factors. In general, however, the mixture is described as somewhere between a very strong liquid wastewater and a very watery solid waste.

To address the issue of how much of this grease-trap material is generated on an ongoing basis, the author conducted a private study in 1984 in the Dallas–Fort Worth area (DFW), as well as other subsequent studies in the Atlanta, Georgia area, and other small-to-medium-sized municipalities. The purpose of the DFW study was to assess the magnitude of the problem as well as to explore the economic and technical feasibility of constructing and operating a private facility to handle these wastes. Based on the number of eating establishments, and mean grease-trap size, as well as projected trap clean-out frequency, an average daily volume of 130,000 gallons was projected. Operating data at full-scale grease-trap waste-processing facilities suggest that the actual volume may be somewhat less, but that is due to less frequent clean-outs than are directed by regulatory agency policies. Therefore the figure is expected to increase to 130,000 gallons per day in the Dallas–Fort Worth (and similar) population areas as enforcement is increased.

## Environmental Agenda and Regulations

### Environmental Agenda

Environmental regulations that address proper grease-trap handling are rapidly changing, particularly at the local level, and the reader is urged to investigate current regulatory positions in a given area of interest.

In terms of understanding all the pertinent issues relative to environmental regulations in general, it is important to note that development of regulations for any given environmental issue, as well as their enforcement, is highly dependent on public perception of environmental impact or threat. For example, there is much good scientific data suggesting that natural radon gas in residential dwellings poses a serious health threat; however, the public at large simply does not perceive this as a real threat. As a result, federal and state environmental regulators moved the issue lower on their agendas in favor of more popular causes that are better understood and funded.

The same argument is true of environmental agendas for corporations. Corporate America has consistently shown that it expends whatever resources are reasonably necessary to maintain compliance with environmental standards. Corporations have also shown that they will go well beyond what is needed to maintain compliance, particularly in response to issues that have public interest.

The environmental agenda of the American public has "recycle" at the very top. We teach it to our children in schools, and we even hear jingles on the radio. Within the food/restaurant industry, the issue of recycling for packaging has caught the fancy and interest of the public. The customer can see and feel the package, and consequently might feel better because the hamburger is wrapped in brown paper with an earthy, rumpled look to it. As a result, the food industry generated a great deal of publicity and committed substantial resources to make recycling happen.

In terms of public perception, the proper handling of waste fat from food preparation is an entirely different matter. Today's health focus is on low fat or no fat. Also, waste fat and grease are essentially messy, sloppy materials with an undesirable odor. There is simply no way to present this to the public in a palatable manner, even in the framework of a recycled or reused product. As such, it has not been as high on the public or corporate environmental agendas. Therefore, environmental regulations and enforcement must be relied on, combined with economic considerations, to determine the proper handling and disposal of these materials. Significant local activity at the municipal enforcement and management level has caused the major fast-food chains to place more emphasis on this issue. Most of these companies now have organized internal programs to address the proper management of grease-trap wastes.

## Regulations

### Recovered Fryer Fats

Recovered fryer fat is collected by renderers and other processors, and transported to their facilities for further processing. In most states, this material is exempt from solid-waste-handling regulations because it is recycled and nonhazardous. However, some states now have manifesting programs in place to track this material.

The processing and recycling facilities themselves are subject to environmental regulations under various state and local ordinances. They must have permits for air emissions (boiler stacks, storage tank vents, dryer stacks, and so forth) and wastewater discharges. Solid waste materials, such as food particle filtrates, must be discarded in permitted solid waste disposal sites.

It is important to note that the collection, recovery, and sale of these fats is a profitable business and therefore requires somewhat less environmental regulation and enforcement to ensure proper procedures.

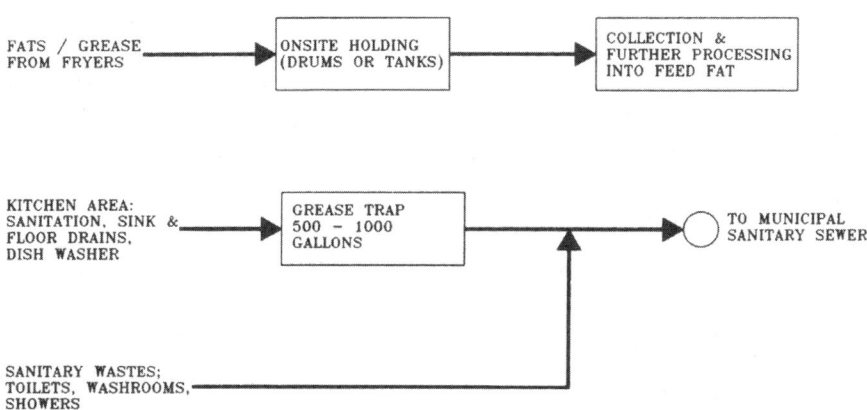

**Fig. 22.1.** Typical institutional/commercial grease-waste and grease-trap arrangement.

*Grease Traps*
Proper handling and disposal of grease-trap waste is not profitable for anyone other than the haulers and waste processing companies who manage the material. As a result, proper management of this material is a much bigger challenge for environmental regulatory programs. Figure 22.1 represents a typical grease-trap arrangement at a restaurant or institutional food-preparation facility. With a few exceptions, virtually every city and county in the United States has ordinances requiring restaurants, cafeterias, and other large food-preparation establishments to have grease traps. Plumbing must be arranged so that sanitary sewage does not pass through the grease trap, although this is not always so in older establishments. Ordinances also require that grease traps be cleaned on a regular basis. This is also somewhat difficult to enforce, given the large number of traps in any given city.

# Recycle Potential and Treatment Facilities

## Recycle Potential
Kitchen fryer fat already gets collected and recycled, and there is a relatively efficient system to handle this. The remainder of this section deals with grease-trap waste. Grease-trap material is another significant source of recycled material. The material can be processed and used for animal-feed fat blending and inedible (nonfeed) purposes, such as fatty acid feedstock, as well as for its fuel value.

*Animal-Feed Fat*
Once recovered in a pure form, grease-trap material is essentially the same as processed cooker fat. It is subject to oxidation and other factors that influence fat chemistry; however, this is essentially reflected in the price paid for the material. Recently, concerns were expressed about other foreign materials that might reach the grease trap, such as drain-unclogging chemicals, pesticide sprays in kitchens, and related materials, entering the food chain. Analysis of the recovered fat by the author did not show this to be a significant problem, and the negative impacts are more perceived than actual. In any case, a facility that processes this material, or purchases it for feed-fat blending, should have an adequate Quality Assurance/Quality Control program to assess these influences on an ongoing basis.

# Inedible (Nonfeed) Uses

There is substantial demand on the open market for lower grade fat materials. The market price is obviously not as good as for feed fat, but it is a positive return, in any case.

## Fuel Use
There is much effort to optimize the use of animal and vegetable based oils for fuel

and energy. These materials have desirable properties for this use. There have been isolated full scale applications of this using deodorizer distillate and other recovered oil fractions from oil refining and processing operations as boiler fuel supplements. The handling characteristics of the hard fat materials have been a major drawback for this.

Using animal and vegetable materials for transportation based fuels in buses, cars, and locomotives is gaining interest at a very rapid pace at this time. Refined soybean oil is a major feedstock for the production of biodiesel. Mixed waste grease and fats are not as well suited for this application due to high FFA levels and melt point characteristics; however, the demand for feedstocks for the large number of biodiesel manufacturing facilities now under construction is creating a demand for these mixed greases for feedstocks as well. Several new biodiesel facilities are being designed with the additional feedstock pretreatment capacity to utilize mixed fats and greases. This application will create demand for this material as well as other materials such as poultry and other animal fats that are currently absorbed into the animal feed fat and other markets.

## *Treatment Recovery Facilities*

Numerous facilities are now in place to receive and process grease-trap waste. Most of them are privately owned and operated facilities, but a few are owned and operated by the local municipal wastewater management authority or entity (2). These facilities are one of two types:

- Facilities designed to treat the waste material and recover grease for marketable recycle uses.
- Facilities that simply treat the entire stream as a waste material.

The second types of facility are usually smaller operations that receive lower quantities of this material. These smaller facilities are focused on dewatering, or otherwise drying, the material to a state in which it can be dealt with as a solid waste and disposed of in local sanitary landfills.

The larger facilities utilize processes and technology directed at recovery of the fat material. The smaller facilities cannot justify the capital investment to recover the fat material, and the larger facilities cannot bear the additional costs of mass handling and disposal of the material as a solid waste.

The actual processing for fat recovery is physical fractionation requiring elevated temperatures. When the temperature of grease-trap content mixture exceeds the melting point of the fat, phase separation occurs, and any solids and water separate to the bottom. The recovered fat mixture forms a relatively pure, dry product. Figure 22.2 shows a typical flow diagram. Some of the problems and challenges involved in operating such a facility are noted.

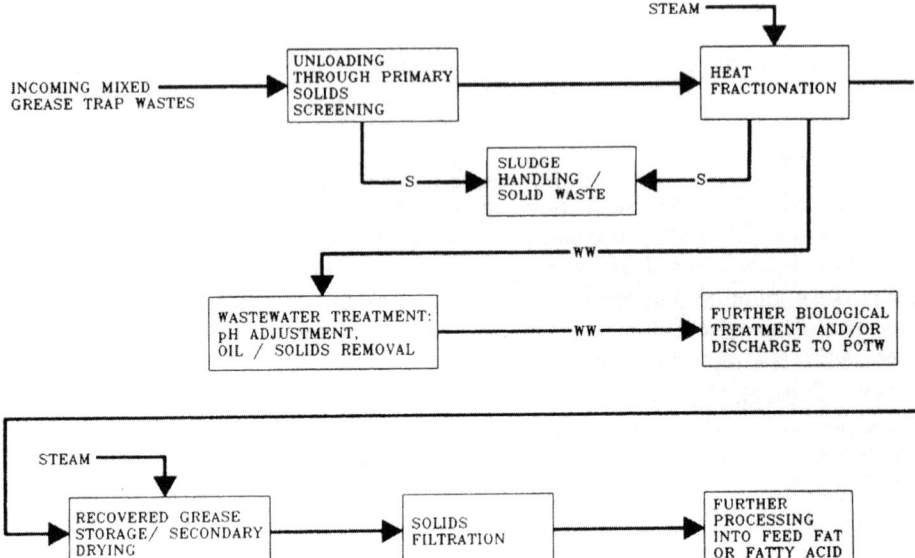

**Fig. 22.2.** Typical grease-trap waste-processing facility.

i)  Obtaining a steady, reliable supply of feedstock is a major problem.

ii) The grease-trap mixture is extremely heterogeneous. Quantities and qualities of fats and oils vary. The mixture contains every possible kind of solid matter from a food-handling operation. The mixture must receive initial screening for larger particles. Handling the screenings is a difficult odor-emitting undertaking. From a mechanical equipment standpoint, a vibrating table screen (Sweeco type) works well.

iii) Pumps and piping are subject to clogging. Pumps must be of the heavy-duty raw-sewage type (not oil pumps). Piping must be insulated and heat-traced, and all parts must be designed for easy disassembly and cleaning. All of this adds to the initial investment costs.

iv) Heat fractionation is effectively accomplished in vertical carbon steel tanks (insulated) with steam coils or sparging.

v)  Water from the tank decant is relatively high in BOD and other contaminants (see Tables 22.1 and 22.2).

vi) The sludge phase volume can be substantial. Also, it is difficult to avoid taking some oil and fat over in the sludge decant if a pure fat stream is desired for recovery/recycle purposes.

vii) Effective sludge-dewatering and sludge-handling faculties are essential, and a high-fat sludge should be anticipated.

viii) The entire operation creates an objectionable odor; therefore, a fail-safe odor control program must be designed into the facilities at the outset. Clearly, this has an impact on site selection as well.

ix) The processor must be committed to the business.

# The Long-Term Perspective

## *Fryer Fat*

As previously noted, this material is already being captured, processed, and recycled in an economically viable marketplace. Increasing regulatory scrutiny is anticipated in the areas of recovered feed-fat quality, and handling of wastewater and sludge in an environmentally acceptable manner. It is also anticipated that the same renderers that collect fryer fat will begin to focus on grease-trap material.

## *Grease Traps*

Grease traps are not a product that will support themselves strictly on the basis of economics. From a regulatory agency perspective, there has been an increasing amount of regulation and enforcement on an area case-by-case basis. This is being supported by better recognition of the problem and management of solutions by the larger fast-food companies. Regarding economics, those cities still accepting grease-trap waste at their municipal treatment works charge $25–50 per vacuum truckload. Private processors/recyclers charge about $250 per load. The absolute value of the recycled product is not significant on a per-load basis. Therefore, the economics favor potentially inappropriate solutions.

## References

1. Water Environment Federation. *Proceedings of the Fats, Oils and Grease (FOG) Management Workshop*; Water Environment Federation: Alexandria, VA, 2005; www.wef.org.
2. Joyce, M.; and B. Donaldson. Fattening up the Bottom Line, Changing Sewer Grease from a Liability to an Asset. *Water Environ. Technol.* **2005,** *17*, 16–19.

# 23

# Options for Reducing/Eliminating *trans* Fatty Acids in Deep Fat Frying and Labeling Implications

**Robert E. Wainwright[a], Denise Fallaw[a], Lynne Morehart[b], Jim Womack[b], and Consuelo Renteria[c]**

*Cargill, [a]5000 South Boulevard, Charlotte, NC 28217, [b]2400 Industrial Drive, Sidney, OH 45365, [c]Apartado Postal M07137, Mexico City, Mexico D.F.*

Throughout the previous chapters, essentially two fundamental methods or categories of deep fat frying have been addressed: foodservice and food processor (or industrial, batch, and continuous). Inherent fryer design differences between the two methods usually warrant different selection criteria when identifying oils offering the best practical no-, or low-, *trans* solutions.

Traditionally, frying fat selection criteria focused only on convenience, economics (cost and availability), and stability. Except in rare instances where the foods fried already catered to diet and health-conscious consumers, the nutritional profile of frying fats and oils was not considered part of the selection criteria, at least not directly.

However, as public awareness grew, oil suppliers began offering "healthier" alternatives, especially when nutritional labeling laws mandated the declaration of total fat, saturated fat, and cholesterol on packaged consumer food products. The initial response by some manufacturers was simply not to make any changes and declare what was in their existing product. However, others saw value in the potential sales and marketing benefit of changing their fat or oil to reflect a "healthier" nutritional statement. Regardless, labeling and its implications now give the fried-food producer more reason, in general, to think of fats and oils in other than traditional terms. Today, the food industry is keenly aware of the potential impact on a product if it contains a fat or oil not generally perceived to be "good for you."

Mandatory labeling of *trans* fat became effective on January 1, 2006, and requires that certain packaged retail foods display total grams of *trans* fat per serving, directly below the saturated fat line, on the Nutrition Facts Panel. This applies to foods manufactured under the jurisdiction of the Food and Drug Administration (FDA) and not to foods manufactured under the U.S. Department of Agriculture

(USDA). Further discussion related to standard serving sizes and reference amounts for compliance follows later in this chapter.

Most foodservice fried foods, and many pre-fried foods sold to the foodservice market, are excluded from this legislation. However, exclusion of this market does not necessarily mean Foodservice companies should not make an effort to improve the nutritional profile of their fried foods. Many companies become subject to the *trans* fat labeling implications indirectly because they have customers, such as re-packers, that, in turn, sell to the retail markets.

From a nutritional perspective, frying remains a subject vulnerable to negative public perceptions, with little expectation of changing significantly in the near future. Therefore, it is critical that food manufacturers and foodservice companies identify practical nutritional goals when first addressing the issue. Fundamental to the decision-making process is first establishing whether the objective is i) removing the term "hydrogenated" from an ingredient statement somewhere, or ii) achieve zero grams *trans* fat per serving without concern for the term "hydrogenated" appearing in an ingredient statement somewhere. The answer to this question determines the capabilities and limitations of a solution. Specifically, "no hydrogenation," by definition, means a liquid, or a semisolid such as palm oil—unhydrogenated oils only. However, if zero grams *trans* fat per serving is the only issue, and the term "hydrogenated" is not an issue, more options become available. For example, blending various liquid (unhydrogenated) oils with fully hydrogenated fats offers far more potential solutions than liquid oils alone. Other practical considerations early on in the decision-making process are:

- Do consumers buy the product for "healthy" reasons? (e.g., "healthy donut"?)
- Are there historical data suggesting current consumers' tolerance for change?
- Is a likely fry-life reduction, and associated cost increase, worth the "healthier" perception?
- Will current customer base remain loyal after the likely cost increase?
- Conversely, what is a reasonable expectation for decreasing market share by changing?
- Is a new product (line extension) more practical than the risk of losing loyal customers?
- Alternatively, does a new product or line risk implying the current brand is less healthy?
- Historically, has the company demonstrated sensitivity to public perception?
- Should/can the change be done all at once or incrementally?
- Would advertising the change be appropriate or not?
- Does the issue carry the same importance, or significance, for all the target markets?
- Even if exempt from labeling laws, is there an opportunity to improve consumer perception?

- Will the likely processing changes be an issue?
- What, if any, specific fats/oils are not options?

Though certainly not a comprehensive list, these are some of the major issues commonly faced by fried food producers when deciding if change is appropriate or necessary.

If a change is determined to be appropriate, the next challenge becomes selecting the best, practical solution. To that end, the following fundamental considerations are suggested:

## Establish Current Nutritional Profiles

An important first step is confirming the nutritional profile of the current fat or oil, especially *trans* and saturated fat. *Trans* and saturated fat information should be available from your oil supplier. In some cases, options for reduction or elimination of *trans* fat mean an unavoidable increase in total saturated fat. Moreover, it may be that your current finished product is already less than 0.5 grams of *trans* fat per serving, depending on absorption, other sources of fat in the matrix and, of course, serving size, therefore a change in formulation may not be necessary. Unless the goal is to completely remove hydrogenated fat from the food matrix, including frying oil, then no change is required if there is less than 0.5 grams of *trans* fat per serving.

Equally important is confirming the nutritional profile of the finished product. There may not be a frying oil option for reducing *trans* fat content to less than 0.5 grams per serving if the *trans* fat in the product before frying is already high. Knowing this in advance will help focus resources more efficiently for both producers and suppliers.

## Typical No- and Low-*trans* Frying Oil Options

Table 23.1 shows common options for reducing *trans* fat in frying oils, and appears in the order of increasing relative stability. (NOTE: The intent is to provide direction only.) "Stability" here refers to the oils' inherent sensitivity to oxidation in general. Also implied are comparative rates of deterioration and the resulting tendency toward flavor reversion, recognizing that type of frying and product fried determines whether oil can be used alone or in combination with another oil listed. Typical approximate saturated and *trans* fat contents are included to help identify potential combinations if the term "hydrogenated" in an ingredient statement is not an issue.

Hydrogenation is not the only potential source of *trans* fat in vegetable oils. The deodorization step required in the processing of all vegetable oils can generate small amounts of *trans*. This occurs because of the necessary high temperatures, and low vacuum, required to produce high-quality, finished oils. However, not all vegetable oils respond to the same deodorization conditions so simply reducing the temperature is not always an option for producing oils with undetectable levels of *trans*.

**TABLE 23.1**
Common Options for No- and Low- *trans* Frying

| Relative Stability | Oil Options—Type | Total SATS[a] | Typical trans[a] | Declare Hydro | Physical Form |
|---|---|---|---|---|---|
| Less Stable | Soybean salad oil (commodity) | 14% | 2% | No | Liquid (Clear) |
| | Canola salad oil (commodity) | 7% | 2% | No | Liquid (Clear) |
| | Corn salad oil (commodity) | 13% | 2% | No | Liquid (Clear) |
| | Cottonseed cooking oil (not winterized cottonseed salad oil) | 26% | 2% | No | Liquid (Clear) |
| | Peanut oil | 17% | 2% | No | Liquid (Clear) |
| | Fluid "economy" frying shortening—soybean salad oil and hydrogenated [b] | 18% | 5% | Yes | Fluid (Opaque) |
| | Hi-oleic canola | 8% | 5% | No | Liquid (Clear) |
| | Nu Sun™—"mid-oleic" sunflower oil | 12% | 2% | No | Liquid (Clear) |
| | Fluid frying shortening—soybean partially hydrogenated and hydrogenated [b] | 18% | 16% | Yes | Fluid (Opaque) |
| | Fluid frying shortening—canola partially hydrogenated and hydrogenated [b] | 11% | 16% | Yes | Fluid (Opaque) |
| | Partially hydrogenated and fractionated—soybean | 14% | 29% | Yes | Liquid (Clear) |
| | All purpose shortening—soybean/cotton | 25% | 31% | Yes | Plastic (Solid) |
| | Palm oil | 49% | 2% | No | Semi-Plastic (Soft solid) |
| | Beef tallow | 50% | 7% | No | Plastic (Solid) |
| More Stable | Heavy-duty frying shortening—soybean | 24% | 47% | Yes | Plastic (Solid) |

[a]*Sources*: *USDA Handbook 8-4*; AOCS Fatty Acid Finder©, Version 1.0b; and private communication.
[b]Hydrogenated = "fully hydrogenated" that's typically <5 Iodine Value and doesn't contribute to total *trans*.

Obviously, "hydrogenated" need not appear in an ingredient statement simply because an oil has detectable *trans*. However, if the objective is zero grams of *trans* fat per serving, it is important to recognize the potential for another source of *trans* beside that which is contributed by partially hydrogenated oils alone.

## Considerations for Snack Frying

The oil used in frying affects three critical issues in the finished product: flavor profile, nutritional profile, and shelf life. Therefore, the nutritional objective should not be to the exclusion of any significant changes in how the product looks, tastes, and smells at the time of frying and during its shelf life. Since a change in the frying oil will likely change the flavor profile of the product, the primary consideration is whether the change is acceptable or not.

Since snack products are usually stored at room temperature, changing to a frying oil for better nutritional profiles usually means reduced oxidative stability, both in the fryer and on the shelf. Changes to manufacturing practices, packaging, or transportation methods may also be necessary in order to keep your current shelf life. Moreover, it is important to understand that making changes in these areas is not a guarantee either. There is always the potential for sacrificing shelf life when switching oils under these circumstances.

## Considerations for Par-Frying

Par-fried foods are only partially fried and, typically, are frozen immediately after this initial frying. Being frozen offers some initial advantages over products that are not fried because the process tends to reduce the oxidation rate caused by frying, which can sometimes effectively prolong its shelf life. However, freezing does not eliminate the potential for autoxidation.

Packaging material is another important factor that can affect oxidation rate. Currently, two different types of packaging are in use, both serving primarily as oxygen barriers: plastic packages (bags) and boxes lined with a thin layer of protective plastic film. The latter is used mostly for par-fried foods sold directly to the consumer who then either fries or bakes the product.

Paper bags lined with thin plastic laminate are typically used for par-fried foods that are sold to foodservice. If changing from a partially hydrogenated fat to a non-hydrogenated oil, then the possibility, if not probability, of bag staining is a real consideration. Though this tends not to become a quality issue, it can be a handling issue for the end user.

Changing to a less stable oil typically translates to processing changes too. It's extremely rare that an option is a "drop-in solution." In some cases, crumb buildup in the equipment (belts, etc.) becomes an issue that, in turn, risks more crumbs in the packaged product. Because liquid oils do not solidify as the product cools, there is more opportunity for greater amounts of crumb material to separate from the finished product. This "shedding" activity, or "shed," remains in the fryer and, in turn, often

**TABLE 23.2**

Reference Amounts Customarily Consumed per Eating Occasion: Foods for Frying—General Food Supply

| Product Category | Reference Amount | Label Statement |
|---|---|---|
| Biscuits, croissants, bagels, tortillas, soft bread sticks, soft pretzels, corn bread, hush puppies | 55 g | _ piece(s) (_ g) |
| Coffee cakes, crumb cakes, doughnuts, Danish, sweet rolls, sweet quick type breads, muffins, toaster pastries | 55 g | _ piece(s) (_ g) for sliced bread and distinct pieces (e.g., doughnut); 2 oz (56 g/visual unit of measure) for bulk products (e.g., unsliced bread) |
| Taco shells, hard | 30 g | _ shell (s) (_ g) |
| Plain or fried fish and shellfish, fish and shellfish cake | 85 g cooked 110 g uncooked | _ piece(s) (_g) for discrete pieces; _ cup(s) (_ g); _ oz (_ g/visual unit of measure) if not measurable by cup |
| Meat, poultry and fish coating mixes, dry; seasoning mixes, dry, e.g., chili seasoning mixes, pasta salad seasoning mixes. | Amount to make one reference amount of final dish | _ tsp(s) (_ g); _ tbsp(s) (_ g) |
| Mixed dishes – not measurable with cup, e.g., burritos, egg rolls, enchiladas, pizza, pizza rolls, quiche, all types of sandwiches | 140 g, add 55 g for products with gravy or sauce topping, e.g., enchilada with cheese sauce, crepe with white sauce | _ piece(s) (_ g) for discrete pieces; _ fractional slice (_ g) for large discrete units |
| French fries, hash browns, skins, or pancakes | 70 g prepared 85 g for frozen unprepared french fries | _ piece(s) (_ g) for large distinct pieces (e.g., patties, skins): 2.5 oz (70 g/_ pieces) for prepared fries: 3 oz (84 g/_ pieces) for unprepared fries |
| Snacks: All varieties, chips, pretzels, popcorn, extruded snacks, fruit-based snacks (e.g., fruit chips) grain-based snack mixes | 30 g | _ cup (s) (_ g) for small pieces (e.g., popcorn) _ piece(s) (_ g) for large pieces (e.g., large pretzels: pressed dried fruit sheet); 1 oz (28 g/visual unit of measure) for bulk products (e.g., potato chips) |

**TABLE 23.2, CONT.**
Reference Amounts Customarily Consumed per Eating Occasion: Foods for Frying—
General Food Supply

| | | |
|---|---|---|
| Entrees without sauce, cuts of meat including marinated, tenderized, injected cuts of meat, beef patty, corn dog, croquettes, fritters, cured ham, dry cured ham, dry cured cappicola, corned beef, pastrami, country ham, pork shoulder picnic, meatballs, pureed adult foods | 85 g ready-to-serve 114 g ready-to-cook | N/A |
| Mixed dishes not measurable with a cup. e.g., burrito, egg roll, enchilada, pizza, pizza roll, quiche, all types of sandwiches, cracker and meat lunch type packages, gyro, stromboli, burger on a bun, frank on a bun, calzone, taco, pockets stuffed with meat, foldovers, stuffed vegetables with meat, shish kabobs, empanada | 140g (plus 55 g for products with sauce toppings) ready-to-serve | N/A |
| Entrees without sauce, poultry cuts, ready to cook poultry cuts, including marinated, tenderized, injected cuts of poultry ham products, adult pureed poultry | 85 g ready-to-serve114 g ready-to-cook | N/A |
| Mixed dishes not measurable with a cup; e.g., poultry burrito, poultry enchiladas, poultry pizza, poultry quiche, all types of poultry sandwiches, cracker and poultry lunch-type packages, poultry gyro, poultry stromboli, poultry frank on a bun, poultry burger on a bun, poultry taco, chicken cordon bleu, poultry calzone, stuffed vegetables with poultry, poultry kabobs | 140 g (plus 55g for products toppings) | N/A |

*Source:* Code of Federal Regulations 21 CFR 101.12 and FISI|9|CFR 317.309.

contributes to premature breakdown of the oil if not removed frequently.

## Considerations for Foodservice Frying

If the current deep-frying shortening is a partially hydrogenated vegetable oil, a change to a zero-grams-of-*trans*-per-serving solution usually requires changes in the restaurant operation too.

The first change, most likely, will be an increase in cost. Canola, corn, and soybean oils tend to be similar in cost but will not last as long in a fryer as heavy-duty fluid, or solid, frying fats. Peanut oil and cottonseed oil have a longer fry-life than canola, corn, or soybean oil, but cost more. Yet the more recently available trait-enhanced oils, such as high-oleic canola, high-oleic sunflower, and low-linolenic soybean cost more than traditional partially hydrogenated soybean and canola products. The basic question here is, "What is an extra day of fry-life worth?"

Better oil and fryer maintenance will help offset the anticipated oil cost increase considerably. This includes: daily oil filtering, frequent fryer temperature calibration to ensure thermostat accuracy, frequent checking of recovery times, and not letting fryers idle at frying temperatures for prolonged periods of time. Ideally, fryers should be turned off during slower periods but that is not always a practical consideration.

## Serving Sizes and Calculation of trans Fat Per Serving

In order to calculate total grams of *trans* fat per serving, it is important to first understand serving sizes and how they are is determined. Serving sizes are based on, but are not identical to, reference amounts established by the FDA in 21CFR101.12 of the Code of Federal Regulations. This section pertains to FDA-regulated foods only, and not foods falling under USDA jurisdiction.

Reference amount is defined as "the amount of food customarily consumed per eating occasion" (1). All reference amounts are based on data obtained from national food consumption surveys. The surveys include population groups that are demographically and socio-economically similar and contain the proper sampling size. The information obtained from these surveys is based on actual conditions of food consumption over a given period. These reference amounts are the basis for establishing serving sizes of food products (Table 23.2).

After the reference amount is known, serving size is then determined by the product's form. Discrete units (e.g., hot dogs, muffins, etc.), products within a multi-serving package and two or more foods packaged with the intention of being consumed together, normally list the number of whole units that closely approximates the reference amount based on the following criteria:

1. If a unit weighs 50% or less of the reference amount, the serving size shall be the number of whole units that most closely approximates the reference amount for the product category.

2.  If a unit weighs more than 50%, but less than 67% of the reference amount, the manufacturer may declare one unit or two units as the serving size.

3.  If a unit weighs 67% or more, but less than 200% of the reference amount, the serving size shall be one unit.

4.  If a unit weighs 200% or more of the reference amount, the manufacturer may declare one unit as the serving size if the whole unit can reasonably be consumed at a single-eating occasion.

5.  For products that have reference amounts of 100 g or more, and are individual units within a multi-serving package, if a unit contains more than 150%, but less than 200%, of the reference amount, the manufacturer may decide whether to declare the individual unit as one or two servings.

6.  The serving size for products that naturally vary in size (e.g., pickles, shellfish, whole fish, and fillet of fish) may be the amount, in ounces, that most closely approximates the reference amount for the product category.

7.  For products consisting of two or more foods packaged and presented for consumption together where the ingredient represented as the main ingredient is in discrete units (e.g., pancakes and syrup), the serving size may be the number of discrete units represented as the main ingredient plus proportioned minor ingredients used to make the reference amount.

8.  For packages containing several individual single-serving containers, each of which is labeled appropriately for individual sale as single-serving containers, the serving size shall be one unit.

9.  For products in large discrete units that are usually divided for consumption (e.g., cake, pie, pizza, melon, and cabbage), for unprepared products where the entire content of the package is used to prepare large discrete units that are usually divided for consumption (e.g., cake mix, pizza kit), and for products which consist of two or more foods packaged and presented to be consumed together where the ingredient represented as the main ingredient is a large discrete unit usually divided for consumption (e.g., prepared cake packaged with a can of frosting), the serving size shall be the fractional slice of the ready-to-eat product (e.g., 1/12 cake, 1/8 pie, 1/4 pizza, 1/4 melon, etc.) that most closely approximates the reference amount for the product category.

For bulk products such as flour, sugar, etc., the serving size shall be the amount in household measurement that approximates the reference amount for the product category (2). Other serving size requirements can be found in 21 CFR 101.9 for the FDA and 9 CFR 317.309 and 381.409 for the FSIS meat and poultry products.

## Calculating trans *Fat per Serving*

When calculating *trans* fat per serving, it is necessary to include all contributing sources of fat in the product. An example of multiple fat sources contributing to total grams of fat per serving is fried, breaded chicken breast. Initially, the product contains moisture that escapes as vapor during frying. As moisture is lost, the product begins to absorb oil. The amount of oil absorbed is primarily a function of frying temperature, the type of product being fried and its surface characteristics. In this example, oil absorption is not the only source contributing to total grams of fat per serving. Breading mixes usually contain some type of fat or oil.

There is no substitute for analyzing the finished product for total *trans* fat per serving size. However, calculating the percentage of *trans* fat per serving of each of the ingredients used in the product, and then adding them together, will give product developers a directional threshold to use when formulating new products or ingredients.

## Example: Breaded and Fried Chicken Breast Strips With 30 Grams Total Fat Per Serving

The average total fat per serving of several, statistically significant, extractions of individual servings is 30 g. Contributors to total fat/serving and their respective *trans* fat content are:

1.  Natural fat content of the chicken breast = 0.0% *trans* fat
2.  Fat in the breading mix = 0.75% *trans* fat
3.  Fat absorbed during frying = 15.0% *trans* fat

Step One: Calculate how much each fat source contributes to total fat/serving.

1.  The raw chicken breast weighs 120 g, has ~2.0% total fat and 0% *trans* fat:
    *Grams of fat per/serving contributed by chicken = 120 g × 0.02 = **2.4 g***

2.  Breading mix adds 15 g per serving, has ~5.0% total fat and ~0.75% *trans* fat:
    *Grams of fat/serving contributed by breading = 15 g × 0.05 = **0.75 g***

3.  Since the remaining fat per serving is contributed by absorbed frying oil, subtract the grams of fat per serving from chicken and breading from total fat per serving (30 grams):

    *30 grams – (2.4 + 0.75) = 26.85 grams fat per serving contributed by frying oil*

Therefore,

*Total grams of* trans *fat per serving* = 0.0 + .006 + 4.03 = **4.036 g**

Step Two: Calculate *trans* fat per serving from each fat source

Knowing how much each fat source in the matrix adds to the total fat per serving, it becomes a matter of multiplying the amount each fat source adds to total fat by the percentage of *trans* fat they contain, and then taking the sum:

1. *Grams* trans/*serving from raw chicken* = (2.4 g fat/serving) × (0.0) = **0.0 g**
2. *Grams* trans/*serving from breading* = (0.75 g fat/serving) × (0.0075) = **0.006 g**
3. *Grams* trans *fat/serving from frying oil* = (26.85 g fat/serving) × (0.15) = **4.027 g**

Therefore,

*Total grams* trans *fat/serving* = 0.0 + .006 + 4.027 = **4.033 g**

As this example shows, the frying oil contributes most of the *trans* fat per serving. To decrease the *trans* level to less than 0.5 g per serving, and declare 0 g of *trans* fat per serving, the frying oil must contain much less than 15% *trans* fat. For directional purposes, however, simply substitute the *trans* fat contained in the new oil directly into the previous calculation to determine its effect on total *trans*:

*Grams* trans *fat/serving from frying oil* = (26.85 × .015) = **0.4027 g**

Therefore, with the new oil:

*Total grams* trans *fat/serving* = 0.0 + 0.006 + 0.4027 = **0.4087 g**

Since food products are often complicated matrices, there is no substitution for gathering accurate data from credible testing sources when evaluating the nutritional profile for label declarations. However, during the product development phase, it is useful to calculate the nutritional components before final product approval or submission of samples for expensive nutritional testing.

## Labeling Regulations and Guidelines in Mexico

Currently, the labeling of *trans* fat is not mandatory in Mexico. However, Mexico does require a nutritional declaration similar to that in the United States and Canada. The Secretariat of Commerce and Industrial Development (Secretaría de Comercio y Fomento Industrial, SCFI) and the Secretariat of Health (Secretaría de Salud, SSA) are the regulating authorities for food. They are responsible for providing the official

**TABLE 23.3**
Compliance requirements established by the Secretariat of Commerce and Industrial Development

| Document | Number |
| --- | --- |
| Pre-packed products, net content and verification methods. | NOM-002-SCFI-1993 |
| Trade information, amount declaration on the label. | NOM-030-SCFI-1993 |
| General system of measure units. | NOM-008-SCFI-2002 |
| Guidelines for redaction, buildup and presentation of Mexican regulations. | NMX-Z-013/1-1977 |
| General specifications for labeling pre-packed non-alcoholic beverages and foods. | NOM-051-SCFI-1994 |
| Industrial-use vegetable shortenings and mixed fats specifications. | NMX-F-009-SCFI-2005 |
| Cottonseed oil specifications. | PROY-NMX-F-004-SCFI-2005 |
| Palm oil specifications. | PROY-NMX-F-019-SCFI-2005 |
| Palm olein specifications. | PROY-NMX-F-020-SCFI-2005 |
| Palm stearine specifications. | PROY-NMX-F-022-SCFI-2005 |
| Vegetable and animal oils and fats. | NMX-F-116-SCFI-2005 |
| Safflower oil specifications. | NMX-F-161-SCFI-2005 |
| Corn oil specifications. | NMX-F-030-SCFI-2005 |
| Edible vegetable oil. | NMX-F-223-SCFI-2005 |
| Soybean oil specifications. | NMX-F-252-SCFI-2005 |
| Sunflower oil specifications. | NMX-F-265-SCFI-2005 |
| Vegetable shortening and edible fat specifications. | MX-F-373-SCFI-2005 |
| Canola oil specifications. | MX-F-475-SCFI-2005 |

regulations for minimum quality standards for edible fats and oils used in foods intended for human consumption and traded in the Mexican territory. Compliance requirements established by the Secretariat of Commerce and Industrial Development are in the following documents: (Table 23.3)

## Definitions for the Labeling of Edible Fats and Oils in Mexico

*"Edible vegetable oil"* A deodorized oil that may contain one or several edible vegetable oils such as soybean, canola, corn, cotton, safflower, sunflower, sesame, etc., that is obtained by solvent extraction and is an amber-colored liquid.

*"Pure oil of...."* (soybean, canola, corn, safflower, sunflower, etc.)
Amber-colored liquid obtained by solvent extraction or mechanically expelled. If

**TABLE 23.3, CONT**
**Compliance requirements established by the Secretariat of Commerce and Industrial Development**

| Test methods. (Metodos de prueba) | Number |
|---|---|
| OSI stability index determination in fats and oils. | MX-F-012-SCFI-2005 |
| Determination of fatty acids by means of gas chromatography. | NMX-F-017-SCFI-2005 |
| Refractive index determination by Abbe refractometer. | NMX-F-074-S-1981 |
| Relative density determination. | MX-F-075-1987 |
| Acidity index determination. | NMX-F-101-1987 |
| Melting point determination. | NMX-F-114-SCFI-2005 |
| Color determination. | NMX-F-116-1987 |
| Iodine Value determination by the cyclohexane-acetic acid method. | NMX-F-152-SCFI-2005 |
| Peroxide value determination. | NMX-F-154-1987 |
| Qualitative determination of mineral oil contents in vegetable or animal fats and oils. | NMX-F-156-1970 |
| Saponification index determination. | NMX-F-174-S-1981 |
| Moisture and volatile matter. | NMX-F-211-1987 |
| Insoluble impurities determination. | NMX-F-215-1987 |
| Undesirable impurities sensorial determination-Odor. | NMX-F-473-1987 |
| Unsaponificable matter determination in vegetable or animal fats and oils. | NMX-K-306-1972 |

Source: (1,2)

purity is at least 99%, it can be labeled "pure."

*"Vegetable shortening"* A semisolid fatty product, obtained by processing oilseeds considered edible by the Secretariat of Health (Secretaría de Salud), which are: sesame, cotton, peanut, canola, safflower, coconut, sunflower, corn, palm, palm kernel, and soybean. The vegetable shortenings can be modified by partially hydrogenation alone, by blending vegetable oils and partial hydrogenated vegetable oils, or by interesterification. Additionally, vegetable shortenings may contain up to 15% maximum of fully hydrogenated oils.

*"Vegetable shortenings and mixed fats"* A semisolid fatty product, obtained by blending edible animal fats (edible tallow, lard, and partially hydrogenated fish oil) and vegetable oils and/or partially hydrogenated vegetable oils. Mixed fats may contain up to 15%

maximum fully hydrogenated oils.

Tables 23.4a–d Oils, vegetable shortenings and mixed fats physical and chemical specifications.

In Mexico, there are eight types of shortenings allowed for commercial use with paramaters defined by the region (Table 23.5) in which they are sold. They are shortening Type 1 for bakery in the Northern Region, Type 2 for bakery in the Southern Region, Type 3 for confectionary, Type 4 for cookies, Type 5 all purpose, Type 6 for cookies, Type 7 for cakes and Type 8 for filling creams. Only Types 7 and 8 have emulsifiers in their formulas and are made from vegetable oils such as sunflower, soybean, canola, palm, etc., in order to comply with the specifications shown in Table 23.4b.

Similarly, shortenings made from mixed fats, blends of vegetable oils and animal fats, are also known as Type 1 bakery—Northern Region, Type 2 bakery—Southern Region, Type 3 bakery—Central Region and used according to the region and its climatic conditions. Type 4 is bakery shortening that requires emulsifiers in the formula for proper functionality.

Anti-foaming agents: dimethylpolysiloxane; max limit: 10 mg/kg (10 ppm)

### *Official Sampling*

Sampling is under the jurisdiction of official Mexican norm NOM-002.SCFI-1993 mentioned previously. It is the accepted methodology for testing physical and chemical properties of fats and oils.

## Packaging and Wrapping

All products subject to these regulations should be packed in containers made from materials resistant to fat absorption, are inert, and protect product stability. Additionally, they must have the ability to prevent contamination and not alter the product's sensory characteristics or quality.

Bulk deliveries require stainless steel cars or tanks. The temperature of the product during its transportation should not exceed 10°C (50°F) above the melting point.

### *Required Labeling Information*

*Packaged Products*
Each unit package should have a label with permanent printing, visible and indelible, in accordance with the official Mexican norm NOM-051-SCFI. The information

**TABLE 23.4a**
**Oils, Vegetable Shortenings, and Mixed Fats: Physical and Chemical Specifications**

| Parameters | Soybean | Sunflower | Canola | Corn | Safflower | Vegetable Oil |
|---|---|---|---|---|---|---|
| FFA as oleic % (max) | 0.05 | 0.05 | 0.05 | 0.05 | 0.05 | 0.05 |
| Moisture and volatile matter % (max) | 0.05 | 0.05 | 0.05 | 0.05 | 0.05 | 0.05 |
| Saponification index mg KOH/g (min) | 189 | 188 | 182 | 187 | 186 | — |
| Color (Lovibond) red (max) | 2 | 1.5 | 2.5 | 4.0 | 1.5 | 3.5 |
| OSI Stability 110°C hrs (min) | 6 | 6 | 8 | 6 | 5 | 5 |
| Peroxide value meq/kg (max) | 2 | 2 | 2 | 2 | 2 | 2 |
| Unsaponificable matter % (max) | 1 | 1 | 1 | 1 | 1 | 1 |

**TABLE 23.4b**
**All Vegetable Shortenings**

| Parameters | Type 1 Bakery Northern Region | Type 2 Bakery Southern Region | Type 3 Confection | Type 4 Cookies | Type 5 All Purpose | Type 6 Cookies | Type 7 Cakes | Type 8 Filling creams |
|---|---|---|---|---|---|---|---|---|
| FFA as oleic % (max) | 0.05 | 0.05 | 0.05 | 0.05 | 0.05 | 0.05 | 0.15 | 0.15 |
| Moisture and volatile matter % (max) | 0.05 | 0.05 | 0.05 | 0.05 | 0.05 | 0.05 | 0.15 | 0.15 |
| Melting Point °C | 46-48 Typ. 50 max. | 50 max. | 50 max. | 50 max. | 50 max. | 50 max. | 50 max. | 50 max. |
| Saponification Index mg KOH/g (min) | 175 | 175 | 175 | 175 | 175 | 175 | 175 | 175 |
| Lovibond Color-Red (max) | 3.5 | 3.5 | 3.5 | 3.5 | 3.5 | 3.5 | 3.5 | 3.5 |
| OSI Stability 110°C Hrs (min) | 20 | 20 | 50 | 30 | 100 | 50 | 30 | 30 |
| Peroxide value meq/kg (max) | 1.5 | 1.5 | 1.5 | 1.5 | 1.5 | 1.5 | 1.5 | 1.5 |
| Unsaponifiable matter % (max) | 1.0 | 1.0 | 1.0 | 1.0 | 1.0 | 1.0 | 1.0 | 1.0 |
| α-mono-glycerides % (max) | — | — | — | — | — | — | 8.0 | 8.0 |

**TABLE 23.4c**
**Mixed Fats – Bakery**

| Parameters | Type 1 Bakery Northern Region | Type 2 Bakery Southern Region | Type 3 Bakery Central Region | Type 4 Bakery Emulsified |
|---|---|---|---|---|
| FFA as oleic % (max) | 0.05 | 0.05 | 0.05 | 0.15 |
| Moisture and volatile matter % (max) | 0.05 | 0.05 | 0.05 | 0.15 |
| Melting point °C | 46-48 Typ. (50 max.) | 50 max. | 50 max. | 50 max. |
| Saponification Index mg KOH/ g (min) | 175 | 175 | 175 | 175 |
| Color (Lovibond) Red (max) | 4.5 | 4.5 | 4.5 | 4.5 |
| OSI Stability 110°C Hrs (min) | 20 | 20 | 50 | 30 |
| Peroxide value meq/kg (max) | 1.5 | 1.5 | 1.5 | 1.5 |
| Unsaponifiable matter % (max) | 1.0 | 1.0 | 1.0 | 1.0 |
| Alpha-monoglycerides % (max) | — | — | — | 8.0 |

**TABLE 23.5**
**Commercial Regions in Mexico**

| Northern Region States | Central Region States | Southern Region States |
|---|---|---|
| Baja California | Nayarit | Guerrero |
| Baja California South | Jalisco | Veracruz |
| Sonora | Colima | Tabasco |
| Sinaloa | Michoacán | Oaxaca |
| Chihuahua | San Luis Potosí | Chiapas |
| Coahuila | Hidalgo | Quintana Roo |
| Nuevo León | Queretaro | Yucatán |
| Durango | Edo de México | |
| Tamaulipas | Guanajuato | |
| Zacatecas | Puebla | |
| | Tlaxcala | |
| | Morelos | |
| | Aguascalientes | |
| | Mexico City | |

should state compliance with applicable sanitary guidelines specified in the Sanitary Control Ruling of the General Health Law. It must also include product size, handling precautions, uses of the product, preferred consumption date (best if used before), and preferred storage conditions.

*Wrapped Products*
The same information indicated for unit packages should be included, as well as

**TABLES 23.6a and b**
**Food Additives and Their Maximum Amounts Allowed Limits by the Secretariat of Health (Secretaría de Salud)**

**TABLE 23.6a**
**Antioxidants**

| Active Ingredient | % Maximum |
|---|---|
| Tocopherols | 0.03 |
| Propyl Gallate (PG) | 0.01 |
| (BHA) | 0.01 |
| (BHT) | 0.02 |
| (TBHQ) | 0.02 |
| Ascorbyl palmitate | 0.02 |

**TABLE 23.6b**
**Synergists to Antioxidants (chelators)**

| Substance | % Maximum |
|---|---|
| Citric Acid (Acido cítrico) | 0.005 |
| Phosphoric Acid (Acido fosfórico) | 0.005 |

precautions for handling and use of wrappings, product code, preferred consumption date (best if used before), and recommended conditions for the storage of the product.

*Bulk Shipments*
All the information for packaged and wrapped products indicated above should appear in the appropriate shipping documents such as bill of lading or invoice.

*Declaration of Nutritional Properties*
Section 3.10: Any text or wording that declares, suggests or implies the nutritional characteristics contained in 100 grams of the packaged products. Also required is total available energy (1 Kjoule = .24 Kcalorie), protein, fats (lipids), carbohydrate and any vitamin or mineral content contained in the product.

Not allowed in the nutritional properties:
• Mention of substances on the list of ingredients with trademarks of the pre-packaged product.
• Declaration (quantitatively or qualitatively) or mention of any nutrient or compound when its addition is mandatory in accordance with applicable official ruling.

4.2.8.3.6 The nutritional information may be presented in the following way:

| Product name: | | |
|---|---|---|
| Nutrition facts Per 100g, per size, per unit: | | |
| Energy Content_____kJ (kcal)<br>Proteins_____g<br>Fats (Lipids) _____g<br>Carbohydrates _____g<br>Sodium _____g | | |
| Additional information | | |

## Storage

The finished product should be stored in facilities that comply with the sanitary standards described by the Secretariat of Health (Secretaría de Salud). The ideal temperature for the conservation of these products should be 20–28°C.

## References

1.  *Code of Federal Regulations*, Title 21, Volume 2, January 1, 2006, Washington, DC.
2.  *Code of Federal Regulations*, Title 9, Volume 2, January 1, 2006, Washington, DC.

# Current and Future Frying Issues

**Robert E. Wainwright[a] and Dan Lampert[b]**

Cargill, [a]5000 South Boulevard, Charlotte, NC 28217; [b]2400 Industrial Drive, Sidney, OH 45365-8952

Fried foods are rarely associated with "healthy" foods. They remain, however, among the most popular foods, not just in the United States, but globally as well. In addition to reliable consumer appeal, restaurants enjoy the benefits of frying because it's quick and convenient.

The basic frying process, involving simultaneous complex and dynamic events, is well understood. The frying medium (fat or oil) is heated to high temperatures, typically 335–400°F, and foods containing various mixes of proteins, carbohydrates, their own lipid material, water, seasonings, salt, etc. cause a nearly infinite number of potential chemical reactions. These reactions, in turn, result in various end products such as polar compounds and polymers, not found in either fresh oil or unfried foods. The other obvious issue with fried foods is in the relatively high-fat/oil content, and with this, the high caloric content of most fried foods.

Thus, when attempting to compile a list of current and future nutritional issues related to fried foods, the most logical place to start is the ever-increasing concern about obesity. Fried foods and sweetened, carbonated beverages are the most frequently referenced foods relating to this subject. The biggest issue with fried foods, of course, is the high percentage of fat in these types of foods. The challenge for food processors is to identify attributes that consumers associate with health and well-being and recognize to be nutritionally sound for their families. Attributes that can be communicated to consumers via label information include reduced caloric density, *trans* fatty acid content, reduced saturates, and amount of hydrogenated fat. Once those criteria are identified, reformulation hurdles can be assessed and potential options developed to improve the nutritional attributes.

Obesity is approaching epidemic proportions in the United States. Children two years old throughout adolescence have shown a ten-fold increase in becoming overweight, and an estimated 115 million people in developing countries are obese (1). Chronic diseases, including heart disease and Type 2 diabetes (e.g., high-glycemic load), top the list of concerns. Type 2 diabetes has increased dramatically in children, and is directly linked to obesity.

Efforts addressing obesity are underway. This includes review of school lunch menus, development of educational programs so people better understand nutritional issues, and the importance of increased physical exercise is being stressed, especially for children.

During the past 30 years, the average percentage of calories from fat has decreased from 36–33%. However, per capita consumption of fat has increased over that same time period because total calorie intake has increased (Dietary Guidelines Advisory Committee Report). For the past decade, the labeling law has emphasized the fat content and discouraged its consumption. The belief was that eliminating, reducing, or controlling fat could have an impact on health. Clearly though, as evidenced by the steady increase in per capita fat consumption, consumers' appetite for fat has not abated. The FDA is currently contemplating a shift in the focus of the Nutrition Facts Panel from "fat" to "total calories."

However, because fat contributes significantly to total calories and furthermore because it is so energy-dense, it would seem prudent to continue to allocate resources toward reducing the fat content of many processed foods provided that this approach can be accomplished without sacrificing the taste. Recently, the FDA has concluded that the current emphasis on fat obscures the message that calories count and that calories, not fat per se, are the main concern in obesity. It is important to note, however, that in addition to calories, fat also supplies essential fatty acids, is a carrier for fat-soluble vitamins, contributes to satiety, and improves palatability of foods.

The U.S. Dept. of Health and Human Services and the USDA have jointly published Dietary Guidelines for Americans every five years since 1980. The guidelines provide authoritative advice for people two years and older about how good dietary habits can promote health and reduce risk for major chronic diseases. The most recent revision was issued on January 12, 2005. The chapter devoted to fat recommends limiting total dietary fat intake to no more than 35% of calories. This compares to a recommendation of no more than 30% in the 2000 Dietary Guidelines. The 2005 update addresses the importance of adequate dietary fat. Minimum intakes for three populations (adults, children 2–3 years of age, and adolescents) have been added. In addition, saturated fat should be limited to no more than 10% of calories (currently 11–13%, an increase of 1–2% over the past 30 years) and cholesterol no more than 300 mg/day. The Advisory Committee to the USDA/DHHS held lengthy discussions about *trans* fat prior to submitting their recommendations and for a period of time the expectation was that the new guidelines would reflect a recommendation that *trans* fat represents no more than 1% of calories. However, this posture was softened somewhat and the published guidelines recommend that the intake of *trans* fat should be as low as possible.

In early 2006, the U.S. Food and Drug Administration unveiled MyPyramid, which represents the latest revision to the Food Guide Pyramid. Contained within the fats and oils portion of My Pyramid are the following recommendations: Most of the fats we consume should be polyunsaturated (PUFA) or monounsaturated (MUFA). These represent the good or healthy oils that do not raise LDL cholesterol. Good sources are fish, nuts, and vegetable oils. Saturated fatty acids (SFA), *trans* fatty acids (TFA), and cholesterol, on the other hand, tend to raise bad LDL cholesterol, which in turn increases the risk for heart disease. To reduce this risk, consumers are advised to cut back on foods that contain SFA, TFA, and cholesterol.

The FDA has been working to develop regulatory strategies whereby science-based labeling can provide a springboard for manufacturers to develop and position healthier dietary choices for consumers. Qualified health claims represent one vehicle available to food processors that desire to offer options that contain healthy lipids. These are steps that can permit consumers to make healthier choices more easily. In addition, potential health claims can motivate processors to formulate products that are more nutritionally desirable and thus use such benefits as a competitive edge.

*Trans* fatty acids are another global issue — and fried foods are a potential primary source of them. The United States, Canada, and a number of other countries are requiring labeling of *trans* fatty acids in retail foods. Denmark and Canada are attempting to mandate a maximum of 2% *trans* in all non-dairy/non-animal fats and oils.

A number of major retail food companies that produce fried or par-fried foods have already reduced the level of *trans* fats in their frying fats/oils to a level of less than 0.5 g per serving of food, allowing the labeling of 0.0 g per serving. Frozen fried chicken, fish, and par-fried french fries (and other frozen potato products) top the list of these types of foods. Most of the other food companies are at least testing various options.

Restaurants and other foodservice applications are not required at this time to label *trans*. Several major restaurant chains have announced intentions to lower *trans*.

## Typical Fatty Acid Compositions

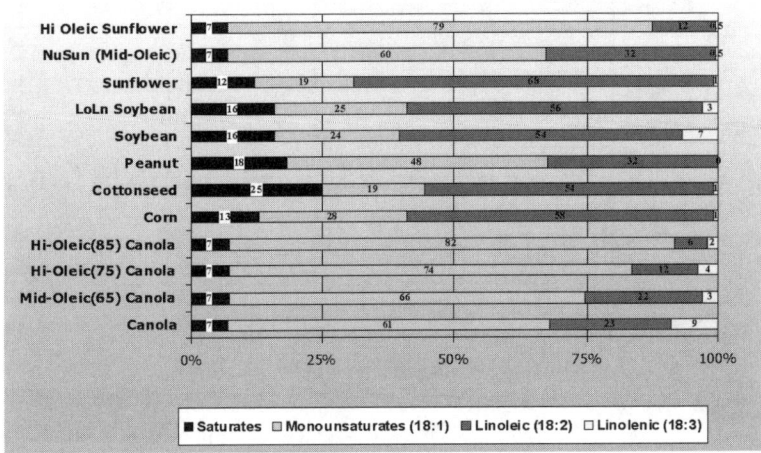

**Fig. 24.1.** Typical fatty acid compositions of higher stability oils.

For example, the city of New York recently challenged restaurants to eliminate *trans* fats from their menus. This brought additional attention and pressure for the entire foodservice industry to reformulate. In both retail and foodservice frying applications, liquid (non-hydrogenated) salad oils found some success, either alone or in blends with other oils and shortenings. Many of the healthy oils are highly unsaturated and as a result can be the victims of deteriorative reactions with oxygen.

These oils range from commodity soybean oil to high-oleic and/or low-linolenic specialty oils such as canola and sunflower. The specialty oils are more stable and more expensive. When compared to the partially hydrogenated oils they replace, commodity oils are usually less stable than these specialty oils. Trait-enhanced oils offer improved stability because their fatty acid profiles are altered in an effort to convey enhanced performance. Several varieties are commercialized. Figure 24.1 and 24.2 shows a number of these benchmarked against traditional varieties in each category.

The key to improved oxidative stability is suppression of linolenic ($C_{18:3}$) and linoleic ($C_{18:2}$) fatty acids. Many more enhancements are in the developmental pipeline, including LL/mid-oleic, LL/mid-oleic/low-saturate, and high omega-3. Oxidation management is becoming more of a challenge as a result of more PUFA oils being used, and thus is an issue that requires adequate support in order for these

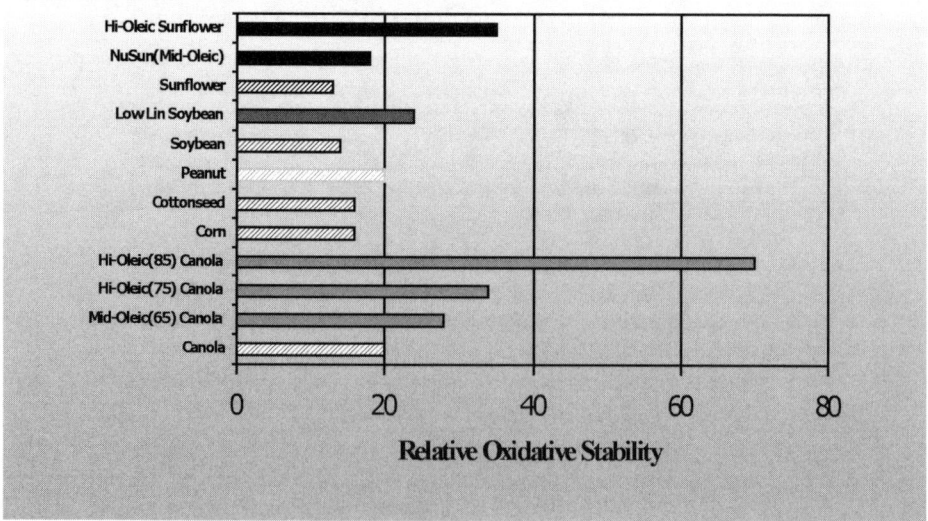

**Fig. 24.2.** Relative stabilities of trait-enhanced oils.

solutions to be successfully implemented in many applications. These oils represent tools whereby omega-3/omega-6 ratios can be better managed and can be conduits toward realizing the benefits of especially omega-3.

Low-linolenic soybeans produce oil that contains less than 3% linolenic acid (compared to 8% in regular varieties). Iowa State University developed a bean that produces oil with a linolenic content of just 1%. Both Monsanto and Pioneer have introduced varieties that produce oil with less than 3% linolenic acid. Monsanto expects to launch a mid-oleic/low-linolenic bean in two to three years, followed by a low saturates trait and high omega-3 variety in five to 10 years.

Traditional sunflower oil has a linoleic acid content of around 68% and generally no more than 1% linolenic acid. Mid-oleic and high-oleic varieties produce oils with significantly reduced linoleic acid contents. Mid-oleic or NuSun™ oil is characterized by a 50% reduction in linoleic acid content, while high-oleic sunflower oil typically contains only about 12% linoleic acid and as much as 80% oleic acid. Higher oleic acid levels are desirable because they improve shelf life and flavor stability of an oil, without the need for hydrogenation. Sunflower oil, regardless of variety, is also quite low in saturates, typically 12% for traditional and less than 9% for trait-enhanced.

A number of trait-enhanced canola varieties, reduced-linolenic/high-oleic, are available. Linolenic acid content is reduced from the 9% typical of commodity canola down to 3%, while oleic acid can be as high as 80% or more. Canola oil has a very healthy profile and the lowest level of saturates, typically only 7%, of any commercially available vegetable oils.

Another issue to deal with, in some applications, is the need for solids to be present to some degree in partially hydrogenated oils. Two basic options for providing some solids are adding either fully hydrogenated fats, which are almost entirely saturated fats, or palm oil, or palm hard fraction (stearine). Although the addition of these options (fully hydrogenated fats or hard fractions) raises the melting point, it also raises total saturates. However, what about saturated fat? Are all saturates equivalent in terms of health and nutrition? The 2005 Dietary Guidelines Advisory Committee concluded that stearic acid has different biological effects compared to other saturates. Many studies have shown it to have a neutral effect on serum total and HDL cholesterol concentrations. However, most work to date has been related to the effects of stearic acid on cardiovascular disease risk factors like blood lipids and lipoproteins. It is important to go beyond these effects, fibrinogen for example, to more completely understand how stearic acid differs from other saturates and whether the apparent neutral effect reported thus far extends to other physiologic processes as well. Clearly, there is much work to be done in this area with the aim of identifying the quantity and types of saturates that are good dietary options.

The last major issue of discussion deals with the chemical compounds formed in frying oil during use. Today, we benefit from this better understanding of frying dynamics which has led to significant improvements in both industrial and foodservice frying systems. Indirect heat to heat the oils, more sensitive temperature control, improved filtration systems and filter aids, all help to slow down the degradation of the frying fat/oil. That being said, the fat/oil still does decompose during its fry-life, but the goal of prolonging the inevitable is achieved.

Again, the primary decomposition compounds, are free fatty acids, polar compounds, and polymers. Fresh oils are non-polar. As the oil breaks down, some components become polar, such as free fatty acids, polymers, and other decomposition products. Polymers normally refer to the group of compounds having higher molecular weight, higher than the original triglycerides (2). As frying oils decompose, they generally become more viscous, due to the presence of many forms of these compounds.

In many European countries, regulatory limits have been established for many of these compounds such as free fatty acids, viscosity, polymers, and/or polar compounds to prevent frying in highly oxidized, or spent, fats/oils, that have long surpassed their discard point. In the future, these compounds may be more closely monitored over the rest of the world.

Before 2002, predicting future frying issues would have been easier. When Swedish scientists accidentally found acrylamide in a number of foods, priorities appeared to change. Scientists are now looking for more carcinogens/potential carcinogens, toxins, etc. in all types of foods.

The list of harmful and potentially harmful substances expands as the means of identification and ability to detect smaller and smaller amounts expand. Today, modern analytical techniques routinely allow detection limits of parts per billion and in some cases parts per trillion. With respect to California's Proposition 65, if a substance known or suspected to be harmful is found at any detectable level in a food, that food has to be identified as potentially harmful. Historically, at least in the United States, the use of high-oleic oils (80% oleic fatty acid content, or higher) for frying was limited. This was due primarily to the high cost, unavailability, and the perception of a different fried food flavor in the finished foods. As some of the issues related to breakdown components become better understood, and/or of more significance, the use of these oils may increase substantially.

With ever-increasing frequency, scientists report something new about foods and about the human condition, so trying to predict the next significant issue, either positive or negative, remains, essentially, a challenge of unpredictable magnitude. Certainly, health-related issues linked to fats and oils will continue, as in the past, to offer opportunities for future scrutiny.

## References

1. Watkins, C. Can Obese People Sue the Food Industry—And Win? *inform* **2004**, *15*, 505–507.
2. Paul, S.; and G.S. Mittal. Regulation of the Use of Degraded Oil/Fat in Deep-Fat/Oil Food Frying. *Crit. Rev. Food Sci. Nutr.* **1997**, *37*, 635–662.

# Index

## A

Absolute filters. *See* Passive filtration
Absorption
  degree of, 262–263
  field fry-tests and, 327
  nonnutritive fats and, 242–243
  polymers and, 193
  thermally oxidized oils/fats and, 177–178
  of *trans* fatty acids, 205–206
Acceptable Macronutrient Distribution Range (AMDR), 168–169
Acid value, laboratory quality control tests and, 374
Acrylamide
  current/future frying issues and, 420
  used oils and, 339–340
Active filtration
  adsorption, 344–345
  phase separation/emulsion breaking, 344
Active Oxygen Method (AOM), 23, 45
Acyl-CoA Oxidase (ACO), CFAM and, 215, 217
Adsorption
  as active filtration, 344–345
  bleaching during crude oil processing, 37
  and extraction of polar compounds from filter cakes, 352–353
  laboratory evaluation of, 349–352
  physical/chemcial properties and, 348
  polar dimers and, 95
  temperature-programmed desorption (TPD), 353–354
Adsorption chromatography
  and analysis of FAME concentrated fractions, 100–101
  and analysis of total polar compounds, 96–97
Aeration (forced) effect on heated oils, 174–175
Alcohols formation, 89
Aldehydes
  analysis of by capillary GC, 102–104
  formation of, 89
  low-weight molecular compounds as model research, 188–189
Alginates, fat substitutes and, 237
Alkaline contaminant materials (ACM), and assessment of heat abuse, 338
Alkane esters. *See Tris* alkane esters as fat substitutes
Alkili/physical refining, 37
Alkoxyl radicals, 88. *See also* Free radicals
Alkyl glycoside FA polyester as fat substitute, 236–237
Alkylmalonic acid, 231–232
American Oil Chemists' Society (AOCS), 35, 363

# F